Michael

IN VITRO METHODS FOR PARASITE CULTIVATION

IN VITRO METHODS FOR PARASITE CULTIVATION

Edited by

A. E. R. TAYLOR

Bureau of Hygiene and Tropical Diseases, London, England

and

J. R. BAKER

Cambridge, England

1987

ACADEMIC PRESS

Harcourt Brace Jovanovich, Publishers

London San Diego New York Berkeley
Boston Sydney Tokyo Toronto

ACADEMIC PRESS LIMITED
24–28 Oval Road
London NW1 7DX

United States Edition published by
ACADEMIC PRESS INC.
Orlando, Florida 32887

British Library Cataloguing in Publication Data
In vitro methods for parasite cultivation.
1. Parasitology —— Technique 2. Cultures (Biology).
I. Taylor, Angela E. R. II. Baker, John R. (John Robin)
574.5'249'0724 QL757
ISBN 0-12-683855-0

Typeset by Colset Private Limited, Singapore
Printed in Great Britain at the University Press, Cambridge

Contributors

C. Arme, *Parasitology Research Laboratory, Department of Biological Sciences, University of Keele, Keele, Staffordshire ST5 5BG, England*

J. R. Baker, *4 Belvoir Road, Cambridge CB4 1JJ, England*

C. G. D. Brown, *Centre for Tropical Veterinary Medicine, University of Edinburgh, Easter Bush, Roslin, Midlothian EH25 9RG, Scotland*

R. Brun, *Swiss Tropical Institute, Socinstrasse 57, CH-4051 Basel, Switzerland*

J. A. Clegg, *Wellcome Research Laboratories, Langley Court, Beckenham, Kent BR3 3BS, England*

G. S. Coleman, *"Chestnuts", 5 Foreman's Road, Thriplow, Royston, Hertfordshire SG8 7RA, England*

D. W. T. Crompton, *Zoology Department, The University, Glasgow GI2 8QQ, Scotland*

Elizabeth U. Canning, *Department of Pure and Applied Biology, Imperial College, London SW7 2AZ, England*

L. S. Diamond, *Laboratory of Parasitic Diseases, National Institutes of Health, Building 5, Room 112, Bethesda, Maryland 20892, USA*

Frank W. Douvres, *Helminthic Diseases Laboratory, Animal Parasitology Institute, Agricultural Research Service, United States Department of Agriculture, Maryland 20705, USA*

David E. Evans, *Department of Medical Protozoology, London School of Hygiene and Tropical Medicine, Keppel Street, London WC1E 7HT, England*

P. R. Gardiner, *International Laboratory for Research on Animal Diseases, PO Box 30709, Nairobi, Kenya*

M. A. Gray, *Kenya Trypanosomiasis Research Institute, PO Box 362, Kikuyu, Kenya*

H. Hirumi, *Kenya Agricultural Research Institute, PO Box 32, Kikuyu, Kenya*

M. R. Hollingdale, *Biomedical Research Institute, 12111 Parklawn Drive, Rockville, Maryland 2082, USA*

M. J. Howell, *Department of Zoology, Australian National University, GPO Box 4, Canberra, ACT 2601, Australia*

L. Jenni, *Swiss Tropical Institute, Socinstrasse 57, CH-4051 Basel, Switzerland*

Olivia L. Lassière, *Zoology Department, The University, Glasgow G12 8QQ, Scotland*

T. Ponnudurai, *Institute of Medical Parasitology, Katholieke Universiteit Nijmegen, Geert Grooteplein zuid 24, 6500 HB Nijmegen, The Netherlands*

Irene Popiel, *Biosis, 1057 East Meadow Circle, Palo Alto, CA 94303, USA*
M. D. Rickard, *University of Melbourne, Veterinary Clinical Centre, Princess Highway, Werribee, Victoria 3030, Australia*
M. A. Smith, *Wellcome Research Laboratories, Langley Court, Beckenham, Kent BR3 3BS, England*
Joseph F. Urban, Jr, *Dept of Agriculture, Beltsville, Maryland 20705, USA*
Caroline M. Winger, *Department of Pure and Applied Biology, Imperial College, London SW7 2AZ, England*

Preface

When, in 1968, we wrote our first book on "The Cultivation of Parasites *in Vitro*" (Taylor and Baker, Blackwell Scientific Publications, Oxford) little did we realize that 20 years later we would be editing our third volume on the subject. Although we have endeavoured to retain the "cook-book" style of the previous volumes, giving full practical details of successful techniques (when given in the original publications), we have also attempted to emphasize the various ways in which *in vitro* culture has been, and will be, applied.

Major advances in this field are reflected in the increased size of not only the 1978 version, which contained four more chapters (11) than the original, but also of this book which now has four chapters (instead of one) on the Kinetoplastida, and in which the Plasmodiidae are divided between two chapters (erythrocytic and exoerythrocytic stages) and new chapters are included on the Babesiidae, Theileriidae, nematodes parasitic in insects and helminth vaccines (16 chapters in all). Thus *in vitro* culture is no longer (and probably never was) an end in itself and now is sufficiently advanced to be applied to the study of the numerous parasitic problems in fields such as immunology, physiology, biochemistry, chemotherapy, diagnostics, molecular biology and the production of vaccines. *In vitro* techniques can be standardized more easily than those *in vitro* thus assisting new-comers to this field and easing the comparison of experiments from different laboratories. Furthermore, such techniques are leading to a reduction in animal experimentation, much needed in this period of constraint on laboratory expenses, and are providing more humane systems.

Lastly, we thank all our contributors, who have not only done an excellent job but who also have borne our editorial interference with remarkable equanimity. We also thank Mark Speight for assistance in preparing the Appendix.

ANGELA E. R. TAYLOR
JOHN R. BAKER

Contents

Chapter 1

Entamoeba, *Giardia* and *Trichomonas*

LOUIS S. DIAMOND

Laboratory of Parasitic Diseases,
NIAID, National Institutes of Health, Bethesda,
Maryland, USA

I. INTRODUCTION

Since publication of the earlier version of this book (Taylor and Baker, 1978), the major technological advances in the cultivation of the lumen-dwelling protozoa discussed in this chapter include (1) development of two media, TYSGM-9 and TYI-S-33, for xenic and axenic[1] cultivation of *Entamoeba* (Diamond, 1982; Diamond *et al.*, 1978); (2) development of two growth media for *Giardia lamblia* (Visvesvara, 1980; Keister, 1983) and two techniques for mass cultivation of this flagellate (Farthing *et al.*, 1982; Wieder *et al.*, 1983); (3) development of soft-agar techniques for cloning *Entamoeba* (Gillin and Diamond, 1978) and *G. lamblia* (Gillin and Diamond, 1980); and (4) development of a medium for cultivation of *Tritrichomonas muris*, the ubiquitous trichomonad of rodents (Saeki *et al.*, 1983).

[1]The nomenclature of the culture systems used in this chapter is based on the terminology of Dougherty (1959). Thus, a culture in which the protozoon is grown in the presence of several associates, usually bacteria, all of which are unidentified, is called xenic; with one known associate, monoxenic; with no associate, axenic. Xenic cultures are often mistakenly referred to as polyxenic: this refers to a culture containing several associates, all of which are known.

Except for *Trichomonas vaginalis*, completely defined synthetic media enabling molecular studies at a level not possible with media containing complex biological ingredients remain elusive goals. Two defined media, DL7 and DL8, are now available for culturing *T. vaginalis* (Linstead, 1981).

A partly defined medium, HUT, for cultivation of *Tri. foetus*, in which all the ingredients except serum are defined (Wang *et al.*, 1984), and a serum-free medium for *G. lamblia* (Wieder *et al.*, 1983), has been developed.

All the lumen-dwelling protozoa considered here have been grown with and without one or more of the microorganisms with which they are associated in their normal habitat within the host. Historically, the majority have been grown first with one or more of these associates, then later in their complete absence.

In the case of *Entamoeba* there are no reports of initiating an axenic culture directly from host material, all having been started using amoebae grown first in xenic or monoxenic culture, or in the presence of a slice of fresh tissue. In contrast, trichomonads, with the exception of *T. tenax* (Diamond, 1962), can be cultured axenically directly from their hosts.

For diagnostic purposes, only xenic culture techniques can be used for the detection of amoebic infections, whereas xenic or axenic techniques can be used to detect trichomonad infections. In fact, the ease with which the three pathogenic species *T. vaginalis*, *T. gallinae* and *Tri. foetus* can be axenized directly from host material has led to the almost exclusive use of this culture system for diagnosis.

Axenic cultures of *Giardia*, especially from humans, can be initiated directly with cysts discharged in faeces (Kasprzak and Majewska, 1983), or with trophozoites obtained by duodenal intubation (Gordts *et al.*, 1984). David Keister, while a member of my laboratory, in collaboration with C. F. T. Mattern, developed an effective but indirect method for initiating axenic cultures of the human parasite (Keister, 1983; Nash *et al.*, 1985). Neonatal mice are inoculated with cysts by percutaneous intragastric injection. After 7 d they are killed and trophozoites obtained from the small intestine are used to initiate the cultures. In all the methods discussed, the microorganisms associated with the flagellates in the host are eliminated through use of various combinations of antibiotics.

The usual procedure in a chapter such as this is to discuss xenic, monoxenic, and axenic culture systems in that order. Today, because the bulk of research is conducted with axenically cultivated organisms, and indeed the newer disciplines of molecular biology and immunology demand the use of such organisms, I shall reverse the usual presentation and assign priority to the culture techniques used in axenic culture. This in no way demeans xenic and monoxenic culture systems. They have played, and will continue to play, important roles in the study of these parasites.

Trypticase, a registered trademark for a pancreatic digest of casein (BBL Microbiology Systems, Inc.), is an essential ingredient of all media to be discussed in this chapter. In 1981 the company that had been manufacturing this particular casein digest, known in the trade as Medo-Peptone, ceased manufacturing it. The casein digest sold today under the name Trypticase™ barely supports the growth of *E. histolytica* or *G. lamblia*. Some lots will not support any growth. During the past four years, with the cooperation of BBL Microbiology Systems, Inc., Sheffield Products Inc. and Deltown Chemurgic Co., my laboratory has tested dozens of casein digests manufactured in the USA and abroad for their ability to support the growth of both parasites. One has been identified, and is distributed by BBL Microbiology Systems as a special order under the name Casein Digest Peptone™, catalogue no. 97023. Unfortunately, it supports only about one-half of the growth of *E. histolytica* obtained with Medo-Peptone. In the following pages Casein Digest Peptone (cat. no. 97023) has been substituted for all original references to Trypticase™. All sera must be heat-inactivated (56°C, 30 min), unless otherwise stated.

II. CULTIVATION OF *ENTAMOEBA*

A. AXENIC CULTIVATION

1° TYI-S-33 medium
(Diamond, Harlow and Cunnick, 1978)

This probably is the most widely used medium for the axenic cultivation of *E. histolytica* and related *Entamoeba* species, including *E. histolytica*-like, Laredo type amoebae of man[2], *E. invadens*, a pathogen of carnivorous reptiles, *E. terrapinae* from the turtle *Chyrsemys elegans*, *E. barretii* of the snapping turtle *Chelydra serpentina*, and *E. moshkovskii* isolated from the effluent of sewage treatment plants.

[2]Characterized by their ability to grow optimally at 24°C and under extreme conditions of hypotonicity (Goldman, 1969).

MEDIUM

1. NUTRIENT BROTH TYI

K_2HPO_4	1.0 g
KH_2PO_4	0.6 g
NaCl	2.0 g
Casein Digest Peptone (BBL, 97023)	20.0 g
Yeast extract (BBL)[3]	10.0 g
Glucose	10.0 g
L-Cysteine-HCl	1.0 g
Ascorbic acid	0.2 g
Ferric ammonium citrate[4]	22.8 mg

Dissolve the ingredients in 600 ml of glass-distilled water, bring final volume to 870 ml and adjust pH to 6.8 with 1N NaOH. Clarify by filtering through one layer of Whatman no. 1 paper. Autoclave (121 °C, 15 min) in units of 87 ml or multiples up to four units. Cool to room temperature. Broth can be stored at − 20 °C for at least 6 months.

2. VITAMIN MIXTURE NO. 13

This mixture is a modification and simplification of the mixture described by Diamond *et al.* (1978).

Prepare the following solutions, each in glass-distilled water, and then combine all five.

(i) *Solution 1.*

(1a) Dissolve niacin, 40 mg, and *p*-aminobenzoic acid, 180 mg, in H_2O, bringing final volume to 125 ml.

(1b) Dissolve nicotinamide, 40 mg; pyridoxal hydrochloride, 40 mg; pyridoxine, 80 mg; calcium pantothenate, 25 mg; choline chloride, 830 mg; *i*-inositol, 125 mg; thiamine hydrochloride, 25 mg; and vitamin B_{12}, 12 mg, in H_2O, bringing final volume to 125 ml.

(1c) Dissolve riboflavin, 25 mg; in H_2O with aid of 1 N NaOH, bringing final volume to 450 ml.

(1d) Dissolve folic acid, 30 mg, in H_2O with aid of 1 N NaOH, bringing final volume to 450 ml.

(1e) Dissolve D-biotin, 30 mg, in H_2O with aid of 1 N NaOH, bringing final volume to 450 ml.

Note. The final pH should be 6.5–7.0. If it is higher, the solution tends to be cloudy and should be discarded; higher pH may result from using

[3]Baker's yeast. Brewer's yeast (Difco) can be substituted
[4]Only the brown form (Mallinckrodt) can be used.

too much NaOH in steps (c), (d) or (e).

(ii) *Solution 2*. Prepare the following solutions, and then combine them, bringing the final volume to 300 ml with glass-distilled water.

(2a) Dissolve DL-6-8-thioctic acid (oxidized form) in 50 ml 95% ethanol.

(2b) Dissolve Tween-80, 5.0 g; menadione sodium bisulphite, 30 mg; and α-tocopherol acetate, 25 mg, in glass-distilled H_2O, bringing final volume to 200 ml.

(iii) Combine solutions 1 and 2, and bring final volume to 2000 ml with glass-distilled H_2O. Filter sterilize (0.22 μm pore size)*. This is vitamin mixture no. 13. Store at − 20 °C.

3. COMPLETE MEDIUM TYI-S-33

(i) Add aseptically to each unit (87 ml) of TYI broth, 10 ml heat-inactivated bovine serum and 2.0 ml of vitamin mixture 13. The final pH is 6.6. The osmolality ranges between 370 and 400 mosmol kg^{-1} depending on the concentrations of the salts in the biological products used in preparing the medium.

(ii) Dispense to 16 × 125 mm glass, screw-capped, culture tubes, filling to 75–80% of total capacity. Store at 40 °C in dark, because of the preparation's extreme sensitivity to light. For mass cultures, plastic tissue culture flasks (Costar no. 3025) can be used. However, growth per millilitre of medium will not equal that obtained in the glass tubes. For maximum growth in plastic flasks, the medium should be dispensed just before it is inoculated with amoebae.

4. REMARKS

(i) Each culture tube or flask should be filled with medium as indicated above; too small or too large a volume of air can be detrimental to the growth of the amoebae.

(ii) Biosate™ Peptone (BBL, cat. no. 11862, special order) can be used instead of the casein digest and yeast extract in TYI-S-33. Biosate™ Peptone is prepared with Casein Digest Peptone (cat. no. 97023).

(iii) Since biological products used in preparing media can vary from lot to lot, I recommend that a sample of each new lot be evaluated

*Dr Diamond seldom uses filters with pore sizes below 0.45 μm; 0.22μm is more reliable for sterilization, but may be difficult to use with solutions containing protein, and may remove some of the protein and, less probably, other constituents [Eds].

rigorously before accepting the entire lot from a distributor (for technique, see Diamond, 1983).

(iv) While the majority of *E. histolytica* strains I have studied require a concentration of 10% serum for axenization and maintenance, a few require concentrations as high as 15–20%. The two strains of *E. histolytica*-like amoebae and all the reptilian species of *Entamoeba* I have studied could be axenized and maintained in 10% serum, excepting *E. moshkovskii* which could be axenized either in the total absence of serum or in a concentration no greater than 1% (Diamond and Bartgis, 1970). However, years after it had been established in axenic culture, I found it grew luxuriantly in the presence of 10% serum (unpublished data).

(v) Bovine serum, especially that obtained from steers between 18 and 24 months old, best supports the growth of *E. histolytica* of several animal sera tested, including goat, horse, human, rabbit and sheep: no growth was obtained with foetal calf serum. Growth in human serum (pooled or unpooled) was very poor, and the morphology of the amoebae was abnormal (Diamond, 1970). All the other species of *Entamoeba* discussed in this chapter grow equally well in bovine or horse serum.

TECHNIQUE

Axenizing *Entamoeba* species, especially *E. histolytica*, is a time-consuming and laborious task, and thus is rarely attempted. The majority of investigators wishing to use axenic cultures in their research are primarily interested in propagating existing cultures in their own laboratories and, therefore, the techniques which follow are concerned with the management of established cultures. For those interested in axenizing their own isolates, detailed directions are given by Diamond (1968a, b), Diamond and Bartgis (1970), Meerovitch (1978) and Diamond (1983).

1. STOCK CULTURES

(a) *Entamoeba histolytica*

(i) These are maintained in borosilicate glass, screw-capped culture tubes incubated at an angle of 5° from the horizontal and at a temperature of 35.5°C. I grow all lumen-dwelling protozoa of homeotherms one or two degrees below the body temperature of the host, as advised by my former teacher Dr David H. Wenrich. With *E. histolytica*, Rees noted, as early as 1928, that incubator temperatures of 35–36°C gave better results than did higher temperatures (37°C). Also, use of lower temperatures reduces the frequency of subculturing.

(ii) The selection of a culture to be used as a source of amoebae for subculturing is vitally important. The amoebae are examined *in situ* with an inverted microscope: having been incubated in a slanted position, cultures are inverted several times to disperse the amoebae; 10–15 min later, after the amoebae have resettled, they are examined with the microscope at a magnification of × 100. Selection of a culture is based on the following criteria: (a) the medium overlying the amoebae is clear, indicating freedom from overt microbial contamination; (b) the majority of amoebae are adherent to the surfaces of the tube or to each other in clumps; (c) the amoebae display active pseudopod formation, though locomotion need not be apparent; (d) few multinucleated giant cells are present (these cells apparently cannot multiply and strains characterized by the presence of numerous giant cells are especially difficult to maintain); (e) dead amoebae, characterized by highly granular, shrunken cytoplasm and irregular edges, are rare.

(iii) Selected cultures are chilled for 5–10 min in an ice-water bath, and inverted several times to disperse the amoebae; portions are transferred, with a pipette, to tubes of fresh medium. The optimum number of amoebae depends on the strain, some need a few hundred cells per ml, others a few thousand. Cultures not selected for subculture are retained (at 33 °C) as reserve cultures until the next subculture is performed. If not needed at that time, they are discarded.

(iv) The majority of strains I have studied reach the end of the logarithmic phase of growth after about 80 h in TYI-S-33. I have transferred cultures serially at 24, 48 and 72 h, and alternately at 72 and 96 h: the best cultures are obtained by serial transfer every 72 h, and in my laboratory all cultures used for experiments, with rare exception, are of this age. On the other hand, for convenience I have carried stock cultures for years by subculturing them at alternate intervals of 72 and 96 h. To do this, the inoculum must be decreased for the longer incubation interval, and increased for the shorter.

(v) Occasionally, after subculture, cultures will not grow as expected. The only way to save these cultures is to stop subculturing and resort to refeeding, i.e. removing the old medium and replacing it with fresh, until the crisis has passed. During this period it is helpful to incubate the culture tubes vertically.

(b) *Entamoeba histolytica*-like Laredo type amoebae and *E. moshkovskii*. As far as I can ascertain, only two strains of the Laredo type amoebae, Huff and Laredo, and two strains of *E. moshkovskii*, FIC and

CST, have been axenized. Axenic cultures are grown at 24 °C. With the Laredo type amoebae, axenic cultures can be grown at 35.5 °C only if a cell-free chick embryo extract has been incorporated into the medium (Diamond, 1968b). Xenic and monoxenic cultures can be grown at both 24 °C and 35.5 °C.

These amoebae need be subcultured only every 10–14 d. Inocula of several thousand amoebae per ml of medium are required for maintenance of the cultures.

(c) *Entamoeba invadens*, *E. terrapinae* and *E. barretii*
These amoebae are much less fastidious than *E. histolytica* in their cultural requirements. For example, a given lot of bovine serum which will not support the human pathogen will often support luxuriant growth of these species.

I culture these amoebae at 24 °C and transfer them biweekly. Only 50–100 amoebae per ml of medium are needed for maintenance.

2. MASS CULTIVATION

Large numbers of amoebae can be obtained by growing them in plastic tissue culture flasks (Costar, cat. no. 3025 or Falcon, cat. no. 3012) filled to within 80% of their capacity with medium (= 33 ml), and inoculated with amoebae in late log-phase obtained from pooled cultures grown in screw-capped culture tubes. From 1000 to 5000 *E. histolytica*, depending on the strain used, are inoculated per ml of medium and harvested at the end of 72 h incubation as follows. Flasks are chilled for 10–15 min in ice-water, inverted several times to suspend the amoebae, and their contents are pooled for centrifugation (10 min, 275*g*). The supernate is removed and replaced with a wash fluid, PBS no. 8 (NaCl, 9.5 g; K_2HPO_4, 3.7 g; KH_2PO_4, 1.1 g, glass-distilled H_2O to make 1000 ml; pH 7.2; osmolality 360 mosmol kg^{-1}, close to that of TYI-S-33). Depending on the strain used, an inoculum of 1000 *E. histolytica* per ml of medium can be expected to yield a 200 to 300-fold increase in 72 h.

All the other species of *Entamoeba* mentioned in this chapter can be mass cultured in the same manner by taking into consideration their rates of multiplication and temperature requirements.

3. COLONIAL GROWTH IN SOFT AGAR MEDIUM (Gillin and Diamond, 1978)

The sensitivity of this technique, and the excellent proportionality between the number of viable cells introduced into the medium and the number of colonies developing make it especially useful for studies of

Entamoeba in which assessment of amoebic survival is critical, e.g. testing of antiamoebic drugs and temperature effects. With care, cloned colonies can be obtained.

(i) Place 5.0 g Difco Bacto agar in a dry screw-capped serum bottle (125 ml capacity), add 95 ml TYI broth (1°1 above), allow it to soak for 30 min, autoclave (121 °C, 15 min) to melt and sterilize the agar and then cool and store it at 4 °C.

(ii) To prepare the medium for use, melt the agar and maintain it at 60 °C in a water bath. Place screw-capped tubes of TYI-S-33 (13 ml) in a second bath (42 °C), and allow to equilibrate for 10 min. Add the required volume of melted agar to tube (see below for amounts), invert tubes several times to mix and immediately add amoebic inoculum. Again invert tubes several times to disperse amoebae. Set agar rapidly by chilling in an ice-water bath for 10 min. Final volume per tube should be 15 ml. NB. Colony formation is dependent on (1) rapid setting of agar, (2) completely filling tubes with medium to reduce oxygen tension, (3) final agar concentration, and (4) the age of the cultures used for the amoebic inoculum. Best results are obtained with amoebae taken from cultures less than 72 h old.

The optimal agar concentration (between 0.3% and 0.8%) must be determined for each species and each strain. The range is very narrow and should be determined by testing increments of 0.05% agar. At suboptimal concentrations, colonies are diffuse, or may fail to form; at concentrations above optimum, colonies are smaller and fewer form.

One hundred cells should be inoculated per tube. More than 100 colonies are difficult to count.

E. histolytica colonies are usually well defined in 4–5 d, reptilian species of *Entamoeba* after 7 d. The *E. histolytica*-like amoebae fail to form colonies visible to the naked eye.

Clones can be obtained with this technique provided that the inoculum is adjusted so that no more than 10 colonies are formed, and that microscopic examination of the newly prepared soft-agar tubes shows the amoebae to be well separated; after the colonies have formed, no amoebae should be wandering between the agar core and walls of the tube. Cloned colonies are picked using a pipette with a finely drawn tip under a dissecting microscope and are transferred to tubes of TYI-S-33 medium. After 3–6 d incubation, the old medium is replaced with fresh medium. The newly cloned cultures are usually ready for subculture 10–14 d after they have been picked.

B. MONOXENIC CULTIVATION

Two systems are available for monoxenic cultivation of *Entamoeba*: one uses a bacterium associate (usually *Fusobacterium symbiosum*[5] = *Bacteroides symbiosus*) and the other a trypanosomatid.

Trypanosoma cruzi (Culbertson strain) (ATCC™ no. 30013), and *Crithidia* sp. (ReF-1: PRR strain) (ATCC™ no. 50083) are those most frequently used. The former system has mostly been used as source material for physiological and biochemical investigations of *Entamoeba*, principally before the advent of axenic cultures (see review by Reeves, 1984), and the latter as a source of amoebae for axenization (Meerovitch, 1978; Diamond, 1983).

At present, the most widely used medium for monoxenic culture of the amoebae with *F. symbiosum* is MS-F medium (Reeves *et al.*, 1957), and for the amoebae with trypanosomatids, TTY-SB medium (Diamond, 1968a). Both of these media are fully described in the earlier versions of this book (Taylor and Baker, 1968, p. 140, and 1978, p. 70, respectively).

Recently, I developed a technique, based on TYI-S-33 medium, for cultivating amoebae with *F. symbiosum* so that only one basic medium is required for monoxenic and axenic growth. I have also modified TTY-SB so that supplementing the medium with frozen and thawed defibrinated rabbit blood is no longer necessary. This medium, designated TTY-SH-2, is more efficient and supports larger (more than 3 times) populations of amoebae than TTY-SB.

When cultivating the amoebae monoxenically, stock cultures of the associates are maintained separately, and combined with the amoebae only when each subculture is made.

1° Monoxenic culture with *Fusobacterium symbiosum*

TECHNIQUE

(i) MAINTENANCE OF *F. SYMBIOSUM* STOCK CULTURES

Place 12 ml of TYI broth (Section II.A.1°.(1)) in 16 × 125 mm screw-capped culture tubes. Incubate (48 h) to test for sterility (the broth can be stored at 4 °C for several weeks). Subcultures are made by transferring one loopful of old culture to a fresh tube of TYI broth and incubating it overnight at 35.5 °C.

[5]Most investigators use the original isolate (ATCC™ 14940) obtained by Shaffer from the NRS flora (Shaffer and Frye, 1948). I use an isolate derived in 1982 from the same flora (see Diamond, 1983, p. 76 for history).

(ii) **MAINTENANCE OF MONOXENIC CULTURES**

Place 11 ml of TYI-S-33 medium (Section II.A.1°) containing 5–10% heat-inactivated bovine serum in screw-capped culture tubes. Add penicillin, 500 iu ml^{-1} of medium, and 2 ml of a 24 h culture of *F. symbiosum* (optical density 0.3–0.45, measured with a Beckman spectrophotometer, model B, set at 558 μm). Inoculate with *E. histolytica*, 500–1000 ml^{-1} of medium, from a previous subculture after chilling for 5 min in an ice-water bath. Incubate at an angle of 10° for 48–72 h. Depending on the strain, yields of 10 000–20 000 ml^{-1} of medium can be expected.

(iii) **REMARKS**

The best amoebic growth is attained using cultures of *F. symbiosum* transferred daily. However, it is possible to obtain amoebic cultures adequate for routine maintenance by using 24 h cultures of the bacterium stored no longer than 48 h at 4°C.

I have grown only *E. histolytica* with this system. However, in view of the number of species of *Entamoeba* which have been cultivated in MS- F medium (Reeves and Bischoff, 1968), I believe all of those discussed in this chapter could be grown in it.

2° TTY-SH-2 medium for monoxenic cultivation with trypanosomatids

MEDIUM

(i) **NUTRIENT BROTH TTY**[6]

K_2HPO_4	1.0 g
KH_2PO_4	1.0 g
NaCl	3.0 g
Tryptose (Difco)	10.0 g
Casein Digest Peptone (BBL, 97023)	10.0 g
Yeast extract (BBL)	10.0 g
Glucose	5.0 g
L-Cysteine-HCl	1.0 g
Ascorbic acid	0.2 g
Haemin[7]	0.8 ml
Glass-distilled H_2O	930.0 ml

Dissolve salts in 600 ml of water. Add tryptose and dissolve (50°C). Add and dissolve remaining ingredients. Bring final volume to 930 ml.

[6]The salt concentrations and final pH of this broth differ from those in TTY-SB (Diamond, 1968a).
[7]See footnote on p. 12.

Adjust pH to 6.7 with 1 N NaOH. Autoclave (121 °C, 15 min) in units of 93 ml, or in multiples of 2 units. Cool to room temperature. The broth can be stored at − 20 °C for at least 3 months.

(ii) COMPLETE MEDIUM

Add aseptically to each unit of broth, 5.0 ml of bovine serum and 2.0 ml vitamin mixture no. 13 (Section II.A.1°). The final pH is 6.9 and the osmolality ranges between 360 and 390 mosmol kg^{-1}.

Dispense in 8.0 ml amounts to 16 × 125 mm screw-capped culture tubes. The medium can be stored at 4 °C for up to 10 d.

1. *Trypanosomatid stock cultures*

These cultures are maintained in LIT no. 2 medium, a modification (Diamond, unpublished) of Yeager's medium as described by D.A. Evans (Taylor and Baker, 1978, p. 73).

MEDIUM

NaCl	4.0 g
KCl	0.4 g
Na_2PHO_4	8.0 g
Glucose	2.0 g
Tryptose (Difco)	5.0 g
Panmede™ liver digest (Paines and Byrne)	1.0 g
Haemin[7]	1.0 ml
Glass-distilled H_2O	900.0 ml

Prepare according to directions given for TTY-SH-2 medium (Section I.B.2°), adjusting pH to 7.4 with 6 N HCl. Autoclave (121 °C, 15 min). Add, aseptically, to each 90 ml unit, 10 ml of heat-inactivated rabbit serum. Dispense in 20 ml amounts to 125 ml screw-capped Erlenmeyer flasks. The broth can stored at − 20 °C for up to 6 months; the complete medium can be stored at 4 °C for at least 1 month.

TECHNIQUE

Inoculate flasks of LIT no. 2 with 10–20 million *Crithidia* sp., or 100 million *T. cruzi*. Tighten caps and incubate at 24 °C. Subculture crithidia three times a week and trypanosomes at weekly intervals.

[8]Supplied from a 2.5 mg ml^{-1} solution in 50% (w/w) aqueous Quadrol (ethylenedinitrilo)-tetra-2-isopropanol), Sigma (Tamburro and Hutner, 1971).

2. Maintenance of E. histolytica *and* E. histolytica*-like Laredo type amoebae A-T (amoeba-trypanosome) or A-C (amoeba-crithidia) cultures*

Subculture as follows: chill selected cultures[8] 5–10 min in ice-water; invert several times to loosen any amoebae adherent to the sides of the tubes; centrifuge (275*g*, 3 min); decant supernate; resuspend pelleted amoebae in 1.0 ml of fresh TTY-SH-2; transfer aliquots to medium freshly inoculated with trypanosomatids; incubate cultures vertically at 35.5 °C.

Remarks. E. histolytica A-C cultures are inoculated with 3–4 million crithidia per ml of medium, while the Laredo type amoebae cultures are inoculated with half that number. A-T cultures are inoculated with 2 million trypanosomes per ml of medium. Subcultures of *E. histolytica* are made alternately at intervals of 48 and 72 h or 72 and 96 h, depending on the strain of amoebae used. A-C cultures inoculated with 10 000–15 000 *E. histolytica* per ml can be expected to yield a 10- to 20-fold increase in the number of cells, and for *E. histolytica*-like amoebae approximately half that number. A-T cultures produce a 4- to 6-fold increase with no difference between *E. histolytica* and *E. histolytica*-like amoebae.

3. Reptilian amoebae and E. moshkovskii

These amoebae are cultured monoxenically with *T. cruzi*. The crithidia cannot be used because they overgrow the amoebae at the temperature (24 °C) used to cultivate these species. Subculture is carried out at 7–10 d intervals.

C. XENIC CULTIVATION

1° TYSGM-9 medium

(Diamond, 1982)

This liquid medium derived from TYI-S-33 medium (Section II.A.1°), has been used to isolate and maintain the following protozoa: *E. histolytica*, *E. coli*, *E. gingivalis* and *Dientamoeba fragilis* (Diamond, 1982); *E. hartmanni*, *Balantidium coli* and *Pentatrichomonas hominis* (G.S. Visvesvara, personal communication). Its relative clarity permits monitoring of the organisms *in situ*, obviating the need to sample cultures between transfers.

[9]For method of selection, see Section II.A.1°, Technique.

MEDIUM

1. NUTRIENT BROTH

K_2HPO_4	2.8 g
KH_2PO_4	0.4 g
NaCl	7.5 g
Casein Digest Peptone (BBL, 97023)	2.0 g
Yeast extract (BBL)	1.0 g
Glass-distilled H_2O	970.0 ml

The broth can be stored for several months at − 20°C.

2. COMPLETE MEDIUM

(i) Place 200 mg gastric mucin (US Biochemical Corporation, cat. no. 16025) in 125 ml screw-capped serum bottles or Erlenmeyer flasks. Add 97 ml of broth to each container and stir vigorously for at least 1 h. The longer the stirring, the clearer the resulting medium. Autoclave (121°C, 15 min) and cool to room temperature.

(ii) Add aseptically 5.0 ml of heat-inactivated bovine serum[9], and 0.1 ml of a 5% solution (v/v) of filter-sterilized (0.22 μm pore size) Tween 80 in distilled water[10].

(iii) Dispense in 8.0 ml portions to 16 × 125-mm screw-capped glass culture tubes.

(iv) Add 2–3 mg of sterile rice starch per ml of medium. The starch is sterilized (dry heat, 150°C, 2.5 h) in 500 mg amounts in screw-capped tubes. To dispense, add 9.5 ml of sterile PBS no. 8 (Section II.A.1°2) to each tube. Bring rice into suspension by vigorous shaking and distribute with a graduated pipette. NB. Purified rice starch prepared by a modification of the method of Griffin and McCarten (1949) (Diamond, 1983, p. 75), and rice starch (British Drug Houses Ltd, and Gailard Schlesinger, Inc.,) are recommended.

(v) The final pH of the medium is 7.2; the osmolality is 320 mosmol kg^{-1}.

TECHNIQUE

Control of the bacterial population is paramount for successful xenic

[10]3% bovine serum was used in the original formula. I have since learned with more experience that some strains of *E. histolytica* are more easily isolated and maintained in 5% serum.

[11]This is another change from the original formula. There is no need to prepare the Tween 80 as a solution in absolute alcohol.

cultivation of lumen-dwelling protozoa, especially *Entamoeba*. Characteristically, media devised for xenic cultivation do not support luxuriant bacterial growth. Nevertheless, in many cases the bacterial flora accompanying the isolate will overgrow the culture unless it is controlled. This is best accomplished through the use of antibiotics, such as penicillin (150–200 iu ml^{-1}) combined with streptomycin (150–200 $\mu g\ ml^{-1}$), erythromycin (100–200 $\mu g\ ml^{-1}$) and/or bacteriostatic dyes such as acriflavin (2.5–10 $\mu g\ ml^{-1}$). These agents can be dispensed with after a balance between the bacteria and amoebae has been reached. Robinson's medium (see below, Section II.C.2°) requires continuous use of erythromycin to control the multiplication of the *Escherichia coli*, which must be added at each subculture.

Initiate cultures with pea-sized bits of faeces, the usual source of patient material of xenic cultures, or with tissue obtained by rectal biopsy. Incubate at 35.5 °C. Subculture three times per week, in one of two ways. Chill 5–10 min in ice-water. If growth is sparse, centrifuge cultures (275*g*, 3 min) and transfer the sediment to tubes of fresh medium. If growth is luxuriant, invert tubes several times to disperse the amoebae and transfer from as little as a few tenths of a millilitre to as much as 1.0 or 2.0 ml. Incubate newly initiated cultures vertically, and established cultures in a slanted position (15 °). The latter position increases yields. Monitor the cultures as follows: gently invert the tube several times to disperse its contents, return it to the incubator, and allow it to remain 1 h at an angle of about 5 °; then remove it from incubator, and examine the entire length of the tube with an inverted microscope at a magnification of × 100. Amoebae, when present, will usually be attached to the glass, sometimes to masses of bacteria and/or starch particles.

2° Robinson's medium

(Robinson, 1968)

This diphasic medium had largely been ignored until its recent revival by Sargeaunt and Williams (1978) for the isolation and maintenance of *Entamoeba* used in their isoenzyme studies. The following lumen-dwelling protozoa of humans have been cultured in it: *E. histolytica*, *E. hartmanni*, *E. coli*, *Endolimax nana*, *Dientamoeba fragilis* (Robinson, 1968); *Iodamoeba buetschlii*, *Pentatrichomonas hominis* and *Balantidium coli* (P.G. Sargeaunt, personal communication). The following description is based on information provided by Peter G. Sargeaunt, the leading exponent of the use of this medium.

MEDIUM

The technique is designed for use with 7.0 ml glass, screw-capped McCartney culture bottles (Gallencamp)[11].

1. **SALINE AGAR SLOPES.** Dissolve 15 g of fine agar powder and 7.0 g of NaCl in 1000 ml of distilled H_2O by heating. Distribute 2.5 ml portions to McCartney bottles. Sterilize by autoclaving (121 °C, 15 min). Remove bottles while agar is still molten. Position bottles at an angle to form a slope, allow to cool.

2. **ERYTHROMYCIN.** Dissolve 0.5 g of pure erythromycin powder (Abbott Laboratories) in 20 ml of 70% ethanol in a sterile container. Allow to stand at 4 °C for 2 h or longer to sterilize the antibiotic, then add 30 ml of sterile distilled H_2O.

3. **DIFCO BACTO-PEPTONE.** Dissolve 20 g in distilled H_2O, autoclave (121 °C, 15 min). NB. Not all brands of peptone will support amoebic growth.

4. **RICE STARCH.** Obtained from British Drug Houses. Use without sterilization.

5. **PHTHALATE SOLUTION.** Dissolve 204 g potassium phthalate in 1800 ml of distilled H_2O by adding 100 ml of 40% NaOH until a pH of 6.3 is reached; add water to bring final volume to 2000 ml, dispense in 10 ml amounts to small bottles, and autoclave (121 °C, 15 min). Dilute this stock solution 1:10 with sterile distilled H_2O to make a 50 mM solution.

6. **SERUM.** The serum can be used inactivated or activated, and from any animal. NB. Robinson (1968) used sheep serum.

7. **DEFINED MEDIUM R FOR *ESCHERICHIA COLI***

(i) *Stock solution*

NaCl	125.0 g
Citric acid, monohydrate	50.0 g
KH_2PO_4	12.5 g
$(NH4)_2SO_4$	25.0 g
$MgSO_4 \cdot 7H_2O$	1.25 g
Lactic acid (90% pure)	100.0 ml
Distilled H_2O	2500.0 ml

Store in unsterilized good quality glass stoppered bottle.

[12]Other culture vessels can be used. Dr E. Meerovitch (personal communication) uses 17 × 100 mm screw-capped plastic culture tubes.

(ii) *Working solution*. Add 100 ml of stock solution 7(i) to 850 ml distilled H_2O. Titrate to pH 7.0 with 40% NaOH. Dispense 25 ml portions to flat-sided 100 ml screw-capped bottles. Autoclave (121°C, 15 min).

8. **BR BASAL AMOEBIC MEDIUM.** Inoculate 25 ml of working solution 7(ii) with *Escherichia coli*, and incubate at 37°C for 48 h with bottle lying flat.

9. **BRS COMPLETE AMOEBIC MEDIUM.** Add an equal volume of serum to BR medium (8) and continue incubating for another 24–48 h.

TECHNIQUE

1. ISOLATION OF AMOEBAE

To a bottle containing a sterile agar slope, add aproximately 10 mg rice starch, 0.12 ml erythromycin solution (item 2), and sufficient BR medium (item 8) to cover slope. Do not fill bottle. Place approximately 50 mg of faeces in bottle, cap and incubate at 37°C for 24 h. Aspirate and discard overlay, leaving only starch and faeces. Replace overlay with sufficient BRS (item 9), diluted 1:4 with the phthalate solution (item 5), to cover slope. Add 0.06 ml erythromycin solution (item 2), 0.06 ml Bacto-peptone solution (item 3), and additional starch. Incubate for 24 h, then remove a drop of faeces and starch from the bottom of slope, mix with a drop of double-strength Lugol's iodine solution, and examine with the aid of an optical microscope for the presence of amoebae. If none is found, add starch if required, and continue incubating culture for an additional 48 h before re-examination.

2. SUBCULTURING

Transfer about 0.1 ml of faecal-starch layer to a fresh agar slope overlaid with BRS diluted 1:4 with phthalate solution, erythromycin, Bacto-peptone and starch as indicated above.

III. CULTIVATION OF *GIARDIA*

A. AXENIC CULTIVATION

Axenic culture systems are the principal ones used for growing *Giardia*. Monoxenic cultures, in which the yeast *Saccharomyces cerevisiae* is the usual associate, have been used primarily to initiate axenic cultures (Meyer, 1970, 1976). These will not be discussed, since axenic cultures can now be initiated directly from host material. *Giardia* have never been grown xenically. In general, cultures of *G. lamblia* are hardier than those of *E. histolytica*.

TP-S-1 (Diamond, 1968b), and TYI-S-33 (Diamond *et al.*, 1978), originally devised for axenic cultivation of *Entamoeba*, support axenic growth of *G. lamblia* provided they are filter-sterilized and *not autoclaved* (Visvesvara, 1980; Gillin and Diamond, 1981). Growth in TYI-S-33 is marginal unless it is supplemented with crude bile and the concentration of cysteine is doubled (Keister, 1983). Both TP-S-1 and the enhanced TYI-S-33 provide heavier yields of *Giardia* in a shorter time than either of Meyer's (1976) media HSP-1 and HSP-2.

1° TP-S-1 Medium

(Diamond, 1968b; Visvesvara, 1980)

MEDIUM

1. **NUTRIENT BROTH TP**

K_2HPO_4	1.0 g
KH_2PO_4	0.6 g
NaCl	5.0 g
Casein Digest Peptone (BBL, 97023)	10.0 g
Panmede™ liver digest (Paines and Byrne)	20.0 g
Glucose	5.0 g
Ascorbic acid	0.2 g
L-Cysteine-HCl	1.0 g
Glass distilled H_2O	880.0 ml

Dissolve ingredients as presented in 600 ml of water. Bring final volume to 880 ml. Adjust pH to 6.8 with 1 N NaOH. Clarify by filtering through one layer of Whatman no. 1 paper. Sterilize by passage through a 0.45 μm membrane filter. Broth can be stored up to 10 days at 4 °C, and up to 6 months at −20 °C provided the cysteine and ascorbic acid are omitted at the time of its preparation. When the medium is preserved frozen, the cysteine and ascorbic acid are added as a concentrated neutralized solution after thawing the broth.

2. **COMPLETE MEDIUM**

To each unit (88 ml) of TP broth, add aseptically 10 ml of heat-inactivated bovine or horse serum and 2.0 ml of vitamin mixture no. 13[12]

[13]Can be substituted for vitamin mixture NCTC 107 used in the original medium.

(Section II.A.1° (2)) for composition. The final pH is 6.6, and the osmolality ranges between 410 and 436 mosmol kg^{-1}.

Dispense 15 ml aliquots to 16 × 125 mm glass, screw-capped culture tubes, filling to 85% of total capacity. Store at 4°C in dark.

TECHNIQUE

The inoculum is prepared from 2-day-old cultures. Parasites are dislodged from the walls of the tube by immersing the latter in ice-water for 5–10 min, then rolling vigorously between the hands before centrifuging it at 1000 rpm (time not stated) and 4°C. The sediment is resuspended in fresh medium, and the parasites in a small sample are counted on a haemacytometer, after fixation in 1% formalin at a dilution of 1:10. The suspension is then diluted with medium so that an inoculum of 0.5 ml will give a final parasite density of 22 000 ml^{-1}. Cultures are incubated at 37°C.

Baveja *et al.* (1984) studied the growth response of a single strain of *G. lamblia* (identified only as a strain obtained from Visvesvara, CDC, Atlanta, Georgia) to human, calf and horse serum. Best growth was obtained with human serum, poorest growth with sheep serum. In view of the problems associated with obtaining large batches of human serum and the number of human viruses potentially contained in such serum, I do not recommend its use.

2° TYI-S-33 medium for *Giardia*
(Keister, 1983)

This medium differs from that devised for the cultivation of *Entamoeba* (Section II.A.1°) as follows: (1) concentration of cysteine is doubled; (2) 500 mg $litre^{-1}$ dehydrated bovine bile (Sigma, cat. no. B 8381) is added; (3) the pH of the enhanced TYI broth is adjusted to 7.0–7.1; (4) broth is filter-sterilized (0.22 μm pore size); (5) vitamin supplement is deleted.

Heat-inactivated bovine serum is used in this medium as it is in TYI-S-33 for cultivation of *Entamoeba*.

3° TYI-33 serum-free medium
(Wieder, Keister and Reiner, 1983)

This medium is recommended for those investigators requiring one free of whole serum. The number of *Giardia* produced is approximately 30% less than that obtained in the presence of whole serum. It differs from Keister's (1983) medium (Section III.A.2°) in that the bovine serum is replaced with bovine serum albumin, Fraction V, 1.0 g $litre^{-1}$ (Sigma, cat. no. 4503) and a lipoprotein cholesterol solution, 4.0 ml $litre^{-1}$ (Miles, cat. no. 82–018–5). I have designated the medium TYI-33 to distinguish it from TYI-S-33, in which the S stands for whole serum.

TECHNIQUE

1. STOCK CULTURES

These are maintained either in modified TP-S-1 (Section III.A.1°) or TYI-S-33 (Section III.A.2°) medium at a temperature of 35.5°C, using similar tubes in a slanted position. During the log-phase of growth, 80–90% of the flagellates are firmly attached to the walls of the tube; during stationary phase this decreases to about 60%. Subcultures are made at alternating intervals of 72 and 96 h. Cultures are selected for transfer in the same way as are those of *Entamoeba* (see Section II.A.1°, Technique, 1 (ii)). These are chilled 10–15 min in ice-water, then inverted several times to loosen the adherent cells. Portions of the suspension are transferred to tubes of fresh medium. The yield of flagellates in TYI-S-33 is approximately twice that obtained in TP-S-1. The generation time in TYI-S-33 is 6–7 h, and cell densities of 2×10^6 are routinely attained.

2. MASS CULTIVATION

For those requiring massive numbers of *Giardia*, two methods are available. One entails the use of specially designed ''outside-in'' roller bottles (Farthing *et al.*, 1982). The other, a rather cumbersome system, uses a 19 litre glass carboy containing 20 hollow glass tubes, 50 mm in diameter, which is connected to a Pellicon Cassette System fitted with a Durapore HVLP 0.45 μm filter (Millipore) (Wieder *et al.*, 1983).

3. COLONIAL GROWTH IN SOFT AGAROSE MEDIUM (Gillin and Diamond, 1980)

This technique, originally devised for *Giardia lamblia*, should be suitable for other species.

(i) *Stock agar suspension* (2.5% w/v). Place 2.5 g Sigma agarose III in a dry screw-capped serum bottle (125 ml capacity). Add 97.5 ml TP (Section III.A.1°) or TYI broth (for culturing *Giardia,* Section III.A.2°). Autoclave (121°C, 15 min) to dissolve and sterilize agarose. Remove from autoclave and place in 60°C water bath.

(ii) *Preparation of soft argarose medium and its use*. Place tubes containing 13 ml of TP-S-1 or enhanced TYI-S-33 medium

(Section III.A.1° and 2°) in a 40–42°C water bath, and equilibrate (15 min). Add appropriate amounts (see below) of agarose stock solution with warmed pipette. Mix contents of tubes by inverting them gently, and immediately add *Giardia*. Again invert tubes several times to disperse the *Giardia*. Set agarose rapidly by chilling in ice-water bath (10 min). The number of colonies formed was directly proportional to the number of *Giardia* inoculated; inocula of fewer than 100 resulted in few or no colonies, whereas inocula of about 750 produced about 150 colonies. Colonies become visible after 5–6 d, and are counted after 7 d. NB. Colony-forming efficiency is dependent upon the concentration of agarose, and its rapid gelling after inoculation with *Giardia*. The optimal concentration of agarose, which is between 0.18% and 0.2%, must be determined for each strain. Colony-forming efficiencies are consistently higher for attached cells than for free-swimming cells. The range for the former is between 20 and 40%. To rid the parent culture of the free-swimming flagellates, the old medium is decanted and replaced with fresh medium. The remaining attached cells are freed from the glass surface of the tube by chilling for 10 min in ice-water.

IV. CULTIVATION OF *TRICHOMONAS*

A. AXENIC CULTIVATION

Most of the trichomonads that have been grown xenically can be grown axenically. The two most commonly used media for accomplishing this are CPLM (Johnson and Trussell, 1943), and TYM (Diamond, 1957) (see Taylor and Baker, 1968, pp. 104 and 84 respectively; Taylor and Baker, 1978; and Diamond, 1983 for descriptions of these and other media).

The most important recent advances in the axenic cultivation of *Trichomonas*, as stated in the Introduction, are the development by Linstead (1981) of two synthetic defined media, DL7 and DL8, for *T. vaginalis*, and a partially defined medium, HUT, by Wang *et al.* (1984) for *T. foetus*.

1° DL8 defined medium for *T. vaginalis*
(Linstead, 1981)

This is the less complicated of the two media devised by Linstead.

MEDIUM

1. BASAL MEDIUM

(i) *Salts*	$litre^{-1}$
NaCl	2000 mg
KCl	2000 mg
KH_2PO_4	1000 mg
K_2HPO_4	500 mg
Ferrous gluconate (Serva)	55 mg
$MgSO_4.H_2O$	750 mg
$CaCl_2.6H_2O$	350 mg
(ii) *Amino acids*	
Glutamine	440 mg
MEM amino acid solution 50 times concentrate (Gibco)	30 ml
MEM non-esssential amino acid solution 100 times concentrate (Gibco)	15 ml
(iii) *Nucleic acid precursors*	
Adenine sulphate	70 mg
Cytidine	50 mg
Uridine	50 mg
Guanosine	30 mg
Thymidine	15 mg
(iv) *Carbohydrates*	
Maltose	5000 mg
DL-glycerophosphate disodium.$6H_2O$	100 mg
D(+)glucosamine-HCl	100 mg
D(+)galactosamine-HCl	50 mg
N-acetyl-D-glucosamine	10 mg
N-acetylneuraminic acid (Sigma type VI)	10 mg
Ascorbic acid	1000 mg

(v) *Trace elements solution (200 times concentrate)*	5.0 ml
	$litre^{-1}$
$FeSO_4.7H_2O$	1000 mg
$ZnSO_4.7H_2O$	1000 mg
$CoSO_4.7H_2O$	500 mg
$CuSO_4.5H_2O$	200 mg
$MnSO_4.4H_2O$	100 mg
$Na_2MoO_4.2H_2O$	100 mg
KI	6.0 mg

Adjust pH to 2.0 with 0.1M H_2SO_4

(vi) *Vitamin solution C (100 times concentrate)*	10 ml
	dl^{-1}
d-Biotin	20 mg
D-pantothenic acid hemi-calcium salt	100 mg
Nicotinamide	50 mg
Pyridoxamine-HCl	50 mg
Riboflavin	10 mg
Thiamine-HCl	10 mg
p-Aminobenzoic acid	5.0 mg
Coenzyme Z_{10} (ethanol solution)	1.0 mg
Choline chloride	10 mg
i-Inositol	100 mg
Glutathione (reduced)	100 mg
Folic acid (dissolved with NaOH)	20 mg
Vitamin B_{12}	10 mg
Deoxyribose	50 mg
Dithiothreitol	100 mg
Tween 80	100 mg

Adjust pH to 7.4 with 1M NaOH.

Adjust final pH of basal medium to 7.1 with 1M NaOH, and sterilize using a 0.22 μm filter. Final osmolality is 220 mosmol kg^{-1}.

Store trace elements solution (v) at 4°C, vitamin solution C (vi) in the dark at −20°C, in 10 ml aliquots, and single-strength basal medium in dark at −20°C, in 100 ml aliquots.

2. ARTIFICIAL SERUM

(i) Cholesterol (chromatographic grade, Sigma). Add at 20 mg ml^{-1} to analytical grade ethanol, warmed to 37°C.

(ii) DL-glyceryl-1-palmitate-2-oleate-3-stearate (GPOS; Calbiochem). Add at 20 mg ml^{-1} to warmed ethanol as in (i).

(iii) Bovine serum albumin (Armour "Fraction V"). Prepare as 5% (w/v) solution in distilled water. Adjust pH to 7.1 with 0.1 M NaOH.

Combine 50 μl of (i) and 50 μl of (ii) in a sterile 20 ml screw-cap bottle. Place 5 ml of (iii) in second sterile bottle. Warm both bottles to 37°C. Using sterile technique, rapidly add contents of bottle containing (iii) to bottle containing (i) and (ii). Mix thoroughly. The resulting emulsion can be stored for several months at 4°C. NB. optimum results depend on carefully following the directions.

3. COMPLETE MEDIUM

Dispense basal medium in 6.0 ml amounts to 8 × 100 mm screw-capped culture tubes. Add 0.5 to 0.6 ml of artificial serum.

Notes

(i) Cell densities of 1×10^6 ml^{-1} and a doubling time of 10 h were routinely obtained for an established culture of Bushby's strain.

(ii) Linstead notes that DL8 supplemented with 8% human serum could be used to isolate *T. vaginalis* from clinical material. No information is provided regarding initiation and routine maintenance of an isolate in DL8 supplemented with the artificial serum.

2° HUT partly defined medium for *Tri. foetus*
(Wang *et al.*, 1984)

HUT was designed to study the purine and pyrimidine metabolism of *Tri. foetus*, and is a modification of Linstead's DL8 medium. Linstead (1981) was able to isolate *Tri. foetus* from host material using DL8 supplemented with 8% human serum. Only the modifications to DL8 will be listed.

MEDIUM

(i) All the nucleic acid precursors of DL8 were replaced with a combination of hypoxanthine (20 mg $litre^{-1}$), uracil (20 mg $litre^{-1}$), and thymidine (0.2 mg $litre^{-1}$).

(ii) Folic acid, *p*-aminobenzoic acid, vitamin B_{12}, *i*-inositol, and deoxyribose were deleted.

(iii) Ferrous gluconate was replaced by ferric ammonium citrate.

(iv) Whole serum was replaced by 10% dialysed, heat-inactivated horse serum.

(v) The final pH was adjusted to 7.5, with 1M NaOH.

HUT is dispensed to plastic tubes or flasks[13]. It can be stored at 0–4 °C for at least 1 month.

TECHNIQUE

Cultures are initiated with flagellates grown in TYM medium (Diamond, 1957, 1983 p. 93; Taylor and Baker, 1968, p. 84; see also 3° below). The initial cell density is kept at 5×10^5 flagellates per ml. The stationary phase population, reached in 24 h, becomes stabilized at $5–7 \times 10^6$ cells per ml after the sixth serial subculture. Thereafter, as few as 50 cells per ml can be used to maintain a culture. Under these conditions, maximum growth is attained in 168 h. Incubation is carried out anaerobically under nitrogen gas at 37 °C[13].

Colonial growth of *Tri. foetus* can be obtained on agar poured plates by adding 0.53% Difco Noble agar to HUT medium.

3° TYM-caecal extract medium for *Tri. muris*

(Saeki *et al.*, 1983)

This is the first report of the cultivation of *Tri. muris* in any culture system. The medium consists of TYM (Diamond, 1957) combined with an extract prepared from caeca of mice.

MEDIUM

1. **NUTRIENT BROTH TYM**

Casein Digest Peptone (BBL) ..	2.0 g
Yeast extract (BBL)	1.0 g
Maltose	0.5 g
L-Cysteine-HCl	0.1 g
Ascorbic acid	0.02 g

2. **CAECAL EXTRACT**

(i) Phosphate buffer

KH_2PO_4	80.0 mg
K_2HPO_4	80.0 mg
Distilled H_2O	90.0 ml

[14]Neither the size of the culture vessels nor the amount of medium used per vessel are given. According to the authors, since incubation is carried out anaerobically under nitrogen gas, the vessels can be filled to any partial level with medium. No mention is made of gassing the culture vessels before placing them in the incubator, an important detail.

(ii) Remove caeca together with their contents from Std-dd mice[14]. Mince and dilute with 25 volumes phosphate buffer (i). Stir 6 h at 4 °C. Centrifuge (3500 rpm, 15 min). Autoclave (121 °C, 15 min).

3. COMPLETE MEDIUM

Dissolve ingredients of TYM (1 above) in 90 ml caecal extract. Adjust pH to 7.0–7.2 with 1N NaOH. Autoclave (10 min, 120 °C). Cool to 48 °C and add 10 ml of heat-inactivated sheep serum sterilized by passage through a Millipore filter (0.22 μm pore size). Add sterilized agar to make a final concentration of 0.05 %. Finally, add potassium penicillin, 100 000 iu; streptomycin sulphate, 100 mg; and Fungizone™, 500 μg. Dispense in 5.0 ml amounts to 11 × 100 mm test tubes plugged with rubber stoppers. Store at 4 °C until used.

TECHNIQUE

1. ISOLATION OF TROPHOZOITES OF *TRI. MURIS*

Suspend caecal contents of infected mice in a large volume of prewarmed Hanks's solution. Filter through 100-mesh, stainless steel sieve. Centrifuge filtrate (1000 rpm, 10 min). Remove supernate with pipette and discard. Resuspend sediment in Hanks's solution. Carefully layer 4 volumes of suspension on 3 volumes Ficol-Conray 800 solution ($\rho = 1.077$, Lymphoprep, Nyegaard). Centrifuge (1200 rpm, 15 min). Collect second layer from top which contains the flagellates and suspend in Hanks's solution to a concentration of 10^6–10^7 cells per ml.

2. INITIATION AND MAINTENANCE OF AXENIC CULTURES

Inoculate tubes of prewarmed medium with 10^5–10^6 parasites contained in 0.1 ml of Hanks's solution. Established cultures yield a 2–8-fold increase in the number of cells after 48 h incubation at 37 °C. Subculturing is performed by centrifuging a thriving culture (1000 rpm, 5 min) and transferring 10^5–10^6 cells to tubes of fresh medium.

B. XENIC CULTIVATION

Practically all culture media devised for xenic cultivation of *Entamoeba* can be used for cultivating trichomonads with a mixed bacterial flora (see Taylor and Baker, 1968).

[15]I presume any mouse strain can be used.

ACKNOWLEDGEMENT

The author thanks Dr Pierre-Marc Daggett for a critical review of the manuscript.

REFERENCES

Baveja, U.K., Jyoti, A.S., Anand, B.S. and Agarwal, D.S. (1984). Optimum conditions for growth of *Giardia lamblia in vitro*. *Journal of Communicable Diseases* **16**, 184–189.

Diamond, L.S. (1957). The establishment of various trichomonads of animals and man in axenic cultures. *Journal of Parasitology* **43**, 488–490.

Diamond, L.S. (1962). Axenic cultivation of *Trichomonas tenax*, the oral flagellate of man. I. Establishment of cultures. *Journal of Protozoology* **9**, 442–444.

Diamond, L.S. (1968a). Improved method for the monoxenic cultivation of *Entamoeba histolytica* Schaudinn, 1903 and *E. histolytica*-like amoebae with trypanosomatids. *Journal of Parasitology* **54**, 715–719.

Diamond, L.S. (1968b). Techniques of axenic cultivation of *Entamoeba histolytica* Schaudinn, 1903 and *E. histolytica*-like amoebae. *Journal of Parasitology* **54**, 1047–1056.

Diamond, L.S. (1970). Serum requirements of axenically cultivated *Entamoeba histolytica*. *Journal of Parasitology* **56**, 79.

Diamond, L.S. (1982). A new liquid medium for xenic cultivation of *Entamoeba histolytica* and other lumen-dwelling protozoa. *Journal of Parasitology* **68**, 958–959.

Diamond, L.S. (1983). Lumen-dwelling protozoa: *Entamoeba*, trichomonads and *Giardia*. *In* "*In Vitro* Cultivation of Protozoan Parasites" (J.B. Jensen, ed.), pp. 65–109. CRC Press, Boca Raton, Florida.

Diamond, L.S. and Bartgis, I.L.B. (1970). *Entamoeba moshkovskii*: axenic cultivation. *Experimental Parasitology* **28**, 171–175.

Diamond, L.S., Harlow, D.R. and Cunnick, C. (1978). A new medium for the axenic cultivation of *Entamoeba histolytica* and other *Entamoeba*. *Transactions of the Royal Society of Tropical Medicine and Hygiene* **72**, 431–432.

Dougherty, E.C. (1959). Introduction to axenic culture of invertebrate metazoa: a goal. *Annals of the New York Academy of Sciences* **77**, 27–54.

Farthing, M.J.G., Pereira, M.E.A. and Keusch, G.T. (1982). *Giardia lamblia*: evaluation of roller bottle cultivation. *Experimental Parasitology* **54**, 410–415.

Gillin, F.D. and Diamond, L.S. (1978). Clonal growth of *Entamoeba histolytica* and other species of *Entamoeba* in agar. *Journal of Protozoology* **25**, 539–543.

Gillin, F.D. and Diamond, L.S. (1980). Clonal growth of *Giardia lamblia* trophozoites in a semisolid agarose medium. *Journal of Parasitology* **66**, 350–352.

Gillin, F.D. and Diamond, L.S. (1981). *Entamoeba histolytica* and *Giardia lamblia*: growth responses to reducing agents. *Experimental Parasitology* **51**, 382–391.

Goldman, M. (1969). *Entamoeba histolytica*-like amebae occurring in man. *Bulletin of the World Health Organization* **40**, 355–364.

Gordts, B., Retoré, P., Cadranel, S., Hemelhof, W., Rahman, M. and Butzler, J.P. (1984). Routine culture of *Giardia lamblia* trophozoites from human duodenal aspirates. *Lancet* **ii**, 137–138.

Griffin, A.M. and McCarten, W.G. (1949). Some methods for the qualitative study of entozoic amoebae in cultures. *Journal of Parasitology* **35**, 193–198.

Johnson, G. and Trussell, R.E. (1943). Experimental basis for the chemotherapy of *Trichomonas vaginalis* infestations. I. *Proceedings of the Society for Experimental Biology and Medicine* **54**, 245–249.

Kasprzak, W. and Majewska, A.C. (1983). Isolation and axenic growth of fresh *Giardia intestinalis* in TPS-1 medium. *Transactions of the Royal Society of Tropical Medicine and Hygiene* **77**, 223–234.

Keister, D.B. (1983). Axenic culture of *Giardia lamblia* in TYI-S-33 medium supplemented with bile. *Transactions of the Royal Society of Tropical Medicine and Hygiene* **77**, 487–488.

Linstead, D. (1981). New defined and semi-defined media for cultivation of the flagellate *Trichomonas vaginalis*. *Parasitology* **83**, 125–137.

Meerovitch, E. (1978). *Entamoeba, Giardia*, and *Trichomonas*. *In* "Methods of Cultivating Parasites *In Vitro*," (A.E.R. Taylor and J.R. Baker, eds), pp. 19–37. Academic Press, London and New York.

Meyer, E.A. (1970). Isolation and axenic cultivation of *Giardia* trophozoites from the rabbit, chinchilla, and cat. *Experimental Parasitology* **27**, 179–183.

Meyer, E.A. (1976). *Giardia lamblia*: isolation and axenic cultivation. *Experimental Parasitology* **39**, 101–105.

Nash, T.E., McCutchan, T., Keister, D., Dame, J.B., Conrad, J.D. and Gillin, F.D. (1985). Restriction-endonuclease analysis of DNA from 15 *Giardia* isolates obtained from humans and animals. *Journal of Infectious Diseases* **152**, 64–73.

Rees, C. (1928). The infectivity and pathogenicity of a starch-fed strain of *Entamoeba histolytica*. *Journal of Parasitology* **15**, 131–134.

Reeves, R.E. (1984). Metabolism of *Entamoeba histolytica* Schaudinn, 1903. *Advances in Parasitology* **23**, 105–142.

Reeves R.E. and Bischoff, J.M. (1968). Classification of *Entamoeba* species by means of electrophoretic properties of amebal enzymes. *Journal of Parasitology* **54**, 594–600.

Reeves, R.E., Meleney, H.E. and Frye, W.W. (1957). A modified Shaffer-Frye technique for the cultivation of *Entamoeba histolytica* and some observations on its carbohydrate requirements. *American Journal of Hygiene* **66**, 56–62.

Robinson, G.L. (1968). The laboratory diagnosis of human parasitic amoebae. *Transactions of the Royal Society of Tropical Medicine and Hygiene* **62**, 285–294.

Saeki, H., Togo, M., Imai, S. and Ishii, T. (1983). A new method for the serial cultivation of *Tritrichomonas muris*. *Japanese Journal of Veterinary Science* **45**, 151–156.

Sargeaunt, P.G. and Williams, J.E. (1978). Electrophoretic isoenzyme patterns of *Entamoeba histolytica* and *Entamoeba coli*. *Transactions of the Royal Society of Tropical Medicine and Hygiene* **72**, 164–166.

Shaffer, J.G. and Frye, W.W. (1948). Studies on the growth requirements of *Entamoeba histolytica*. I. Maintenance of a strain of *E. histolytica* though one hundred transplants in the absence of an actively multiplying bacterial flora. *American Journal of Hygiene* **47**, 214–221.

Tamburro, K.M. and Hutner, S.H. (1971). Carbohydrate-free media for *Crithidia*. *Journal of Protozoology* **18**, 667–672.

Taylor, A.E.R. and Baker, J.R. (1968). "Cultivation of Parasites *In Vitro*." Blackwell Scientific Publications, Oxford.

Taylor, A.E.R. and Baker, J.R. (eds) (1978). "Methods of Cultivating Parasites *In Vitro*." Academic Press, London and New York.

Visvesvara, G.S. (1980). Axenic growth of *Giardia lamblia* in Diamond's TP-S-1 medium. *Transactions of the Royal Society of Tropical Medicine and Hygiene* **74**, 213–215.

Wang, C.C., Wang, A.L. and Rice, A. (1984). *Tritrichomonas foetus*: partly defined cultivation medium for study of the purine and pyrimidine metabolism. *Experimental Parasitology* **57**, 68–75.

Wieder, S.C., Keister, D.B. and Reiner, D.S. (1983). Mass cultivation of *Giardia lamblia* in a serum-free medium. *Journal of Parasitology* **69**, 1181–1182.

Chapter 2

Rumen Entodiniomorphid Protozoa

G.S. COLEMAN

Biochemistry Department, Agricultural and Food Research Council, Institute of Animal Physiology, Babraham, Cambridge, England

I. INTRODUCTION

There are two main types of ciliate protozoa present in the rumen, those of the order Trichostomatida (commonly known as holotrich protozoa) usually represented only by *Isotricha intestinalis*, *I. prostoma* and *Dasytricha ruminantium*, and those of the order Entodiniomorphida of which there are many genera and species*. Of the holotrich protozoa only *D. ruminantium* has been cultured for longer than 60 d *in vitro* (Clarke

*In the classification proposed by Small and Lynn (1985), *Isotricha* and *Dasytricha* are placed in the order Vestibuliferida which, with the Entodiniomorphida, constitutes the subclass Trichostomatia [Eds].

and Hungate, 1966); since it was necessary to carry out two daily manipulations separated by an interval of 2–4 h, the method was very time-consuming and tedious. In contrast, all the entodiniomorphid protozoa tested have been cultured *in vitro* with varying degrees of success, although it is still necessary to feed the cultures every day. The methods available will be considered below. The rationale of these is based on reproducing the conditions in the rumen, i.e. a constant warm temperature and an anaerobic environment deficient in readily metabolizable compounds but rich in many particles such as bacteria, starch grains, chloroplasts and cellulose fibres. The media used, therefore, are all oxygen-free, contain viable bacteria but none of the usual constituents, such as peptone, of bacteriological media, are incubated at 39 ± 2°C and have the minimum amount (necessary for protozoal growth) of particulate food such as starch grains added to them each day. As these food particles are readily engulfed by the protozoa, but are comparatively resistant to bacterial attack, they promote protozoal growth and minimize that of bacteria. No rumen ciliate protozoon has been grown axenically for more than a few days (see Section IV.E).

II. IDENTIFICATION AND CLASSIFICATION

As rumen protozoa cannot be freeze-dried, and because there is no type collection of living organisms, anyone wishing to culture them must start with crude rumen contents obtained from the host animal by stomach tube or rumen cannula. The investigator must therefore be able to identify the protozoon he wishes to isolate; this is not always easy, especially with protozoa containing many food particles, which make it impossible to see any internal detail.

The literature on the classification of these protozoa is difficult to follow, as some organisms have had five or six names since they were first recognized. The descriptive papers and books are unfortunately not always easy to obtain. The beginner is advised to consult Kofoid and MacLennan (1930, 1932, 1933), Dogiel (1927), Ogimoto and Imai (1981) and Coleman (1979). The first four contain many line drawings; the fourth is written in German. Ogimoto and Imai (1981, pp. 32–49) provide over 130 black-and-white photomicrographs of rumen ciliates that have been separated by treatment with 30% sucrose and stained with methyl green. The shape of some of the protozoa is slightly distorted and the skeletal plates are not as obvious as in iodine-stained preparations. Coleman (1979) gives a few photomicrographs of commonly found entodiniomorphid protozoa stained with iodine. Any beginner trying to identify rumen ciliates is advised to begin by staining starved protozoa with iodine. This will show up the highly characteristic skeletal plates if any are present. The presence, size and distribution of caudal spines is very variable in clone cultures of rumen ciliate protozoa and is a poor taxonomic character. Both Hungate (1966) and I believe that there are many fewer species than have been described and that each species has several variant forms.

III. MATERIALS AND GENERAL METHODS

A. BASAL MINERAL SALTS SOLUTIONS

1. *Hungate-type*

This salts medium is based on that of Hungate (1942) and contains (per litre): NaCl, 5.0 g; $CH_3.COONa$, 1.5 g; K_2HPO_4, 1.0 g; KH_2PO_4, 0.3 g. Before use the medium (30 ml) is gassed vigorously with 95% (v/v) N_2 + 5% (v/v) CO_2 for 2 min before sealing the flask with a rubber bung.

2. *Coleman-type*

This salts solution is based on that of Coleman (1958) and is usually made at double the final concentration. The concentrated solution contains (per litre): K_2HPO_4, 12.7 g; KH_2PO_4, 10.0 g; NaCl, 1.3 g; $CaCl_2$ (dried), 0.09 g; $MgSO_4.7H_2O$, 0.18 g. When this solution is prepared, a precipitate will form unless care is taken not to add the calcium salt (dissolved in at least 10% of the H_2O) until the phosphates have been dissolved in most of the water. This concentrated solution can be made up in bulk, as it will keep indefinitely.

3. *Caudatum-type*

This consists of 50 ml concentrated Coleman-type salts solution, 50 ml H_2O, 0.5 ml 15% (w/v) $CH_3.COONa$, and 1.0 ml 2% (w/v) L-cysteine hydrochloride (neutralized with 1 M NaOH, 0.6 ml per 0.1 g cysteine HCl, immediately before use). If required, the medium without cysteine can be autoclaved (115 °C, 20 min) without a precipitate forming. Before use the complete medium is gassed vigorously with 95% (v/v) N_2 + 5% (v/v) CO_2 (or N_2 in established cultures — see Section III. G) for 2 min before sealing the flask with a rubber bung.

4. Simplex-type

This consists of 40 ml concentrated Coleman-type salts solution, 60 ml H_2O, 15 ml 5% (w/v) $NaHCO_3$ and 1.1 ml 2% (w/v) L-cysteine hydrochloride (neutralized as described in para. 3). CO_2 is bubbled vigorously through the mixture for 2 min before sealing the flask with a rubber bung.

B. RUMEN FLUID FOR MEDIA

All rumen fluid for media is taken from sheep (although other ruminants could presumably be used), via a rumen cannula, 1–5 h after feeding, and strained through one layer of muslin to remove the larger food particles. This strained fluid is centrifuged (500*g*, 5 min) to remove the protozoa. The supernatant fluid should be stored at 4 °C for 24 h before use, to kill any residual protozoa. In an emergency, the fluid can be filtered through a Ford "Sterimat CC/01" to remove these protozoa, but the method is not completely reliable. Freshly prepared rumen fluid can be stored for up to 2 weeks at 4 °C. Freshly prepared rumen fluid can be autoclaved (115 °C, 20 min) under N_2 in "medical flat" bottles.

C. DRIED GRASS

Dried grass is prepared from lawn mowings that have been air-dried in a thin layer on the bench for 3 d. They are then ground in a Lee Attrition Mill (Lee Engineering) to produce a coarse powder containing particles up to 5 mm long ("coarse dried grass") or in a ball mill for 16 h to produce a fine powder ("powdered dried grass"). These preparations can be stored indefinitely in the dark but, if sterilized in an oven (160 °C, 1 h), no longer support protozoal growth. Dried grass should be added to cultures from the end of a small spatula: with practice, the correct amount can be judged fairly accurately. In my laboratory, where grass has to be added as quickly as possible to over 200 tubes each day, experience has shown that, even when adding powdered dried grass alone, the amount added to each 50 ml tube in any one day can vary from 1–20 mg without affecting the protozoal population density, although over several days the amount added should approximate to the optimum.

D. STARCH

Rice starch (British Drug Houses) or wholemeal flour (W. Prewett, or an equivalent product from other suppliers) is added as a 1.5% (w/v) suspension in water. The wholemeal flour suspension (0.1–0.5 ml) must be added to the medium rapidly (6 × 0.15 ml samples in 3 s) through a 1 ml calibrated pipette with a large hole at the tip, as the solid matter tends to sink quickly to the bottom of the pipette if pipetting is carried out in the conventional way. In cultures of *Entodinium caudatum*, the use of rice starch sterilized by heating dry grains (160 °C, 1 h) resulted in death of the protozoa. However, the protozoa grew well in the presence of rice starch that had been heated dry (120 °C, 24 h) before use.

1. *Preparation of sterol-coated starches*

Sterol-coated starches are prepared by a method based on that of Onodera and Henderson (1980), in which 20 ml of a 1% (w/v) ethanolic solution of the sterol is mixed with 20 g of rice starch or wholemeal flour and then evaporated to dryness at 39 °C. The cycle of addition of sterol solution and evaporation is then repeated once more to produce the coated starch.

E. STRAW

Straw, as supplied by any farm, is chopped into short lengths (2–5 cm) and then put through a cutter mill (Baird and Tatlock) in which the material is sheared between rotating and stationary knives until it is small enough to pass through a screen at the bottom. This material, which is 1–2 mm long, is then ground to powder in a ball mill.

F. HAEMIN

Use a 1% (w/v) solution of haemin (Sigma — from bovine blood) in 0.01 M NaOH.

G. GASES

"Medical quality" CO_2, N_2 (oxygen-free) and 95% N_2 + 5% CO_2 (British Oxygen) can be used without prior treatment. Removal of traces of O_2 by red-hot copper or chromous sulphate solution does not improve the growth of *En. caudatum*. As 95% N_2 + 5% CO_2 is expensive compared with oxygen-free N_2 or CO_2, the caudatum-type salts medium required for the dilution of *established* cultures can be bubbled instead with oxygen-free N_2. After feeding cultures, CO_2 can be used to gas over the top of the medium, without harming the protozoa. This modification of previously published methods (see, e.g., Coleman, 1960) is satisfactory provided that it is not used during the initial isolation of the protozoa or with cultures in which the protozoa are growing poorly.

H. CULTURE VESSELS

The cultures are normally maintained in 50, 100 or 250 ml centrifuge tubes almost completely filled with medium. In these, the protozoa are present as a loose pellet on the bottom of the tube and are therefore unaffected by any leakage of air into the vessel due to accidental displacement of the rubber bung. For the safe continuation of a culture, it is essential to divide it between at least six tubes as occasionally all the protozoa in some tubes die for no obvious reason.

I. FIXATION AND COUNTING OF PROTOZOA

The best fixative for entodiniomorphid protozoa is iodine, which also stains any starch-like material in the organisms and makes them easier to count: protozoa fixed with formalin frequently disintegrate after 24 h. Provided that sufficient protozoa are present, the best method is to add 0.5 ml protozoal culture to 4.5 ml 0.1 M I_2. This fixed material will keep indefinitely, stoppered, in the dark; the final product must be well coloured with iodine.

To sample a protozoal culture, it should be well mixed and then, using a 1.0 ml graduated pipette with a large hole at the tip, the material should be sucked up and blown into the iodine as quickly as possible. To count all the protozoa, except for *Polyplastron multivesiculatum* and *Metadinium medium*, 0.1 ml of this mixture is expelled on to a microscope slide from a 0.2 ml graduated pipette with the tip cut off (the bore of a 0.1 ml pipette being too narrow). After a cover slip is put on, all the protozoa are counted under the microscope using a ×4 or ×10 objective by tracking back and forth across the area of the liquid. With the large protozoa (*P. multivesiculatum* and *M. medium*), when fewer than 100 organisms ml^{-1} are present in culture, all the protozoa in the original fixed material are counted by placing the tube (preferably 100 × 12 mm) horizontally on the stage of an inverted microscope (Olympus Optical, model CK) and moving the tube across the field of vision over a ×4 objective. Only those protozoa showing no sign of disintegration should be counted.

IV. SPECIAL TECHNIQUES

A. INITIAL ISOLATION OF A PROTOZOAL SPECIES

1. Picking protozoa from a mixed population

No selective medium is available, except for *En. caudatum*. It is therefore necessary, in order to isolate most entodiniomorphid protozoa, to pick out individual protozoa from rumen contents or a mixed culture *in vitro*. Provided that the required protozoon will compete successfully with the other protozoa present, it is usually easier to begin by inoculating crude rumen contents into culture medium and then, when the protozoa are growing well, to pick individuals out from this. If this approach is adopted, it is recommended that the crude rumen contents be inoculated into caudatum and simplex salts in the presence and absence of freshly prepared rumen fluid and the protozoa fed on grass and wholemeal flour or on grass alone. This has the added advantage of providing information on the culture conditions most likely to favour the survival and growth of the selected organism. A suitably diluted drop of the material is placed on a slide under an inverted microscope and individual protozoa sucked up through a micropipette attached to a LeFonbrune suction and force syringe (Beaudouin). The position of the micropipette is adjusted by attaching it to a Brinkman CP5 micromanipulator (Camlab).

The two main difficulties, apart from the manual dexterity involved, are (a) the necessity for speed, as the protozoa are strict anaerobes and are exposed to air during the procedure, and (b) the risk of removing small, transparent *Entodinium* spp. along with the larger protozoon that it is hoped to isolate. Three drops of caudatum-type salts medium containing 10% prepared fresh rumen fluid are placed on a slide next to the crude material. The required protozoa are then removed from the crude material without worrying overmuch whether other protozoa are picked up as well. All the material is then placed in one of the drops of medium. This process is continued through the drops of medium until individuals of just one protozoal species can be picked up easily in the pipette. Less than 5 min should elapse between putting crude material on the slide and placing the protozoa in a tube of growth medium. If it takes longer, the whole process should be begun again as the protozoa are unlikely to survive.

The chance of successfully culturing a protozoal species from individuals picked out in this way is improved if 5–10 protozoa can be inoculated into the medium. This is a safe procedure if the wanted protozoon is morphologically distinct from the others present, e.g. when the protozoon is the only *Ophryoscolex* species present, but it may result in a mixed culture when trying to grow one *Diplodinium* species morphologically similar to others present.

TABLE I. Conditions for the isolation and cultivation of entodiniomorphid protozoa

	Isolation method	Preferred source of starch[a]	Growth on grass alone in absence of starch[b]	Growth on straw alone in absence of starch[c]
1. *Diplodinium monacanthum*	Single cell	WF 0.05	+	?
2. *Diplodinium pentacanthum*	Single cell	(WF 0.05)	++	++
3. *Diploplastron affine*	Single cell	WF 0.05	(++)	–
4. *Enoploplastron triloricatum*	Single cell	(WF 0.05)	++	?
5. *Entodinium bursa*	Single cell	WF 0.03	–	–
6. *Entodinium caudatum*	Enrichment	RS 0.15	–	–
7. *Entodinium furca bilobum*	Single cell	WF 0.05	–	–
8. *Entodinium longinucleatum*	Single cell	WF 0.1	–	–
9. *Entodinium simplex*	Differential centrifugation	WF 0.02	–	–
10. *Epidinium ecaudatum caudatum* (bovine)	Single cell	WF 0.1	?	–
11. *Epidinium ecaudatum caudatum* (ovine)	Single cell	WF 0.1	+	–
12. *Epidinium ecaudatum tricaudatum*	Single cell	WF 0.1	?	–
13. *Eremoplastron bilobum*	Single cell	WF 0.05	++	?
14. *Eremoplastron bovis*	Single cell	WF 0.05	++	?
15. *Eudiplodinium maggii*	Single cell	(WF 0.03) (RS 0.03)	++	+
16. *Metadinium medium*	Single cell	(WF 0.03)	++	–
17. *Ophryoscolex caudatus*	Single cell	WF 0.25	–	–
18. *Ostracodinium obtusum bilobum*	Single cell	WF 0.05 (RS 0.05)	++	++
19. *Polyplastron multivesiculatum*(+)	Single cell	WF 0.05		–
20. *Polyplastron multivesiculatum*(–)	Single cell	WF 0.05 RS 0.05	(++)	+

[a] WF = wholemeal flour; RS = rice starch; the figures are the optimum amount of the material (mg ml^{-1}) to be added each day; parentheses indicate that, although the protozoon will grow for a limited time on starch, this is not the preferred substrate.

[b] + = protozoon will grow poorly on grass alone; ++ = protozoon will grow well on grass alone; ? = not determined. Symbols in parentheses do not apply to all strains.

[c] + = protozoon will grow for a limited time on straw alone; ++ = protozoon will grow well for over a year on straw alone; ? = not determined.

2. *Treatment of newly inoculated cultures*

The newly picked protozoa are inoculated into 3.0 ml of the appropriate salts medium (see Table I under individual species) containing 10%

TABLE I. *Continued*

Stimulation of growth of established cultures by rumen fluid[d]	Salts medium preferred	Method of increasing culture	Requirement for other protozoa[f]	Maximum population density (protozoa ml^{-1})		Maximum life of culture[g]
				Starch & grass	Grass alone	
—	Simplex	Dilution	–			13 months
(FRF)	Caudatum or simplex	Dilution	–	400	800	14 months
(FRF)	Caudatum	Dilution	–	100	200	27 months
—	Simplex	Dilution	–		250	24 months
FRF	Caudatum	Dilution	+	150		12 months
FRF,ARF	Caudatum + CAP[e]	Replace medium and dilution	–	32 000		>26 years
FRF	Simplex	Dilution	–			3 months
FRF,ARF	Caudatum	Dilution	–	2 100		15 months
FRF	Simplex	Dilution	–	27 000		8 years
—	Caudatum or simplex	Dilution	–	500		15 months
—	Caudatum or simplex	Dilution	–	1 200	20	21 months
—	Caudatum or simplex	Dilution	–	1 000		18 months
—	Simplex	Dilution	–			>10 months
(FRF)	Caudatum	Dilution	–		1000	15 months
—	Caudatum	Dilution	–	330	350	2 years
—	Caudatum	Dilution	–		2	6 months
FRF	Caudatum	Dilution	–	120		29 months
(FRF)	Simplex	Dilution	–	200	100	> 2 years
(FRF)	Caudatum	Dilution	(+)	50		3 years
(FRF)	(Caudatum) Hungate	Dilution	–	100	20	12 months

[d] FRF = prepared fresh rumen fluid; ARF = autoclaved rumen fluid. Abbreviations in parentheses do not apply to all strains.
[e] CAP = chloramphenicol, 48 $\mu g\ ml^{-1}$.
[f] Symbols in parentheses do not apply to all strains.
[g] The symbol > indicates that at the time of writing a culture was still alive and growing well.

freshly prepared rumen fluid in a 125 mm × 12 mm tube. For protozoa which use starch, one drop of a dilute suspension of wholemeal flour (obtained by allowing a 1.5% w/v suspension of wholemeal flour to settle for 1–2 min and removing the liquid from near the surface) is added together with a few mg of coarse dried grass. It is essential that most of the grass sinks to the bottom of the tube or the protozoa will not survive. For cellulolytic protozoa (whose preferred substrate is grass alone), add only a few mg of powdered dried grass. The tubes are then gassed with the appropriate gas, sealed with a rubber bung and incubated at 39°C.

The tubes are opened daily, starch and/or grass is added as before and the tubes resealed. After a week the protozoa that survived the isolation process should have increased in numbers sufficiently to be easily visible under the microscope. If an inverted microscope is used, the protozoa can be observed without opening the tubes. If no protozoon is visible, the process should be continued for a further week and then, if there is still none present, the tubes discarded. In my experience the success rate varies from nil to 100%, depending on the protozoon, but is usually about 50%. A week after inoculation, the cultures should be diluted with an equal volume of the same medium and the daily feeding continued for a further week. If the protozoa repeatedly refuse to grow, it is worthwhile repeating the procedure but replacing the fresh medium by spent medium from a crude culture from which the protozoa have been removed by centrifugation, by incubating the culture at 4°C for 24 h, or by ceasing to feed it.

After this time the procedure adopted will depend on how well the protozoa are growing. If they are growing well (if over 50 active protozoa can be seen at first glance under the microscope), the contents of the tube can be poured into 30 ml of the same medium in a 50 ml centrifuge tube and bi-weekly dilution of the cultures can be carried out as described below. If the protozoa are growing less well (if 10–50 protozoa are visible), pour the contents of the tube into an equal volume of medium in a 15 ml centrifuge tube and a week later dilute this again with an equal volume of fresh medium. If only a few protozoa are visible, pipette off the top 80% of the culture medium and replace it with an equal volume of fresh medium; then treat as above when sufficient protozoa are present. It must be emphasized that, during all these procedures, daily feeding must be continued, and that the amount added, especially of starch, should be increased as the number of protozoa increases.

Unfortunately, at any stage in the first month in the life of an isolate, the protozoa may suddenly all die. If several apparently successful isolates have been made, it is inadvisable to discard any until the chosen one is established in at least six tubes.

Until the protozoa are growing well and reproducibly it is advisable to continue to add 10% freshly prepared rumen fluid to the media for 1–3 months, depending on the species; this can then be omitted with those species where it is known not to be stimulatory (Table I). Similarly, with those species in which autoclaved rumen fluid is stimulatory to growth, the change from fresh rumen fluid can be made at this time.

3. *Contamination*

As it is usual when isolating individual protozoa to inoculate a series of tubes, care must be taken when feeding the cultures each day not to transfer protozoa from one tube to another. If this precaution is not taken it is easy for a vigorously growing contaminant protozoon such as *En. simplex* to be transferred from one tube, into which it was accidentally inoculated originally, to all the others in a batch. The necessary precautions include always using the same bung for a tube, flaming the pipette used to gas the tubes before insertion into each, and *never* allowing the pipette used to add the starch suspension to touch the side of a tube.

4. *Temperature*

The optimum temperature for all species is 39°C; they will not grow above 42°C or below 35°C. A few individuals in a culture of *En. caudatum* will still be alive after 12 h at room temperature, but this treatment kills all other species. Growth of *En. caudatum* at 41°C produces distorted protozoa that do not separate normally during binary fission and which grow in chains.

B. MAINTENANCE OF ESTABLISHED CULTURES

Each day, every tube of culture is opened and food is added. If the protozoon is growing on wholemeal flour or rice starch, the amount given in Table I is added (as a 1.5% w/v suspension in water) together, twice a week, with a few mg coarse dried grass per tube. If the substrate is powdered dried grass, 0.2 mg ml^{-1} culture is added. The tubes are then gassed with CO_2, sealed with a rubber bung and returned to the incubator (39°C) as soon as possible. Once or twice a week the number of tubes containing each protozoal species is doubled by division of the culture. Except for *En. caudatum* (section IV.D.6), the culture is mixed, half is poured into another tube, and each tube is made up to the original volume with fresh medium. Food is then added as usual and the tubes are gassed and sealed with a rubber bung.

When protozoa are required for metabolic experiments, the additional tubes produced at the bi-weekly division can be used for this purpose.

C. LIFE OF ESTABLISHED CULTURES

Except for *Entodinium caudatum* that has been cultured on rice starch in the presence of chloramphenicol for over 26 years, most of the other species have a finite life of 1–2 years *in vitro*. I have successfully reversed this "loss of vigour" by inoculating protozoa from a 12–15 month old culture into the rumen of a sheep that had been rendered protozoon-free by treatment with manoxol (dioctyl sodium sulphosuccinate: BDH) (Coleman and Sandford, 1980). Not all cultures will grow in the sheep if inoculated in this way, but if one does, on re-isolation after only 2 weeks in the rumen, the new culture will survive for the full 1–2 years.

D. MAINTENANCE OF INDIVIDUAL PROTOZOAL SPECIES

The basic methods for the isolation and subsequent growth and maintenance of these protozoa are given above and in Table I. The following notes are intended to supplement these and emphasize points of technique relevant to a particular genus or species. References to published work with the organism are given at the end of the section.

1. Diplodinium monacanthum

Diplodinium monacanthum has been isolated only once (on wholemeal flour and coarse dried grass) and proved difficult to culture. However much care was taken, some tubes in a batch always contained very few organisms. Once established, it grew better on powdered dried grass than on wholemeal flour and coarse dried grass (6000 compared with 2000 protozoa ml^{-1}) (Coleman *et al.*, 1976).

2. Diplodinium pentacanthum

Diplodinium pentacanthum can be isolated equally well on powdered dried grass or wholemeal flour and coarse dried grass and grows well on both substrates, although some strains tend to grow better on one than on the other. Population densities of up to 800 ml^{-1} have been found although 100–200 ml^{-1} are more common. This protozoon is one of the few that grows (50 ml^{-1}) for prolonged periods on straw as sole added source of food (Coleman *et al.*, 1976).

3. *Diploplastron affine*

Diploplastron affine shows marked strain differences in its growth requirements. Some isolates will grow well on powdered dried grass alone, without freshly prepared rumen fluid, whereas others will grow only on wholemeal flour in the presence of prepared fresh rumen fluid (Coleman *et al.*, 1976).

4. *Enoploplastron triloricatum*

Enoploplastron triloricatum must be isolated on simplex-type salts medium containing powdered dried grass or wholemeal flour and coarse dried grass. However, once established it grows equally well on caudatum- or simplex-type salts media and better without freshly prepared rumen fluid than with it. It grows better and lives longer in the presence of powdered dried grass than on wholemeal flour (Coleman *et al.*, 1976).

5. *Entodinium bursa*

Entodinium bursa engulfs *En. caudatum* and will not grow or survive *in vitro* except in the presence of living protozoa from a culture of *En. caudatum* grown as described below. It is isolated by transferring protozoa from crude rumen contents into caudatum-type salts medium containing 10% freshly prepared rumen fluid and fed initially, and daily thereafter, with wholemeal flour, coarse dried grass and a drop of *En. caudatum* culture. As the *En. bursa* grows, and provided that most of the *En. caudatum* have disappeared, it is essential to add fresh and increasing amounts of *En. caudatum* every day, even though a few are still apparently present in the culture, because after 1–3 weeks exposure to *En. bursa* those *En. caudatum* that survive engulfment grow caudal spines. These caudate protozoa are engulfed much less readily than the normal culture form and, although *En. bursa* survives in their presence, the population density is much lower. *En. bursa* will not grow on simplex-type salts medium or in the absence of rumen fluid.

To obtain large numbers of *En. bursa*, 8.0 ml of a stock culture is inoculated into 100 ml of a culture of *En. caudatum* obtained 3 d previously from the division of a culture and containing approximately

20 000 protozoa ml^{-1}. The inoculated culture is fed daily with 0.075 mg rice starch and 0.015 mg wholemeal flour ml^{-1}. After 4 d (occasionally 3 or 5 d) all the *En. caudatum* have disappeared and about 900 *En. bursa* organisms ml^{-1} are present. Occasionally, under these conditions, the size of *En. caudatum* increases to such an extent that the *En. bursa* cannot engulf them. If this happens the culture should be fed only 0.05 mg wholemeal flour ml^{-1} (Coleman *et al.*, 1977).

6. *Entodinium caudatum*

Entodinium caudatum is isolated by inoculating 0.3% v/v rumen contents into caudatum-type salts medium containing 10% prepared fresh rumen fluid and 48 μg ml^{-1} chloramphenicol (in a 50 ml tube); rice starch (0.03 mg ml^{-1}) and a few mg dried grass are added daily to each tube. Twice a week 80% of the supernatant fluid is removed and replaced by fresh medium. This process is continued until sufficient protozoa are present (20 000 ml^{-1}) for the culture to be divided into two, by first replacing the supernatant fluid and then, after mixing, pouring half into another tube and making both up to the original volume with fresh medium. This procedure always works well if the starting population is of the "A-type" (Eadie, 1962) but on one occasion, using a "B-type" population, I found that some of the larger protozoa also used the rice starch and a mixed culture resulted. In this event, *En. caudatum* must be picked out individually.

During the first week after inoculation, both the number of protozoal species present and the total number of protozoa decline until few can be seen. The number of *En. caudatum* then increases steadily and after 2 weeks the amount of rice starch added each day can be increased to 0.25 mg ml^{-1}. After several months in culture, maintenance can be facilitated by the replacement of the fresh rumen fluid in the medium by autoclaved rumen fluid, provided that the daily addition of rice starch is decreased to 0.15 mg ml^{-1}. With some strains, the rather complicated manipulations described for the division of a culture cannot be altered, whereas, with others, simple dilution will suffice. The presence of chloramphenicol is essential during the initial isolation in order to kill some of the other protozoa present, and it cannot be omitted from established cultures without the *En. caudatum* dying. The protozoa in cultures that have been maintained for over 10 years metabolize starch more slowly than freshly isolated organisms and need be fed only every 2 d. The requirement for dried grass also ceases and this can be omitted completely. The characteristic caudal spines of *En. caudatum* become shorter and disappear after a few months in culture; they never reappear unless the protozoa are exposed to *En. bursa*.

TABLE II. Steady-state population densities of some entodiniomorphid protozoa grown in the presence or absence of sterols and haemin

	Dilution frequency (week^{-1})	Haemin[b]	Population density[a] with sterol indicated					
			None[c]	β-Sitosterol	Stigmasterol	Ergosterol	α-Spinasterol	Three sterols[d]
Ophryoscolex caudatus[e]	Once	−	124	226	295	250		
		+	91	286				
	Twice	−	56	110	37	37	62	125
		+	63	181	92	97	94	136
Entodinium caudatum	Once	−		7700	4150		5220	
		+		6500	7300			
Epidinium ecaudatum caudatum	Once	−	155	730	770	640	555	935
		+	200	850			945	1120
	Twice	−		590				
		+		510				
Polyplastron multivesiculatum	Once	−	8	23	25	24		22
		+	10	25				

[a] Protozoa ml^{-1}.
[b] 20 μg ml^{-1} once a week.
[c] Plus 5.0 mg dried grass twice a week.
[d] β-Sitosterol, stigmasterol and ergosterol each at one-third of the standard concentration.
[e] Grown in the presence of freshly prepared rumen fluid (10% v/v).

Based on Coleman and Reynolds (1982).

It is possible to omit the rumen fluid if the rice starch used is coated with β-sitosterol or stigmasterol, although the population density is then lower than normal (Table II); it should be noted that rice starch as supplied by the manufacturer contains choline, which is essential for protozoal growth (Coleman, 1958, 1960; Coleman and Reynolds, 1982; Broad and Dawson, 1976).

7. *Entodinium longinucleatum*

Entodinium longinucleatum can be isolated equally well on wholemeal flour or rice starch in the presence of coarse dried grass, but established cultures grow more reliably on the former. The freshly prepared rumen fluid added for isolation can be replaced by autoclaved rumen fluid without change in the population density; omission of all rumen fluid results in the death of the culture (Owen and Coleman, 1976).

8. *Entodinium simplex*

A number of small *Entodinium* spp. (under 45 μm long) without caudal spines, e.g. *En. exiguum*, *En. nanellum*, *En. simplex* and *En. parvum*, were described by Dogiel (1927); the only one apparently present in Clun Forest sheep at Babraham is *En. simplex*. In this circumstance it can be separated from all the larger protozoa by differential centrifugation. Some rumen contents are removed 6 h after feeding the animal and strained through one layer of muslin to remove the larger food particles. The fluid is then allowed to stand in a 200 mm × 25 mm tube at 39°C for 2 h to allow the larger protozoa to fall to the bottom. The fluid from the top of the tube is then centrifuged at 600 *g* for about 3 min or until, on microscopic examination, *En. simplex* is the only protozoon present in a drop of the supernatant fluid. One drop of this fluid is then inoculated into the medium. If at any stage a protozoon with a caudal spine is observed, that culture must be discarded. As all rumen ciliates in culture tend to lose their caudal spines, it is possible to be certain that the culture is not that of another species that has lost its spine only if care is taken to ensure that no caudate protozoon has ever been present.

The fresh rumen fluid in the isolation medium can be replaced by autoclaved rumen fluid without change in the population density; omission of rumen fluid decreases the density to 10–15% (Jarvis and Hungate, 1968; Coleman, 1969).

9. Epidinium ecaudatum caudatum (bovine)

Epidinium ecaudatum caudatum, found only in cattle, has a single, straight and heavy caudal spine that tends to get smaller and shorter with time in culture (Coleman *et al.*, 1972).

10. Epidinium ecaudatum caudatum (ovine)

This subspecies, found only in sheep with a ''B-type'' population (Eadie, 1962, 1967), has a single, short, curved caudal spine that disappears after 4–5 months in culture. The form of the spines in this and the bovine form must be determined in living material as they become distorted on fixation. This protozoon is one of the easiest to isolate by picking individual organisms, as it is the least sensitive to exposure to air. It grows equally well on caudatum- or simplex-type salts medium, in the presence or absence of freshly prepared rumen fluid, when fed with wholemeal flour and coarse dried grass. *Epidinium* species attack grass particles more rapidly than other protozoa and it is essential to ensure that some grass is always present at the bottom of the tubes. However, if starch grains coated with β-sitosterol, stigmasterol, or α-spinasterol are used, the grass can be omitted and the population density increases (Table II). Haemin (20 μg ml^{-1}) is also stimulatory (Table II). There are very marked strain differences in the vigour with which *Epidinium* isolates from different sheep will grow under standard conditions, with population densities ranging from 150 to 1200 ml^{-1}.

This protozoon will also grow on powdered dried grass alone, but only at a population density of 20 ml^{-1} (Coleman *et al.*, 1972; Coleman and Reynolds, 1982).

11. Epidinium ecaudatum tricaudatum

Epidinium ecaudatum tricaudatum has three caudal spines which gradually disappear in culture, but at an intermediate stage both caudate and non-caudate organisms are present.

12. Eremoplastron bovis

Eremoplastron bovis is normally found in cattle but will also grow in the sheep rumen and I have maintained it for several years in one sheep as the only protozoon present. At intervals over 8 years individuals were removed and cultured *in vitro*. The various isolates showed very different growth requirements. When originally isolated, the organisms grew well on simplex-type salts in the absence of rumen fluid with a daily feed of powdered grass, whereas later isolates preferred caudatum-type salts, freshly prepared fluid and wholemeal flour.

13. Eudiplodinium maggii

Different strains of *Eudiplodinium maggii* behave differently with respect to their preferences for caudatum- or simplex-type salts media. Most can be isolated in simplex-type salts medium. However, some will never grow on simplex-type salts medium and must be isolated and grown subsequently on caudatum-type salts medium. This protozoon always grows well on powdered dried grass, but some strains grow better on powdered dried grass and wholemeal flour. However, unlike many protozoal species, it is essential to add the grass each day: addition only twice a week results in death of a culture. With some strains, the wholemeal flour is replacable by rice starch with no decline in protozoal numbers. The growth of most established strains is not stimulated by the addition of 10% freshly prepared rumen fluid to the medium. *Eudiplodinium maggii* will grow for a limited period (3 weeks) on powdered straw in the presence of freshly prepared rumen fluid (Coleman *et al.*, 1976).

14. Metadinium medium

I have isolated *Metadinium medium* only once, from the rumen of a cow. It grew best in caudatum-type salts medium in the absence of rumen fluid on powdered dried grass, but proved difficult to keep alive and died after 6 months.

15. Ophryoscolex caudatus

Established cultures of *Ophryoscolex caudatus* can be grown for 3–4 weeks in the absence of rumen fluid, but for reliable maintenance over long periods it is essential to add 10% freshly prepared rumen fluid. This can be replaced by autoclaved rumen fluid; but growth is less consistent under these conditions. The population density is increased in the presence of sterols (coated on the starch grain); β-sitosterol, stigmasterol, ergosterol and α-spinasterol are all stimulatory under some conditions (Table II). Haemin has a small further stimulatory effect.

The morphology of *Op. caudatus* changes markedly after 2–4 weeks in culture, the long main caudal spine becoming much shorter and twisted so that it is at right angles to, rather than parallel to, the long axis of the protozoon (for photograph of culture form, see Coleman, 1971). This morphological change is also associated with some physiological change, because some strains that had grown well after initial isolation suddenly died at this time. If they survive this change, they grow well for 2 years (Coleman and Reynolds, 1982).

16. *Ostracodinium obtusum bilobum*

All isolates of *Ostracodinium obtusum bilobum* grow well with the daily addition of powdered dried grass, but some grow better (population density increased from 100 to 200 ml^{-1}) when wholemeal flour also is added daily. The growth of some isolates can improve in the presence of freshly prepared rumen fluid. *Os. obtusum bilobum* also grows for long periods (> 1 year) on powdered straw as the sole added source of food, but only in the presence of freshly prepared rumen fluid and at low population density (50 ml^{-1}). After a few weeks in culture, this protozoon loses its two caudal lobes and comes to resemble *Os. obtusum obtusum*.

17. *Polyplastron multivesiculatum* (with *Epidinium* spp.)

Cultures of this protozoon are initiated by inoculating single organisms into 3.0 ml of a culture of *Ep. ecaudatum caudatum* (ovine) growing on caudatum-type salts medium containing 10% prepared fresh rumen fluid and fed daily, before and after the inoculation, with wholemeal flour and coarse dried grass. Some strains of *P. multivesiculatum* have, when grown on caudatum-type salts medium, an obligate requirement for one of a number of other species of rumen protozoa, e.g. *Ep. ecaudatum caudatum* (ovine or bovine), *Ep. e. tricaudatum*, *Eu. maggii* or *Diplodinium monacanthum* but not *Entodinium* spp. or *Diploplastron affine*. *Ep. e. caudatum*, which is the best food protozoon, is engulfed by *P. multivesiculatum* at a rate of 1–10 organisms per protozoon per day. When those *Epidinium* spp. initially present in the culture have disappeared, it is essential to add more daily (as 1.0 ml of a culture of *Epidinium* spp. per 30 ml *P. multivesiculatum* culture). About one *Epidinium* per *Polyplastron* per day gives good growth of *Polyplastron*. Prepared fresh rumen fluid can be omitted from established cultures, but this protozoon will not grow on simplex-type salts medium.

Some cultures of *P. multivesiculatum* will engulf only protozoa from the strain of *Ep. e. caudatum* on which they were isolated, and die when this strain dies. However, others can engulf *Epidinium* sp. from a new isolate initiated when the original dies, although in my experience no culture of *P. multivesiculatum* has ever survived the death of a second isolate of *Epidinium* sp. The maximum life of a culture (Table I) therefore depends on the life of the food protozoon (Coleman *et al.*, 1972).

18. Polyplastron multivesiculatum (without *Epidinium* spp.)

Some strains of *P. multivesiculatum* will grow *in vitro* in the absence of *Epidinium* only when individuals are placed in Hungate-type salts medium (inoculated the previous day with a drop of freshly prepared rumen fluid) and fed daily with wholemeal flour and coarse dried grass. As the maximum number of protozoa found under these conditions is never more than 5 ml^{-1}, the amount of wholemeal flour added each day must be very carefully controlled and must never exceed 0.03 mg ml^{-1}.

Other strains will grow well if individuals are inoculated into caudatum-type salts medium containing 10% freshly prepared rumen fluid to which wholemeal flour and powdered dried grass, or grass alone, are added daily. Once the cultures are established, the rumen fluid can be omitted, although the population density declines (from 100 to 60 ml^{-1} in presence of wholemeal flour). With *P. multivesiculatum*, unlike most other species, the wholemeal flour can be replaced by rice starch, though with a 40% drop in population density (Coleman *et al.*, 1972).

E. AXENIC CULTURE

All entodiniomorphid protozoa contain viable bacteria in vesicles in their endoplasm and therefore can be rendered bacteria-free only by treatment with antibiotics. Several people have successfully rendered the protozoa "bacteria-free", but nobody has persuaded such protozoa to grow under axenic conditions, although Hino and Kametaka (1977) maintained *Entodinium caudatum* alive in steadily decreasing numbers for 3 weeks. Other relevent papers are those by Coleman (1962) and Bonhomme *et al.* (1982a, b).

F. CONTINUOUS CULTURE

It is not possible to grow entodiniomorphid protozoa in a conventional continuous culture apparatus, owing, in part, to the inability of the protozoa to engulf starch grains in continuously stirred medium. Even if the stirring is stopped for a short time to allow the protozoa to feed, they still do not thrive. However, the problem has been overcome by the provision of a large dead space in which the protozoa can sequester. The successful apparatus consists of a fermentation vessel separated by a membrane (0.4 μm pore size) from a dialysing chamber through which medium is passed comparatively rapidly. Inside the fermentation vessel is a hollow perforated piston which is moved slowly up and down (8 strokes min^{-1}). Inside the piston are a number of nylon bags (245 μm mesh size) containing substrate such as crushed barley and hay. Half the bags are replaced each day by new ones containing fresh substrate so that each bag remains *in situ* for 2 d (Weller and Pilgrim, 1974; Nakamura and Kurihara, 1978). The system has now been further improved by the addition of pieces of sponge (Abe and Kurihara, 1984). Under these conditions, and at a dilution rate of 1.0 d^{-1}, a normal mixed protozoal fauna has been maintained for over 2 months with *Entodinium* population densities of 10^6 and 2×10^5 ml^{-1} in the bags and the effluent respectively. In the author's experience, the use of nylon bags containing only sponge was unsuccessful, as the protozoa enter the bags to find food and not just to shelter from the moving liquid.

REFERENCES

Abe, M. and Kurihara, Y. (1984). Long term cultivation of certain rumen protozoa in a continuous fermentation system supplemented with sponge materials. *Journal of Applied Bacteriology* **56**, 201–213.

Bonhomme, A., Fonty, G. and Senaud, J. (1982a). Essai d' obtention et de survie de ciliés Entodiniomorphes du rumen en cultures axéniques. *Annales de Microbiologie* **133B**, 335–341.

Bonhomme, A., Fonty, G. and Senaud, J. (1982b). Obtention de *Polyplastron multivesiculatum* (cilié Entodiniomorphe du rumen) en condition axénique. *Journal of Protozoology* **29**, 231–233.

Broad, T.E. and Dawson, R.M.C. (1976). Role of choline in the nutrition of the rumen protozoon *Entodinium caudatum*. *Journal of General Microbiology* **92**, 391–397.

Clarke, R.T.J. and Hungate, R.E. (1966). Culture of the rumen holotrich ciliate *Dasytricha ruminantium* Schuberg. *Applied Microbiology* **14**, 340–345.

Coleman, G.S. (1958). Maintenance of oligotrich protozoa from the sheep rumen *in vitro*. *Nature* **182**, 1104–1105.

Coleman, G.S. (1960). The cultivation of sheep oligotrich protozoa *in vitro*. *Journal of General Microbiology* **22**, 555–563.

Coleman, G.S. (1962). The preparation and survival of almost bacteria-free suspensions of *Entodinium caudatum*. *Journal of General Microbiology* **28**, 271–281.

Coleman, G.S. (1969). The cultivation of the rumen ciliate *Entodinium simplex*. *Journal of General Microbiology* **57**, 81–90.

Coleman, G.S. (1971). The cultivation of rumen entodiniomorphid protozoa. *In* "Isolation of Anaerobes" (D.A. Shapton and R.G. Board, eds), pp. 159–176. Academic Press, London and New York.

Coleman, G.S. (1979). Rumen ciliate protozoa. *In* "Biochemistry and Physiology of

Protozoa" (M. Levandowsky and S.H. Hutner, eds), Vol. 2, pp. 381–408. Academic Press, New York and London.

Coleman, G.S. and Reynolds, D.J. (1982). The effect of sterols and haemin on the growth of the rumen ciliate *Ophryoscolex caudatus* and some other entodiniomorphid protozoa. *Journal of Applied Bacteriology* **52**, 129–134.

Coleman, G.S. and Sandford, D.C. (1980). The uptake and metabolism of bacteria, amino acids, glucose and starch by the spined and spineless forms of the rumen ciliate *Entodinium caudatum*. *Journal of General Microbiology* **117**, 411–418.

Coleman, G.S., Davies, J.I. and Cash, M.A. (1972). The cultivation of the rumen ciliates *Epidinium ecaudatum caudatum* and *Polyplastron multivesiculatum in vitro*. *Journal of General Microbiology* **73**, 509–521.

Coleman, G.S., Laurie, J.I., Bailey, J.E. and Holdgate, S.A. (1976). The cultivation of cellulolytic protozoa isolated from the rumen. *Journal of General Microbiology* **95**, 144–150.

Coleman, G.S., Laurie, J.I. and Bailey, J.E. (1977). The cultivation of the rumen ciliate *Entodinium bursa* in the presence of *Entodinium caudatum*. *Journal of General Microbiology* **101**, 253–258.

Dogiel, V.A. (1927). Monographie der Familie Ophryoscolecidae. *Archiv für Protistenkunde* **59**, 1–227.

Eadie, J.M. (1962). Inter-relationships between certain rumen ciliate protozoa. *Journal of General Microbiology* **29**, 579–588.

Eadie, J.M. (1967). Studies on the ecology of certain rumen ciliate protozoa. *Journal of General Microbiology* **49**, 175–194.

Hino, J. and Kametaka, M. (1977). Gnotobiotic and axenic cultures of a rumen protozoon, *Entodinium caudatum*. *Journal of General and Applied Microbiology* **23**, 37–48.

Hungate, R.E. (1942). The culture of *Eudiplodinium neglectum*, with experiments on the digestion of cellulose. *Biological Bulletin, Marine Biological Laboratory, Woods Hole* **83**, 303–319.

Hungate, R.E. (1966). "The Rumen and its Microbes". Academic Press, London and New York.

Jarvis, B.D.W. and Hungate, R.E. (1968). Factors influencing agnotobiotic cultures of the rumen ciliate *Entodinium simplex*. *Applied Microbiology* **16**, 1044–1052.

Kofoid, C.A. and MacLennan, R.F. (1930). Ciliates from *Bos indicus* Linn. 1. The genus *Entodinium* Stein. *University of California Publications in Zoology* **33**, 471–544.

Kofoid, C.A. and MacLennan, R.F. (1932). Ciliates from *Bos indicus* Linn. 2. A revision of *Diplodinium* Schuberg. *University of California Publications in Zoology* **37**, 53–152.

Kofoid, C.A. and MacLennan, R.F. (1933). Ciliates from *Bos indicus* Linn. 3. *Epidinium* Crawley, *Epiplastron* gen. nov. and *Ophryoscolex* Stein. *University of California Publications in Zoology* **39**, 1–33.

Nakamura, F. and Kurihara, Y. (1978). Maintenance of a certain rumen protozoal population in a continuous *in vitro* fermentation system. *Applied and Environmental Microbiology* **35**, 500–506.

Ogimoto, K. and Imai, S. (1981). "Atlas of Rumen Microbiology." Japan Scientific Societies Press, Tokyo.

Onodera, R. and Henderson, C. (1980). Growth factors of bacterial origin for the culture of the rumen oligotrich protozoon, *Entodinium caudatum*. *Journal of Applied Bacteriology* **48**, 125–134.

Owen, R.W. and Coleman, G.S. (1976). The cultivation of the rumen ciliate *Entodinium*

longinucleatum. *Journal of Applied Bacteriology* **41**, 341–344.
Small, E.B. and Lynn, D.H. (1985). Phylum Ciliophora Doflein, 1901. *In* "An Illustrated Guide to the Protozoa" (J.J. Lee, S.H. Hutner and E.C. Bovee, eds), pp. 393–575. Society of Protozoologists, Lawrence, Kansas.
Weller, R.A. and Pilgrim, A.F. (1974). Passage of protozoa and volatile fatty acids from the rumen of the sheep and from a continuous *in vitro* fermentation system. *British Journal of Nutrition* **32**, 341–351.

Chapter 3

Leishmania

DAVID A. EVANS

Department of Medical Protozoology,
London School of Hygiene and Tropical Medicine,
Keppel Street, London, England

I. INTRODUCTION

In vitro cultivation of the causative organism probably plays a more important role in the study and treatment of the leishmaniases than in any other group of diseases caused by protozoa. It is hard to think of an aspect of leishmaniasis in which cultivation of the organism does not have a role. Parasitological diagnosis, identification and characterization of the organism, biochemical, cell biological, immunological and chemotherapeutic studies usually involve *in vitro* cultivation of the parasites in one form or another.

Two distinct morphological forms of the organism occur during the digenetic life-cycle of *Leishmania*, the almost entirely intracellular amastigotes found in the macrophage and macrophage-like cells of the parasites' vertebrate hosts, and the extracellular flagellated promastigotes that occur in the alimentary tract of the sandfly vector. Both amastigote and promastigote forms can be cultivated *in vitro*, but for a variety of reasons it is the promastigote that is more usually cultivated.

In the past most ''review''-type articles on the *in vitro* cultivation of *Leishmania* have tended to concentrate on culture media formulations without much reference to the techniques of cultivation (e.g. Hendricks and Childs, 1980; Chang and Fish, 1983; Jaffe *et al.*, 1984). The aim of this chapter is to try to redress the balance somewhat by devoting more space to cultivation *per se*, and limiting the media recipes to those in common use that are known to perform well.

II. CULTURE METHODS

A. PROMASTIGOTES

Unfortunately a ''universal'' culture medium in which all the different leishmanias will grow is still lacking, and it is almost imposssible to predict which medium will be best suited for the growth of a particular isolate of *Leishmania*. This difficulty becomes especially apparent in the initial isolation of the organism from its vertebrate host, as certain culture media, although quite satisfactory for the growth of established promastigote cultures, are singularly useless for the isolation and establishment of the organisms *in vitro*. The processes of isolation and initial culture will therefore be dealt with separately from the propagation of established promastigote cultures.

1. *Isolation and initial establishment in culture*

Isolation and initial establishment are without doubt the most important steps in the cultivation of any leishmanial organism and are those which cause the most trouble and loss of organisms. It is here that the choice of culture medium is of vital importance and where so many media perform especially poorly. As well as having to support promastigote division, an isolation medium must permit efficient transformation from amastigote to viable promastigote form. Almost as important as the choice of isolation medium is the method by which tissue containing amastigotes is obtained and inoculated into the medium. Generally speaking, the greater the number of amastigotes inoculated, the greater are the chances of obtaining a viable culture of *Leishmania*. It is inappropriate here to describe the different methods that are employed to obtain tissue samples from cutaneous or visceral sites that will serve as the source of leishmanial amastigotes, as they are described in detail elsewhere (e.g. Peters and Killick-Kendrick, 1987). The inoculation process does, however, differ somewhat between cutaneous and visceral material and these will be dealt with separately.

(a) Isolation from cutaneous sites. Biopsy material from obvious skin lesions is the usual source of the inoculum, but occasionally biopsies of apparently normal skin are used, especially when potential animal reservoirs of leishmaniasis are under investigation. The inoculum usually consists of a piece of skin (epidermis plus the shallow portion of the dermis) from the edge of a cutaneous lesion, or an aspirate or scraping taken from this site.

Before taking the biopsy, the skin must be sterilized. With human skin, swab the edge of the lesion thoroughly with cotton wool soaked in 70% ethanol, taking care not to get alcohol into any open part. Very dirty skin must be swabbed several times using fresh swabs until the cotton wool is no longer dirtied by the process. Allow the skin to dry before taking the sample. Avoid using iodine-alcohol as this is very leishmanicidal; if it is used then every trace of iodine *must* be removed by swabbing with 70% ethanol before the sample is taken.

Sterilizing the skin may be difficult with a live wild-caught animal, which will usually have to be tranquillized. Wash the area of skin to be sampled with soap and water, rinse with clean water, shave the skin, swab very thoroughly with 70% ethanol as above, and allow the skin to dry before taking the sample. It is important to ensure that the inoculum is not excessively bloody, as most bloods contain serum proteins highly inhibitory to the growth of leishmanial promastigotes (Evans and Smith, 1986). Tissue samples containing very low numbers of amastigotes, especially those taken from old healing lesions, are especially difficult to culture. With such material fairly large pieces of tissue should be inoculated into the culture medium. With material taken from human skin, there is usually no need to macerate or homogenize the material before its inoculation, and simply pressing the skin against the wall of the culture vessel with a sterile needle will release sufficient amastigotes into the medium. The skin of some animals, particularly edentates such as sloths and arboreal ant-eaters, is so tough and rubbery that homogenization or grinding before inoculation is necessary.

(b) Isolation from visceral sites. Bone marrow juice or, more rarely, splenic aspirate material are the usual sources of amastigotes from cases of human visceral leishmaniasis; additionally, lymph node aspirates or pieces of liver or spleen are employed when wild and domestic animals are under investigation. Here, again, it is important to avoid the transfer of more than small amounts of whole blood or serum components into the culture medium along with the inoculum. This is especially important when inoculating bone marrow juice. It is extremely tempting to inoculate substantial amounts of such material into culture, especially if direct examinations of marrow smears have failed to demonstrate the presence of parasites. It is much safer to inoculate one or two drops (about 0.1 ml) of marrow juice into several tubes of culture medium rather than to inoculate it all into one tube. Alternatively, the marrow juice, which must be taken into an anticoagulant, may be centrifuged to sediment the cellular components; the supernatant fluid is discarded and the pellet, after washing by resuspension in isotonic salts solution and

recentrifugation, is used as inoculum for one or two tubes of culture medium.

(c) Choice of culture medium. Ideally, several different media should be inoculated in the hope that at least one will prove suitable for the growth of the organism. More often than not, a single medium is used, usually one the worker has faith in or one that has proved successful in the past with *Leishmania* from the particular area of study.

Faced with an unknown organism from an unfamiliar geographical area, I would always choose one of the traditional biphasic blood-agar media in preference to a tissue culture medium supplemented with foetal calf serum.

(d) Temperature of incubation. Many workers favour a temperature between 25 and 27 °C for culturing promastigotes. I prefer 21–23 °C, at which temperature I have found that most leishmanias grow well; at 27 °C some seem to grow less well. Indeed, it is easier to kill leishmanial promastigotes by heat than by cold, so if any doubt exists about the temperature of incubation for a particular culture, err on the cold side.

(e) Use of antibiotics. Although I have always been against the routine inclusion of antibiotics in culture media (Evans, 1978), one exception I do make is when using media for the isolation of *Leishmania*, as the chances of introducing contaminating microorganisms along with the inoculum are high. I have found gentamycin, at a final concentration of 50–100 $\mu g\ ml^{-1}$ of medium, gives reliable broad-spectrum antibacterial protection and has the advantage of far greater stability than the often-used penicillin–streptomycin combination.

(f) Culture vessels for isolation purposes. This is a very personal choice and there are really two schools of thought: (i) those, including myself, who prefer small, screw-capped universal glass bottles of ¼ fluid ounce (about 5.0 ml) capacity (''bijou'' bottles); and (ii) those who use 15 mm × 150 mm (approx.) round glass tubes, either screw-capped or sealed by a rubber bung. Both kinds of culture vessel seem to function equally well, but I find the ''bijou'' bottles easier to handle and they take up considerably less space in an incubator. Agar slopes in ''bijou'' bottles require 1.5 ml of solid phase and those in 15 mm × 150 mm culture tubes need 4.0 ml.

Some workers prefer to use disposable plastic culture tubes; these present no special problems. The method of making blood-agar slopes does require some modification as it is impossible to autoclave disposable plastic tubes without destroying them. The simple modification consists of autoclaving the agar base in a capped flask or bottle and cooling it to about 55 °C before adding the blood, mixing, and dispensing (using aseptic precautions) into the plastic tubes.

(g) Cultivation technique for isolation. Inoculate the material from which the isolation is to be attempted into the liquid phase of a biphasic medium or directly into a liquid medium, using strict aseptic technique. Incubate the inoculated medium at between 20 and 25 °C. Inspect the culture every 2 or 3 d by removing a small drop and examining this as a wet preparation under the low power of, preferably, a phase-contrast microscope. As soon as active promastigotes are seen in the primary culture, subinoculate one or two drops into fresh culture medium. As growth proceeds in the cultures, add fresh liquid phase if modified Tobie's medium or USMARU medium is being used (see below), taking care not to add more than the volume of liquid already present. This process may be continued until the liquid phase extends about half-way up the blood-agar slope, after which the liquid phase of the culture should be inoculated into larger culture vessels.

(h) Initial management of organisms of the L. donovani *group*. Quite often, when isolating *Leishmania* from cases of visceral leishmaniasis, the primary culture in a medium such as USMARU or modified Tobie's is positive but subsequent subcultures into medium of identical composition fail to grow and the organisms are lost. One way of salvaging such cultures is to make the first subculture into a less rich medium such as NNN or one of the semi-solid blood-agar media described below. The semi-solid media in particular seem to have the ability to revive ailing cultures, and after one passage through such a medium the promastigotes invariably grow quite satisfactorily in any of the more usual biphasic or liquid media.

2. *Media suitable for the initial isolation of* Leishmania

(a) Biphasic blood agar-based media

1° NNN medium
(see Wenyon, 1926)

Nicolle's simplification of Novy and MacNeal's original medium, although now some seventy years old, still has its place. Beware, however, of the incorrect use of the name of NNN or "3N" for any biphasic medium based on blood-agar, as has regrettably become common practice; many so-called NNN media have little in common with true NNN other than that they are biphasic and contain rabbit blood.

(i) **SOLID PHASE**

Agar (plain, non-nutrient)	1.4 g
NaCl	0.6 g
Distilled water	90 ml

Heat all the ingredients together in a flask until the agar melts, keeping the contents well mixed, otherwise the agar will burn on the bottom of the flask. A magnetic stirrer fitted with a hot-plate is useful. Transfer the appropriate amounts of molten agar directly into the culture vessels and sterilize by autoclaving (121 °C, 15 min). Allow the agar to cool to about 50 °C and add aseptically collected defibrinated rabbit blood to a final concentration of approximately 15 %. Mix the blood and agar by rolling the tubes in an upright position between the hands. Place the tubes in a sloped position until the agar has set, then stand them upright and transfer to a refrigerator at 4 °C or into iced water.

(ii) **LIQUID PHASE**

This consists of the water that condenses at the bottom of the slopes, no additional liquid phase being added. The rapid cooling of freshly made slopes by transfer to a refrigerator or iced water greatly increases the amount of water of condensation that will accumulate.

2° USMARU medium (Difco blood-agar medium)

(Walton *et al.*, 1977)

A very much richer medium than NNN, especially useful in the isolation of nutritionally more fastidious organisms such as those of the *L. braziliensis* group.

(i) **SOLID PHASE**

"Bacto" blood-agar base (Difco) ..	4 g
Distilled water	100 ml

Dissolve and dispense as described for NNN medium.

(ii) **LIQUID PHASE**

As for NNN medium; if additional liquid is required, a few drops of sterile distilled water may be added.

3° Evans' modified Tobie's medium
(Evans *et al.*, 1984)

A rich biphasic medium that has been used successfully for the isolation of a great variety of leishmanias from both Old and New World sources.

(i) **SOLID PHASE**

Beef extract (Oxoid ''Lab-Lemco'' L29)	0.3 g
Bacteriological peptone (Oxoid L37)	0.5 g
NaCl	0.8 g
Agar (Oxoid purified)	2.0 g
Distilled water	100 ml

Mix and heat the ingredients in a flask as for NNN medium. Transfer the molten agar into culture tubes and autoclave (121 °C, 15 min). Cool the sterilized agar to about 55 °C, then add defibrinated horse blood (inactivated by heating at 56 °C for 30 min) to give a final concentration of approximately 15%. Mix and slope as for NNN medium.

(ii) **LIQUID PHASE**

A liquid phase other than the simple water of condensation is used — proline-containing balanced salts solution (PBSS). As PBSS has extremely good keeping properties, it is suggested that at least 1 litre is prepared at a time.

KCl	0.4 g
$Na_2HPO_4 \cdot 12H_2O$	0.06 g
KH_2PO_4	0.06 g
$CaCl_2 \cdot 2H_2O$	0.185 g
$MgSO_4 \cdot 7H_2O$	0.1 g
$MgCl_2 \cdot 6H_2O$	0.1 g
NaCl	8.0 g
L-Proline	1.0 g
Phenol red	(q.s.)*
Distilled water	1000 ml

*Quantum sufficiat.

Dissolve the ingredients one at a time in about 750 ml of distilled water. Adjust the pH to 7.2 with solid Tris (tris(hydroxymethyl) aminomethane), make up the volume to 1000 ml, dispense into convenient screw-capped containers and autoclave (121 °C, 15 min). Add 0.2–0.3 ml of the liquid phase to the agar slope immediately before inoculation.

(*b*) *Liquid medium*

1° Schneider's *Drosophila* medium
(see Hendricks and Wright, 1979)

Schneider's *Drosophila* medium (revised)
with L-glutamine (Gibco) .. 100 ml
Heat-inactivated (56°, 30 min) foetal calf
serum to 10, 20 or 30% (v/v)

This is a commercially available, liquid insect tissue culture medium supplemented with foetal calf serum (FCS), which has been widely used for the isolation and bulk cultivation of *Leishmania* spp. It is a very expensive medium and somewhat variable in performance. I have found it considerably inferior to the richer biphasic blood-agar for the isolation of members of the *L. braziliensis* group of organisms. Not more than 1.0 ml per culture tube should be used for isolation.

(*c*) *Foetal calf serum substitute*. The majority of non-defined, liquid culture media for the cultivation of *Leishmania* have FCS as one of their essential ingredients. Without doubt, FCS will be the single most expensive ingredient, and in some parts of the world it is extremely difficult to obtain reliable supplies. A variety of other sera have been tried as substitutes for FCS, but invariably the promastigotes fail to grow and simply agglutinate and die in the presence of these sera. Gamma-globulin-free sera, used to replace FCS, seem to support excellent promastigote growth (Evans, 1986), but such sera are difficult to obtain and are even more expensive than FCS. However, a cheap, easily obtainable substitute for FCS can be simply made from any easily available mammalian serum (Evans, 1986). This substitute is solid and therefore turns the previously liquid medium containing FCS into a biphasic medium, albeit a very much less messy one than a medium based on blood-agar. The best results are obtained when 10% FCS is replaced by 15–20% of the solid substitute.

A great variety of different leishmanias, including all the WHO recommended reference strains, have been successfully cultivated in a mixture of minimum essential medium (MEM, Gibco; 100 parts), Evans' blood lysate broth (EBLB; 5 parts) and FCS substitute (20 parts) (Evans, 1986). FCS substitute can also replace FCS in a variety of other media used for the cultivation of *Leishmania* spp., such as Schneider's medium, RPMI 1640, tissue culture medium 199 (see Taylor and Baker, 1968, pp. 355–356), etc.

Preparation from serum collected aseptically. Transfer the required amount of serum (15–20% of the total volume of culture medium to be contained in each culture vessel) into sterile culture vessels. If the vessels are plastic, they must be capable of withstanding heating to 86°C. Place the capped vessels upright in a water bath maintained at 80–86°C for 3 h. The heating coagulates the serum, producing "inspissated" serum. After cooling, add the remaining liquid ingredients of the complete culture medium to the vessels and inoculate as usual.

Preparation from serum not collected aseptically. The method is the same as that described above, with two additional heatings of the serum. On the day of collection, heat the vessels containing the serum to 80–86°C for 3 h; on the next two days, reheat to the same temperature for 30 min on each occasion.

(d) The use of blood other than rabbit's blood in biphasic media. Quite often, rabbit blood is not easily available for inclusion in biphasic media such as NNN or USMARU; mammalian bloods other than rabbit may then be used. Sheep, horse or human bloods have all been used, but it is worth experimenting with whatever bloods are easily available. Bloods other than rabbit's should be used either defibrinated or with an anticoagulant, but *always* heat-inactivate them (56°C, 30 min) and increase the concentration of agar in the medium to 2%.

(e) Sterility checking of blood-agar. It is advisable to incubate freshly made blood-agar media at 37°C for 24 h and to examine the surface of the slope for signs of bacterial growth; any medium showing such signs is immediately discarded. In practice, such sterility checks are often omitted.

3. *Maintenance and bulk cultivation of* Leishmania.

(a) Culture vessels. This is a matter of personal choice, and a great variety of sizes and shapes of glass or plastic bottles or flasks is used. A range of vessels with different capacities should be available, as, when scaling-up the volume of the cultures, it is important not to inoculate large volumes of fresh medium with small inocula. A convenient range of sizes would be ¼ fluid ounce "bijou" bottles (5.0 ml capacity), 1 fluid ounce "universal" bottles (25 ml) and 125 ml "medical flat" bottles, all filled to about ⅓ capacity with liquid culture medium. With biphasic media, a similar range of vessels may be used. The solid phase is sloped when tubes and cylindrical bottles are used, but in the larger culture

flasks and "medical flats" the solid phase is cast on the base or on one of the two large flat inner surfaces of the container. The liquid portion is added and the vessel is incubated so that the solid phase is at the bottom with the liquid covering its surface.

The contents of the smallest bottle are used to inoculate the middle-size containers, and their content is subsequently used to inoculate the large vessels. The contents of the latter can then be used to inoculate three more large vessels. I avoid having more than 50 ml of medium in any one vessel; when greater volumes are required, I use several flasks in preference to one very large one. Very large vessels are generally more difficult to handle, and a contaminant in one large flask will result in the loss of a large volume of culture.

(b) Liquid media. Liquid media are more convenient for large volume cultures than are the biphasic ones.

1° MEM:FCS:EBLB medium
(Evans *et al.*, 1984)

A nutritionally rich liquid medium suitable for growth (but not isolation) of almost any leishmania.

MEM medium with Earle's salts modified, autoclavable; obtained from Gibco	100 ml
$NaHCO_3$ solution (7.5% w/v)	3.0 ml
Evans' blood lysate broth (EBLB)	5.0 ml
Foetal calf serum (heat-inactivated)	10 ml

The original method of preparing EBLB (Evans, 1978) is lengthy and requires several membrane filtrations. A much simpler method of producing essentially the same medium ("pseudo-EBLB"; Evans, 1986), which performs equally well, is given below.

Tryptose (Oxoid L47)	1.5 g
Casein hydrolysate (Oxoid L41)	1.0 g
Liver digest (Oxoid L27)	1.0 g
L-Proline	0.15 g
KH_2PO_4	0.68 g
NaOH	0.17 g
Distilled water	100 ml

Final pH 7.3–7.4 (adjust if necessary with either 1 = HCl or 1 = NaOH)

(i) Dissolve the solid ingredients in the distilled water and sterilize by autoclaving (121°C, 15 min).

(ii) After cooling, add 15 ml of blood lysate prepared aseptically from whole blood (human, equine, ovine, caprine or rabbit bloods all seem to be equally good) aseptically collected either into an anti-coagulant or defibrinated. Sediment the blood cells by centrifugation at approximately 3000*g* for 10 min, discard the liquid portion (serum or plasma) and wash the packed blood cells twice by resuspension in an equal volume of either sterile isotonic saline or PBSS (see above) and recentrifuge (3000*g*, 10 min). Lyse the washed blood cells by adding sterile distilled water equal to the volume of packed cells, and mixing thoroughly.

(iii) The medium will be very cloudy at this point owing to the cellular debris added in the blood lysate. It may be clarified by aseptic centrifugation (15 000*g* at least 30 min). Decant the clear supernatant, taking care not to disturb the pellet, and bottle and store it at 4°C.

2° HO-MEM MEDIUM

(Berens *et al.*, 1976)

This is a widely used medium which supports good growth of a variety of leishmanias. Although transformation of amastigotes to promastigotes will take place in the medium, it is best suited as a liquid medium for the growth of established promastigote cultures.

Minimal essential medium with Spinner's salts (SMEM; Gibco)	10.6 g
MEM amino acids 50 times concentrate (Gibco)	10.0 ml
MEM non-essential amino acids 100 times concentrate (Gibco)	10.0 ml
Sodium pyruvate 100 mm (Gibco)	11.0 ml
$NaHCO_3$	2.2 g
D-Glucose	2.0 g
Biotin	0.1 mg
p-Aminobenzoic acid	1.0 mg
Distilled water	to 1000 ml

Mix the solid and liquid ingredients in a flask, using only 900 ml of the water. When all have dissolved, adjust the pH (if necessary) to 7.2–7.4 with 1 = HCl or 1 = NaOH, make up the volume to 1000 ml with distilled water and sterilize by filtration. The medium is completed by the addition of heat-inactivated FCS to a final concentration of 10%

(15–20% FCS substitute may be used as an alternative — see Section II.A.2.(c)).

3° Other tissue culture media

Most of the commercially available tissue culture media, usually with the addition of a minimum of 10% heat-inactivated FCS, have been used with varying degrees of success for the cultivation of leishmanial promastigotes. Of these FCS-supplemented tissue culture media, Schneider's *Drosophila* medium seems to be one of the most useful for long-term cultivation. Tissue culture medium 199, MEM and RPMI 1640 often support growth of some of the more fastidious organisms such as *L. braziliensis* or *L. donovani* for only 2 or 3 passages before growth falls away rapidly and the cultures die.

4. Semi-solid media

These are extremely valuable as transport media and for reviving ailing cultures. I have yet to find a leishmania that can be grown in another culture medium that will not grow in one of the semi-solid media. Leishmanial cultures survive for very long periods in these media, and on two occasions I have received cultures through the post in "sloppy Evans" medium that had travelled for 6 weeks and were still viable. However, such media are not normally used for routine cultivation, because they are somewhat messy and slightly difficult to examine, as the promastigotes are always found amongst masses of blood cells.

1° "Sloppy Evans" medium
(Evans *et al.*, 1984)

PBSS (see Section II.A.-2.(a)3°)	80 ml
Bacteriological peptone (Oxoid) ..	0.1 g
Beef extract (Oxoid, "Lab-Lemco")	0.03 g
Washed packed horse blood cells..	10 ml
FCS (heat-inactivated)	10 ml
Agar (Oxoid, purified)	0.3 ml

(i) Harvest blood cells from aseptically collected defibrinated horse blood by centrifugation (3000*g*, 10 min); decant the supernatant, wash the packed cells by resuspension in at least an equal volume of PBSS, recentrifuge (3000*g*, 10 min) and repeat the washing process twice.

(ii) Mix the ingredients (omitting the packed cells and FCS) in a flask or screw-capped bottle. Sterilize by autoclaving (121 °C, 15 min), cool to about 50 °C and add the packed cells and FCS; mix well and dispense whilst still molten into suitable sterile culture tubes. The packed cell:FCS mixture can be replaced by 20 ml of defibrinated rabbit blood.

TECHNIQUE

See medium 2° below for description of techniques.

2° Semi-solid Locke-blood-agar medium

(Modified by L. Schnur [personal communication] from Adler and Theodor, 1926)

LOCKE'S SOLUTION

NaCl	9.2 g
$CaCl_2$	0.24 g
$NaHCO_3$	0.15 g
KCl	0.42 g
D-Glucose	1.0 g
Distilled water	1000 ml

Blood-agar base:

Agar (plain non-nutrient)	2.5 g
Bacteriological peptone	1.0 g
NaCl	0.5 g
Distilled water	100 ml

Mix 7 parts of Locke's solution with 1 part of molten blood-agar base, adjust the pH to 7.4 with 1 = HCl or 1 = NaOH, and autoclave to sterilize. Allow to cool to about 50 °C and add defribinated rabbit blood to a concentration of approximately 10%. Mix thoroughly and dispense into sterile culture vessels.

TECHNIQUE

(i) Small culture tubes or bottles are the most convenient vessels for cultivation in semi-solid media. If the media are to be used for transport purposes then ¼ fluid ounce "bijou" bottles (about 5 ml) or plastic tubes such as Nunc 80 × 12.5 mm polypropylene screw-capped cryotubes are ideal, as both are tough and easily packed.

(ii) Inoculate deeply into the medium with a Pasteur pipette. If used as a transport medium it is best to use a small inoculum (about 0.1 ml), so that the culture does not overgrow if there is a delay in reaching its destination.

5. Cloning

There is an increasing demand, especially from immunologists and molecular biologists, for cultures of *Leishmania* spp. derived from single promastigotes. To meet this demand, several methods for cloning promastigotes have been described. These include those which rely on direct observation, selection and cultivation of single organisms (Alekseev and Saf'janova, 1977; Gradoni *et al.*, 1983; Evans and Smith, 1986), those involving limiting dilution (Handman *et al.*, 1983), and plating methods (Jaffe *et al.*, 1984; Hill *et al.*, 1983); with the two latter techniques, it is only assumed that the resultant cultures have arisen from single organisms. For obvious reasons, most workers will accept only those cultures derived from directly observed single organisms as true clones. The method of Alekseev and Saf'janova (1977) involves the use of a micromanipulator and a great deal of skill; that of Gradoni *et al.* (1983) requires serial dilution, direct observation and cultivation in microtitre plates; the method of Evans and Smith (1986) includes isolation into microdrops, observation in hanging drops and cultivation in capillary tubes. This method, which is described below, is probably the most rapid of the three and has produced a 100% success rate on some occasions.

1° Cloning technique
(Evans and Smith, 1986)

APPARATUS

(i) Humid chamber (Fig. 1(a)) made by cementing, with epoxy resin adhesive, 4 mm wide × 1 mm thick glass strips to 76 × 26 mm glass microscope slides.
(ii) Glass capillary tubes 100 mm long × 0.5 mm internal diameter with one end drawn out in a flame to an even finer capillary (Fig. 1(b)).
(iii) 22 × 22 mm glass coverslips.
(iv) 1.0 ml disposable plastic syringes.
(v) "Pipetteman" automatic pipette with autoclavable plastic tips (smallest size).

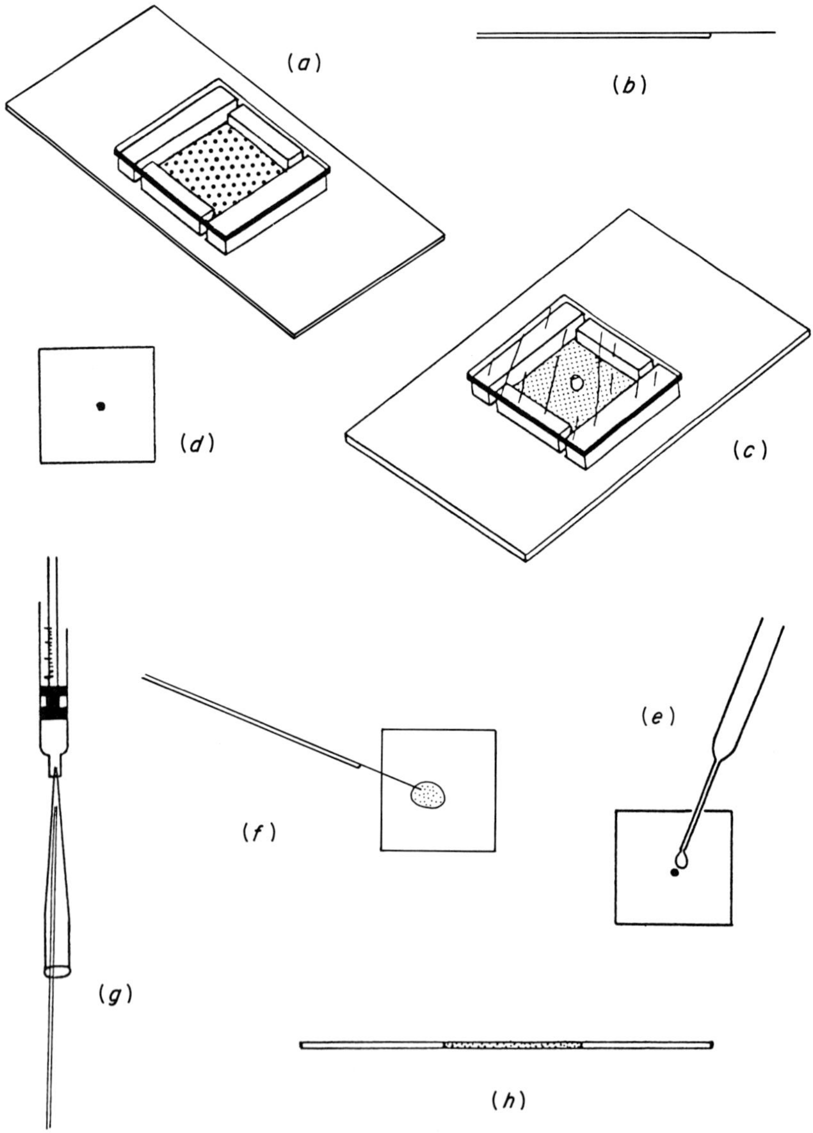

Fig. 1. *(a–h)* Steps in the cloning technique of Evans and Smith (1986); see text for detailed explanation.

TECHNIQUE

The following operations (except for microscopic examination) must be carried out in a laminar flow or similar sterile cabinet using strict aseptic technique and sterile apparatus.

(i) Dilute a mid-log-phase liquid culture of the organism to be cloned with fresh culture medium so that one microscope field ($\times$ 10 eyepieces, $\times$ 10 objective) contains about two promastigotes.

(ii) Transfer a minute drop (about 0.2–0.5 μl) of the diluted culture on to a sterile coverslip (Fig. 1(d)); immediately invert the coverslip and place it over the humid chamber.

(iii) Examine the drop under a $\times$ 10 objective using $\times$ 10 eyepieces and select only those drops that can be seen within a single field of view. Look for drops that contain a single active promastigote and make more dilutions if no such drop is found. When a drop with a single promastigote is found, have this confirmed by at least one (preferably two) independent observer(s).

(iv) Remove the coverslip from the humid chamber, turn it so that the drop is uppermost and immediately add a drop (approx 20 μl) of fresh culture medium to the microdrop on the slide (Fig. 1(e)).

(v) Push the finely drawn out end of a sterile capillary tube into the drop on the coverslip and allow the tube to take up the drop by capillary attraction (Fig. 1(f)). If the capillary tube is less than ⅓ filled, add a further drop of culture medium to the coverslip and continue to fill the capillary until it is about ⅓ full.

(vi) Insert the wide end of the tube into the automatic pipette tip (Fig. 1(g)), the narrow end of which is inserted into the nozzle of a 1.0 ml syringe, and gently withdraw the plunger of the syringe until the culture medium is drawn into the centre of the capillary tube (Fig. 1(h)).

(vii) Seal both ends of the capillary by heating, then transfer it to an incubator at 22°C.

(viii) Make regular microscopical examinations of the contents of the capillary by placing the intact tube on a microscope slide and focusing on both menisci between the liquid and air phases in the tube. The first signs of growth are usually seen at these interfaces and are not usually apparent for at least 7 d.

(ix) When good growth of the organisms is seen spreading through the tube, score both ends with a glass cutter, swab the outside of the tube with 70% ethanol, allow to dry, and break off both ends of the tube; expel its contents into about 0.5 ml of medium contained in a small culture bottle.

6. *Defined media for leishmanial promastigotes*

When the previous version of this book (Taylor and Baker, 1978) was prepared, only two defined media had been described that would support the growth of *Leishmania*. Trager's defined medium C (Trager, 1957), which was useful for growing only *"Leishmania" tarentolae* — an organism which almost certainly is a trypanosome and not a leishmania (Wallbanks *et al.*, 1985), and the "RE" series of media (Steiger and Steiger, 1976, 1977). Since then two additional defined media have appeared, HOSMEM II (Berens and Marr, 1978) and, more recently, MD-29 (Melo *et al.*, 1985). The compositions of media C, RE-III and HOSMEM II are detailed by Chang and Fish (1983) and are not repeated here.

1° Defined medium MD-29
(Melo *et al.*, 1985)

	mg litre^{-1}		
Fructose	2000	Pyridoxine-HCl	0.125
$Na_3PO_4.12H_2O$	5000	Pyridoxal	10
L-α-Glycerophosphate, Na salt	20 000	Thiamine	10
Sodium acetate. $3H_2O$	540	Ascorbic acid	0.25
NaCl	4000	Choline	12
KCl	400	L-Alanine	285
EDTA, Na salt	80	L-Arginine	550
Trisodium citrate $2H_2O$	600	L-Aspartic acid	100
$FeNO_3.9H_2O$	0.0000005	L-Cysteine-HCl	100
Sodium succinate $6H_2O$	270	L-Glutamic acid	1045
Adenosine	20	L-Glycine	350
Adenine-HCl	50	L-Histidine	350
Guanosine	20	L-Isoleucine	260
Guanine-HCl	1.5	L-Leucine	760
Hypoxanthine	1.5	L-Lysine-HCl	750
Xanthine	1.5	L-Methionine	175
Folic acid	30	L-Phenylalanine	373
D-α-Tocopherol succinate	4.0	L-Proline	780
DL-α-Lipoic acid	0.4	L-Hydroxyproline	50
Nicotinic acid	0.125	L-Serine	285
Nicotinamide	10	L-Threonine	270
D-Biotin	10	L-Tryptophan	140
Riboflavin	1.0	L-Tyrosine ethyl ester	360
Inositol	2.0	L-Valine	485
D-Ca pantothenate	10	Haemin	1.0

The amino acid ingredients can be stored deep-frozen as a single stock solution in double-distilled water; the other chemicals, with the exception of haemin and fructose, are added in dry form, dissolved in double-distilled water; the pH is adjusted to 7.5 with 1 = HCl and the medium is sterilized by filtration. The haemin (presumably dissolved in a solution of NaOH) is autoclaved and added to the medium. The final addition is the fructose; the original paper does not give details, but presumably it was added as either a filter-sterilized or mildly autoclaved (115 °C, 10 min) solution.

The medium which, unlike RE III and HOSMEM II, contains no bovine serum albumin serum fraction V, is reported to grow at least 19 different stocks and strains of *Leishmania* to concentrations up to 6×10^7 ml^{-1} through numerous passages. Amastigote-to-promastigote transformation will occur in the medium and it has been used for initial isolation of amastigotes from experimentally infected laboratory animals.

7. *Control of culture contamination*

Even in the best-ordered laboratories there will be occasions when cultures become contaminated with microorganisms other than leishmanias. Gross bacterial and fungal contamination are difficult to deal with, but a recently described chromatographic method (Okot-Kotber, 1985) promises to be very helpful in "cleaning" such cultures. Lesser degrees of bacterial contamination can usually be controlled and eventually eliminated by the inclusion of antibiotics in the culture medium. Gentamycin, penicillin and streptomycin, singly or in combination, seem to deal with most of the common bacterial contaminants. Occasionally, bacteria resistant to these antibiotics will be found and alternative antibacterial agents will need to be used. If time permits, it is helpful to determine the antibiotic sensitivity of the contaminating organism(s), but in practice the choice of antibiotic is usually made empirically.

Fungal contamination has always presented more of a problem. Most antifungal agents are also antileishmanial — e.g., fungizone (amphotericin B) and ketoconazole — but one, 5-fluorocytosine, is quite useful in controlling yeast contamination (Kimber *et al.*, 1981); at the concentrations employed (500 $\mu g\ ml^{-1}$ or less) it does not inhibit the growth of leishmanial promastigotes. 5-Fluorocytosine does not, however, seem to be useful in controlling contamination caused by filamentous fungi. The method of Okot-Kotber (1985) is likely to prove the most useful for eradicating fungal contamination. It is based on that originally described by Lanham and Godfrey (1970), for the separation of salivarian trypanosomes from blood cells, and involves passing the contaminated leishmanias through a column of diethylaminoethyl (DEAE) cellulose.

1° Method for freeing *Leishmania* cultures from fungal contamination

(Adapted from Okot-Kotber, 1985)

(i) PHOSPHATE-SALINE BUFFER (PS) (pH = 8.0, I = 0.362)

Na_2HPO_4 (anhydrous)	13.48 g
$NaH_2PO_4.2H_2O$	0.78 g
NaCl	4.25 g
Distilled water to	1000 ml

(ii) EQUILIBRATION OF DEAE CELLULOSE

Suspend the required amount of DE52 cellulose (Whatman) in PS and add orthophosphoric acid (1/20 dilution of the concentrated acid), stirring constantly until the pH reaches 8.0; about 12.5 ml of acid is needed for 100 g of DE52. If the DE52 becomes too acid, readjust with 0.5 M NaOH. Allow the cellulose to settle, then pour off the supernatant. Add more PS (about twice the volume of the cellulose), stir and allow to settle. Repeat this washing process at least five more times. Check the pH of the supernatant buffer, which should still be 8.0; if it is not, readjust with acid or alkali as above, wash at least twice more and check the pH again. Sterilize the equilibrated DE52 cellulose by autoclaving (121°C, 15 min).

(iii) ION-EXCHANGE CHROMATOGRAPHY

The chromatography column consists of a sterile hypodermic syringe barrel (2.0, 5.0 or 10 ml depending on the volume of the culture to be treated). In a laminar flow or similar sterile cabinet, introduce a sterile polyvinyl sponge (permeable to leishmanial promastigotes) into the bottom of the barrel, then pour sterile equilibrated DE52 into the barrel so that the cellulose, when packed down, comes to within about 15 mm of the top. Place a sterile lid over the top of the column and allow the cellulose to settle for about 30 min. Lower a disc of sterile Whatman no. 41 filter paper gently on to the top of the cellulose.

The contaminated culture, made up in sterile PS buffer, is added to the top of the column. PS buffer is then run on to the column from an appropriate reservoir, until 5 times the volume of the DE52 has passed through. The eluant buffer is collected in a sterile container and centrifuged (about 1500*g*, 10 min). The supernatant is discarded and the pellet is resuspended in a suitable culture medium, preferably the same as it was growing in previously. Antibiotics should be included in the medium (see Section II.A.1(e)).

B. AMASTIGOTES

Leishmanial amastigotes normally parasitize mononuclear phagocytic cells of the mammalian hosts, and it is usually in such cells that they are cultivated *in vitro*. The ability to grow these forms *in vitro* enables one to investigate processes such as macrophage-parasite interactions and to screen potential anti-leishmanial compounds relatively quickly, without the use of large numbers of laboratory animals. Recently considerable progress has been made towards the cultivation of amastigotes in cell-free media (Pan, 1984), but as yet only a very limited number of strains can be grown extracellularly and there remains some doubt as to how closely such amastigotes resemble those found inside cells.

1. *Intracellular amastigotes*

A detailed account of the various procedures available for the cultivation of these forms is given by Chang and Fish (1983), so only the commonly used methods will be referred to here. Basically, two main systems are used: (a) cultivation in peritoneally-derived macrophages and (b) cultivation in macrophage cell lines.

Peritoneally-derived macrophages are usually good hosts for leishmanial amastigotes and are chiefly used for relatively short experiments. As these macrophages normally do not divide, any amastigote culture using such cells is self-limiting since, although multiplication of the parasites will take place, the host cells will not divide. While such a system is perfectly satisfactory for most purposes, occasions do arise where long-term cultivation to produce large numbers of parasites is required. For this purpose, macrophage cell lines, the cells of which do divide, such as J774G8 may be used as host cells (Chang, 1980). The infections may be initiated with either amastigotes or promastigotes, but the former produce more reliable infections.

2. *Extracellular amastigotes*

It has been known for some time that "amastigotes" will appear occasionally in cultures of leishmanial promastigotes (Row, 1922) and several attempts have been made to cultivate these amastigotes in cell-free media (e.g. by Trager, 1953, and Lemma and Schiller, 1964), without much success. More recently Pan (1984) succeeded in cultivating amastigotes of a stock of *L. mexicana pifanoi* in a cell-free medium. There is still some question about how closely these amastigotes relate to their intracellular counterparts, but they do have several important features in common. They are rapidly phagocytosed by macrophage cell line cells and divide within the cells; they are morphologically and ultrastructurally similar; they transform into promastigotes; and their surface membranes are antigenically more similar to each other's than either of them are to promastigotes.

1° JH-30 medium (for the serial cultivation of intracellular stages of *L.m.pifanoi*)

(Pan, 1984)

(i) BASAL SOLUTION

	ml
Glucose (8%)	2.5
Medium 199 (10 times concentrate)	10.0
Trypticase (5%)	10.0
Vitamin mix (see below)	2.0
Biotin (4 mg dl^{-1} in 0.083% (w/v)HCl)	2.0
Folic acid (10 mg dl^{-1} in 0.008% (w/v) NaOH)	5.0
Vitamin B_{12} (1 mg dl^{-1} ml)	2.0
Sodium pyruvate (0.1 M)	3.0
Methyl cellulose (4.0%)	5.0
Penicillin (100 000 iu ml^{-1})	0.1
Haemin (50 mg dl^{-1} in 0.04% (w/v) NaOH)	5.0
Nucleotide mix (see below)	10.0
FCS (heat-inactivated)	33.5
L-Glutamine (3.0%)	2.0

The solutions of trypticase and methyl cellulose are sterilized by autoclaving (121°C, 15 min), those of the remaining ingredients by membrane filtration (0.22 μm pore size). Stock solutions of glucose, biotin, folic acid, vitamin B_{12}, sodium pyruvate, penicillin, haemin, nucleotide mix and L-glutamine are stored at – 20°C; those of medium 199, trypticase, vitamin mix, sodium pyruvate, methyl cellulose, haemin and sodium bicarbonate are stored at 4°C.

(ii) VITAMIN MIX

	mg
p-Aminobenzoic acid	30
D-Calcium pantothenate	40

Choline chloride	30
Iso-inositol	30
Nicotinamide	50
Nicotinic acid	20
Pyridoxal-HCl	20
Pyridoxine-HCl	20
Pyridoxamine-2HCl	20
Riboflavin-5-phosphate-Na.$2H_2O$	4
Thiamine-HCl	20

All dissolved in 100 ml double-distilled water.

(iii) NUCLEOTIDE MIX

	mg
Adenosine triphosphate	200
Adenosine diphosphate	100
Adenosine monophosphate	100
Glutathione (reduced)	20
L-Cysteine-HCl	20
Ascorbic acid	20

All dissolved in 100 ml double-distilled water.

(iv) COMPLETE MEDIUM JH-30

	ml
Basal solution	92.1
HEPES buffer (1.0 M)	1.0
$NaHCO_3$ (4.2%)	3.0
Double-distilled water	39.5

TECHNIQUE

Medium JH-30 was inoculated with promastigotes of a cloned strain of *L. m. pifanoi* known to transform into amastigotes in cell-free medium. The population density was adjusted to about 10^6 organisms ml^{-1}, the culture was incubated at 33 °C and subcultured into fresh medium every 96 h. By the time of the second subculture, 99% of the organisms were amastigotes which could be subcultured indefinitely. Population densities of over 2.5×10^7 ml^{-1} were obtained. It was also reported that two other leishmanial stocks, one of *L. braziliensis* and the other of the *L. mexicana* group, had been successfully cultivated as amastigotes in this medium.

REFERENCES

Adler, S. and Theodor, O. (1926). The identity of *Leishmania tropica* Wright, 1903 and *Herpetomonas papatasi* Adler, 1925. *Annals of Tropical Medicine and Parasitology* **20**, 355–364.

Alekseev, A.N. and Saf'janova, V.M. (1977). Cloning *Leishmania* at the promastigote stage by Vonbrun micromanipulator. *Parazitologiia* **11**, 158–161. [In Russian.]

Berens, R. and Marr, J. (1978). An easily prepared defined medium for cultivation of *Leishmania donovani* promastigotes. *Journal of Parasitology* **64**, 160.

Berens, R.L., Brun, R. and Krassner, S.M. (1976). A simple, monophasic medium for axenic cultivation of hemoflagellates. *Journal of Parasitology* **62**, 360–365.

Chang, K.-P. (1980). Interactions of leishmanias with mammalian cells *in vitro*. *In* "The *In Vitro* Cultivation of the Pathogens of Tropical Diseases", *Tropical Disease Research Series* **3**, pp. 275–276. Schwabe, Basel.

Chang, K.-P. and Fish, W.R. (1983). *Leishmania*. *In* "In Vitro Cultivation of Protozoan Parasites" (J.B. Jensen, ed.), pp. 111–153. CRC Press, Boca Raton, Florida.

Evans, D.A. (1978). Kinetoplastida. *In* "Methods of Cultivating Parasites *in Vitro*" (A.E.R. Taylor and J.R. Baker, eds), pp. 55–88. Academic Press, London and New York.

Evans, D.A. (1986). An inexpensive easily available replacement for foetal calf serum in media for the *in vitro* cultivation of *Leishmania* spp. *Zeitschrift für Parasitenkunde* **72**, 567–572.

Evans, D.A. and Smith, V. (1986). A simple method for cloning leishmanial promastigote. *Zeitschrift für Parasitenkunde* **72**, 573–576.

Evans, D.A., Lanham, S.M., Baldwin, C.I. and Peters, W. (1984). The isolation and isoenzyme characterization of *Leishmania braziliensis* subsp. from patients with cutaneous leishmaniasis acquired in Belize. *Transactions of the Royal Society of Tropical Medicine and Hygiene* **78**, 35–42.

Gradoni, L., Gramiccia, M., Maroli, M., Pozio, E. and Bettini, S. (1983). A contribution to the technique of cloning *Leishmania*. *Annales de Parasitologie Humaine et Comparée* **58**, 301–303.

Handman, E., Hocking, R.E., Mitchell, G.F. and Spithill, T.W. (1983). Isolation and characterization of infective and non-infective clones of *Leishmania tropica*. *Molecular Biochemistry and Parasitology* 7, 111–126.

Hendricks, L.D. and Childs, G.E. (1980). Present knowledge of the *in vitro* cultivation of *Leishmania*. *In* "The *In Vitro* Cultivation of the Pathogens of Tropical Diseases", *Tropical Disease Research Series* **3**, Schwabe, Basel.

Hendricks, L. and Wright, N. (1979). Diagnosis of cutaneous leishmaniasis by *in vitro* cultivation of saline aspirates in Schneider's *Drosophila* medium. *American Journal of Tropical Medicine and Hygiene* **28**, 962–964.

Hill, J.O., North, R.J. and Collins, F.M. (1983). Advantages of measuring changes in the number of viable parasites in murine models of experimental cutaneous leishmaniasis. *Infection and Immunity* **39**, 1087–1094.

Jaffe, C.L., Grimaldi, G. and McMahon-Pratt, D. (1984). The cultivation and cloning of *Leishmania*. *In* "Genes and Antigens of Parasites: a Laboratory Manual (C.M. Morel, ed.), pp. 47–91. Instituto Oswaldo Cruz, Rio de Janeiro.

Kimber, C.D., Evans, D.A., Robinson, B.L. and Peters, W. (1981). Control of yeast contamination with 5-fluorocytosine in the *in vitro* cultivation of *Leishmania* spp. *Annals of Tropical Medicine and Parasitology* **75**, 453–454.

Lanham, S.M. and Godfrey, D.G. (1970). Isolation of salivarian trypanosomes from man and other mammals using DEAE-cellulose. *Experimental Parasitology* **28**, 521–534.

Lemma, A. and Schiller, E.L. (1964). Extracellular cultivation of the leishmanial bodies of species belonging to the protozoan genus *Leishmania*. *Experimental Parasitology* **15**, 503–513.

Melo, N.M., De Azevedo, H.P., Roitman, I. and Mayrink, W. (1985). A new defined medium for cultivating *Leishmania* promastigotes. *Acta Tropica* **42**, 137–141.

Okot-Kotber, B.M. (1985). A rapid chromatographic method for elimination of fungal contamination in *in vitro* cultures of *Leishmania* spp. *Parasitology* **91**, 1–7.

Pan, A.A. (1984). *Leishmania mexicana*: serial cultivation of intracellular stages in cell-free medium. *Experimental Parasitology*, **58**, 72–80.

Peters, W. and Killick-Kendrick, R. (eds) (1987). "The Leishmaniases in Biology and Medicine". Academic Press, London and New York.

Row, R. (1922). On the reversion of the flagellate form of *Leishmania donovani* and *Leishmania tropica*. *Indian Journal of Medical Research* **10**, 476–481.

Steiger, R.F. and Steiger, E. (1976). A defined medium for cultivating *Leishmania donovani* and *Leishmania braziliensis*. *Journal of Parasitology* **62**, 1010–1011.

Steiger, R.F. and Steiger, E. (1977). Cultivation of *Leishmania donovani* and *Leishmania braziliensis* in defined media: nutritional requirements. *Journal of Protozoology* **24**, 437–441.

Taylor, A.E.R. and Baker, J.R. (1968). "The Cultivation of Parasites *in Vitro*". Blackwell Scientific Publications, Oxford.

Taylor, A.E.R. and Baker, J.R. (eds). (1978). "Methods of Cultivating Parasites *In Vitro*". Academic Press, London and New York.

Trager, W. (1953). The development of *Leishmania donovani in vitro* at 37°C. Effects of the kind of serum. *Journal of Experimental Medicine* **97**, 177–189.

Trager, W. (1957). Nutrition of a hemoflagellate (*Leishmania tarentolae*) having an interchangeable requirement of choline or pyridoxal. *Journal of Protozoology* **4**, 269–276.

Wallbanks, K.R., Maazoun, R., Canning, E.U. and Rioux, J.A. (1985). The identity of *Leishmania tarentolae* Wenyon 1921. *Parasitology* **90**, 67–78.

Walton, B.C. Shaw, J.J. and Lainson, R. (1977). Observations on the *in vitro* cultivation of *Leishmania braziliensis*. *Journal of Parasitology* **63**, 1118–1119.

Wenyon, C.M. (1926). "Protozoology", vol. 2, pp. 1300–1303. Baillière, Tindall and Cox, London.

Chapter 4

Trypanosoma cruzi and other Stercorarian Trypanosomes

J.R. BAKER

4, Belvoir Road, Cambridge, England

I. INTRODUCTION

The major advances in the cultivation of stercorarian trypanosomes since our last book was published (Taylor and Baker, 1978) have concerned the development of cell culture systems for growing *in vitro* the entire vertebrate section of the life cycle of *Trypanosoma cruzi*. In the previous book, Ryley and Wilson (1978) devoted one page to the topic, describing the method developed by Gutteridge *et al.* (1969) using a line of human heart cells. A wide range of host cells has been found suitable for the intracellular growth of *T. cruzi* as amastigotes, with subsequent differentiation into trypomastigotes; these include macrophages (e.g., Dvorak and Schmunis, 1972; Nogueira and Cohn, 1977), HeLa cells (e.g., Dvorak and Hyde, 1973; Hyde and Dvorak, 1973; Dvorak and Crane, 1981), monkey Vero cells (e.g., Bertelli *et al.*, 1977; Henriquez *et al.*, 1981), human skin-muscle cells (Pan, 1978b), chick embryo muscle cells (Henriquez *et al.*, 1981), bovine embryo skeletal muscle cells (Dvorak and Hyde, 1973), canine kidney cells (Hyde and Dvorak, 1973), human diploid cells with a finite life *in vitro* (Sanderson *et al.*, 1980), fibrosarcoma cells and murine thigh muscle tumour cells (Hudson *et al.*, 1984). Glauert *et al.* (1982) successfully grew the closely related parasite *T. dionisii* (a species of the subgenus *Schizotrypanum* which parasitizes Chiroptera) in a commercially available "immortal" cell line derived from *Bison bison* (so-called "buffalo" lung or BL cells, Bu(IMR-31), ATCC no. CCL40, obtainable from Flow Laboratories, catalogue no. 05541); it is likely that this cell line would be equally satisfactory for *T. cruzi*.

Many experimental studies have been made on *T. cruzi* in these various cell culture systems, particularly those involving macrophages (e.g., Dvorak and Schmunis, 1972; Milder and Kloetzel, 1980; Meirelles *et al.*, 1982; and — with emphasis on cell-mediated immunity — Nogueira and Cohn, 1977). The kinetics of entry into, and subsequent development in, non-phagocytic cells (or those not "professionally" phagocytic) have been studied by Dvorak and Hyde (1973) and Hyde and Dvorak (1973), using an apparatus designed by Dvorak (1980) to permit continuous microscopical observation of a cell culture containing trypanosomes. The effects of temperature on intracellular development were investigated by Bertelli *et al.* (1977) and Bogliolo *et al.* (1980), and the role of the host cell membrane in parasite entry was studied by Henriquez *et al.* (1981). Villalta and Kierszenbaum (1984) found that entry of *T. cruzi* into cultured macrophages and myoblasts was related to galactose residues on the cell surface: prior treatment of potential host cells with α-galactosidase depressed entry, whereas similar treatment of the parasites enhanced it. Dvorak and Crane (1981) found that the susceptibility of HeLa cells to *T. cruzi* was influenced by the stage of the host cell's cycle. Hudson *et al.* (1984) developed a system of continuously cycling cell cultures infected with *T. cruzi* which permitted the continual harvesting of liberated amastigotes over periods of up to 18 months (Section III.B.1°). Hanson (1980) briefly reported the use of roller bottles with 670 cm^2 surface area for large-scale growth of Vero cells, infected with *T. cruzi*, in 100 ml Eagle's minimum essential medium (MEM; see Taylor and Baker, 1968, p. 355) supplemented with 2% v/v foetal bovine serum; this permitted the daily harvesting of amastigotes and trypomastigotes, in variable proportions, for 2–3 weeks, with yields of 3 to 4 $\times 10^7$ ml^{-1}.

Schmatz and Murray (1982) found that cell cultures of rat myoblasts (L-6) permitted much more synchronous intracellular development of *T. cruzi* if the cultures were irradiated (3000 rad from a cobalt-76 source) before infection with *T. cruzi*.

Lanar (1979) conducted an interesting study on *T. cruzi* cultivated in medium ML-15H (see Section III.B.2°) in the presence of an insect cell line (*Triatoma infestans* embryo cells) at 28°C. Bloodstream trypomastigotes transformed into "staphylomastigotes" (grape-like clusters of amastigotes) in the supernatant medium and then — but only in the presence of the cells — re-differentiated into trypomastigotes. Lanar

(1979) suggested that the insect cells produced a soluble factor essential for this second differentiation. Baker and Selden (1978) had concluded similarly that mouse peritoneal macrophages produced a soluble factor *in vitro* which was essential for the maintenance of extracellular *T. dionisii* at 37 °C in medium 199 plus 20% v/v calf serum; the cells were not essential for survival of trypanosomes at 28 °C. Growth of *T. cruzi* as extracellular clusters of amastigotes (presumably equivalent to Lanar's staphylomastigotes) in L4N medium (see Evans, 1978, pp. 75–76) at 28 °C had also been reported by Baker and Price (1973).

Abrahamson and Kloetzel (1980) demonstrated the presence of trypanosomal antigen on the surface of cells (epithelioid diploid cells) in cultures infected with *T. cruzi*; both parasitized and unparasitized cells were antigen-coated, an observation of possible significance in understanding the pathogenesis of Chagas disease. A similar observation had been made by Baker and Green (1977) concerning macrophages cultured with *T. dionisii*.

Several other interesting studies, more peripherally related to the *in vitro* cultivation of *T. cruzi*, have been made. These include cryopreservation of the bloodstream trypomastigotes (Filardi and Brener, 1975). The use of DEAE cellulose columns to purify trypomastigotes derived from cell cultures was discussed by Al-Abassy *et al.* (1972) and Villalta and Leon (1979); the latter showed that the procedure might reduce the infectivity of the trypanosomes. However, metacyclic trypomastigotes derived from cultures in cell-free media may be purified without apparent damage by centrifugation in a discontinuous gradient of percoll (Castanys *et al.*, 1984), and amastigotes liberated from cell cultures can be similarly purified with the aid of a discontinuous metrizamide gradient (Villalta and Kierszenbaum, 1982 — see Section III.B.2° — and De Carvalho and De Souza, 1983). The stability of epimastigotes and bloodstream trypomastigotes in different washing solutions was investigated by Katzin *et al.* (1977); epimastigotes survived well in simple solutions such as 0.85% w/v (0.15 M) sodium chloride solution, trypomastigotes needed the addition of 1% v/v serum, albumin or soluble casein to ensure survival. Schmatz *et al.* (1983) reported a method of separating broad bloodstream trypomastigotes from slender forms by the use of a continuous gradient of hypaque (12.5–25.0%), to enable the differential inoculation of cell cultures. These authors found that, whereas both forms were equally capable of infecting macrophages (presumably by phagocytosis), slender forms were much more infective to muscle cells (presumably by active penetration).

Goldberg and Chiari (1980) reported the cloning of *T. cruzi* by limiting dilution followed by plating on to solid medium and the subsequent isolation of single colonies, and Eyzaguirre (1983) found that colonies would grow also on blood-free agar.

The differentiation of cultured epimastigotes into metacyclic trypomastigotes (metacyclogenesis) *in vitro* has been the subject of several studies. Sullivan (1982) found that this will occur in Grace's insect tissue culture medium (see Taylor and Baker, 1968, p. 312) plus 10% (v/v) foetal bovine serum, but that decreasing or eliminating the serum content depressed the level of transformation. Lopetegui and Miatello (1982), using similar media, found that metacyclogenesis occurred after exponential growth had ceased, and was inhibited by the fall in pH to 5.5 that accompanied continued exponential growth in richer media. Isola *et al.* (1983) reported that metacyclogenesis was apparently stimulated by the presence in Grace's medium of haemolymph from *Triatoma infestans*. Ucros *et al.* (1983) observed that metacyclogenesis did not occur in the absence of serum (newborn calf serum; the possible effect of insect haemolymph was not investigated) and was stimulated by the presence in the medium of certain sugars (glucose, fructose, sucrose and possibly trehalose). Metacyclogenesis was enhanced also by an alkaline environment (pH 9.0) and inhibited by acid conditions (pH 5.5 to 6.0); it occurred only in microtitre culture plates and not in test tube cultures, suggesting that gaseous exchange played a role.

It is clear from these studies that the factors inducing and influencing metacyclogenesis are by no means fully understood. It does, however, seem legitimate to conclude, from these and other studies, that continued exponential growth *in vitro* and metacyclogenesis are mutually exclusive. This may be related to the natural occurrence of the process in the insect vector's hindgut, where nutritional conditions are presumably poor and a low pH prevails: it is a matter of "divide *or* differentiate".

Brun and Jenni (1985) have provided a useful, fairly recent, review of all aspects of the cultivation *in vitro* of *T. cruzi* (and other species).

II. CULTURE OF *T. CRUZI* INSECT HOST STAGES

The stages of *T. cruzi* which normally develop in its vector insects (Triatominae) grow readily in a range of liquid and biphasic media (see Evans, 1978), and would probably do so in any of the media described in Chapter 3 of this book. Cultivation temperature is not critical, but should not be above 28°C. A "classical" favourite medium for *T. cruzi* is LIT (see Evans, 1978, p. 73), and Engel *et al.* (1982) described the growth kinetics of 19 clones of *T. cruzi* in LIT medium.

Hendricks *et al.* (1978) found that *T. cruzi* (and some other haemoflagellates, including *T. rangeli*) could be cultivated in tissue culture medium 199 and in Grace's, and Schneider's, insect tissue culture media, all three of which can be obtained from commercial suppliers such as Flow Laboratories and Gibco Bio-Cult, amongst others (the composition of the first two is given by Taylor and Baker, 1968, pp. 355–356 and 312–313 respectively). For the cultivation of trypanosomes, the media are supplemented with 30% v/v foetal bovine serum, inactivated by heating at 56°C for 30 min (see also Section IV). This is a convenient, though expensive, means of cultivating suitable species; its discovery led to a brief debate in print on the need of trypanosomatids for haem, which is thought to be absolute (Baker, 1984; Hutner, 1984). Arévalo *et al.* (1985) showed that haem was indeed required for the differentiation *in vitro* of *T. cruzi* epimastigotes to metacyclic trypomastigotes (see also Roitman, 1980 and Ezyaguirre, 1983).

Defined media for *T. cruzi* have been devised by Azevedo and Roitman (1977; see Evans, 1978, pp. 68–69) Pan (1978c) and Avila *et al.* (1979, 1983); details of the last three of these are given in 1°, 2° and 3° below. Williams and Hudson (1982) described the mass culture of *T. cruzi* in LIT or Warren's medium with 10% heat-inactivated newborn calf serum, by means of a continuous flow system in a modified bacterial fermenter.

1° Defined medium F-84 for *T. cruzi*
(Pan, 1978c)

COMPOSITION

	ml
Amino acid solution (see below)	15.0
L-Tyrosine (40 mg dl^{-1}, aqueous)	10.0
Glucose (0.8% w/v, aqueous)	25.0
Medium 199 (tenfold concentrate)	10.0
Sodium pyruvate (0.1 M)	2.0
Vitamins solution (see below)	2.0
Biotin (2.0 mg dl^{-1}, aqueous)	2.0
Folic acid (10 mg dl^{-1}, aqueous)	5.0
Vitamin B_{12} (1.0 mg dl^{-1}, aqueous)	1.0
Metals solution (see below)	1.0
Methyl cellulose (4% w/v)	4.0
Haemin (Teichmann's crystals, 50 mg dl^{-1} of 0.01 N NaOH)	4.0
L-Glutamine (200 mM)	4.0
Antibiotics mixture (see below)	0.1
Adenosine phosphates solution (see below)	10.0
Sodium bicarbonate (4.2% w/v, aqueous)	5.0

AMINO ACID SOLUTION
(Trager, 1957, without tyrosine)

	mg
L-Arginine-HCl	120
L-Histidine	60
L-Isoleucine	240
L-Leucine	600

L-Lysine-HCl	500
L-Methionine	120
L-Phenylalanine	160
L-Threonine	200
L-Tryptophane	80
L-Valine	200
L-Alanine	280
L-Aspartic acid	480
Glycine	40
L-Proline	200
L-Serine	160
L-Glutamic acid	760
Distilled water	100 ml

VITAMINS SOLUTION

	mg
p-Aminobenzoic acid	30
D-Calcium pantothenate	40
Choline chloride	30
i-Inositol	30
Nicotinamide	50
Nicotinic acid	20
Pyridoxal-HCl	20
Pyridoxine-HCl	20
Pyridoxamine-2HCl	20
Riboflavin-5-phosphate-sodium	2.0
Thiamine	20
Distilled water	100 ml

METALS SOLUTION

	mg
$ZnSO_4 \cdot 7H_2O$	220
H_3BO_3	1.0
$CuSO_4$	5.0
$MnSO_4 \cdot 4H_20$	30
$CoSO_4 \cdot 7H_20$	4.0
Distilled water	100 ml

ANTIBIOTICS MIXTURE

	ml^{-1}
Penicillin G	10^5 iu
Streptomycin	10^5 μg

ADENOSINE PHOSPHATES SOLUTION

	mg
L-Cysteine-HCl	5.0
Glutathione (reduced)	20
Ascorbic acid	5.0
Adenosine triphosphate	50
Adenosine diphosphate	10
Adenosine monophosphate	10
Distilled water	100 ml

All solutions except trypticase, methyl cellulose and metals were sterilized by filtration (0.45 μm pore size). Trypticase, methyl cellulose and metals solution were sterilized by autoclaving (121 °C, 15 min).

TECHNIQUE

The initial inoculum consisted of washed parasites (details not given) harvested from cultures in a serum-free semi-synthetic medium (F-81), itself based on F-69 with serum omitted and a higher concentration of vitamins (see Pan, 1978a, c). The medium was used in 5.0 ml amounts contained in screw-capped tubes (16 × 125 mm), incubated at 27 °C. Serial transfers were made at approximately monthly intervals, using an inoculum of 1.0 ml per tube. The organisms were almost entirely epimastigotes.

2° Defined medium for *T. cruzi*
(Avila *et al.*, 1979)

COMPOSITION

SOLUTION A

	mg litre^{-1}
NaCl	6800
KCl	400
$MgSO_4 \cdot 7H_2O$	200
$NaH_2PO_4 \cdot H_2O$	140
$CaCl_2$	200
$NaHCO_3$	2200

SOLUTION B

L-Arginine-HCl	126
L-Cystine-2HCl	31.3
L-Histidine-HCl	42
L-Glutamine	292
L-Isoleucine	52
L-Leucine	52
L-Lysine-HCl	72.5
L-Methionine	15
L-Phenylalanine	32
L-Threonine	48
L-Tryptophan	10
L-Tyrosine Na_2	52.1
L-Valine	46

SOLUTION C

HEPES	2340

SOLUTION D (see note below)

Choline Cl	2.0
Folic acid	2.0
i-Inositol	4.0
Nicotinamide	2.0
D-Ca pantothenate	2.0
Pyridoxal-HCl	2.0
Riboflavin	0.2
Thiamine-HCl	2.0

SOLUTION E

Glucose	1000

SOLUTION F

Phenol red	10

SOLUTION G (see note below)

2-Deoxyinosine-5'-triphosphate	0.02
2-Deoxyadenosine-5'-triphosphate	0.02
2-Deoxyuridine-5'-triphosphate	0.02
Adenosine triphosphate	2.0
Deoxyribose	0.5

SOLUTION H

Streptomycin	10^5 iu
Penicillin	10^5 iu

Note. Solutions D and G were prepared 100 times concentrated and stored frozen; other solutions (presumably) were prepared to give concentrations indicated in a total volume of 1.0 litre of deionized water; pH adjusted to 6.8–7.0 with 1 N HCl. It is not stated how the medium was sterilized: presumably by filtration (0.45 μm pore size). This medium was referred to by Avila *et al.* (1983) as MCRM. For use, the medium was supplemented with between 1.0 and 3.0 mg ml^{-1} of bovine liver catalase (Sigma, C 40). This was dialysed against culture medium for 24 h at 4 °C after being dissolved in deionized water and sterilized by filtration

(0.45 μm pore size). It was then added (presumably as an equal volume) to sterile plastic tubes already containing sterile twofold concentrated culture medium.

TECHNIQUE

The inoculum routinely consisted of 5 × 10^5 trypanosomes ml^{-1} of culture. Parasites were obtained from cultures in liquid medium. Medium was used in 10 ml aliquots in 25 cm^3 plastic tissue culture flasks (Falcon, 10-126-2B), incubated at 27 °C. Subcultures were made about every 10 d. Growth up to 2 × 10^7 parasites ml^{-1} (95% epimastigotes) was achieved. The medium did *not* support growth of *Leishmania*, although all five strains of *T. cruzi* tested grew satisfactorily.

3° Minimal defined medium TMM for *T. cruzi*
(Avila *et al.*, 1983)

COMPOSITION

	g $litre^{-1}$		
NaCl	6.8	HEPES (*N*-2-hydroxyethylpiperazine-*N*'-2-hydroxypropanesulphonic acid)	2.26
KCl	0.4	EPPS (*N*-[2-hydroxyethyl]-piperazine-*N*'-3-propanesulphonic acid)	2.34
$MgSO_4 \cdot 7H_2O$	0.2	Phenol red	0.01
$NaH_2PO_4 \cdot H_20$	0.14	Streptomycin	0.1
$CaCl_2$	0.2	Penicillin	100 000 iu
Choline Cl (see note below)	0.002	Bovine liver catalase (Sigma, C 40)	3.0
Folic acid (see note below)	0.002		
Glucose	1.0		
$NaHCO_3$	2.2		

Note. Choline chloride and folic acid were added from a 100 times concentrated stock solution. Components were dissolved in deionized water and the pH was adjusted to 6.4 with 1 M HCl. No detail is given concerning sterilization, which was presumably by filtration (0.45 μm pore size).

TECHNIQUE

The medium was used in 10 ml aliquots in plastic tissue culture flasks (Falcon, 3012), incubated at 28 °C. Subcultures were made every 10 d, using an inoculum of 5 × 10^6 parasites contained in 0.3–0.8 ml. Parasites were obtained from cultures in the defined medium described above (Section II.2°). Growth up to about 9 × 10^6 trypanosomes ml^{-1}

was achieved. Infectivity for mice was retained until at least the ninth passage.

III. CULTURE OF *T. CRUZI* VERTEBRATE HOST STAGES

Evans (1978, pp. 74–75) listed two media (F-29 and F-32) devised by Pan (1971) for the extracellular cultivation of amastigotes of *T. cruzi*, which were thought to represent the stage normally grown within cells; however, cultures in these media would not survive at temperatures above 35 °C (Pan, 1971). Subsequently, Pan (1978a) modified the medium to support growth of amastigotes at 37 °C. Details of the method are given below.

A. CELL-FREE MEDIUM

1° Medium F-69 for *T. cruzi* amastigotes
(Pan, 1978a)

COMPOSITION

	ml
Distilled water	22.0
Medium 199 (tenfold concentrate)	10.0
Glucose (0.8% w/v, aqueous)	25.0
Sodium pyruvate (0.1 M)	2.0
Trypticase (5%, w/v)	10.0
Vitamins mixture (see note below)	1.0
Biotin (2.0 mg dl^{-1}, aqueous)	1.0
Folic acid (10 mg dl^{-1}, aqueous)	5.0
Vitamin B_{12} (1.0 mg dl^{-1}, aqueous)	1.0
Methyl cellulose (4% w/v, aqueous)	5.0
Haemin (Teichmann's crystals, 50 mg dl^{-1} of 0.01 N NaOH)	5.0
Antibiotics mixture (see note below)	0.1
Adenosine phosphates solution (see note below)	10.0
Sodium bicarbonate (4.2% w/v, aqueous)	3.0

Note. The vitamins, antibiotics and adenosine phosphates solutions are the same as those given under medium 1° in Section II, except that riboflavin-5-phosphate-sodium.$2H_2O$ (4.0 mg) was used in the vitamins mixture for this medium.

Distilled water, trypticase and methyl cellulose solutions were sterilized by autoclaving (121 °C, 15 min) and stored at 4 °C. All other solutions were sterilized by filtration (0.45 μm pore size) and stored at − 20 °C in 10 ml aliquots. All solutions were prepared with deionized water. The osmotic pressure of the medium was 357 mosmol.

TECHNIQUE

For use, the base was supplemented with 10 ml litre^{-1} foetal bovine serum, not inactivated (though inactivation at 56 °C for 30 min was said to do no harm). The medium was used in 5.0 ml aliquots contained in screw-capped tubes (16 × 125 mm), gassed with 4 to 5 % CO_2 in air and incubated at 37 ± 0.5 °C. Serial transfers of 1.0 ml of culture were made every 96 h. After eight subcultures, organisms were ''predominantly'' (about 95 %) amastigotes, the remainder being promastigotes.

B. CELL CULTURES

As already stated (Section I), intracellular *T. cruzi* have been grown in a wide range of vertebrate cell types *in vitro*; in many of these systems, the intracellular life cycle is completed, with the final emergence of (presumed) bloodstream-type trypomastigotes from the host cells. These cell cultures can be infected either with trypanosomes from biphasic or liquid medium cultures of insect stages, when it is likely that it is metacyclic trypomastigotes which are the infective stages (Hudson *et al.*, 1984; see also Glauert *et al.*, 1982), or with bloodstream trypomastigotes. Hudson *et al.* (1984) achieved continually cycling cultures, in which *T. cruzi* completed its intracellular life cycle every 4–6 d and the emerging trypomastigotes reinfected the host cells; such cultures have been continuously maintained for up to 18 months. Details of their method, and of another which permits the harvesting of pure amastigotes, are given below.

1° Continuous intracellular culture of *T. cruzi*
(Hudson *et al.*, 1984)

MEDIUM

Dulbecco's modification of Eagle's minimal essential medium (DMEM), supplemented with 10% v/v heat-inactivated foetal bovine serum, penicillin (200 iu ml^{-1}) and streptomycin (100 μg ml^{-1}), buffered with bicarbonate [presumably $NaHCO_3$] (22 g litre^{-1}) and 5% CO_2 in air (using phenol red as indicator).

CELL LINE

S2 cells, which originated in 1970 from the thigh muscle tumour of a BALB/c mouse injected with murine sarcoma virus (Harvey strain). The cell line contained a complex of the defective murine sarcoma virus and its helper murine leukaemia virus.

PARASITE INOCULUM

T. cruzi Y strain, clone Wel Tryp A2, was harvested from irradiated (900 rad whole-body irradiation) CBA/T6 mice by cardiac puncture at peak parasitaemia, using heparin as anticoagulant. The blood was allowed to sediment under gravity for 30 min at room temperature, the trypomastigotes were removed with the plasma, washed thrice by suspension in DMEM followed by centrifugation (150*g*, 10 min, room temperature), and counted in a haemocytometer.

TECHNIQUE

One million S2 cells were plated into 75 cm^2 plastic tissue culture flasks, culture volume 20 ml (Costar, catalogue no. 3075). After 24 h, 10^7 trypomastigotes were added to each flask. After a further 16 h, the supernatant medium was removed, the adherent cells were washed three times with medium, and 20 ml fresh medium were added. All incubation was at 37°C.

Parasites were harvested by removal of the supernatant medium and separated from it by centrifugation (250*g*, 15 min, room temperature). If necessary, amastigotes were freed from contaminating trypomastigotes or murine cells by density gradient centrifugation (33.9*g* Ficoll (Pharmacia) + 114.5 g Triosil (Nyegaard) in 471 ml water mixed with up to 50% v/v phosphate-buffered saline; 1.0 ml parasite suspension was layered on to 9.0 ml gradient and centrifuged (800*g*, 20 min, 4°C).

Usually, the removal of infected, non-adherent murine cells in the supernatant medium was sufficient to prevent overgrowth of the cell culture, which was maintained simply by adding fresh medium to a final volume of 20 ml. If necessary, the cell culture could be split after brief trypsinization (0.125% w/v trypsin in citrate buffer + 0.05% EDTA, at a ratio of 1 : 60). Stock cultures of uninfected murine cells were split in this way every 7 d.

The amastigotes prepared from these cell cultures were closely similar morphologically, biochemically and immunologically to those derived from infected rodents *in vivo*, and were infective to mice.

2° Medium ML-15HA for *T. cruzi* amastigotes

(Villalta and Kierszenbaum, 1982)

COMPOSITION

	ml dl^{-1}
Tissue culture medium L-15 (Flow Laboratories)	79

Tryptose phosphate broth with glucose (Difco)	10
Foetal bovine serum (Flow Laboratories)	10
Glutamine, 5% solution (Difco)	0.5
Haemin, 2.5 mg ml^{-1} solution (Sigma)	0.2

Note. The pH was adjusted to 7.3 with sterile 2% v/v HCl. This constituted medium ML-15H (Lanar, 1979). To prepare medium ML-15HA, it was supplemented with foetal bovine serum (heat-inactivated), 10% v/v; penicillin, 100 iu ml^{-1}; streptomycin, 100 μg ml^{-1}.

TECHNIQUE

Vero cells (fibroblast-like cells derived from the kidney of *Cercopithecus aethiops*; obtainable from commercial suppliers including Flow Laboratories and Gibco-Bio Cult) were maintained in 25 cm^2 plastic tissue culture flasks (Dow Corning) in Eagle's minimum essential medium (Gibco; see Taylor and Baker, 1968, p. 355) supplemented with serum and antibiotics as given above and incubated at 37°C. Bloodstream trypomastigotes of *T. cruzi* were harvested aseptically from mice by cardiac puncture (anticoagulant not stated) at maximum parasitaemia and 10^6 were added to Vero cell cultures in which Eagle's medium had been replaced by ML-15HA. After 48 h the supernatant medium was replaced by fresh ML-15HA; similarly after a further 48 h. Parasites were harvested after another 24–48 h (6 d in all) by removal of the supernatant medium, and washed twice by centrifugation (460*g*, 10 min, 4°C). Amastigotes, liberated from infected cells, were separated from trypomastigotes (if required) by centrifugation (800*g*, 20 min, 4°C) through a discontinuous metrizamide gradient (8% metrizamide (Nyegaard), relative density 1.048, at the top; 16% metrizamide, relative density 1.06, at the bottom). Amastigotes (5–10 × 10^7 per culture) collected at the bottom of the centrifuge tube. These amastigotes (about 10^6 ml^{-1}) could be used to inoculate fresh Vero cell cultures instead of bloodstream trypomastigotes, if desired.

Kierszenbaum *et al.* (1983) described a method for separating infected mastocytoma (P815) cells from uninfected cells in *T. cruzi* cultures by centrifugation of cells washed twice with Tyrode's solution (Taylor and Baker, 1968, p. 350) by centrifugation 450*g*; temperature and time not stated, presumably 4°C and 15 min): washed cells were layered on to a discontinuous gradient of 2.0 ml solutions of metrizamide of relative densities 1.120 (lower layer), 1.100 and 1.090 in 15 ml plastic centrifuge tubes (Corning; catalogue no. 25310) and centrifuged (450*g*, 15 min,

4 °C). The pellet consisted of about 80% infected cells. The method could presumably be adapted to other cell culture systems.

IV. CULTURE OF OTHER STERCORARIAN TRYPANOSOMES

Many species of stercorarian trypanosomes of mammals and other vertebrates will grow readily, predominantly as epimastigotes, in a wide range of undefined media at temperatures not higher than 28 °C. Some may require lower temperatures — *T. rajae*, for example, isolated from the ray *Raja clavata* grows well in 4N medium (Evans, 1978, p. 67) only if kept at 15 °C or below (T.J. Preston, personal communication). Details of media suitable for the cultivation of stercorarian trypanosomes have been given by Taylor and Baker (1968) and Evans (1978). There have been few significant developments in this subject within the last 10 years, apart from the introduction of Grace's insect tissue culture medium as a suitable medium for the cultivation of an increasing range of stercorarian species, including *T. rangeli* (Hendricks *et al.*, 1978; Pennick and Paul, 1983; Mohamed and Molyneux, 1987). The possibility of using this medium with other species of *Trypanosoma* clearly should be explored; it is convenient (if expensive) and is free of blood or other organic components, except for the foetal bovine serum which must be added to support trypanosome growth. Schneider's *Drosophila* medium is equally suitable for at least some stercorarian trypanosomes, if supplemented with serum (Mohamed and Molyneux, 1987).

The cultivation of species of the subgenus *Herpetosoma* has recently been reviewed by Mohamed and Molyneux (1987). Most members of the subgenus can be grown fairly easily in conventional blood-agar media (mainly NNN; see Chapter 3), though an exception appears to be *T. nabiasi* of rabbits *(Oryctolagus cuniculi*). Mohamed and Molyneux (1987) succeeded in serially culturing *T. microti*, *T. evotomys*, *T. grosi*, *T. musculi* and *T. lewisi* in both Schneider's and Grace's media (obtained from Gibco Bio-Cult), supplemented with 20% v/v foetal bovine serum (inactivated at 56 °C for 30 min) and kanamycin (500 μg ml^{-1}) and incubated at 27 °C. Parasites were isolated from infected hosts directly into the media and, after an establishment period of 3 weeks, subcultures were made weekly; 5.0 ml volumes of the media were contained in T-25 plastic tissue culture flasks. Metacyclic trypomastigotes developed, which were infective to the appropriate mammalian hosts. Only transient growth of *T. nabiasi* was obtained in Grace's medium, and none in Schneider's.

In my laboratory, Grace's medium (supplemented with 30% v/v heat-inactivated foetal bovine serum) has been used for serial culture of several species of the subgenus *Schizotrypanum* isolated from bats (Chiroptera), for *T. scyllii* from the dogfish *Scyliorhinus canicula* (kindly provided by Dr Ann Pulsford) and, with marginal success only, for *T. rajae* (at temperatures below 15 °C). *T. scyllii* has also been grown in 4N medium (see Evans, 1978, p. 67); 3% w/v urea added to the overlay improved growth (Pulsford, 1984).

V. PROSPECTS

It seems likely, if one dares to prophesy, that future developments in the cultivation of stercorarian trypanosomes, and *T. cruzi* in particular, will be directed towards the production of different life-cycle stages, and especially those normally developing in the vertebrate host, in increased yields and of increased purity. Thus suitable material should become available for furthering the pursuit of the "philosopher's stone" — a usable vaccine against Chagas disease. At present this search is bedevilled by the fact that the pathogenesis of Chagas disease may be autoimmunological in nature. Perhaps the ability to purify selected antigens from suitable stages, and to clone their genes in an appropriate vector, could eventually lead to the selection of an antigen which induces only protective antibodies in the recipient.

More immediately, the ability to cultivate more species in large numbers in relatively "clean" media, such as Grace's insect tissue culture medium, facilitates the production of monoclonal antibodies (see Chao and Dusanic, 1985) and the application of the current battery of biochemical and immunological procedures which are so markedly clarifying the taxonomic interrelationships of trypanosomes and other parasites. Thus, the application of isoenzyme electrophoresis to a number of species of the subgenus *Herpetosoma* (Mohamed *et al.*, 1987) became possible only after means of cultivating them in Grace's medium had been developed (Mohamed and Molyneux, 1987).

REFERENCES

Abrahamson, I.A. and Kloetzel, J.K. (1980). Presence of *Trypanosoma cruzi* antigen on the surface of both infected and uninfected cells in tissue culture. *Parasitology* **80**, 147–152.

Al-Abassy, S.N., Seed, T.M. and Kreier, J.P. (1972). Isolation of the trypomastigote form of *Trypanosoma cruzi* from a mixture of the trypomastigote and epimastigote forms of the parasite by use of a DEAE-cellulose column. *Journal of Parasitology* **58**, 631–632.

Arévalo, J., Panebra, A. and Santa Cruz, C. (1985). Relevance of hemin for *in vitro* differentiation of *Trypanosoma cruzi*. *Journal of Protozoology* **32**, 553–555.

Avila, J.L., Bretaña, A., Casanova, M.A., Avila, A. and Rodríguez, F. (1979). *Trypanosoma cruzi*: defined medium for continuous cultivation of virulent parasites. *Experimental Parasitology* **48**, 27–35.

Avila, J.L., Pérez-Kepp, R. and Bretaña, A. (1983). A minimal medium for the cultivation of infective *Trypanosoma cruzi* epimastigotes. *Journal of General Microbiology* **129**, 285–291.

Azevedo, H.P. and Roitman, I. (1977). Growth of the Y strain of *Trypanosoma cruzi* in an HX 25-modified defined medium. *Journal of Parasitology* **63**, 485.

Baker, J.R. (1984). Haem and haemoflagellates. *Transactions of the Royal Society of Tropical Medicine and Hygiene* **78**, 142.

Baker, J.R. and Green, S.M. (1977). Trypanosomal antigen on macrophages cultivated *in vitro* with *Trypanosoma (Schizotrypanum) dionisii*. *Annals of Tropical Medicine and Parasitology* **71**, 141–145.

Baker, J.R. and Price, J. (1973). Growth *in vitro* of *Trypanosoma cruzi* as amastigotes at temperatures below 37°C. *International Journal for Parasitology* **3**, 549–551.

Baker, J.R. and Selden, L.F. (1978). *Trypanosoma (Schizotrypanum) dionisii*: influence of mouse peritoneal macrophages and calf sera on extracellular growth *in vitro* at 37°C. *Journal of General Microbiology* **106**, 27–32.

Bertelli, M.S.M., Golgher, R.R. and Brener, Z. (1977). Intraspecific variation in *Trypanosoma cruzi*: effect of temperature on the intracellular differentiation in tissue culture. *Journal of Parasitology* **63**, 434–437.

Bogliolo, A.R., Guimarães, R.C., Brener, Z., Bertelli, M.S.M. and Golgher, R.R. (1980). *Trypanosoma cruzi*–host cell interaction and the temperature-sensitive phenotype of CL strain. *Ciência e Cultura* **32**, 742–749.

Brun, R. and Jenni, L. (1985). Cultivation of African and South American trypanosomes of medical or veterinary importance. *British Medical Bulletin* **41**, 122–129.

Castanys, S., Osuna, A., Gamarro, F. and Ruiz-Pérez, L.M. (1984). Purification of metacyclic forms of *Trypanosoma cruzi* by Percoll discontinuous gradient centrifugation. *Zeitschrift für Parasitenkunde* **70**, 443–449.

Chao, D. and Dusanic, D.G. (1985). Monoclonal antibodies to metacyclic stage antigens of *Trypanosoma cruzi*. *American Journal of Tropical Medicine and Hygiene* **34**, 694–701.

De Carvalho, T.U. and de Souza, W. (1983). Separation of amastigotes and trypomastigotes of *Trypanosoma cruzi* from cultured cells. *Zeitschrift für Parasitenkunde* **69**, 571–575. [*Protozoological Abstracts* 1984, **8**, abstr. 900.]

Dvorak, J.A. (1980). Trends in the use of *in vitro* cell cultures for *Trypanosoma cruzi* research. *In* "The *In Vitro* Cultivation of the Pathogens of Tropical Diseases", *Tropical Diseases Research Series* **3**, pp. 223–236. Schwabe, Basel.

Dvorak, J.A. and Crane, M. St J. (1981). Vertebrate cell cycle modulates infection by protozoan parasites. *Science* **214**, 1034–1036.

Dvorak, J.A. and Hyde, T.P. (1973). *Trypanosoma cruzi*: interaction with vertebrate cells *in vitro* I. Individual interactions at the cellular and subcellular levels. *Experimental Parasitology* **34**, 268–283.

Dvorak, J.A. and Schmunis, G.A. (1972). *Trypanosoma cruzi*: interaction with mouse peritoneal macrophages. *Experimental Parasitology* **32**, 289–300.

Engel, J.C., Dvorak, J.A., Segura, E.L. and Crane, M. St J. (1982). *Trypanosoma cruzi*: biological characterization of 19 clones derived from two chronic Chagasic patients. I. Growth kinetics in liquid medium. *Journal of Protozoology* **29**, 555–560.

Evans, D. (1978). Kinetoplastida. *In* "Methods of cultivating Parasites *In Vitro*" (A.E.R. Taylor and J.R. Baker, eds), pp. 55–88. Academic Press, London and New York.

Eyzaguirre, N.L. (1983). Culture of *Trypanosoma cruzi* on agar plates free of blood. *Transactions of the Royal Society of Tropical Medicine and Hygiene* **77**, 740–741.

Filardi, L.S. and Brener, Z. (1975). Cryopreservation of *Trypanosoma cruzi* bloodstream forms. *Journal of Protozoology* **22**, 398–401.

Glauert, A.M., Baker, J.R. and Selden, L.F. (1982). Mechanism of entry and development of *Trypanosoma dionisii* in non-phagocytic cells. *Journal of Cell Science* **56**, 371–387.

Goldberg, S.S. and Chiari, E. (1980). Growth and isolation of single colonies of *Trypanosoma cruzi* on solid medium. *Journal of Parasitology,* **66**, 677–679.

Gutteridge, W.E., Knowler, J. and Coombes, J.D. (1969). Growth of *Trypanosoma cruzi* in human heart tissue cells and effects of aminonucleosides of puromycin, trypacidin and aminopterin. *Journal of Protozoology* **16**, 521–525.

Hanson, W.L. (1980). Mass production of trypomastigote and amastigote stages of *Trypanosoma cruzi* in cell culture. *In* "The *In Vitro* Cultivation of the Pathogens of Tropical Diseases", *Tropical Diseases Research Series* **3**, pp. 245–246. Schwabe, Basel.

Hendricks, L.D., Wood, D.E. and Hajduk, M.E. (1978). Haemoflagellates: commercially available liquid media for rapid cultivation. *Parasitology* **76**, 309–316.

Henriquez, D., Piras, R. and Piras, M.M. (1981). The effect of surface membrane modifications of fibroblastic cells on the entry process of *Trypanosoma cruzi* trypomastigotes. *Molecular and Biochemical Parasitology* **2**, 359–366.

Hudson, L., Snary, D. and Morgan, S.-J. (1984). *Trypanosoma cruzi*: continuous cultivation with murine cell lines. *Parasitology* **88**, 283–294.

Hutner, S.H. (1984). Haem and haemoflagellates: a reply to J.R. Baker. *Transactions of the Royal Society of Tropical Medicine and Hygiene* **78**, 564–565.

Hyde, T.P. and Dvorak, J.A. (1973). *Trypanosoma cruzi*: interaction with vertebrate cells *in vitro*. II. Quantitative analysis of the penetration phase. *Experimental Parasitology* **34**, 284–294.

Isola, E.L.D., Lammel, E.M. and González Cappa, S.M. (1983). Influencia de la hemolinfa de *Triatoma infestans* en la morfogénesis del *Trypanosoma cruzi. Revista Argentina de Microbiología* **15**, 181–185. [*Protozoological Abstracts* 1984, **8**, abstr. 2970.]

Katzin, A.M., Lajmanovich, S. and González Cappa, S.M. (1977). Comparative stability of epimastigotes and trypomastigotes in different wash solutions. *Journal of Parasitology* **63**, 925–927.

Kierszenbaum, F., Villalta, F. and Pankratz, H.S. (1983). Separation of *Trypanosoma cruzi*-infected from non-infected P815 cells. *Acta Tropica* **40**, 85–87.

Lanar, D.E. (1979). Growth and differentiation of *Trypanosoma cruzi* cultivated with a *Triatoma infestans* embryo cell line. *Journal of Protozoology* **26**, 457–462.

Lopetegui, R. and Miatello, C.S. (1982). Desarrollo y diferenciación de *Trypanosoma cruzi* en medios liquidos libres de células. *Revista Latinoamericana de Microbiología* **24**, 125–132. [*Protozoological Abstracts* 1984, **8**, abstr. 2385.]

Meirelles, M.N.L., Chiari, E. and de Souza, W. (1982). Interaction of bloodstream, tissue culture-derived and axenic culture derived trypomastigotes of *Trypanosoma cruzi* with macrophages. *Acta Tropica* **39**, 195–203.

Milder, R. and Kloetzel, J. (1980). The development of *Trypanosoma cruzi* in macrophages *in vitro*. Interactions with lysosomes and host cell fate. *Parasitology* **80**, 139–145.

Mohamed, H.A. and Molyneux, D.H. (1987). *In vitro* cultivation of *Herpetosoma* trypanosomes in insect cell tissue culture media. *Parasitology Research* **73**, 9–14.

Mohamed, H.A., Molyneux, D.H. and Scott, C.M. (1987). Isoenzyme characterization of trypanosomes of the subgenus *Herpetosoma. Parasitology* **94**, 39–48.

Nogueira, N. and Cohn, Z. (1977). *Trypanosoma cruzi*: uptake and intracellular fate in normal and activated cells. *American Journal of Tropical Medicine and Hygiene* **26**, 194–202.

Pan, S.C. (1971). Cultivation and morphogenesis of *Trypanosoma cruzi* in improved liquid media. *Journal of Protozoology* **18**, 556–560.

Pan, S.C. (1978a). *Trypanosoma cruzi*: intracellular stages grown in a cell-free medium at 37°C. *Experimental Parasitology* **45**, 215–224.

Pan, S.C. (1978b). *Trypanosoma cruzi*: *in vitro* interactions between cultured amastigotes and human skin-muscle cells. *Experimental Parasitology* **45**, 274–286.

Pan, S.C. (1978c). *Trypanosoma cruzi*: cultivation in macromolecule-free semisynthetic and synthetic media. *Experimental Parasitology* **46**, 108–112.

Pennick, N.C. and Paul, T.R. (1983). Alternative culture media for trypanosomes. *Annual Report, Institute of Terrestrial Ecology, 1982,* pp. 93–94. Institute of Terrestrial Ecology, Cambridge.

Pulsford, A. (1984). Preliminary studies on trypanosomes from the dogfish, *Scyliorhinus canicula* L. *Journal of Fish Biology* **24**, 671–682.

Roitman, I. (1980). Nutritional requirements of *Trypanosoma cruzi. In* "The *In Vitro* Cultivation of the Pathogens of Tropical Diseases", *Tropical Diseases Research Series* **3**, pp. 247–248. Schwabe, Basel.

Ryley, J.F. and Wilson, R.G. (1978). Cell and tissue culture. *In* "Methods of Cultivating Parasites *In Vitro*" (A.E.R. Taylor and J.R. Baker, eds), pp. 111–128. Academic Press, London and New York.

Sanderson, C.J., Thomas, J.A. and Twomey, C.E. (1980). The growth of *Trypanosoma cruzi* in human diploid cells for the production of trypomastigotes. *Parasitology* **80**, 153–162.

Schmatz, D.M. and Murray, P.K. (1982). Cultivation of *Trypanosoma cruzi* in irradiated muscle cells: improved synchronization and enhanced trypomastigote production. *Parasitology* **85**, 115–125.

Schmatz, D.M., Boltz, R.C. and Murray, P.K. (1983). *Trypanosoma cruzi*: separation of broad and slender trypomastigotes using a continuous hypaque gradient. *Parasitology* **87**, 219–227.

Sullivan, J.L. (1982). Metacyclogenesis of *Trypanosoma cruzi in vitro*: a simplified procedure. *Transactions of the Royal Society of Tropical Medicine and Hygiene* **76**, 300–303.

Taylor, A.E.R. and Baker, J.R. (1968). "The Cultivation of Parasites *in vitro*." Blackwell, Oxford and Edinburgh.

Taylor, A.E.R. and Baker, J.R. (eds) (1978). "Methods of Cultivating Parasites *in vitro*". Academic Press, London and New York.

Trager, W. (1957). Nutrition of a hemoflagellate (*Leishmania tarentolae*) having an interchangeable requirement for choline or pyridoxal. *Journal of Protozoology* **4**, 269–276.

Ucros, H., Granger, B. and Krassner, S.M. (1983). *Trypanosoma cruzi*: effect of pH on *in vitro* formation of metacyclic trypomastigotes. *Acta Tropica* **40**, 105–112.

Villalta, F. and Kierszenbaum, F. (1982). Growth of isolated amastigotes of *Trypanosoma cruzi* in cell-free medium. *Journal of Protozoology* **29**, 570–576.

Villalta, F. and Kierszenbaum, F. (1984). Host-cell invasion by *Trypanosoma cruzi*: role of cell surface galactose residues. *Biochemical and Biophysical Research Communications* **119**, 228–235. [*Protozoological Abstracts* 1984, **8**, abstr. 2379.]
Villalta, F.V. and Leon, W. (1979). Effect of purification by DEAE-cellulose column on infectivity of *Trypanosoma cruzi* blood forms. *Journal of Parasitology* **65**, 188–189.
Williams, G.H. and Hudson, L. (1982). Growth of *Trypanosoma cruzi in vitro*: development and application of a continuous-flow culture system. *Parasitology* **84**, 511–526.

Chapter 5

Salivarian Trypanosomes: Bloodstream Forms (Trypomastigotes)

R. BRUN and L. JENNI

Swiss Tropical Institute, Socinstrasse 57, CH-4051 Basel, Switzerland

I. INTRODUCTION

Salivarian trypanosomes proliferate in mammalian hosts as trypomastigote forms in the bloodstream and also in extravascular sites such as the lymphatic system, the central nervous system and tissue of various organs. The extravascular distribution is more pronounced in the *Trypanozoon* subspecies, whereas *T. congolense* and *T. vivax* are located predominantly in the bloodstream. The parasites are covered by a surface coat of variable surface glycoproteins (VSG), which protects them from the lytic action of the mammalian plasma and allows them to evade elimination by the host's immune response.

Until 1977 no method was available to grow the bloodstream stages in culture. All previous attempts to cultivate the mammalian forms resulted either in the death of the trypanosomes or their transformation to vector forms that were no longer infective for mammalian hosts. The application of tissue culture techniques finally led to a breakthrough. Hirumi *et al.* (1977a, b) were the first workers to succeed in growing bloodstream forms continuously, using a feeder layer of mammalian cells in combination with a commercial tissue culture medium. The basic idea of growing bloodstream forms in co-cultivation with mammalian cells had been tried 10 years earlier by LePage (1967); using mouse L-cells and NCTC 109 medium, bloodstream forms could be maintained for 3 d.

The subsequent rapid development of culture methods for the bloodstream stages of salivarian trypanosomes has been reviewed by Hirumi (1979), Hill and Hirumi (1983) and Brun and Jenni (1985).

II. *TRYPANOSOMA (TRYPANOZOON) BRUCEI*

A. FEEDER LAYER SYSTEMS

Bloodstream forms of African trypanosomes can be successfully cultivated *in vitro* in the presence of a mammalian cell feeder layer. The critical elements of established culture systems are summarized in Table I, and discussed in more detail below.

TABLE I. Critical elements of feeder layer culture systems

Feeder layer	origin of cell line
	cell type
	growth rate of cell line
Medium	suitability to support growth of feeder cells and trypanosomes
	pH, tonicity
Serum	origin
	preparation, filtration and storage
	normal or heat-inactivated
Gas phase	5% CO_2 required
	reduced O_2 favourable
Culture vessel	volume of medium related to surface area
	open or closed vessel
Medium change	interval and portion of volume

1. *Feeder layer cells*

Many different mammalian cell lines of different cell type (''finite'' lines with a limited life-span as well as continuous lines) have been used as feeder layers. Hirumi *et al.* (1980a) used several lines of fibroblast or epithelial cell type isolated from bovine blood or embryonic tissue. Two commercially available cell lines, buffalo lung and Chinese hamster lung, have been used by Hill *et al.* (1978 a, b). Table II summarizes some of the major cell lines used as feeder layers. Most of them belong to the fibroblast-like cell type that seems to favour the growth of *Trypanozoon* subspecies. In fibroblast monolayers, intercellular clusters of parasites are

TABLE II. Cell lines used as feeder layer for the cultivation of *Trypanosoma (T.)brucei* bloodstream forms

Origin	Type[a]	Trypanosome[b]	References
Bovine adult blood	fibr.	*T.b.*	Hirumi *et al.* (1977a,b)
Bovine embryonic tissue	fibr.	*T.b.*	Hirumi *et al.* (1977a,b)
Bovine embryonic pericardium	epith.	*T.b.*	Hirumi *et al.* (1977a,b)
New Zealand White rabbit whole embryo	fibr.	*T.b.*, *T.r.*	Brun *et al.* (1979, 1981)
Microtus montanus whole embryo	fibr.	*T.b.*, *T.r.*, *T.g.*, *T.e.*	Brun *et al.* (1979, 1981)
Goat embryonic kidney	fibr.	*T.b.*, *T.r.*	Brun *et al.* (1984b)
Eland embryonic lung	fibr.	*T.b.*, *T.r.*	Brun *et al.* (1984b)
Grant's gazelle embryonic lung	fibr.	*T.r.*	Brun *et al.* (1984b)
Impala embryonic heart	fibr.	*T.b.*, *T.r.*	Brun *et al.* (1984b)
Human embryonic tissue	fibr.	*T.r.*, *T.g.*	Brun and Jenni (1983)
Foetal rat lung	fibr.	*T.b.* (dyskinetoplastic)	Stuart (1980)
Mouse bone marrow		*T.b.*, *T.r.*, *T.g.*	Balber (1983, 1984)
New Zealand White rabbit diaphragm	fibr.	*T.e.*	Zweygarth *et al.* (1983)
Buffalo lung[c]	fibr.	*T.b.*, *T.r.*	Hill *et al.* (1978a,b)
Chinese hamster lung[c]	fibr.	*T.b.*, *T.r.*	Hill *et al.* (1978a,b)
Mouse L-cells[c]	fibr.	*T.b.*	Hirumi *et al.* (1980a)
Flow 2002, human embryonic lung[c]	fibr.	*T.r.*, *T.g.*	Brun *et al.* (1981)

[a]fibr. = fibroblast-like cells; epith. = epithelioid cells.
[b]*T.b.* = *Trypanosoma b. brucei*, *T.r.* = *Trypanosoma b. rhodesiense*, *T.g.* = *Trypanosoma b. gambiense*, *T.e.* = *Trypanosoma evansi*.
[c]commercially available cell line.

a prerequisite for establishment in culture. These intercellular trypanosomes possibly represent tissue forms, and in fact their ultrastructural composition is similar to that of extravascular forms found in the rat (Hecker and Brun, 1982).

In culture systems using epithelial cells, intercellular clusters are not formed, trypanosomes growing solely in the supernatant medium (Hirumi *et al.*, 1980a). A completely new approach was chosen by Balber (1983, 1984), who introduced a feeder layer of murine bone marrow cells. These primary cultures contained adipocytes as the growth supporting elements and the trypanosomes were located mainly free in the culture supernate.

From the work so far published, it becomes obvious that several cell lines from different mammalian hosts can be used as feeder layers. However, the cell line has to be carefully selected and several criteria have to be considered.

(i) *Growth properties*. A slow-growing cell line is preferable to a fast-growing line. The latter depletes the medium of nutrients and accumulates metabolites very quickly and thus is in competition with the trypanosomes.

(ii) *Origin of the cells*. Cell lines from natural or laboratory hosts provide "logical" culture systems, especially in combination with a serum supplement from the same animal species, or another species which represents a potential host for *Trypanozoon* trypanosomes.

(iii) *Serum incompatibility*. A cell line which tolerates heterologous sera is an advantage since it can be used together with inexpensive or easily prepared sera of appropriate suitability.

(iv) *Cell type*. Fibroblast-like cell lines, in opposition to epithelial cell lines, allow intercellular growth of trypanosomes.

A cell line which fulfils most of these criteria is MEF, *Microtus montanus* whole embryo fibroblasts (Brun *et al.*, 1981). *Microtus montanus*, a North American vole, is a suitable laboratory host for *T.b. gambiense* (Seed and Negus, 1970). MEF were originally isolated for the *in vitro* propagation of *T. brucei gambiense* stocks, but it turned out that they were also ideal for growing *T.b. brucei* and *T.b. rhodesiense* (Brun *et al.*, 1981). One of the most favourable properties of this cell line is its compatibility with most mammalian sera, e.g. human serum. Also it is slow growing and forms a thin monolayer with intercellular spaces that can easily be occupied by the trypanosomes.

For the human infective *Trypanozoon* subspecies*, *T.b. rhodesiense*

*Throughout this and the subsequent chapter, the term "subspecies" and the names *T. brucei rhodesiense* and *T. brucei gambiense* are used for convenience, without necessarily indicating the taxonomic status of the populations concerned [Eds].

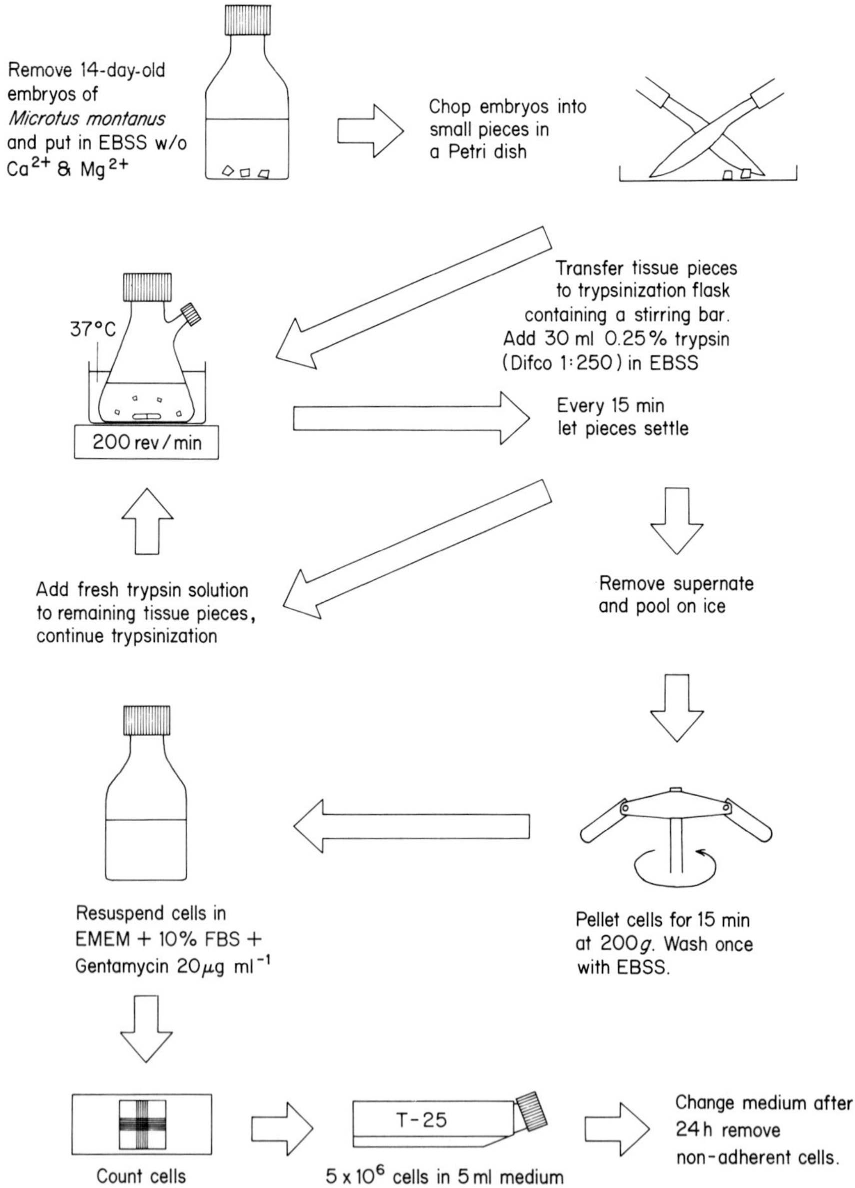

Fig. 1. Isolation of cell line from whole embryos of *Microtus montanus*.

and *T.b. gambiense*, human embryonic fibroblasts are also ideal feeder layers. Cell lines isolated from human embryonic tissues as well as a commercial cell line, Flow 2002, supported growth of *T.b. rhodesiense* and *T.b. gambiense* very efficiently (Brun and Jenni, 1983).

The isolation of cell lines has been described by Hirumi *et al.* (1977 a, b) and Brun *et al.* (1981). Procedures for the isolation of cell lines can be found in most of the currently used handbooks on culture techniques for animal cells (e.g. Freshney, 1983). Different methods exist for isolating fibroblast-like cell lines, of which enzymatic disaggregation is commonly used (Fig. 1).

2. *Culture media*

Two different tissue culture media are currently used for growing the bloodstream forms of salivarian trypanosomes in the presence of a feeder layer: Roswell Park Memorial Institute (RPMI) 1640 medium and Eagle's minimum essential medium with Earle's salts (EMEM: see Taylor and Baker, 1968, p. 355). Both media can be bought in liquid or powder form from various suppliers and are used with L-glutamine and buffered with 25 mM HEPES.

EMEM solution can be purchased already supplemented with L-glutamine and buffered with 25 mM HEPES. To this solution, 1% of a stock solution of glucose (100 g litre^{-1}) and 1% MEM non-essential amino acids, 100 times concentrate, can be added aseptically without the need to sterilize the final medium by filtration. If EMEM is prepared from the powder, HEPES (25 mM) and $NaHCO_3$ (2.2 g litre^{-1}) must be added and the pH adjusted with 4 N NaOH to 7.4; the medium must then be sterilized by filtration (0.22 μm pore size filter).

The medium can be stored frozen for several months, or at 4 °C for at least 4 weeks. Before use, it is supplemented with serum (10–20% v/v) and, optionally, with antibiotics; we routinely use gentamycin (10 μg ml^{-1}) for cultures of both mammalian cells and trypanosomes.

3. *Serum*

Serum is an obligatory addition to the culture medium, not yet replaceable by defined components. A variety of mammalian sera can be used to grow the mammalian stages of *Trypanozoon* subspecies. Hirumi *et al.* (1977 a, b) used 20% heat-inactivated (HI) foetal bovine serum (FBS) when they first achieved continuous growth of bloodstream forms *in vitro*. These authors warned of the considerable variation among different batches of FBS with regard to growth promotion or inhibition and considered that batches should be tested before ordering (Hirumi *et al.*, 1980a; Hill and Hirumi, 1983).

In our laboratory we had difficulty in finding suitable batches of FBS for mammalian form cultures, especially when working with fresh pleomorphic isolates, and therefore we use self-prepared sera from various mammalian hosts. The blood is allowed to coagulate at room temperature for 2 h, then left for another 2 h, or even overnight, at 4°C. The separated serum is sterilized by filtration through a 0.22 μm filter, and stored frozen in suitable aliquots at −28°C until use. The change from commercial to self-prepared sera immediately gave better growth of bloodstream forms in our cultures. Almost all sera prepared from the blood of various mammalian hosts supported growth of *Trypanozoon* subspecies to some degree. Table III lists some of the sera with good growth-supporting properties leading to the formation of intercellular trypanosome clusters and final parasite densities above 2×10^5 ml^{-1}.

Whether a serum should be used heat-inactivated (HI) depends on the compatibility of the feeder cells with a given serum, and on the purpose of the *in vitro* experiment. Incompatibility between cells and serum has to be expected with normal serum whenever feeder cells and serum are not derived from the same animal species. The reaction of mammalian feeder layer cells to heterologous serum comprises vacuolization, detachment and finally destruction of the cells. This normally occurs within 24 h and terminates the culture. Heat-inactivation (56°C, 30 min) before use may help to overcome this problem.

TABLE III. Mammalian sera that support good growth of *Trypanosoma (T.) brucei* bloodstream forms *in vitro*[a]

Serum[b]	*Trypanozoon* "subspecies"[c]
Rabbit	*T.b., T.r., T.g.*
Rat	*T.b., T.r.*
Horse	*T.b., T.r., T.g., T.e.*
Foal	*T.b., T.e.*
Goat	*T.b., T.r.*
Sheep	*T.b., T.r.*
Impala	*T.b., T.r.*
Human	*T.r., T.g.*

[a]EMEM was supplemented with 15–20% serum (normally heat-inactivated) and used in the presence of a mammalian cell feeder layer (Brun *et al.*, 1979, 1981, 1984b; Zweygarth *et al.*, 1983).

[b]All sera were prepared in the laboratory.

[c]*T.b.* = *Trypanosoma b. brucei*, *T.r.* = *Trypanosoma b. rhodesiense*, *T.g.* = *Trypanosoma b. gambiense*, *T.e.* = *Trypanosoma evansi*.

Heat-inactivated rabbit serum, prepared from blood of New Zealand White rabbits, is excellent for growing all *Trypanozoon* subspecies (Brun *et al.*, 1979, 1981). Commercial rabbit serum from various suppliers supported no, or only very little, growth of bloodstream forms *in vitro*. It was first suspected that antibiotics or other additions to the rabbit diet could be responsible for this. However, most suppliers do not use such additives in rabbit food, but commercial sera are often sterilized by multiple filtration. With each filtration step, proteins are lost by binding to the filter membrane. Another reason for the loss of growth-supporting activity could be careless shipping from the producer to the customer. Sera often thaw in transit and are refrozen at their destination. This freeze–thawing is extremely detrimental to the sera.

The low cost of self-prepared sera should also be considered, especially when blood can be obtained from a slaughter-house. Commercial sera are expensive and proper shipment, especially overseas, may almost double the price.

4. *Culture vessels*

Various disposable plastic vessels (tissue culture quality) of different shape and size can be used to grow bloodstream forms depending on the number of trypanosomes required and on the purpose of the experiment (Table IV). Our laboratory normally uses T-25 and T-75 flasks, or C-24 multiwell plates. For large-scale cultures the Nunc cell factory, closed culture plates with 600 cm^2 surface area, can be used.

TABLE IV. Production of trypanosomes in different culture vessels[a]

Culture vessel	Surface area (cm^2)	Amount of medium (ml)	Daily output[b]
C-24 cluster plate, per well	2	1	10^6
T-25 flask	25	7–10	10^7
T-75 flask	75	20–30	3×10^7
Nunc cell factory, single tray	600	200	2.5×10^8

[a]Culture system with fibroblast feeder layer.
[b]Realistic number of culture forms per vessel that can be expected with *T. (T.)brucei* stocks adapted to culture.

5. Detailed techniques

1° Feeder cell-layer technique for *T. brucei*

MEDIUM

(i) Either RPMI 1640 with HEPES and L-glutamine, or EMEM with HEPES, L-glutamine, glucose (g litre^{-1}) and MEM non-essential amino acids (100 times concentrate, 1% v/v) (see Section II.A.2 above).

(ii) Serum 10–20% (see Section II.A.3 above).

(iii) Antibiotics if required (e.g., gentamycin 10 μg ml^{-1}).

FEEDER CELLS

Feeder cells are subcultured by trypsinization about 48 h before inoculation of the culture with trypanosomes. Stock cultures of the feeder cell lines are grown in EMEM with 10% HI-FBS at 37°C in 5% CO_2 in air. Immediately before inoculating the culture system with trypanosomes, the medium is completely changed and HI-FBS replaced by a suitable mammalian serum (see Section II.A.3).

TECHNIQUE

(i) *Initiation of cultures*. The following trypanosome forms may be used to initiate an *in vitro* culture: metacyclic forms from tsetse flies; bloodstream forms from an infected animal, separated from blood cells by differential centrifugation, anion exchange chromatography (Lanham and Godfrey, 1970) or gradient centrifugation; bloodstream forms from a cryopreserved stabilate; lymph forms from lymph nodes, lymph fluid, brain or other tissue. There is no requirement for extensive washing of the trypanosomes after isolation. On the contrary, the more gentle and quick the isolation procedure, the better the *in vitro* propagation of the trypanosome population. Slight contamination of the inoculum by host cells (e.g., red blood cells) does not affect the initiation of a culture unless experiments require the complete absence of host cells. Although there is no minimum number of trypanosomes for inoculation, (single trypanosomes can be cloned directly *in vitro*; see below) the initial trypanosome density should not exceed 2 × 10^5 ml^{-1} culture medium, or the trypanosomes may die within 24 h owing to overcrowding.

Metacyclic forms from infected tsetse flies are ideal for starting cultures. Tsetse flies are induced to probe into a drop of warmed medium on a lymphocyte migration plate (Sterilin). The metacyclics are counted and put into one well of a 24-well cluster plate containing a feeder layer. They transform, within 8–10 h, to slender bloodstream forms (Brun *et al.*, 1984a).

After inoculation the cultures are incubated at 37 °C. Open flasks or plates require a humidified incubator with 5 % CO_2 in air. Flasks can also be gassed with 5 % CO_2 in air, closed air-tight and then incubated in a conventional incubator. When open systems are used, a modular incubation chamber (Flow Laboratories) can replace an expensive CO_2 incubator, if gassed from a cylinder of 5 % CO_2 in air.

When a feeder layer of fibroblast-like cells is used, the trypanosomes rapidly invade the intercellular spaces (a few minutes after inoculation) and eventually form clusters (Fig. 2). In our experience, these intercellular forms are important for the establishment and continuity of cultures. The supernatant medium also becomes populated with free trypanosomes exhibiting a slight degree of pleomorphism, in contrast to the intercellular forms which appear to be uniformly of the slender-intermediate form.

(ii) *Maintenance*. The cultures should be examined daily with an inverted phase-contrast microscope. As soon as the trypanosome density reaches 10^6 ml^{-1} the medium has to be changed daily by removing most or all of the used medium and replacing it by the same volume of fresh medium. If the used medium is gently mixed before removal, most of the free trypanosomes are suspended in it and can thus easily be harvested. The intercellular forms in the feeder layer remain and repopulate the fresh supernatant medium. The daily production of bloodstream forms in culture vessels of different size depends on the surface area of the feeder layer (Table IV). If the culture was inoculated with few

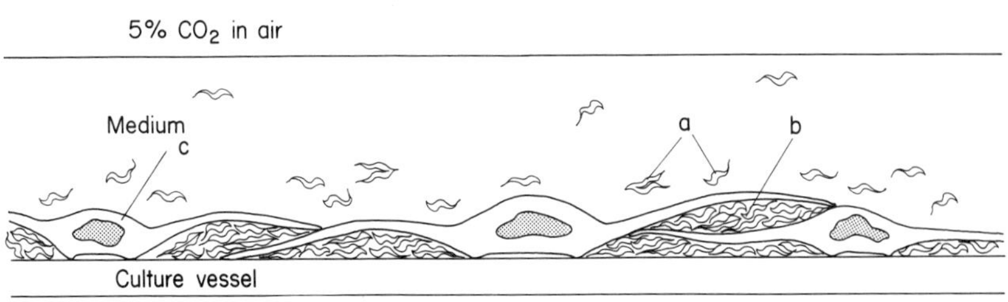

Fig. 2. Schematic drawing of cross section through established feeder layer culture of *T. (T.) brucei* bloodstream forms: a, trypanosomes in the supernatant medium; b, intercellular clusters of bloodstream forms; c, fibroblasts attached to culture vessel.

trypanosomes ($< 10^4$ ml^{-1}), the medium need not be changed for the first 48 h. Afterwards, part of the medium (50–70%) should be replaced daily until the final trypanosome density of about 2×10^6 ml^{-1} is reached. At this density growth of the trypanosomes stops, probably owing to glycolysis-inhibiting metabolites (pyruvate and lactate) accumulating in the medium. The same feeder layer can be used for 2–3 weeks or longer. Subculture is simply performed by transferring a few drops of medium from an established culture on to a fresh feeder layer.

2° *In vitro* cloning by the microdrop technique

In vitro cloning can be carried out using a microdrop technique with bloodstream forms from mice or culture, or with metacyclic forms from tsetse flies (Hirumi *et al.*, 1980b; Brun *et al.*, 1981; Hill and Hirumi, 1983). A short description of the procedure as done in our laboratory is given below.

(i) All peripheral wells of a flat-bottomed microtitre plate are filled with medium or sterile distilled water to help reduce evaporation.

(ii) A suspension of metacyclics or bloodstream forms is prepared in medium containing serum. The trypanosome concentration must be adjusted so that, on average, one trypanosome is obtained per microdrop.

(iii) To create a "wet chamber effect" in the empty wells, a drop of 30 μl of medium is placed at the edge of each well located in the centre area of the plate.

(iv) Using a sterile piece of wire, dipped into the trypanosome suspension, a microdrop is placed beside the 30 μl drop. The plate is immediately covered.

(v) The microdrop is examined using an inverted phase-contrast microscope. If two observers can each detect only one trypanosome, 70 μl of medium containing serum is added to the well.

(vi) The contents of the well, now containing one trypanosome in 100 μl medium, are transferred into a well on another microtitre plate containing a 2-day-old feeder layer.

(vii) The second plate is incubated at 37 °C in a humidified incubator with 5% CO_2 in air; the medium is changed every 2 d.

(viii) Once the clone reaches a population density of 10^5 ml^{-1} it can be transferred on to a fresh feeder layer in a larger culture vessel.

B. CELL-FREE CULTURE SYSTEMS

Recently, two culture media have been developed that allow growth of bloodstream forms in the absence of a mammalian feeder layer. Both media are based on minimum essential medium (MEM) supplemented with 10% serum. The key substances are 2-mercaptoethanol in one medium (Baltz *et al.*, 1985), and L-cysteine in the other (Duszenko *et al.*, 1985).

1° Cell-free semi-defined medium for bloodstream forms of *Trypanosoma (T.) brucei*
(Baltz *et al.*, 1985)

This medium is based on EMEM that is used for feeder layer culture systems. The feeder layer can be replaced by 2-mercaptoethanol or thioglycerol: Na-pyruvate, hypoxanthine and thymidine are further supplements of this culture medium (see below).

MEDIUM

MEM with Earle's salts, with L-glutamine, without $NaHCO_3$, powder (package for 1 litre).	
MEM non-essential amino acids (100 times concentrate	10 ml
HEPES	5.95 g
Glucose	1.0 g
$NaHCO_3$	2.2 g
Double-distilled water	1100 ml

Adjust pH to 7.3 with 5 N NaOH, sterilize by filtration (0.22 μm), and store (up to 10 d, 4°C).

Before use the medium is further supplemented as follows.

	Stock solution	Final concentration
Kanamycin	10 mg ml^{-1}	30 μg ml^{-1}
2-Mercaptoethanol	20 mM	0.2 mM
Na-pyruvate	1 M	2.0 mM
Hypoxanthine	10 mM	0.1 mM
Thymidine*	1.6 mM	0.016 mM
Rabbit, horse, rat or human serum (heat-inactivated)		10–15 %

*Possibly not essential.

TECHNIQUE

After initiation of the cultures an adaptation stage of 4–8 d was evident, and this was more reproducible in the presence of a macrophage feeder layer. After adaptation, it was possible to subculture without feeder cells by transferring 50–100 μl into 1.0 ml of fresh medium in C-24 multiwell plates. From then on, subcultures were maintained at a trypanosome density of 5×10^4–10^6 ml^{-1}. The highest trypanosome density reached, with *T.b. rhodesiense* ETat 2, was 4×10^6 ml^{-1}, with a generation time of 6.1 h.

Out of 11 *Trypanozoon* stocks, Baltz *et al.* were able to grow eight stocks axenically, six of them directly from mouse derived bloodstream forms and two of them after an adaptation phase over a macrophage feeder layer for 4–8 d. Three other *T.b. gambiense* stocks could not be grown. Different sera (all HI) were cross-checked with three differently supplemented media; various combinations resulted in different growth support, with or without the need for an adaptation phase over a macrophage feeder layer. It was concluded that a feeder layer should be used for the initial adaptation phase, and that several sera should be tested, in particular rabbit, horse, and foetal bovine sera (all HI). For *T.b. gambiense*, rat and human sera were recommended. EMEM with mercaptoethanol, hypoxanthine and Na-pyruvate gave the best results.

2° Cell-free semi-defined medium for bloodstream forms of *Trypanosoma b.brucei* strain 427
(Duszenko *et al.*, 1985)

Bloodstream forms of the monomorphic *Trypanosoma b. brucei* stock 427 were axenically cultivated in a modified MEM supplemented with L-cysteine. The cysteine-free MEM, prepared in the laboratory, was supplemented with L-ornithine, myristic acid/BSA complex and 10% HI-FBS. The key substance in this medium was obviously L-cysteine, which can replace the mammalian feeder layer. Owing to the fact that L-cysteine is oxidized within 4 h and that concentrations over about 10 mg litre^{-1} are toxic, this amino acid has to be added every 12 h in small amounts, depending on the trypanosome density in the culture. Omission of L-cysteine at any subculture eliminated the growth support of the medium.

MEDIUM

		ml
EMEM (with Earle's salts): see Taylor and Baker, 1968, p. 355) without cysteine and $NaHCO_3$, with added adenosine (12 mg litre^{-1} HEPES (7.14 g litre^{-1} and glucose (33 mM), self-prepared	.. ..	84.5
$NaHCO_3$ 5% (w/v)	.. ..	3.6
L-Ornithine (10 mg litre^{-1})	.. ..	0.1
Myristic acid/BSA complex (see note (a))	.. .. ..	2.0
FBS, heat-inactivated	.. .. ..	10.0
Cysteine (see note (b))		mg litre^{-1}
	added after inoculation	0.6
	added after 6 h	2.0
	added after 12 h	4.0
	added after 24 h	4.0
	and so on.	

Notes:

(a) Prepared in a molar ratio of 7 : 1, using myristic acid (1.2 mg ml^{-1}) and defatted BSA (50 mg ml^{-1}) as described by Cross and Manning (1973).

(b) A 1000 times concentrate cysteine stock was prepared in EMEM and sterilized by filtration each time immediately before use.

TECHNIQUE

Cultures were initiated with bloodstream forms from feeder layer cultures, starting with a trypanosome density of 10^5 ml^{-1} and an L-cysteine concentration of 0.6 mg litre^{-1}. After 6 h at 37 °C in 5% CO_2 in air, fresh L-cysteine was added to a final concentration of 2.0 mg litre^{-1}, and after another 12 h to 4.0 mg litre^{-1}. This concentration was maintained by adding 4.0 mg litre^{-1} L-cysteine every 12 h as a freshly prepared 1000 times concentrate stock solution in EMEM. Cultures could also be initiated with trypanosomes isolated from infected mouse blood; 1 mM Na-pyruvate or catalase (1.75 μg ml^{-1}) was then added to the medium to prevent the toxic effect of H_2O_2 produced by autoxidation of L-cysteine. The final trypanosome density reached in this medium was about 2 × 10^6 ml^{-1}. This value is in the same range as those attained in Baltz medium (Baltz *et al.*, 1985) or in the presence of a mammalian feeder layer (Brun *et al.*, 1981; Hill and Hirumi, 1983).

Experience with these two axenic culture media is still very limited. It remains to be established how easily other *Trypanozoon* stocks can be adapted to them, and whether the cultivated forms have the same characteristics as those in the mammalian host.

III. *TRYPANOSOMA CONGOLENSE*

The first, limited, success in the cultivation of bloodstream forms of *T. congolense* was achieved by incubating pieces of trypanosome infected calf or rabbit skin in EMEM or RPMI 1640 buffered with HEPES and supplemented with 20% HI-FBS (Gray *et al.*, 1979). Trypomastigotes, infective for mice, could be maintained for 3 weeks.

This system was improved by replacing the dermal explant with a feeder layer of a bovine aorta endothelial (BAE) cell line (Gray *et al.*, 1985; Ross *et al.*, 1985). A description of the isolation and maintenance of BAE cells is given in Chapter 6 (section II.C.2° (iv), (v)). Bloodstream trypomastigotes could be continuously cultivated at temperatures of 28 and 37°C. The cultures were initiated either with metacyclic forms from infected tsetse flies or from culture, or with bloodstream forms from mice, prepared with the aid of DEAE cellulose (Lanham and Godfrey, 1970). If normal bovine serum was replaced by FBS, the bloodstream forms transformed to procyclic epimastigotes, which eventually started to develop into metacyclic forms. The bloodstream trypomastigotes were attached to the feeder cells by their anterior ends, but could be harvested or subcultured by gently washing the cells with medium. The final trypanosome density in culture depended on the *T. congolense* stocks used, and was in the range of 10^5–10^6 ml^{-1}.

A. DETAILED TECHNIQUES

1° Cultivation of bloodstream forms of *T. congolense*
(Gray *et al.*, 1985)

MEDIUM

EMEM plus 25 mM HEPES and 20% normal bovine serum.

TECHNIQUE

4.0 ml aliquots of medium were used in T-25 flasks, gassed with 5% CO_2 in air and incubated at 28 or 37°C, containing a feeder layer of bovine aorta endothelial (BAE) cells. This cell line was isolated from bovine aorta by incubating pieces in collagenase solution and removing the endothelial cells by gentle pipetting (see Chapter 6, Section II.C.1°C (iv)). The inoculum was either metacyclic forms from tsetse or culture, or trypomastigotes isolated from mouse blood by the method of Lanham and Godfrey (1970). The yield was $5 \times 10^5 - 3 \times 10^6$ ml^{-1}.

A culture system similar to that described by Gray *et al.* (1985) has been developed by Hirumi and Hirumi (1984). The major difference is the medium — RPMI instead of EMEM. The need for foetal goat serum in this medium impedes its wider use. However, it is possible that this component could be replaced by another mammalian serum. Culture conditions and maintenance were similar to those described by Gray *et al.* (1985).

2° Cultivation of bloodstream forms of *T. congolense*
(Hirumi and Hirumi, 1984)

MEDIUM

RPMI 1640 plus 25 mM HEPES, 16% goat serum (HI) and 4% foetal goat serum (both sera prepared in the laboratory from blood of East African Galla crossbred goats).

TECHNIQUE

4.0 ml aliquots of medium were used in T-25 flasks, gassed with 5% CO_2 in air and incubated at 37°C, containing a feeder layer of bovine aorta endothelial (BAE) cells. The inoculum was 5×10^5 trypomastigotes isolated from mouse blood by differential centrifugation (200*g*, 10 min, temperature not stated; supernatant recentrifuged at 1500*g*, 10 min, to sediment trypanosomes). The yield was 500–1000 trypanosomes per microscope field (total magnification × 100).

IV. *TRYPANOSOMA VIVAX*

T. vivax isolates from East and West Africa differ from each other with regard to virulence in domestic and laboratory animals. Whereas East African stocks predominantly infect bovines and do not grow in laboratory rodents, some West African stocks are less virulent for bovines but are virulent in goats and often easy to propagate in small rodents. Such rodent-adapted stocks are more convenient to work with in the laboratory.

A culture system was developed for West African *T. vivax* (stock Zaria Y 486) by Brun and Moloo (1982). Since the goat is a very susceptible host for this West African *T. vivax* isolate, goat feeder cells and goat serum were used for their culture.

1° Culture system for West African stocks of *T. vivax*
(After Brun and Moloo, 1982)

MEDIUM

EMEM with L-glutamine and $NaHCO_3$, buffered with 25 mM HEPES and supplemented with additional glucose (1 g $litre^{-1}$), 1% MEM non-essential amino acids (100 times concentrate), and either 20% v/v goat serum (HI) or 10% v/v goat serum (HI) plus 10% v/v FBS (HI). (Goat sera prepared in the laboratory from the blood of East African Galla crossbred, or European, goats.)

Growth was improved by further supplementing the medium with 0.2 mM 2-mercaptoethanol, 2 mM Na-pyruvate and 0.1 mM hypoxanthine according to Baltz *et al.* (1985).

TECHNIQUE

Cultures were initiated in C-24 cluster plates or in T-25 flasks containing an almost confluent feeder layer of either MEF or goat embryonic lung (GEL) cells, gassed with 5% CO_2 in air and incubated at 37°C. Bloodstream forms were isolated from the blood of infected mice (by differential centrifugation or DEAE 52 cellulose separation according to Lanham and Godfrey, 1970), or goats (Brun and Moloo, 1982). Cultures were inoculated to give an initial density of 3×10^4 to 1.5×10^5 trypanosomes ml^{-1}. The preparation and maintenance of *T. vivax* cultures were similar to those described for *T. (T.) brucei* stocks. In contrast to *T. (T.) brucei*, *T. vivax* bloodstream forms tended to attach by their flagella to the feeder cells, and did not normally enter intercellular spaces. Trypanosomes were also found free in the supernatant medium, where they reached a density of 1×10^6–2×10^6 ml^{-1}. The same feeder layer could be used for at least 2 weeks before a subculture was prepared by transferring a few drops of culture supernate on to a fresh feeder layer with fresh medium.

East African *T. vivax* stocks did not grow in this culture system. Recently, workers in the laboratory of Dr. H. Hirumi (ILRAD, Nairobi, Kenya) succeeded in growing an East African *T. vivax* stock (IL 2133) as bloodstream forms *in vitro*. A feeder layer of glial cells from foetal bovine brain was used in EMEM, supplemented with 10–15% adult bovine plasma and 5–10% HI-FBS (Dr H. Hirumi, personal communication). Cloning of single bloodstream forms derived from cultures could also be achieved in this system.

V. CURRENT AND POTENTIAL USE OF CULTURES

Culture techniques are currently used for various purposes in the fields of biochemistry, immunology and molecular biology. One major use of *in vitro* cultures is the production of parasites without the mammalian host. With the existing culture systems for trypanosome bloodstream forms, however, the trypanosome yield is very modest. One major goal of future research has to be the improvement of parasite density in culture (see Section VI.B). Only then will it become possible to produce the high numbers of these trypanosome forms which at present can be obtained only from laboratory rodents and which are necessary for analytical work. The replacement of laboratory animals by available alternative methods should be an ethical obligation to any scientist doing experimental work.

Cultures are a real alternative to laboratory animals in the case of those trypanosome species or stocks that are difficult to propagate *in vivo*, such as *T. b. gambiense*. Culture methods may allow the isolation and growth of stocks from human patients in short times (Brun and Jenni, 1983), whereas laboratory animals are either refractory or require repeated subpassages for trypanosome densities to become detectable.

Cultures are also an ideal tool for studying the metabolism and morphogenesis of trypanosomes under controlled conditions. New safe and effective drugs against African trypanosomiasis are urgently needed. For empirical screening of compounds *in vivo*, large numbers of animals have to be used. In future, rational drug design based on the detection of metabolic peculiarities in the parasite's metabolism may be achieved with the help of *in vitro* techniques.

Continuous culture systems for bloodstream forms of *Trypanozoon* subspecies have recently been used for screening trypanosomicidal drugs (Balber *et al.*, 1985; Borowy *et al.*, 1985a, b, c; Croft, 1986; Fish *et al.*, 1985). The introduction of cell-free culture media will add a new dimension to drug screening. In fact, the axenic culture system developed by Baltz *et al.* (1985) has now been used to test the trypanosomicidal effect of a new nitroimidazole compound (Borowy *et al.*, in press). In that study, *T. vivax* and *T. congolense* were included as well as *Trypanozoon* subspecies.

The phenomenon of antigenic variation has been demonstrated in cultures of *T. b. brucei* in the absence of an immune response (Doyle *et al.*, 1980) and with *T. congolense* stocks (Luckins *et al.*, 1986). While one stock retained the metacyclic variable antigenic types (VATs) during continuous cultivation as bloodstream forms for 4 months, another stock underwent antigenic variation within 2 weeks after the initiation of bloodstream form cultures with metacyclic forms. Another important contribution to the study of antigenic variation has been made by Baltz *et al.* (1986), using their axenic culture system. These authors demonstrated that two variable surface glycoproteins can be expressed simultaneously on the surface of the same *T. equiperdum* bloodstream form in culture.

Bloodstream forms of *T. b. brucei* derived from metacyclics *in vitro*, which still expressed metacyclic VATs, have been used after irradiation to immunize mice. The mice were protected against infective bites to tsetse flies harbouring the same *T. b. brucei* stock (Jenni and Brun, 1981). Bloodstream forms from culture can also be used as antigen for routine serodiagnostic purposes and for immunological studies.

To distinguish between *T. b. brucei*, which is sensitive to human serum by definition, and the human pathogenic "subspecies" *T. b. rhodesiense* and *T. b. gambiense* (both resistant to human serum), an *in vitro* test has been developed (Jenni and Brun, 1982) that is based on a feeder layer culture system using human serum as a medium supplement. The test allows the detection of single forms resistant to human serum in a trypanosome population.

In future, when simpler and refined *in vitro* techniques become available, cultures will have a wide range of applications in most laboratories studying salivarian trypanosomes. One field in which culture systems are potentially of vital importance is in the genetic analysis of trypanosomes. Genetic reassortment between parasite stages of compatible clones can be studied *in vitro* by simultaneous cultivation. This may allow a detailed analysis of the genetic hybridization events in trypanosomes which can occur in the tsetse fly (Jenni *et al.*, 1986).

VI. UNSOLVED PROBLEMS

A. REPLACEMENT OF FEEDER LAYER CELLS

Since the first continuous co-cultivation of mammalian forms with a mammalian cell feeder layer (Hirumi *et al.*, 1977a, b,), replacement of feeder cells has been a major goal. With the recent development of two cell-free culture media (Baltz *et al.*, 1985; Duszenko *et al.*, 1985) this long-standing problem seemed to be solved. However, some of the trypanosome stocks used by Baltz *et al.* (1985) required an adaptation phase on a macrophage feeder layer, and others could not be grown at all. Such problems are not unexpected to workers accustomed to growing trypanosomes *in vitro*. Both Baltz *et al.* (1985) and Duszenko *et al.* (1985) claimed that the cell-free media they had developed enabled the growth of the trypanosomes by providing reducing conditions. However, the feeder layer cells support trypanosome growth in a complex manner. Mammalian cells in culture probably release factors that stimulate the growth of trypanosomes; they may also detoxify serum or medium components or metabolites released by the parasites. Another important feature of feeder cells could be the modulation of the gas phase. In the body fluids of mammalian hosts O_2 tension is reduced and CO_2 concentration elevated, compared with atmospheric conditions. This has been taken into consideration for the cultivation of erythrocytic stages of *Plasmodium falciparum* (Trager and Jensen, 1976), which are grown in 5% CO_2, 5% O_2 and 90% N_2, and also for filariids (Weinstein, 1986). However, trypanosomes are cultivated in 5% CO_2 in air, i.e. under atmospheric O_2 tension. Additionally, feeder cells create a microclimate enabling short-range interactions that mediate the growth of trypanosomes (Tanner, 1980). None of these properties of feeder cells can yet be simulated in cell-free culture media, with the exception of the reducing conditions induced by 2-mercaptoethanol or L-cysteine.

B. IMPROVEMENT OF TRYPANOSOME DENSITY

The maximum cell density, about 2×10^6 trypanosomes ml^{-1}, at present reached in cultures is very modest compared with that found in susceptible laboratory animals (up to 10^9 ml^{-1}). Bloodstream forms use large amounts of glucose, which they convert to pyruvate, or pyruvate and glycerol, under anaerobic conditions. The accumulated pyruvate is partly converted to lactate by the lactate dehydrogenase present in the serum. Pyruvate and lactate may reach levels in the culture medium of 15 mM or higher (Brun *et al.*, 1981).

When fresh medium containing 15 mM or 30 mM Na-pyruvate was added to young cultures of bloodstream forms with intercellular parasites in the feeder layer, growth of the trypanosomes stopped. The slender forms died within 48 h, while the stumpy forms survived up to 96 h (our unpublished observations). On the other hand, low concentrations of Na-pyruvate (1 mM) have been found to have a favourable effect on growth of bloodstream forms in culture (Tanner *et al.*, 1980). The feeder cells do not metabolize the pyruvate produced by the trypanosomes and are themselves releasing lactate. In order to increase significantly the trypanosome density in culture, methods have to be found to remove the pyruvate produced by the parasites. This could eventually be achieved by degradation of the released pyruvate by enzymes. Another possible method of removing pyruvate from the medium is to co-cultivate microorganisms that metabolize pyruvate to CO_2 in a dialysis bag. If removal of this growth-inhibiting metabolite could be achieved, mass cultivation of bloodstream forms might be realized. At present the existing culture systems for bloodstream forms hardly satisfy the needs of biochemists and molecular biologists. For some studies the procyclic culture form represents an alternative; these can be grown *in vitro* with final parasite densities of up to 5×10^7 trypanosomes ml^{-1} (see Chapter 6).

C. REPLACEMENT OF SERUM

The replacement of the serum component in culture media is undoubtedly another major goal. The development of a serum-free medium for bloodstream forms, however, seems to be extremely difficult. During evolution of the host–parasite relationship, trypanosomes have adapted their biochemical requirements to the environment of the vector and to the mammalian host. The bloodstream stages have acquired metabolic pathways enabling them to grow in the plasma and other body fluids of their mammalian hosts. In the bloodstream, trypanosomes are surrounded by all the nutrients they need for survival, reproduction and differentiation and it has to be assumed that a variety of molecules that are present in the host's serum is required by the parasite. Such factors, necessary for supporting growth of the bloodstream forms, may be many and the variety of molecules offered by the serum almost countless. It therefore seems very unlikely that serum-free cultivation of a parasite that has adapted its metabolism, over a long evolutionary period, to grow in the plasma of its mammalian host can be realized in the near future. Above all, we have to keep in mind that retention of the parasite's natural characteristics *in vitro* is more important than attempting their axenic or serum-free cultivation (or both).

REFERENCES

Balber, A.E. (1983). Primary murine bone marrow cultures support continuous growth of infectious human trypanosomes. *Science* **220**, 421–423.

Balber, A.E. (1984). Growth of *Trypanosoma brucei gambiense, T. b. rhodesiense*, and *T. congolense* in bone marrow cultures of genetically anemic SI/SI^d mice. *Journal of Parasitology* **70**, 179–180.

Balber, A.E., Gonias, S.L. and Pizzo, S.V. (1985). *Trypanosoma brucei gambiense*: growth *in vitro* of bloodstream forms inhibited by dichlorodiamineplatinum (II) compounds and hypolipidemic drugs. *Experimental Parasitology* **59**, 74–80.

Baltz, T., Baltz, D., Giroud, C. and Crockett, J. (1985). Cultivation in a semi-defined medium of animal infective forms of *Trypanosoma brucei, T. equiperdum, T. evansi, T. rhodesiense* and *T. gambiense*. *EMBO Journal* **4**, 1273–1277.

Baltz, T., Giroud, C., Baltz, D., Roth, C., Raibaud, A. and Eisen, H. (1986). Stable expression of two variable surface glycoproteins by cloned *Trypanosoma equiperdum*. *Nature* **319**, 602–604.

Borowy, N.K., Fink, E. and Hirumi, H. (1985a). *In vitro* activity of the trypanocidal diamidine DAPI on animal-infective *Trypanosoma brucei brucei*. *Acta Tropica* **42**, 287–298.

Borowy, N.K., Fink, E. and Hirumi, H. (1985b). *Trypanosoma brucei*: five commonly used trypanocides assayed *in vitro* with a mammalian feeder layer system for cultivation of bloodstream forms. *Experimental Parasitology* **60**, 323–330.

Borowy, N.K., Hirumi, H., Waithaka, H.K. and Mkoji, G. (1985c). An assay for screening drugs against animal-infective bloodstream forms of *Trypanosoma brucei brucei in vitro*. *Drugs under Experimental Clinical Research* **XI**, 155–161.

Borowy, N.K., Brun, R., Nelson, R.T., Waithaka, H.K., Polak, A. and Hirumi, H. (1987). Ro 15–0216: a nitroimidazole compound active *in vitro* against human and animal pathogenic African trypanosomes. *Annals of Tropical Medicine and Hygiene* (in press).

Brun, R. and Jenni, L. (1983). *In vivo* and *in vitro* production of *Trypanosoma b. gambiense*. *In* "Proceedings of Symposium on the Diagnosis of African sleeping Sickness due to *T. gambiense* (P.J. Crooy, ed.), pp. 33–35. Smith Kline RIT, Antwerp.

Brun, R. and Jenni, L. (1985). Cultivation of African and South American trypanosomes of medical or veterinary importance. *British Medical Bulletin* **41**, 122–129.

Brun, R. and Moloo, S.K. (1982). *In vitro* cultivation of animal-infective forms of a West African *Trypanosoma vivax* stock. *Acta Tropica* **39**, 135–141.

Brun, R., Jenni, L., Tanner, M., Schönenberger, M. and Schell, K.-F. (1979). Cultivation of vertebrate-infective forms derived from metacyclic forms of pleomorphic *Trypanosoma brucei* stocks. *Acta Tropica* **36**, 387–390.

Brun, R., Jenni, L., Schönenberger, M. and Schell, K.-F. (1981). *In vitro* cultivation of bloodstream forms of *Trypanosoma brucei, T. rhodesiense* and *T. gambiense*. *Journal of Protozoology* **28**, 470–479.

Brun, R., Hecker, H., Jenni, L. and Moloo, S.K. (1984a). A quantitative ultrastructural study on the transformation of *Trypanosoma brucei brucei* metacyclic to bloodstream forms *in vitro*. *Acta Tropica* **41**, 117–129.

Brun, R., Grootenhuis, J.G., Kunz, C. and Schönenberger, M. (1984b). *In vitro* cultivation of *Trypanosoma (T.) brucei* bloodstream forms using cell lines and sera from African wild Bovidae. *Journal of Parasitology* **70**, 836–837.

Croft, S.L. (1986). *In vitro* screens in the experimental chemotherapy of leishmaniasis and trypanosomiasis. *Parasitology Today* **2**, 64–69.

Cross, G.A.M. and Manning, J.C. (1973). Cultivation of *Trypanosoma brucei* sspp. in semi-defined and defined media. *Parasitology* **67**, 315–331.

Doyle, J.J., Hirumi, H., Hirumi, K., Lupton, E.N. and Cross, G.A.M. (1980). Antigenic variation in clones of animal-infective *Trypanosoma brucei* derived and maintained *in vitro*. *Parasitology* **80**, 356–369.

Duszenko, M., Ferguson, M.A.J., Lamont, G.S., Rifkin, M.R. and Cross, G.A.M.

(1985). Cysteine eliminates the feeder cell requirement for cultivation of *Trypanosoma brucei* bloodstream forms *in vitro. Journal of Experimental Medicine* **162**, 1256–1263.

Fish, W.R., Marr, J.J., Berens, R.L., Looker, D.L., Nelson, D.J., Laton, S.W. and Balber, A.E. (1985). Purine analogues as chemotherapeutic agents in the African trypanosomiases: efficacy and metabolism. *Antimicrobial Agents and Chemotherapy* **27**, 33–36.

Freshney, R.I. (1983). ''Culture of Animal Cells. A Manual of Basic Technique''. Alan R. Liss, New York.

Gray, M.A., Brown, C.G.D., Luckins, A.G. and Gray, A.R. (1979). Maintenance of infectivity of *Trypanosoma congolense in vitro* with explants of infected skin at 37°C. *Transactions of the Royal Society of Tropical Medicine and Hygiene* **73**, 406–408.

Gray, M.A., Ross, C.A., Taylor, A.M., Tetley, L. and Luckins, A.G. (1985). *In vitro* cultivation of *Trypanosoma congolense*: the production of infective forms from metacyclic trypanosomes cultured on bovine endothelial cell monolayers. *Acta Tropica* **42**, 99–111.

Hecker, H. and Brun, R. (1982). Comparative morphometric analysis of bloodstream and lymph forms of *Trypanosoma (T.) brucei brucei* grown *in vitro* and *in vivo. Transactions of the Royal Society of Tropical Medicine and Hygiene* **76**, 692–697.

Hill, G.C. and Hirumi, H. (1983). African trypanosomes. *In* ''*In vitro* Cultivation of Protozoan Parasites'' (J.B. Jensen, ed.), pp. 193–219. CRC Press, Boca Raton, Florida.

Hill, G.C., Shimer, S.P., Caughey, B. and Sauer, L.S. (1978a). Growth of infective forms of *Trypanosoma rhodesiense in vitro*, the causative agent of African trypanosomiasis. *Science* **202**, 763–765.

Hill, G.C., Shimer, S., Caughey, B. and Sauer, S. (1978b). Growth of infective forms of *Trypanosoma (T.) brucei* on buffalo lung and Chinese hamster lung tissue culture cells. *Acta Tropica* **35**, 201–207.

Hirumi, H. (1979). Cultivation of salivarian trypanosomes: application to *in vitro* studies of African trypanosomiasis. *In* ''Practical Tissue Culture Applications'' (K. Maramorosch and H. Hirumi, eds), pp. 309–329. Academic Press, New York and London.

Hirumi, H. and Hirumi, K. (1984). Continuous cultivation of animal-infective bloodstream forms of an East African *Trypanosoma congolense* stock. *Annals of Tropical Medicine and Parasitology* **78**, 327–330.

Hirumi, H., Doyle, J.J. and Hirumi, K. (1977a). African trypanosomes: cultivation of animal-infective *Trypanosoma brucei in vitro. Science* **196**, 992–994.

Hirumi, H., Doyle, J.J. and Hirumi, K. (1977b). Cultivation of bloodstream *Trypanosoma brucei. Bulletin of the World Health Organization* **55**, 405–409.

Hirumi, H., Hirumi, K., Nelson, R.T. and Bwayo, J.J. (1980a). Present status of *in vitro* cultivation of animal-infective African trypanosomes. *In* ''The *In Vitro* Cultivation of the Pathogens of Tropical Diseases'', *WHO Tropical Diseases Research Series* **3**, pp. 165–200. Schwabe, Basel.

Hirumi, H., Hirumi, K., Doyle, J.J. and Cross, G.A.M. (1980b). *In vitro* cloning of animal-infective bloodstream forms of *Trypanosoma brucei. Parasitology* **80**, 371–382.

Jenni, L. and Brun, R. (1981). *In vitro* cultivation of pleomorphic *Trypanosoma brucei* stocks: a possible source of variable antigens for immunization studies. *Transactions of the Royal Society of Tropical Medicine and Hygiene* **75**, 150–151.

Jenni, L. and Brun, R. (1982). A new *in vitro* test for human serum resistance of *Trypano-*

soma (T.) brucei. Acta Tropica **39**, 281–284.

Jenni, L., Marti, S., Schweizer, J., Betschart, B., Le Page, R.W.F., Wells, J.M., Tait, A., Paindavoine, P., Pays, E. and Steinert, M. (1986). Hybrid formation between African trypanosomes during cyclical transmission. *Nature* **322**, 173–175.

Lanham, S.M. and Godfrey, D.G. (1970). Isolation of salivarian trypanosomes from man and other mammals using DEAE-cellulose. *Experimental Parasitology* **28**, 521–534.

Le Page, R.W.F. (1967). Short-term cultivation of *Trypanosoma brucei in vitro* at 37°C. *Nature* **216**, 1141–1142.

Luckins, A.G., Frame, I.A., Gray, M.A., Crowe, J.S. and Ross, C.A. (1986). Analysis of trypanosome variable antigen types in cultures of metacyclic and mammalian forms of *Trypanosoma congolense. Parasitology* **93**, 99–109.

Ross, C.A., Gray, M.A., Taylor, A.M. and Luckins, A.G. (1985). *In vitro* cultivation of *Trypanosoma congolense*: establishment of infective mammalian forms in continuous culture after isolation from the blood of infective mice. *Acta Tropica* **42**, 113–122.

Seed, J.R. and Negus, N.C. (1970). Susceptibility of *Microtus montanus* to infection by *Trypanosoma gambiense. Laboratory Animal Care* **20**, 657–661.

Stuart, K. (1980). Cultivation of dyskinetoplastic *Trypanosoma brucei. Journal of Parasitology* **66**, 1060–1061.

Tanner, M. (1980). Studies on the mechanisms supporting the continuous growth of *Trypanosoma (Trypanozoon) brucei* as bloodstream-like form *in vitro. Acta Tropica* **37**, 203–220.

Tanner, M., Brun, R. and Jenni, L. (1980). Pyruvate as a factor supporting survival of bloodstream forms of *Trypanosoma brucei* strain 427 *in vitro. In* "The *In Vitro* Cultivation of the Pathogens of Tropical Diseases", *WHO Tropical Diseases Research Series* **3**, pp. 211–212. Schwabe, Basel.

Taylor, A.E.R. and Baker, J.R. (1968). "The Cultivation of Parasites *In Vitro*". Blackwell Scientific Publications, Oxford.

Trager, W. and Jensen, J.B. (1976). Human malaria parasites in continuous culture. *Science* **193**, 673–675.

Weinstein, P.P. (1986). Filariasis: problems and challenges. *American Journal of Tropical Medicine and Hygiene* **35**, 221–233.

Zweygarth, E., Ahmed, J.S. and Rehbein, G. (1983). Cultivation of infective forms of *Trypanosoma (T.) brucei evansi* in a continuous culture system. *Zeitschrift für Parasitenkunde* **69**, 131–133.

Chapter 6

Salivarian Trypanosomes: Insect Forms

M.A. GRAY*, H. HIRUMI† AND P.R. GARDINER†

**Centre for Tropical Veterinary Medicine, University of Edinburgh, Easter Bush, Roslin, Scotland*

and

†*International Laboratory for Research on Animal Diseases, Nairobi, Kenya*

I. INTRODUCTION

Since the publication of the predecessor of this book (Taylor and Baker, 1978), many advances have been made in the cultivation *in vitro* of several of the developmental stages in the life-cycles of the salivarian trypanosomes (reviewed by Hirumi *et al.*, 1980a; Hill and Hirumi, 1983;

Brun and Jenni, 1985; Hawke, 1985). These advances have been accompanied by greater knowledge of the morphology and metabolism of some of these stages in both their insect vectors and mammalian hosts (e.g., see Vickerman, 1985; Opperdoes, 1985). This parallel progress has substantiated the premise that the study of the forms maintained or generated in culture will lead to a genuine understanding of trypanosome biology. Cultured trypanosomes hold out hope for improvements in our knowledge of trypanosome biochemistry, for determining the therapeutic potential of drugs, for the study of factors affecting differentiation, and for the production of antigens for use as immunogens or in serodiagnosis (see Section V).

All the morphological forms of trypanosomes that have been described in the tsetse fly vector (*Glossina* spp.) have been obtained *in vitro*. However, the numbers of any one developmental stage that can be produced (and separated from other life-cycle stages to near purity) depends upon the species and stock of trypanosome being grown, and the culture system employed. A method for staining the different life-cycle stages of trypanosomes from culture is given in Section VI.

The media in current use for *in vitro* culture of salivarian trypanosomes are designated as complex, semi-defined or, rarely, defined. Some *in vitro* systems include tissue or organ culture. As more is learned about the individual developmental stages of trypanosomes and, particularly, the role the substratum plays in epimastigote attachment and development, it is hoped that sufficient numbers of each form will ultimately be cultivatable in defined media. The culture media and techniques now used generally for the cultivation of vector form trypanosomes are detailed in this chapter for each of the major tsetse-transmitted species. First, however, we shall describe some of the important factors thought to govern the differentiation of these forms in culture.

The first distinct developmental stage of *T. brucei* and *T. congolense* found in the tsetse fly, about 48 h after the uptake of bloodstream form trypanosomes with a blood meal, is the midgut form, which, in culture, is represented by the procyclic trypomastigote (Brown *et al.*, 1973). Procyclic forms differentiate from bloodstream forms in culture in a similar time and show similar morphology and metabolism to their vector counterparts (Brown *et al.*, 1973). In the tsetse fly, the bloodstream form loses its surface coat and develops a functional mitochondrion to allow aerobic respiration (Vickerman, 1971). When bloodstream form trypanosomes are introduced into cultures in which the normal host serum has been replaced by foetal bovine serum a similar organellar differentiation occurs. In *T. brucei* it appears that the stumpy form of the bloodstream parasite, which is in a stationary phase with regard to DNA synthesis

(Shapiro *et al.*, 1984), and which already has a partially developed mitochondrion (Vickerman, 1971), is the stage which differentiates into the midgut form in the tsetse fly. Bloodstream trypomastigotes that may be functionally (but not morphologically) equivalent to the stumpy forms of *T. brucei* have recently been recognized in the life-cycles of *T. congolense* (Nantulya *et al.*, 1978) and *T. vivax* (Shapiro *et al.*, 1984). It is difficult to initiate cultures leading to the development of insect forms, or to infect tsetse flies, unless the stumpy, or stationary phase, stages are present. Since the differentiation of bloodstream forms to stumpy forms seems to occur less readily following frequent non-cyclical passage of trypanosome stocks in rodents, *recently isolated stocks are recommended for the initiation of cultures*.

It has been found that the frequency of passage can affect trypanosome differentiation in culture. If procyclic cultures are passaged every 3–4 d they remain in logarithmic growth as pure cultures of procyclics. However, if procyclics of *T. congolense* are left in the same culture flask, the population will cease to divide and will differentiate into proventricular forms (Reichenow, 1934), which can be harvested after 8–10 d or more. In procyclic cultures of *T. brucei*, non-dividing forms have also been detected in which the oxidative pathway has changed from the procyclic pattern to that of the proventricular stage found in the tsetse fly (Evans and Brown, 1972).

The procyclic and the proventricular trypanosomes appear to be insensitive to the presence of antibiotics (penicillin plus streptomycin or gentamycin) and these antibiotics can be included in media for their culture. However, if antibiotics are present, the proventricular form can differentiate no further. In the absence of antibiotics, and in the presence of a suitable substratum, the proventricular form of *T. congolense* attaches and eventually differentiates into the (antibiotic-sensitive) epimastigote stage. We stress, in agreement with Evans (1978, and Chapter 3) the importance of working without antibiotics for, if antibiotics had been present in the system used to culture *T. congolense* (Gray *et al.*, 1981), infective metacyclics would never have been obtained. Similar findings have been reported for *T. brucei* (Cunningham *et al.*, 1981). Other stages in the life-cycle appear not to be similarly affected.

Attachment to substrata via the flagellar tip, a process involving the elaboration of hemidesmosomes, is a special feature of the kinetoplastid flagellates (e.g., see Hommel and Robertson, 1976). In salivarian trypanosomes, it appears to be a necessary stage for the generation of infective metacyclics. Epimastigotes of *T. congolense* and *T. vivax* form attachments to substrates in culture and in the tsetse fly (Thevenaz and Hecker, 1980; Gray *et al.*, 1981; Vickerman, 1973; H. Hirumi, unpublished observations). Tetley and Vickerman (1985) have recently shown that epimastigotes of *T. brucei* develop into metacyclics while attached to microvilli of the tsetse fly salivary gland; the use of a substratum of living cells with microvilli has not been sufficiently explored in culture systems for this organism.

The mammal-infective metacyclic, the final developmental stage found in tsetse, is produced as a free-swimming trypomastigote in culture media. It has re-acquired the proteinaceous coat of bloodstream forms and can exist in the presence of normal serum from its mammalian host, and in the presence of the antibiotics already mentioned. The metacyclic forms of *T. congolense* grown in culture have been shown to have the same set of variant antigen types as metacyclics of the same strain produced by cyclical passage through *Glossina* (Crowe *et al.*, 1983). Metacyclics, whether produced *in vivo* or *in vitro*, can produce chancres in rabbits and ruminants (Roberts *et al.*, 1969; Luckins *et al.*, 1981). Metacyclic trypanosomes appear not to divide, and will differentiate no further, unless mammalian feeder layer cells are added to the culture system (Brun and Jenni, 1985). In the presence of feeder layers, the metacyclic forms appear to differentiate into bloodstream forms, even at temperatures of 25–28 °C (the usual temperature for insect-form cultures). With *T. congolense*, these forms continue to exhibit metacyclic surface antigens for 2–16 weeks, even though dividing as differentiated bloodstream forms. This supports the view that temperature may not be a critical factor in differentiation if all other necessary culture conditions are met, as does the observation of procyclic forms in mammalian cell cultures at 37 °C in which foetal bovine serum replaced the normal mammalian host serum (Ferrante and Allison, 1983).

Although all the forms of trypanosomes that arise in tsetse can thus be produced (and many separated as pure populations, see below) it should be borne in mind that *in vitro* cultivation is a discriminatory process, selecting only those trypanosomes in a population that are capable of adaptation to the culture system. Similarly, extensive passaging *in vitro* may cause alterations in the virulence of many types of microorganism (Hirumi, 1979; Hirumi *et al.*, 1980a).

Abbreviations used in this chapter are listed in Section VI.F.

II. *TRYPANOSOMA (NANNOMONAS) CONGOLENSE*

A. PROCYCLIC CULTURES

For the continuous culture of the procyclic stage of *T. congolense* we describe two semi-defined media, SM-77 and 109-C, that are being used to produce pure populations of this stage for metabolic, biochemical and immunological studies. Details of complex and diphasic media that can be used to grow this stage were given by Evans (1978, pp. 81–83).

1° Medium SM-77
(Cunningham, 1977)

MEDIUM

	mg dl^{-1}
Inorganic salts	
NaH_2PO_4	53
$MgCl_2 \cdot 6H_2O$	304
$MgSO_4 \cdot 7H_2O$	370
KCl (anhyd.)	298
$CaCl_2 \cdot 2H_2O$	15
Sugars	
Glucose	70
Fructose	40
Sucrose	40
Organic acids	
L-Malic	67
α-Ketoglutaric	37
Fumaric	5.5
Succinic	6.0
Amino acids	
β-Alanine	200
DL-Alanine	109
L-Arginine	44
L-Asparagine	24
L-Aspartic acid	11
L-Cysteine-HCl	8.0
L-Cystine	3.0
L-Glutamic acid	25
L-Glutamine	164
Glycine	12
L-Histidine	16
DL-Isoleucine	9.0
L-Leucine	9.0
L-Lysine	15
DL-Methionine	20
L-Phenylalanine	20
L-Proline	690
DL-Serine	20
L-Taurine	27
DL-Threonine	10
L-Tryptophan	10
L-Tyrosine	20
DL-Valine	21
Vitamin mixture BME (100 times concentrate)	0.2ml dl^{-1}
Phenol red (0.5%)	0.4

The medium is prepared by dissolving the compounds in deionized distilled water as follows: (a) inorganic salts, except for $CaCl_2$, in 10 ml; $CaCl_2$ in 5.0 ml; (b) sugars in 10 ml; (c) organic acids in 5.0 ml; (d) amino acids in 65 ml. The several solutions are mixed and the pH is adjusted to 7.4 with 2 N NaOH. The mixture is sterilized by filtration, using a Millipore Sterifil apparatus with a 0.22 μm filter, and then divided into convenient volumes and stored at −20°C until needed. Before use, the medium is thawed and foetal bovine serum (FBS) is added (with or without heat-inactivation) to a final concentration of 20% v/v. The addition of 50 mM HEPES aids the buffering of this medium during the growth of procyclic *T. congolense* (R. Brun, personal communication).

TECHNIQUE

Cunningham (1977) described the continuous culture of *T. congolense* in SM-77, but the presence of *Glossina* organs was required for the transformation of bloodstream trypomastigotes to procyclic forms. T.W. Pearson (personal communication) found that, by employing a large initial inoculum of bloodstream trypanosomes, transformation into procyclic trypanosomes took place in the absence of *Glossina* organs.

The initial trypanosomes in mouse blood are mixed with an equal volume of phosphate saline glucose (PSG) at pH 8.0, containing 5% foetal bovine serum, and then passed down a DE-52 anion exchange column (Lanham and Godfrey, 1970) to separate trypanosomes from red blood cells. The separated trypanosomes are suspended in PSG at a concentration of 10^7 ml^{-1}; 1.0 ml of this suspension is used to inoculate 4.0 ml of SM-77 in a T-25 flask and the flask is incubated horizontally at 26°C. The medium (half the volume at first) is changed every 3–6 d. Once established, the cultures are passaged every 3–4 d. For initial subcultures, an inoculum of 10^7 is used but, once established, an inoculum of 10^6 is sufficient for subculture. Such cultures produce 1–4.5 $\times$ 10^7 ml^{-1} of procyclic trypanosomes. The medium not only supported long-term maintenance of tsetse organs, but was also suitable for the establishment and growth of several trypanosomatids besides *T. congolense*.

2° Medium 109-C
(Brun, 1982)

MEDIUM

	$litre^{-1}$
Minimum essential medium (MEM) with Earle's salts without L-glutamine (Gibco cat. no. 320–1090)	950 ml
MEM amino acid solution (50 times concentrate)	20 ml
MEM non-essential amino acid solution (100 times concentrate)	20 ml
HEPES (60 mM)	14.3 g
L-glutamine	0.3 g
L-proline	0.3 g
Glucose	1.0 g
Adenosine	20 mg
Tonicity	355 mosmol

The solid components are dissolved in the 950 ml of MEM and the pH is readjusted to 7.4 with 4 N NaOH. The volume is made up to 1000 ml and filter-sterilized (0.22 μm). Before use, the medium is supplemented with 10% inactivated foetal bovine serum and 0.1% of haemin stock solution (see below). Antibiotics may be used as appropriate (100 iu ml^{-1} penicillin, 100 $\mu g\ ml^{-1}$ streptomycin).

Haemin stock solution. 100 mg haemin chloride (chlorohaemin — Fluka) is dissolved in 25 ml 0.05 N NaOH by heating to boiling point. The volume is made up to 50 ml with distilled water before sterilization by autoclaving (121 °C, 15 min).

TECHNIQUE

Brun (1982) described this semi-defined medium for the continuous culture of *T. congolense* procyclics (and has since used it for *T. simiae* procyclics, personal communication). This was the first report of a semi-defined medium that would support the transformation of bloodstream trypomastigotes of *T. congolense* to procyclic trypomastigotes. For the initiation of cultures, bloodstream trypomastigotes in citrated heart blood of an infected mouse are employed. To this blood an equal volume of medium 109-C is added and the suspension is centrifuged (50*g*, 15 min). The supernatant, containing the trypanosomes, is then transferred to another test tube and centrifuged (1000*g*, 7 min). The pelleted trypanosomes are resuspended in medium 109-C containing 10% heat-inactivated FBS, 200 iu ml^{-1} penicillin and 200 $\mu g\ ml^{-1}$ streptomycin, to give a final concentration of 2–5 $\times$ 10^6 trypanosomes ml^{-1}. 2 ml aliquots of the trypanosome suspension are dispensed in Falcon tubes and incubated at 28 °C. The medium is changed every second day following centrifugation (1000*g*, 7 min) of the trypanosomes. After 1–2 weeks, the culture can be transferred to a culture flask, when the volume of medium can be increased. As soon as the culture reaches stationary phase, as judged by acidification of the pH indicator, the organisms are subcultured, employing a starting inoculum of 3 $\times$ 10^6 ml^{-1}. After establishment of the culture (5–6 weeks) yields of about 4.5 $\times$ 10^7 trypanosomes ml^{-1} can be generated, and additional subcultures can be made with smaller inocula of about 10^6 ml^{-1}.

B. "PROVENTRICULAR" CULTURES

To date, no one has described a culture medium to grow pure populations of "proventricular" forms of *T. congolense*. This form has been described in culture by Reichenow (1934), Tobie (1958), Cunningham (1977) and Steiger *et al.* (1977). Steiger *et al.* (1977) described the appearance of these forms between 7 and 9 d after passage in their continuous culture system. If, by avoiding subculture, procyclic cultures are allowed to go into stationary phase, the numbers of proventricular forms increase. It might be possible to separate the two life-cycle stages, for instance on gradients of Percoll.

C. EPIMASTIGOTE CULTURES

Cultures containing epimastigotes of *T. congolense* were described by Steiger *et al.* (1977). This culture system contained a continuous line of *Glossina morsitans* cells (Schneider, 1979), in Falcon T-25 flasks containing 4.0 ml of medium. The cultures were passaged every 2 weeks, and it is possible that if they had been kept longer the epimastigotes would have adhered and differentiated into metacyclic forms.

Gray *et al.* (1981) obtained cultures of nearly pure epimastigote forms which, once established, could be continuously passaged in semi-defined medium alone. The original technique employed an infected proboscis of *G. morsitans* placed beside a bovine collagen explant. The technique was simplified by using bloodstream trypanosomes and replacing the bovine dermal collagen explant with commercially purified bovine dermal collagen, ''Vitrogen'' (Hirumi *et al.*, 1982; see Section II.C.1° (iii)). Several investigators (W. Fish, and C.A. Ross, personal communications) have now found that, with some stocks, attachment of epimastigotes could be obtained in the medium alone. This attachment appears to be aided by the use of Falcon ''Primaria'' flasks. The appearance of attached epimastigotes occurred later in cultures in which the bovine skin explant was replaced by ''Vitrogen'' (see 1° below).

The bovine dermal collagen explant can also be replaced by a bovine endothelial cell monolayer (Gray *et al.*, 1985; Ross *et al.*, 1985). The choice of culture system will depend on the material and time available. The proventricular form from the tsetse fly, if available, is one of the best forms for the initiation of cultures.

1° Epimastigotes of *T. congolense*

MEDIUM

The best semi-defined medium used to date has been Eagle's minimum essential medium (MEM) with Earle's salts (Gray *et al.*, 1984); see Section VI. MEM can be bought commercially in liquid or powder form. Various laboratories have their own preference. (i) Liquid MEM (Eagle) with Earle's salts, with 25 mM HEPES buffer without L-glutamine (Gibco cat.

no. 041–2360). This is supplemented with 292 mg litre^{-1} of L-glutamine and 20% (v/v) of heat-inactivated FBS added just before use. (ii) Either: liquid MEM with Earle's salts, with L-glutamine (Gibco cat. no. 041–1095); or powdered MEM (Eagle) with Earle's salts, with L-glutamine and non-essential amino acids, without sodium bicarbonate (Gibco cat. no. 072–1500).

To 500 ml of either of the above the following components are added:

Glutamine (4 mM)	146 mg
HEPES (1 M)	2.4 g
$NaHCO_3$	1.1 g
FBS	20% (v/v)

The FBS is inactivated before use by heating for 30 min at 56°C. The pH of the solution is then adjusted with NaOH to 7.4 and the solution is sterilized by filtration.

If antibiotics are to be included for the initial isolation of the trypanosomes, penicillin (100 iu ml^{-1}), streptomycin (100 μg ml^{-1}) or gentamycin (50 μg ml^{-1}) can be used.

TECHNIQUE

(i) Most commercial tissue culture flasks and microplates are suitable for this system, but initiation of cultures is best in 25 cm^2 (50 ml) plastic tissue culture flasks (Nunclon (Nune), Falcon (Primaria, cat. no. 38813), Lux (supplied by Flow Laboratories), Costar).

(ii) To prepare bovine dermal collagen explants, skin is removed from the gluteal or flank region of a calf ($<$ 1 year old) under anaesthesia. The area to be used is previously shaved free of hair, washed with soap and water, and then surface-sterilized with 70% ethanol. Areas of skin (7 $\times$ 5 cm) are removed aseptically and placed, epidermis down, in a Petri dish. Leaving the epidermis intact, the collagen layer is removed with a scalpel and placed in MEM/20% FBS containing antibiotics. The collagen is then cut into appropriately sized pieces (about 5 mm diameter) and stored at 4°C until use (up to 2–3 months).

(iii) Commercial collagen is obtainable as Vitrogen™ (from Collagen Corporation). A stock solution is prepared on ice, by adding 1.0 ml of 0.1 M NaOH and 1.0 ml of MEM (Eagle) 10 times concentrate (Gibco cat. no. 042–1430) to 8.0 ml of Vitrogen solution. For use, 1.0 ml of stock solution is added to 9.0 ml of MEM/FBS. About 3.7–3.8 ml of this solution is added to a Leighton tube containing a plastic coverslip (Costar, cat. no. 3393) so that the medium comes just to the heel of the tube.

The tube is then incubated (horizontally) at 37 °C overnight. When required, the coverslip covered in collagen is placed in the middle of a T-25 flask on the bottom, covered with 4 ml MEM/FBS and gassed with 5% CO_2 in air.

(iv) For initiation of bovine endothelial cell monolayers (after Gray *et al.*, 1985), the surface endothelial cells are scraped from the aortic intima of a *Bos taurus* calf (l year or under) into 4.0 ml MEM/FBS in T-25 flasks, gassed with 5% CO_2 in air, and then incubated at 37 °C. The islands of endothelial-like cells (having no fibroblast contamination) are removed after 1 week, with a silicone rubber "policeman", into a fresh flask with medium. After four such passages, at weekly intervals, a uniform monolayer of endothelial-like cells should be obtained that can then be subcultured weekly using trypsin/EDTA solution as described below. A stabilate of endothelial-like cells can be prepared at the 6th or 7th passage in medium containing 10% dimethyl sulphoxide (DMSO), and stored in liquid N_2. These cells can be used up to passage 20 for trypanosome culture.

(v) For subculture of bovine endothelial monolayers, the 4.0 ml supernatant is removed and the confluent monolayer in the T-25 flask is rinsed twice with sterile phosphate buffered saline (PBS: Oxoid, Dulbecco A, pH 7.3). 2.0 ml of a mixture of EDTA (0.01%) and trypsin (0.01%) in PBS is added. When the cells start rounding up, the excess liquid is removed, and the cells dislodged by striking the flask sharply. 10 ml of MEM/FBS is added to the flask, the medium containing the cells is centrifuged (500*g*, 10 min), and the pellet resuspended in 5.0 ml of fresh MEM/FBS. 0.5–1.0 ml volumes are then dispensed in T-25 flasks to which MEM/FBS is added to give a final volume of 4.0 ml. The flasks are gassed with 5% CO_2 in air and incubated at 37 °C. The cultures remain for one week at a stable pH of 7.3 following a medium change, and can be used for trypanosome culture once confluent (3–4 d after subculture). The temperature is then lowered to 28 °C before inoculation. It is possible to subculture young lines, when the cells are not so adherent, by washing alone without trypsinization.

(vi) To cryopreserve bovine endothelial monolayers, the supernatant is discarded from the monolayer and the cells are removed with EDTA/trypsin as above, centrifuged, and the pellet resuspended in 10 ml MEM/FBS containing 10% DMSO. This is dispensed in 2.0 ml aliquots in plastic freezing ampoules and cooled slowly

($1\,^\circ C\ min^{-1}$) to $-70\,^\circ C$. The ampoules are then stored in liquid nitrogen.

To resuscitate, an ampoule is thawed rapidly in warm water (37 °C) and 1.0 ml of the stabilate is added to 9.0 ml MEM/FBS at 4 °C. This suspension is centrifuged (500*g*, 10 min) and the pellet resuspended in 4.0 ml MEM/FBS in a T-25 flask, gassed with 5% CO_2 in air, and incubated at 37 °C.

(vii) Trypanosomes for initiation of epimastigote cultures can be obtained from an infected tsetse fly which is induced to probe into a pool of 50–100 μl of MEM/FBS containing antibiotics in a container (such as a leucocyte migration plate) warmed to 37 °C. The pool is examined for trypanosomes and, if any are present, transferred with a sterile pipette to the culture flask. If the flask contains dermal collagen explant, the supernatant is first removed and the pool of trypanosomes placed at the edge of the explant and left for 30 min before 4.0 ml of fresh MEM/FBS is added. The same procedure is followed for "Vitrogen" and endothelial cells to allow close contact between the trypanosomes and the endothelial cells or collagen.

Alternatively, the tsetse fly is immobilized (by brief cooling at −20 °C) and the proboscis is removed in a pool of sterile MEM/FBS containing antibiotics, contained in a Petri dish. The proventriculus is then cut from the thorax in a second pool of medium, and, lastly, the midgut is removed in a further pool. The individual organ, containing the trypanosomes, is placed beside any of the substrates previously described, and 4.0 ml of MEM/FBS, containing antibiotics, is added.

(viii) Bloodstream trypanosomes for initiating cultures can be obtained by taking blood from an infected animal into either heparin (to give 5.0 iu heparin ml^{-1}) or 0.3% citrate solution. The trypanosomes are then separated from blood components by anion exchange chromatography or by centrifugation as described for Brun's 109-C medium (Section II.A). The trypanosomes are adjusted to $10^7\ ml^{-1}$ in MEM/FBS. 1 ml of the suspension is added to a culture flask prepared as described, and the volume of medium is gradually increased over the next 2–4 h.

(ix) Cultures can be incubated at 24°–32 °C, but 28 °C has been the temperature most frequently used. Initially only half the supernatant is replaced by fresh medium, without antibiotics, every 48–72 h. Once the trypanosomes have become established, the entire supernatant is replaced by fresh medium every 48 h (sufficient trypanosomes remain on the flask). Once a layer of adhering epimastigotes forms and spreads across the bottom of the flask some may have to be scraped, or washed, off so that areas are freed for new establishment of attached epimastigotes in the culture. Up to 2–4 × 10^7 trypanosomes ml^{-1} are found in the supernatant in addition to the attached layer of epimastigotes.

(x) Once an adhering layer of epimastigotes has been formed, enough epimastigotes are usually present in the supernatant to passage the culture to a fresh flask (in which supporting dermal collagen, "Vitrogen" or bovine aortic endothelial cells are not required). 1 ml containing 10^7 trypanosomes is transferred to 3.0 ml of MEM/FBS in a T-25 flask which is gassed with 5% CO_2 in air and incubated at 37°C.

Mature cultures of nearly pure epimastigotes can be obtained by omitting glutamine or proline from the medium (C. Ross, personal communication). These cultures can be increased in volume in T-75 flasks, but they are harder to maintain as it is difficult to wash off enough adhering epimastigotes to maintain media changes at an interval as great as 48 h. If such cultures overgrow, the trypanosomes tend to die. However, this type of culture has been maintained in one flask for over a year, or by continuous subculture for 2–3 years.

D. METACYCLIC CULTURES

Infective metacyclics are obtained in culture by allowing the adhering epimastigote layers to remain in a culture flask long enough to differentiate into the metacyclic form. Depending upon the stock and the culture conditions, this can take from 7 to 35 d for primary cultures. However, following passage to a fresh culture flask, the interval before the appearance of metacyclics can be 8–12 d, depending on the stock. Metacyclics have remained as stable, infective populations in cultures maintained in single flasks, or in cultures which have been continuously passaged, for over a year.

Metacyclic trypanosomes can be separated from epimastigote cultures either by means of DE-52 anion exchange columns (see Section VI.D.2) or by agglutination with media containing 10% adult bovine plasma, and filtration through a 5.0 μm Millipore filter. More metacyclics appear to be collected by the latter method, although this may be due to the fact that only those with a complete surface coat are collected after passage through an ion-exchange column.

III. *TRYPANOSOMA (DUTTONELLA) VIVAX*

A. INSECT-FORM CULTURES

Insect forms of *T. vivax* were first described in culture by Trager (1959); details are given in Taylor and Baker (1968, pp.42–44), and so are not repeated here. Trager (1975) changed his culture system to employ Falcon culture dishes (catalogue number cult 38–7) containing 0.6 ml of Eagle's MEM with $NaHCO_3$ and PIPES and BES buffers supplemented with 20% sheep serum. Tsetse organs excised from puparia were included in the medium. *T. vivax* blood trypomastigotes, from an infected sheep, were inoculated into the culture maintained at 31°C, and 2–3 d later elongated dividing trypanosomes were observed. After 4 d, small rosettes of trypanosomes were observed which had developed by 7 d to dense layers of epimastigotes attached to the tsetse organs. Transitional forms akin to those described in the hypopharynx of tsetse flies (Vickerman, 1973) were also observed. Infectivity trials with these cultured trypanosomes were not performed. A similar culture technique was employed using separated bloodstream trypomastigotes from an infected sheep, and immature whole tsetse flies from puparia contained in 4.0 ml of MEM/20% FBS or RPMI 1640/20% FBS in plastic T-25 flasks. The proboscis and thoracic tissue of the tsetse material introduced into the culture were favoured as developmental sites, and clusters of trypanosomes were observed on the tsetse tissue and attached to the flask in their immediate vicinity (CTVM, 1980).

Cultures were passaged twice, at 14 and 28 d, into fresh medium containing whole tsetse puparia, and a sheep was infected by trypanosomes from one such culture. Only a small percentage of these cultures was successful in producing infective trypanosomes, possibly, in the reported instance, because of difficulty with the puparia. The successful infection of a sheep with cultured trypanosomes followed the development of a West African stock of *T. vivax* for 53 d *in vitro*.

Hirumi *et al.* (1983, 1984, 1985b) described a technique (employing fibroblast feeder layers and matrix gel green-A beads) by which the whole life cycle of *T. vivax* can be reproduced.

We describe here in detail the technique presently employed to grow the insect forms of *T. vivax*. So far, one West African and three East African stocks have been grown successfully. The cultures contain the various trypanosome stages found in the tsetse vector, from which the metacyclic forms can be separated (Hirumi *et al.*, 1985b).

1° Insect form cultures of *T. vivax*
(Hirumi *et al.*, 1983, 1984, 1985b)

MEDIA

RPMI 1640 with 25 mM HEPES (Gibco cat. no. 041–2400) or MEM (Eagle) with Earle's Salts with 25 mM HEPES (Gibco cat. no. 041–2360) supplemented with 292 mg L-glutamine litre^{-1} before use. Both media are supplemented with 20% FBS (v/v).

TECHNIQUE

(i) Plastic T-25 (25 cm^2; 50 ml) culture flasks (Nunclon (Nune), Falcon, Lux (obtainable from Flow Laboratories), Costar) are suitable.

(ii) Fibroblast feeder layers can be established with cells from 14–16 d embryonic tissues of *Microtus montanus* (Brun *et al.*, 1981), obtained by trypsinization with 0.5% Trypsin-EDTA (Gibco cat. no. 043–5400) in Earle's (Gibco cat. no. 041–4150) or Hanks' buffered salts solution (BBS; Gibco cat. no. 041–4175). The numbers of live cells (estimated by the trypan blue exclusion test) are enumerated by haemocytometer and 5 × 10^6 viable cells are added to each T-25 flask containing 5.0 ml MEM/20% FBS, gassed with 5% CO_2 in air, and incubated at 37 °C. The medium is replaced after 24 and 48 h to remove debris and unattached cells. Cultures can be passaged after 4–6 d into three fresh flasks, but at higher passage numbers this may be reduced to a ratio of one culture to two new flasks. Once established, the cultures are passaged weekly by trypsinization (e.g., with 0.05% bovine trypsin/EDTA), and require a change of medium between passages. Otherwise, the procedure is the same as for bovine endothelial cell monolayers (Section II.C.1°). Stabilates can be prepared and stored in liquid nitrogen.

Alternatively, cells for feeder layers can be obtained by trypsinization of foetal bovine thymus pieces, with stirring (as above). This procedure releases thymus cells which are then grown in 5.0 ml volumes of medium (e.g. L-15, but the other media mentioned here will support growth) and passaged once weekly by trypsinization (as described for bovine endothelial cell monolayers, Section II.C.1°). A stabilate can be prepared from about passage 4 and stored in liquid nitrogen as described in Section II.C.1°. Such a cell line can be used for trypanosome culture up to the 20th or 30th passage.

(iii) Matrix Gel Green-A beads (Amicon cat. no. 19311) sterilized by washing 5.0 ml quantities several times in equal volumes of 70% ethanol. A 50% stock suspension is then prepared by adding 5.0 ml of the appropriate medium. The beads are used at a final concentration of 2–5% suspended in culture medium.

(iv) To initiate cultures from the blood of an infected mammalian host, blood is collected from the infected animal into anticoagulant (heparin) at peak parasitaemia. Trypanosomes are separated from the blood by centrifugation as described for *T. congolense* procyclic cultures (Section II.A) and resuspended in medium to give 2.5×10^6 trypanosomes ml^{-1}. 4 ml of this suspension are added to a T-25 flask containing a fibroblast feeder layer, 0.1–0.4 ml of the washed 50% suspension of Matrix Gel Green-A beads is added carefully, and the flasks are incubated at 25°C.

(v) To initiate cultures from an infected tsetse fly, the proboscis of the fly is removed (as described for the establishment of epimastigote cultures of *T. congolense*, Section II.C) and placed in a pool of medium. The proboscis is then cut into three or four pieces with fine scissors or a scalpel and the pieces are introduced by sterile pipette into a T-25 flask containing a mammalian fibroblast feeder layer with 4.0 ml of medium. The flasks are incubated at 25°C.

(vi) Half the medium is changed every 2–4 d, depending on the growth of the trypanosomes. When the cultures are initiated from blood trypomastigotes, trypanosomes attach to the beads by the anterior ends of their flagella and transform into uncoated, elongated trypanosomes within 3 d. These uncoated trypomastigotes then transform into attached epimastigotes which are at first elongated but subsequently shorten within 5–7 d. At 10 d of culture, patches on the surface of the flask are scraped free of the mammalian cell feeder layer, allowing epimastigote attachment to the exposed surface of the flask. The epimastigotes then multiply. From 12–21 d a small number (0.5–1%) of short, slender trypomastigotes appears amongst the epimastigotes, and supernatants from such cultures become infective for a host mammal. The supernatant trypanosomes can similarly be used to initiate cultures of *T. vivax* mammalian trypomastigote forms for growth at 37°C (Hirumi *et al.*, 1983). When the cultures are initiated from an infected tsetse proboscis, development to infective forms takes longer. During the initial 4 weeks, epimastigotes increase in number within and around the proboscis, also aggregating at the surface of the medium. By the 4th week, metacyclic forms are seen amongst the epimastigotes.

(vii) A culture initiated from bloodstream trypomastigotes can be passaged from about the 10th day of culture onwards, while one initiated from an infected proboscis requires to be left for about 4 weeks before the first passage. Subpassages are made by transferring supernatants containing trypanosomes to a new T-25 flask containing a mammalian cell feeder layer (without beads) and the volume is made up to 4.0 ml with fresh medium. A primary culture can continue to produce infective trypanosomes for over 100 d. It should be born in mind that the metacyclic forms of *T. vivax*, like those of *T. brucei* (Brun *et al.*, 1979) and *T. congolense* (Gray *et al.*, 1985), when cultured in the presence of mammalian cells, probably differentiate and divide as blood forms even at 28 °C.

(viii) For separation of trypomastigotes from mixed cultures, the mixed population is incubated in medium containing 10% bovine plasma for 1–2 h at 25–28 °C. The epimastigotes are aggregated into clumps while the trypomastigotes remain swimming freely and are separated by passing the trypanosome suspension through a Millipore filter of 5.0 μm pore size at 1*g* (Hirumi *et al.*, 1985b).

Hirumi *et al.* (1985b) described two morphologically different forms of metacyclics after separation as described in (viii); "immature metacyclics" having a tapered posterior end and lacking a surface coat, and "mature metacyclics", being shorter with a rounded posterior end, terminal kinetoplast and possessing a surface coat. Hirumi *et al.* (1985b) have grown a West African stock and three East African stocks of *T. vivax* in culture and have found that the West African stock grew better in RPMI 1640 with FBS over a bovine thymus fibroblast feeder layer containing about a 2–5% suspension of Matrix Gel Green-A beads for initiation, while the East African stocks preferred MEM with FBS over a *Microtus* fibroblast feeder layer containing a 1–2% suspension of beads for initiation.

IV. *TRYPANOSOMA (TRYPANOZOON) BRUCEI*

The insect midgut forms of *T. brucei** were the first recorded salivarian trypanosomes to be grown in culture (Novy and MacNeal, 1903).

Most early research into salivarian trypanosomiasis was carried out with this species of trypanosome because it can grow, and be easily passaged, in laboratory animals. Evans (1978) detailed the separation of *T. brucei* trypomastigotes from the blood of infected laboratory animals for the initiation of procyclic cultures in complex media — EBLB (Evans, 1978, pp. 81–83), or a modification of Tobie's biphasic medium; in semi-defined media — HX12 and HX12V (Cross and Manning, 1973; see Evans, 1978, p. 80) and SDM-77 (Brun and Jenni, 1977; see Evans, 1978, pp. 80–81); and also in defined medium — HX25 M (Cross *et al.*, 1975; see Evans, 1978, pp. 78–79). Evans (1978, pp. 85–86) also described Cunningham and Honigberg's technique (1977) for the production of infective metacyclic trypanosomes in culture, in which the midgut-like insect forms growing in SM medium differentiate finally to yield infective metacyclic forms. Thus, here, we limit our discussion to the new media described for better growth of cultures of *T. brucei* insect forms and describe modifications that have been developed for these culture systems.

A. PROCYCLIC CULTURES

Many stocks of *T. brucei* can now be transformed from bloodstream trypomastigotes to procyclic trypomastigotes and established in continuous culture using semi-defined media. Brun and Schonenberger (1979) modified SDM-77 (Brun and Jenni, 1977; see Evans, 1978, pp. 80–81), calling the new formulation SDM-79, and included citrate or *cis*-aconitate for the more successful transformation of bloodstream trypomastigotes to procyclic trypomastigotes (Brun and Schonenberger, 1981). Details of this medium are given below.

Recently Czichos *et al.* (1986) have elegantly demonstrated that a 10 °C reduction in temperature and the presence of *cis*-aconitate are the two essential factors required for the efficient transformation of bloodstream forms to procyclic forms. Recognition of these two factors has allowed the establishment *in vitro* of procyclic trypanosomes even of monomorphic bloodstream form clones of *T. brucei* (Czichos *et al.* 1986).

*Throughout this and the preceding chapter, the term "subspecies" and the names *T. brucei rhodesiense* and *T. brucei gambiense* are used for convenience without necessarily indicating the taxonomic status of the populations concerned [Eds].

1° Medium SDM-79
(Brun and Schonenberger, 1981)

MEDIUM

	litre^{-1}		
MEM, powder (Gibco F-14)	7.0 g	L-Arginine	100 mg
Medium 199 powder (KC Biological DM-310), with Earle's salts	2.0 g	L-Glutamine	300 mg
MEM amino acids (50 ×), without L-glutamine (Gibco)	8.0 ml	L-Methionine	70 mg
MEM non-essential amino acids (100 times concentrate) (Gibco)	6.0 ml	L-Phenylalanine	80 mg
Glucose	1.0 g	L-Proline	600 mg
HEPES	8.0 g†	L-Serine	60 mg
MOPS	5.0 g†	L-Taurine	160 mg
$NaHCO_3$	2.0 g	L-Threonine	350 mg
tri-Na-Citrate ($C_6H_5Na_3O_7.2H_2O$)	588 mg	L-Tyrosine	100 mg
Na-pyruvate	100 mg	Adenosine	10 mg
L-Alanine	200 mg	Guanosine	10 mg
		Glucosamine-HCl	50 mg
		Folic acid	4.0 mg
		p-Aminobenzoic acid	2.0 mg
		Biotin	0.2 mg

*33.6 mM.
†24.0 mM.

The pH is adjusted, with 2 N NaOH, to 7.3 and the medium filter-sterilized with a 0.22 μm Millipore filter. Bovine haemin stock solution (2.5 mg ml^{-1} in 0.05 N NaOH) is autoclaved and added at a final concentration of 7.5 mg litre^{-1}. Before use the medium is supplemented with 2.5–10% inactivated foetal bovine serum (FBS; Gibco). Tonicity of the complete medium is 330 mosmol.

TECHNIQUE

Separated, washed, pleomorphic bloodstream forms of *T. brucei* are incubated in 5.0 ml of this medium for 48 h in T-25 flasks at 27°C to effect transformation to procyclic forms. Thereafter the procyclic forms are cultured in SDM-79, producing about 4–5 × 10^7 trypanosomes ml^{-1} after establishment (within 3–6 weeks). The cultures are passaged every 2–3 d.

Cunningham's (1977) SM-77 medium (Section II.A.1°) is widely used for the establishment from bloodstream forms, and continuous culture, of procyclics of *T. brucei*. T.W. Pearson (personal communication) has recently established procyclics of three stocks of *T. b. brucei*, four stocks of *T. brucei rhodesiense* and one stock of *T. b. gambiense** in SM-77 medium using the technique described in Section II.A.1° for *T.*

congolense in SM-77 medium. Such cultures produce 1–4 × 10^7 procyclics ml^{-1} medium.

When a stock of *T. brucei* is difficult to transform from the bloodstream trypomastigote to the procyclic trypomastigote stage in one of the above semi-defined media, a complex medium such as EBLB or modified Tobie's biphasic medium (Evans, 1978, pp. 81–83) can be tried.

Where only a very few bloodstream trypomastigotes are obtainable from the blood of the host. e.g. for some *T. b. gambiense* stocks, procyclic cultures can be initiated from the midgut forms of an infected tsetse fly (Brun and Jenni, 1985).

Defined media are required for biochemical studies of procyclic forms, but only a limited number of stocks will grow in such media. Medium HX25M (Cross *et al.*, 1975; see Evans, 1978, pp. 78–79) is still used for biochemical studies of procyclic forms. Recently Seebeck and Kurath (1985) described two simple media (ME-83 and HHP-84) for biochemical experimentation with cultured procyclic *T. brucei*. ME-83 is a semi-defined medium based on SDM-79 and facilitates the selective omission of individual compounds, e.g. amino acids, whereas HHP-84 is a fully defined minimal medium which supports cell motility but not cell proliferation. Details of these media are given below.

2° and 3° Media ME-83 and HHP-84

(Seebeck and Kurath, 1985)

MEDIA

	ME-83 mg $litre^{-1}$	HHP-84 mg $litre^{-1}$		ME-83	HHP-84
			Haemin	50	—
NaCl	8000	8000	Bovine serum albumin	—	0.1%
KCl	400	400	Proline	1012	1012
$MgSO_4.7H_2O$	100	100	Alanine	9.0	—
Na_2HPO_4 anhydrous	48	48	Asparagine	15	—
$CaCl_2.2H_2O$	185	185	Aspartate	13	—
$NaHCO_3$	350	350	Glutamate	15	—
KH_2PO_4	60	60	Glycine	8.0	—
$MgCl_2.6H_2O$	100	100	Serine	11	—
Glucose	1000	1000	Arginine	128	—
HEPES	2383	2383	Cysteine	24	—
Glutamine	292	—	Threonine	48	—
Histidine	44	—	Tryptophan	12	—
Isoleucine	96	—	Tyrosine	40	—
Leucine	96	—	Valine	48	—

*See footnote on p. 134.

	ME-83	HHP-84		ME-83	HHP-84
Lysine	72	—	Penicillin/		
Methionine	16	—	streptomycin		
Phenylalanine	32	—	(Gibco)	100 iu ml^{-1}	100 iu ml^{-1}

TECHNIQUE

(i) A 500 ml batch of ME-83 is prepared from the following components: 50 ml of 10 times concentrate Hanks' BSS (Gibco); 50 ml of 0.1 M HEPES, pH 7.3; 5.0 ml of 100 times concentrate non-essential amino acids (Gibco); 5.0 ml of proline (100 mg ml^{-1} in H_2O); 20 ml of tyrosine (0.2 mg ml^{-1} in distilled H_2O); 10 ml each of the following amino acids stock solutions (mg ml^{-1} in H_2O): arginine, 6.4; cysteine, 1.2; glutamine, 14.6; histidine, 2.2; isoleucine, 4.8; leucine, 4.8; lysine, 3.6; methionine, 0.8; phenylalanine, 1.6; threonine, 2.4; tryptophan, 0.6; valine, 2.4. Distilled water is added to 500 ml and the medium is sterilized by filtration. Before use, the medium is completed by adding 5.0 ml haemin solution (5.0 mg ml^{-1} in 1% NaOH), 5.0 ml penicillin/streptomycin solution (Gibco; 10 000 iu ml^{-1}) and 25 ml heat-inactivated FBS.

(ii) A 500 ml batch of HHP-84 is prepared from the following ingredients: 50 ml of 10 times concentrate Hanks' BSS (Gibco); 50 ml of 0.1 M HEPES, pH 7.3; 5.0 ml proline (100 mg ml^{-1} in H_2O); and distilled H_2O to 500 ml. The medium is sterilized by filtration and stored at 4°C. Before use 5.0 ml haemin solution (as above), 0.5 g bovine serum albumin and 5.0 ml penicillin/streptomycin solution are added. Since HHP-84 is mostly used for short-term experiments, the antibiotics can be omitted.

(iii) Trypanosomes are grown in 5.0 ml quantities in plastic tissue culture flasks at 26°C and are passaged twice a week. As with SDM-79, the extent of cell growth is dependent on the batch of FBS and the quantity added.

B. "PROVENTRICULAR" CULTURES

Proventricular-like forms of *T. brucei* were described by Thomson and Sinton (1912) from NNN blood-agar slants and similar media. These forms develop when cultures of procyclic trypanosomes are allowed to go into stationary phase. Trager (1959; see Taylor and Baker, 1968, pp. 42–44) and Cunningham and Honigberg (1977; see Evans, 1978, pp. 85–86) described proventricular forms in their cultures which included tsetse organs, as did Hirumi *et al.* (1980a) in their "cyclic development" culture system. Evans and Brown (1972) investigated the utilization of glucose and proline by stationary phase procyclic culture forms and found that they used glucose in the same manner as did proventricular forms from tsetse.

C. EPIMASTIGOTE CULTURES

Only a few epimastigotes of *T. brucei* have been described in culture and those have occurred in very small numbers. They have, however, been observed in cultures which have eventually produced infective metacyclic forms (Cunningham and Honigberg, 1977; Hirumi, 1979; Hirumi *et al.*, 1980a; Nyindo *et al.*, 1978a). The yield of metacyclics in these same cultures was very low compared with the yield of metacyclics differentiating from the dense layers of epimastigotes found in the salivary glands of the infective tsetse fly. If a suitable substratum (which may be living cells, since the epimastigotes of *T. brucei* are intimately associated with the cells in the tsetse salivary gland; Tetley and Vickerman, 1985) could be found, perhaps adhering layers of epimastigotes of *T. brucei* could be grown in culture, as can those of *T. congolense* and *T. vivax*.

D. METACYCLIC CULTURES

Metacyclic forms of *T. brucei* in culture were first described by Cunningham and Honigberg (1977) and details of their culture technique, which included "head and salivary glands" from tsetse-fly puparia in SM medium (Cunningham, 1977), are given by Evans (1978, pp. 85–86; see also Section VI). Subsequently, this culture system has been used to grow further stocks of *T. b. brucei* to the infective metacyclic stage (Cunningham and Taylor, 1979), and to grow stocks of the other subgroups of *T. brucei*: *T. b. gambiense* (Jones *et al.*, 1981) and *T. b. rhodesiense* (Cunningham, 1985). Cunningham *et al.* (1981) also described a monomorphic stock of *T. brucei* which produced sporadic infections in mice after being cultured in this system: they found the "heads and salivary glands" could be replaced by alimentary canal and abdominal wall tissues from the tsetse fly, but not by skeletal muscle or skin explants from mouse embryos. This suggests that the environment of the salivary gland lumen is not essential for epimastigote to metacyclic maturation to take place. Cunningham (1985) further found that tsetse thoracic tissue was one of the most successful tissues for culturing *T. b. rhodesiense*. Only the thoracic muscle from one tsetse fly was required in 3.5 ml of SM medium, whereas 13–24 "heads and salivary glands" had been required in Cunningham and Honigberg's (1977) original culture system. Occasionally, supernatant trypanosomes from control cultures grown in SM medium alone give sporadic infections in mice. This has been reported by other authors using different media (reviewed by Cunningham and Taylor, 1979). The yield of metacyclics produced in the culture system of Cunningham and her colleagues is very low (0.01–3% of the total trypanosome population), but the infective trypanosomes produced in this system appear to be metacyclics that neither divide nor differentiate to bloodstream forms. These metacyclic trypanosomes can be separated from the remaining, uncoated trypanosomes by means of DE-52 columns (Gardiner *et al.*, 1980a); after injection into mice or rabbits, they develop into bloodstream forms, the variant surface glycoproteins of which correspond to those produced by bloodstream populations of rodents infected by tsetse bites with the same stock of *T. brucei* (Gardiner *et al.*, 1980b). Cunningham and Honigberg's (1977) culture system, although laborious in requiring organs and tissue from tsetse puparia, and producing rather low yields of metacyclic forms has been the only culture system in which the metacyclics of *T. brucei* produced *in vitro* could be separated to obtain a pure suspension. Recently, however, Cunningham and Kaminsky have found that the tsetse explants could be replaced by explants of *Phormia regina* (blowfly). Furthermore, a cell line from *Anopheles gambiae* will support differentiation of procyclics through to metacyclic forms. When transformed procyclics of three stocks of *T. b. brucei* and one stock of *T. b. rhodesiense* were grown at 27 °C in 25 cm^2 flasks containing an *Anopheles gambiae* monolayer 2.8×10^5 metacyclic forms per ml of culture supernatant could be produced (Kaminsky, personal communication).

E. COMPLETE CYCLICAL DEVELOPMENT OF *T. BRUCEI* IN CULTURE

Two other culture systems have been described in which *T. brucei* metacyclic trypanosomes are produced (Hirumi, 1979; Hirumi *et al.*, 1980a and Nyindo *et al.*, 1978a, 1979), but these systems include mammalian feeder layer cells in the presence of which metacyclic trypanosomes can further differentiate to the infective dividing forms found in the mammalian host (Brun *et al.*, 1979). Hirumi (1979) and Hirumi *et al.*, (1980a) described the complete cyclical development of *T. brucei* in culture employing a culture system which included feeder layer cells in RPMI 1640 media with 25 mM HEPES and supplemented with 20% FBS as detailed below.

1° Medium for complete cyclical development of *T. brucei*

(Hirumi, 1979; Hirumi *et al.*, 1980a)

MEDIUM

RPMI 1640 with 25 mM HEPES supplemented with 20% heat-inactivated FBS.

TECHNIQUE

(i) Plastic 25 cm^2 T-25 flasks (Nunclon, Lux, Falcon, Costar) are used.

(ii) Feeder layer cells are grown from the buffy coat cells of normal bovine blood: blood from a normal bovine is bled into heparin, to give 50 iu ml^{-1} of heparin, in a siliconized centrifuge tube and centrifuged (2000*g*, 20 min, 4 °C). The plasma is discarded and the buffy coat layer transferred (with as few red cells as possible) into 3 times its volume of phosphate buffered saline (PBS) at pH 6.8, after which the temperature is kept at 20–25 °C. The buffy coat cells, suspended in PBS, are layered onto "Lymphoprep" (Nyegaard) in a siliconized centrifuge tube and centrifuged (400*g*, 25 min). The supernatant is discarded and the cells at the interface (taking care not to include erythrocytes) are removed into 5 times their volume of PBS and centrifuged (160*g*, 10 min) to deposit mononuclear cells and leave the platelets in suspension. The cell pellet is used to seed T-25 flasks containing 6.0 ml RPMI 1640 with 20% FBS. Fibroblast cells will adhere and overgrow non-adhering cells. Enrichment of the fibroblast component can be achieved by subculturing the freshly isolated buffy coat cells after 18 h, leaving the majority of adherent macrophages in the original flask. This can be repeated once more before establishing the monolayers (Hirumi *et al.*, 1977a). The medium is changed twice weekly, and the fibroblast feeder layer cells are passaged as described for bovine endothelial cell monolayers (Section II.C.1°).

(iii) Bloodstream trypanosomes (3×10^6) from culture (or from the infected blood of a host) are used to inoculate cultures containing bovine fibroblast feeder layers in 6.0 ml of medium contained in a T-25 flask. The cultures are then incubated at 25 °C.

(iv) At 2 d and on d 4 or 5, 3–4 ml of the culture supernatant is replaced by the same volume of fresh medium. During this period many of the trypanosomes may die leaving about 10^5 trypanosomes ml^{-1}.

(v) The cultures are then maintained at 25 °C with no further medium change until d 21. During this time procyclic forms, which start to make their appearance at d 5, increase rapidly in number to approximately 10^7 trypanosomes ml^{-1} and dominate the culture by d 8. Later, slender proventricular forms appear and, in turn, dominate the culture population by d 20. However, in parallel with the increase in number of proventricular forms, the overall trypanosome population density drops to 10^6 trypanosomes ml^{-1}. The cultures are not infective for mice at this stage.

(vi) From d 21 the culture medium is changed every 2–4 d, during which time the proventricular forms change to short trypomastigotes and the trypanosome density rapidly increases to 10^7 trypanosomes ml^{-1}. A few epimastigote forms are seen, amongst various abnormal forms, by d 30. About 24 h after the first appearance of the epimastigote forms, a small number of metacyclic forms are seen.

(vii) From 37 d the incubation temperature can be raised to 37 °C and the medium changed every 1–2 d. The cultures then contain increasing numbers of dividing haematozoic (= bloodstream) trypanosomes (Hirumi *et al.*, 1977b).

Nyindo *et al.* (1978a, 1978b, 1979) described a culture system producing trypanosomes infective for the mammalian host. They combined vector organ culture and mammalian feeder layer cells in RPMI 1640 with 25 mM HEPES supplemented with 20% heat-inactivated FBS, 5% lactalbumin hydrolysate and antibiotics. When such cultures were initiated with alimentary tracts of tsetse flies infected with *T. brucei* trypanosomes (Nyindo *et al.*, 1978a), procyclic forms, proventricular forms and a few epimastigotes are discernible before the cultures became infective for mice. Alternatively, when the cultures were initiated from salivary glands infected with *T. brucei*, the growth of infective bloodstream forms was apparent (Nyindo *et al.*, 1978b, 1979). Since this culture system is suitable for the differentiation of metacyclics to dividing forms, as found in the mammalian host, it is presumed that metacyclics, either from salivary glands or those that have differentiated from epimastigotes in culture, will further differentiate into the forms found in the mammalian host. Nyindo and Rurangirwa (1981) demonstrated that the infective forms found in their continuous culture system stayed as a stable antigenic population for 7 weeks, and corresponded to the metacyclic populations of the same strain produced in the tsetse fly. It has been found that *T. congolense* metacyclics will differentiate and multiply as the infective mammalian forms at 28 °C and yet retain a stable antigenic profile, similar to that represented amongst the metacyclic populations produced by a fly. However, the infective mammalian forms may change their antigenic profile within 2–16 weeks, depending on the stock grown in culture (Hirumi *et al.*, 1980b; Doyle *et al.*, 1980; Luckins *et al.*, 1986). Nyindo (1985), however, assumed that the population of *T. brucei* in his continuous culture system would continue to remain antigenically stable and used trypanosomes that had been in continuous culture for 385 d to immunize rats. The rats were not protected from infection following the bites of tsetse flies infected with the homologous stock of *T. brucei*, presumably because the infective trypanosome forms used to immunize the rats had changed their antigenic nature after such extended culture. Similar cultures of trypanosomes, incubated at 37°C, changed their antigenic profile before the end of the 7-week observation period.

V. APPLICATIONS

The main aim of growing trypanosomes in culture is to be able to produce large numbers of a particular stage in the life cycle of the trypanosome and thus be able to study the trypanosome away from the complicated physiological and defence systems of its insect vector and mammalian host.

Tsetse fly colonies are difficult and expensive to handle and many stocks of trypanosomes have a very low infection rate in tsetse fly. Tsetse may also produce very low numbers of the trypanosome stage required. Being able to propagate the insect stages of trypanosomes in culture means that many laboratories, that are not able to handle tsetse flies because of technical or legal restrictions, can now study these stages of the trypanosome life-cycle.

Cultured trypanosomes have provided material for serology that would be impossible or extremely laborious to obtain by other means. For instance, *T.b. gambiense* is notoriously difficult to work with in the laboratory owing to the inability of many stocks to infect laboratory animals or, when they do, to produce adequate parasitaemias to infect tsetse flies. Jones *et al.* (1981) used the metacyclics of *T.b. gambiense* from culture to compare the antigenic serotypes of different stocks. Recently Brun and Jenni (1985) have used the procyclic forms of this species, which can be propagated *in vitro* in relatively large numbers, to infect tsetse flies.

Tsetse flies infected with *T. congolense* produce very few metacyclic trypanosomes, so Crowe *et al.* (1983) used metacyclics of *T. congolense* from culture to prepare monoclonal antibodies to the metacyclic variable antigen types (mVATs). These monoclonal antibodies were then used to identify the mVAT repertoire of *T. congolense* metacyclics from tsetse flies as well as from culture. In this manner, the authors were able to show that the mVAT repertoire of the *T. congolense* stock under study was limited to 12 VATs. Hirumi *et al.* (1985a) used *T. congolense* metacyclics produced in culture to immunize goats, and were able to demonstrate that killed metacyclics which had been propagated *in vitro* could induce protective immunity in the ruminant host against tsetse challenge with the homologous serodeme.

Cultured procyclic trypanosomes have been used to investigate isoenzyme patterns for the characterization and differentiation of trypanosome stocks (Kilgour, 1980; Betschart *et al.*, 1983). They have been used to prepare stage- and species-specific monoclonal antibodies (Richardson *et al.*, 1986) and as a source of antigens for the development of immunodiagnostic tests for African sleeping sickness (T.W. Pearson, personal communication). Procyclic trypanosomes from culture have been used as antigens in immunofluorescence and simple agglutination assays to detect antibodies in the sera of vervet monkeys infected with *T.b. rhodesiense*. This test has recently been applied to the diagnosis of sleeping sickness in humans (Pearson *et al.*, 1986). Other species-specific monoclonal antibodies prepared against procyclic culture forms are being used in the indirect immunofluorescence test, enzyme-linked immunosorbent assay (ELISA), and the sandwich ELISA for the diagnosis of African trypanosomiasis and for species characterization (Nantulya *et al.*, 1985; Katende *et al.*, 1987). Monoclonal antibodies raised against *T. congolense* procyclic culture forms recognize known *T. congolense* stocks but not *T. simiae*, and thus may be useful in epidemiological studies (T.W. Pearson and R.P. Beecroft, personal communication).

Maudlin *et al.* (1984) immunized rabbits with procyclic forms of *T. congolense* produced *in vitro* and concluded that vaccination against a single procyclic *T. congolense* stock could reduce transmission of both homologous and heterologous *T. congolense* stocks which might, in certain epidemiological conditions, provide an additional control method for African trypanosomiasis. Murray *et al.* (1985) also found that they could suppress trypanosome infection rates in tsetse by maintaining infected flies on goats immunized against procyclic trypanosomes propagated *in vitro*. They found this effect to be species-specific, and within a species it was independent of the trypanosome stock used for immunization. However, they questioned the usefulness of this approach as a practicable method of vaccination of livestock, because of the range of hosts on which tsetse feed.

Genetic recombination has been shown to occur in *T. brucei* during its cycle in the tsetse fly (Jenni *et al.*, 1986). Recombination of VAT repertoires, and isoenzyme characteristics, occurred between two stocks used simultaneously to infect tsetse. This might have the effect of extending or altering mVAT and bloodstream VAT repertoires in the field. Thus, present work with cultures of the insect forms of *T. congolense* is being pursued to elucidate whether and when sexual reproduction occurs in the life-cycle of the trypanosome in the vector (C. Ross, personal communication).

Many bacterial and viral diseases are controlled by attenuated vaccines. Hirumi *et al.* (1980a) indicated that possible attenuation of the virulence of the bloodstream form of trypanosomes might occur after continuous culture. If the same were true of the metacyclics produced in continuous culture, then the possibility of live vaccines for trypanosomiasis could be re-examined.

As well as the studies of serology or trypanosome antigenicity outlined above, procyclic trypanosomes have been used as a convenient source of organisms for the investigation of trypanosome biochemistry (Opperdoes and Steiger, 1981; Opperdoes and van Roy, 1982; Hart *et al.*, 1984; Clayton, 1985). The morphological differentiation (Bohringer and Hecker, 1975; Ghiotto *et al.*, 1979; Barry and Vickerman, 1979; Hecker, 1980) and mitochondrial biogenesis (Bienen *et al.*, 1981), which accompany the transformation of bloodstream forms to procyclics *in vitro*, have already been studied in some detail. Further, the fact that procyclic trypanosomes lose the surface coat prominent on mammalian forms (Barry and Vickerman, 1979) and cease antigen gene expression (Overath *et al.*, 1983) will continue to place this *in vitro* transformation system at the centre of molecular biological studies of antigenic variation in the African trypanosomes.

VI. APPENDIX

No two stocks of *Trypanosoma* spp. appear to have identical growth characteristics *in vitro* from the time of their isolation to their establishment and production of the metacyclic stage.

Batches of medium, even from the same supplier, vary considerably in their ability to support the growth of trypanosomes in culture. This is also true of different batches of FBS. If one is starting to isolate and establish trypanosome stocks in culture for the first time, it is advisable to obtain a stock with known growth characteristics and use it to familiarize oneself with the culture of trypanosomes *in vitro*. It would be preferable also to obtain batches of medium and FBS (from the same laboratory as the stock was obtained) that are known to support good growth of these organisms. Samples of media and FBS obtained from commercial firms can then be compared and the best selected and stored for future use with other trypanosome stocks. It is generally found that a good batch of medium and FBS will support the growth of many different stocks of *Trypanosoma* spp., although there are exceptions.

A. MEDIA

Each batch of medium requires to be tested before use to ensure that it will support good growth of the stages of the trypanosomes' life-cycle that are required. It should not be used beyond the expiry date, and unstable components such as glutamine should be added just before use. It is best to acquire at least two samples of freshly prepared batches from the supplier for testing. Subpassage of trypanosome cultures into duplicate cultures of each of the new medium samples supplemented with the same batch of FBS is recommended. The cultures should be observed for the growth pattern of the trypanosomes for about 2 weeks. When a suitable batch of medium has been selected, sufficient quantities to last for about 3–4 months should be ordered to allow time for testing further samples and for the continuation of long-term cultures.

B. FOETAL BOVINE SERUM

Batches of FBS tend to vary even more than batches of media, some actually having a lethal effect on the trypanosomes, or being unable to support the differentiation of one trypanosome form to another. It is best to test at least 3–5 sample batches if possible, as it can sometimes be difficult to acquire a good batch of FBS for trypanosome culture. As in the selection of suitable media, a trypanosome culture should be subcultured for comparison into duplicate cultures containing the new FBS in the same batch of medium. Observation should continue for at least 2 weeks. FBS can be stored at $-20\,^{\circ}C$ for a year or more as long as repeated freezing and thawing is avoided. Some laboratories have found that dialysis of FBS against medium before it is used improves its ability to support the growth of trypanosomes in culture.

C. STAINING TECHNIQUE

Cultures should be continually observed for changes in trypanosome morphology by preparing stained smears on slides at each medium change. The staining technique described below eliminates the inclusion bodies often found in the insect stages of trypanosomes in culture, and so enables a clear picture to be obtained of the cell membrane, flagellum and position of the kinetoplast in relation to the nucleus (Gray *et al.*, 1984).

A smear of the culture supernatant is quickly dried under a flow of air (e.g. in a laminar-flow hood). The smear is then fixed with methanol (1 min) after which it is hydrolysed at 60°C in 1 N HCl (9 min). It is then washed in phosphate buffer (pH 7.0) and allowed to stand in fresh phosphate buffer (5 min) before staining with 5% Giemsa (Gurr's improved R66 solution; Hopkins and Williams) in phosphate buffer for about 60 min at 37°C. The slide is then rinsed in phosphate buffer and blotted dry ready for examination or mounting.

D. ANION-EXCHANGE COLUMNS

The method for separating trypanosomes with surface coats from blood components or culture supernatants is essentially that of Lanham and Godfrey (1970). Sterile disposable plastic syringes are useful in carrying out this technique. About 9.0 ml of diethylaminoethyl cellulose (Whatman DE-52; Whatman Lab Sales) equilibrated with phosphate buffered saline at pH 8.1 and containing 1% w/v glucose (PSG) are packed into the barrel of a 10 ml syringe and supported on a 2.0 mm depth of packed glass wool. (Recently, some laboratories have been using PSG at pH 7.4). 4–8 ml of the culture supernatant (or blood diluted in PSG) containing the trypanosomes are poured on to the column. 10 ml of PSG are used to elute the trypanosomes with a surface coat from the DE-52. The procedures are carried out aseptically so that the trypanosomes can be subsequently cultured. If required, the trypanosomes can be concentrated by centrifugation (1000*g*, 7 min).

E. SURFACE STERILIZATION OF TSETSE PUPAE

Where tissues or tsetse organs are required, they are usually taken from the puparial stage just before emergence after surface sterilization in White's solution (Trager, 1959) or 10% v/v "Roccal" (Bayer) (Cunningham, 1977), and then washed several times in an insect balanced salt solution (or culture medium) containing 100 iu of penicillin, 100 μg streptomycin and 10 μg "Fungizone" (Squibb) ml^{-1}

before dissection in insect balanced salt solution (Cunningham and Honigberg, 1977), or culture medium.

F. ABBREVIATIONS USED

BES:	*N*,*N*-bis(2-hydroxyethyl)-2-aminoethane-sulphonic acid
BSS:	balanced salt solution
DMSO:	dimethyl sulphoxide
EDTA:	ethylenediaminetetraacetic acid
ELISA:	enzyme-linked immunosorbent assay
FBS:	foetal bovine serum
HEPES:	*N*-2-hydroxyethylpiperazine-*N*′-2-ethanesulphonic acid
MEM:	minimum essential medium (Eagle)
MOPS:	3-(*N*-morpholino)propanesulphonic acid
PBS:	phosphate buffered saline
PIPES:	piperazine-*N*,*N*′-bis(2-ethanesulphonic acid)
PSG:	phosphate-buffered saline glucose
RPMI:	Roswell Park Memorial Institute

ACKNOWLEDGEMENTS

We are grateful to colleagues from several institutes for communicating, and allowing us to quote, unpublished work, and for critical reviews of the manuscript.

This is ILRAD publication No. 451.

REFERENCES

Barry, J.D. and Vickerman, K. (1979). *Trypanosoma brucei*: loss of variable antigens during transformation from bloodstream to procyclic forms *in vitro*. *Experimental Parasitology* **48**, 313–324.

Betschart, B., Wyler, R. and Jenni, L. (1983). Characterisation of *Trypanozoon* stocks by isoelectric focusing and isoenzyme analysis. *Acta Tropica* **40**, 25–28.

Bienen, E.J., Hammadi, E. and Hill, G.C. (1981). *Trypanosoma brucei*: biochemical and morphological changes during *in vitro* transformation of bloodstream to procyclic trypomastigotes. *Experimental Parasitology* **51**, 408–417.

Bohringer, S. and Hecker, H. (1975). Quantitative ultrastructural investigations of the life cycle of *Trypanosoma brucei*: a morphometric analysis. *Journal of Protozoology* **22**, 463–467.

Brown. R.C., Evans, D.A. and Vickerman, K. (1973). Changes in oxidative metabolism and ultrastructure accompanying differentiation of the mitochondrion in *Trypanosoma brucei*. *International Journal for Parasitology* **3**, 691–704.

Brun, R. (1982). Cultivation of procyclic trypomastigotes of *Trypanosoma congolense* in a semi-defined medium with direct adaption from bloodstream forms. *Zeitschrift für Parasitenkunde* **67**, 129–135.

Brun, R. and Jenni, L. (1977). A new semi-defined medium for *Trypanosoma brucei* spp. *Acta Tropica* **34**, 21–33.

Brun, R. and Jenni, L. (1985). Cultivation of African and South American trypanosomes of medical or veterinary importance. *British Medical Bulletin* **41**, 122–129.

Brun, R. and Schonenberger, M. (1979). Cultivation and *in vitro* cloning of procyclic culture forms of *Trypanosoma brucei* in a semi-defined medium. *Acta Tropica* **36**, 289–292.

Brun, R. and Schonenberger, M. (1981). Stimulating effect of citrate and *cis*-aconitate on the transformation of *Trypanosoma brucei* bloodstream forms to procyclic forms *in vitro*. *Zeitschrift für Parasitenkunde* **66**, 17–24.

Brun, R., Jenni, L., Tanner, M., Schonenberger, M. and Schell, K.F. (1979). Cultivation of vertebrate infective forms derived from metacyclic forms of pleomorphic *Trypanosoma* stocks. *Acta Tropica* **36**, 387–390.

Brun, R., Jenni, L., Schonenberger, M. and Schell, K.F. (1981). *In vitro* cultivation of bloodstream forms of *Trypanosoma brucei*, *T. rhodesiense* and *T. gambiense*. *Journal of Protozoology* **28**, 470–479.

Clayton, C.E. (1985). Structure and regulated expression of genes encoding fructose biphosphate aldolase in *Trypanosoma brucei*. *EMBO Journal* **4**, 2997–3003.

Cross, G.A.M. and Manning, J.C. (1973). Cultivation of *Trypanosoma brucei* sspp. in semi-defined and defined media. *Parasitology* **67**, 315–331.

Cross, G.A.M., Klein, R.A. and Linstead, D.J. (1975). Utilisation of amino acids by *Trypanosoma brucei* in culture: L-threonine as a precursor for acetate. *Parasitology* **71**, 311–326.

Crowe, J.S., Barry, J.D., Luckins, A.G., Ross, C.A. and Vickerman, K. (1983). All metacyclic variable antigen types of *Trypanosoma congolense* identified using monoclonal antigens. *Nature* **306**, 389–391.

CTVM (1980). Centre for Tropical Veterinary Medicine, Annual Report 1979-1980. Edinburgh.

Cunningham, I. (1977). New culture medium for maintenance of tsetse tissues and growth of trypanosomatids. *Journal of Protozoology* **24**, 325–329.

Cunningham, I. (1985). Infectivity of *Trypanosoma brucei rhodesiense* cultivated at 28 °C with tsetse fly tissue. *Proceedings, VII International Congress of Protozoology, 22–29 June, 1985*, Abstract No. 47. Nairobi, Kenya.

Cunningham, I. and Honigberg, B.M. (1977). Infectivity reacquisition by *Trypanosoma brucei brucei* cultivated with tsetse salivary glands. *Science* **197**, 1279–1282.

Cunningham, I. and Taylor, A.M. (1979). Infectivity of *Trypanosoma brucei* cultivated at 28 °C with tsetse fly salivary glands. *Journal of Protozoology* **26**, 428–432.

Cunningham, I., Honigberg, B.M. and Taylor, A.M. (1981). Infectivity of monomorphic and pleomorphic *Trypanosoma brucei* stocks cultivated at 28 °C with various tsetse fly tissues. *Journal of Parasitology* **67**, 391–397.

Czichos, J., Nonnengaesser, C. and Overath, P. (1986). *Trypanosoma brucei*: cis-aconitate and temperature reduction as triggers of synchronous transformation of bloodstream to procyclic trypomastigotes *in vitro*. *Experimental Parasitology* **62**, 283–291.

Doyle, J.J., Hirumi, H., Hirumi, K., Lupton, E.N. and Cross, G.A.M. (1980). Antigenic variation in clones of animal-infective *Trypanosoma brucei* derived and maintained *in vitro*. *Parasitology* **80**, 359–369.

Evans, D.A. (1978). Kinetoplastida. *In* "Methods of Cultivating Parasites *In Vitro*"

(A.E.R. Taylor and J.R. Baker, eds), pp. 55–88. Academic Press, New York and London.

Evans, D.A. and Brown, R.C. (1972). The utilisation of glucose and proline by culture forms of *Trypanosoma brucei*. *Journal of Protozoology* **19**, 686–690.

Ferrante, A. and Allison, A.C. (1983). Alternative pathway activation of complement by African trypanosomes lacking a glycoprotein coat. *Parasite Immunology* **5**, 491–498.

Gardiner, P.R., Lamont, L.C., Jones, T.W. and Cunningham, I. (1980a). The separation and structure of infective trypanosomes from cultures of *Trypanosoma brucei* grown in association with tsetse salivary glands, *Journal of Protozoology* **27**, 182–185.

Gardiner, P.R., Jones, T.W. and Cunningham, I. (1980b). Antigenic analysis by immunofluorescence of *in vitro*-produced metacyclics of *Trypanosoma brucei* and their infections in mice. *Journal of Protozoology* **27**, 316–320.

Ghiotto, V., Brun, R., Jenni, L. and Hecker, H. (1979). *Trypanosoma brucei*: morphometric changes and loss of infectivity during transformation of bloodstream forms to procyclic culture forms *in vitro*. *Experimental Parasitology* **48**, 447–456.

Gray, M.A., Cunningham, I., Gardiner, P.R., Taylor, A.M. and Luckins, A.G. (1981). Cultivation of infective forms of *Trypanosoma congolense* at 28°C from trypanosomes in the proboscis of *Glossina morsitans*. *Parasitology* **82**, 81–95.

Gray, M.A., Ross, C.A., Taylor, A.M. and Luckins, A.G. (1984). *In vitro* cultivation of *Trypanosoma congolense*: the production of infective metacyclic trypanosomes in cultures initiated from cloned stocks. *Acta Tropica* **41**, 343–353.

Gray, M.A., Ross, C.A., Taylor, A.M., Tetley, L. and Luckins, A.G. (1985). *In vitro* cultivation of *Trypanosoma congolense*: the production of infective forms from metacyclic trypanosomes cultured on bovine endothelial cell monolayers. *Acta Tropica* **42**, 99–111.

Hart, D.T., Misset, O., Edwards, S.W. and Opperdoes, F.R. (1984). A comparison of the glycosomes (microbodies) isolated from *Trypanosoma brucei* bloodstream form and cultured procyclic trypomastigotes. *Molecular and Biochemical Parasitology* **12**, 25–35.

Hawke, C.J. (1985). Advances in trypanosome culture. *Parasitology Today* **1**, 30–31.

Hecker, H. (1980). Application of morphometry to pathogenic trypanosomes (Protozoa, Mastigophora). *Pathology Research and Practice* **166**, 203–217.

Hill, G.C. and Hirumi, H. (1983). African trypanosomes. *In* "*In Vitro* Cultivation of Protozoan Parasites" (J.B. Jensen, ed.), pp. 193–219. CRC Press, Boca Raton, Florida.

Hirumi, H. (1979). Cultivation of salivarian trypanosomes: application to *in vitro* studies of African trypanosomes. *In* "Practical Tissue Culture Applications" (K. Maramorosch and H. Hirumi, eds), pp. 309–329. Academic Press, New York and London.

Hirumi, H., Doyle, J.J. and Hirumi, K. (1977a). Cultivation of bloodstream *Trypanosoma brucei*. *Bulletin of the World Health Organization* **55**, 405–409.

Hirumi, H., Doyle, J.J. and Hirumi, K. (1977b). African trypanosomes: cultivation of animal-infective *Trypanosoma brucei in vitro*. *Science* **196**, 992–994.

Hirumi, H., Hirumi, K., Nelson, R.T. and Bwayo, J.J. (1980a). Present status of *in vitro* cultivation of animal-infective African trypanosomes. *In* "The *In Vitro* Cultivation of the Pathogens of Tropical Disease", *Tropical Diseases Research Series* **3**, pp. 165–200. Schwabe, Basel.

Hirumi, H., Hirumi, K., Doyle, J.J. and Cross, G.A.M. (1980b). *In vitro* cloning of animal-infective bloodstream forms of *Trypanosoma brucei*. *Parasitology* **80**, 371–382.

Hirumi, H., Hirumi, K. and Moloo, S.K. (1982). Simplified methods for propagating *Trypanosoma congolense* metacyclics *in vitro*. *Journal of Protozoology* **29**, 480.

Hirumi, H., Nelson, R.T. and Hirumi, K. (1983). Complete cyclic development of *Trypanosoma vivax in vitro*. *Journal of Protozoology* **30**, 6A.

Hirumi, H., Hirumi, K., Nelson, R.T. and Moloo, S.K. (1984). Propagation of metacyclic-like forms of East African *Trypanosoma vivax*. *Journal of Protozoology* **31**, 7A.

Hirumi, H., Hirumi, K., Nantulya, V., Moloo, S.K. and Musoke, A. (1985a). Immunisation of goats with *in vitro*-propagated killed metacyclics of *Trypanosoma congolense* against infective tsetse challenge. *Proceedings, VII International Congress of Protozoology, 22–29 June, 1985*, Abstract No. 175. Nairobi, Kenya.

Hirumi, K., Nelson, R.T., Hirumi, H. and Moloo, S.K. (1985b). Characterisation of *Trypanosoma vivax* metacyclic trypomastigotes propagated *in vitro*. *Proceedings, VII International Congress of Protozoology, 22–29 June, 1985*, Abstract No. 46. Nairobi, Kenya.

Hommel, M. and Robertson, E. (1976). *In vitro* attachment of trypanosomes to plastic. *Experientia* **32**, 464–465.

Jenni, L., Marti, S., Schweizer, J., Betschart, B., LePage, R.F.W., Wells, J.M., Tait, A., Paindavoine, P., Pays, E. and Steinert, M. (1986). Hybrid formation between African trypanosomes during cyclical transmission. *Nature* **322**, 173–175.

Jones, T.W., Cunningham, I., Taylor, A.M. and Gray, A.R. (1981). The use of culture-derived metacyclic trypanosomes in studies on the serological relationships of stocks of *Trypanosoma brucei gambiense*. *Transactions of the Royal Society of Tropical Medicine and Hygiene* **75**, 560–565.

Katende, J.M., Nantulya, V.M. and Musoke, A.J. (1987). An improved indirect immunofluorescent antibody test for diagnosis of human trypanosomiasis (sleeping sickness). *Transactions of the Royal Society of Tropical Medicine and Hygiene* (in press).

Kilgour, V. (1980). The electrophoretic mobilities and activities of eleven enzymes of bloodstream and culture forms of *Trypanosoma brucei* compared. *Molecular and Biochemical Parasitology*, **2**, 51–62.

Lanham, S.M. and Godfrey, D.G. (1970). Isolation of salivarian trypanosomes from man and other mammals using DEAE-cellulose. *Experimental Parasitology* **28**, 521–534.

Luckins, A.G., Rae, P.and Gray, M.A. (1981). Development of local skin reactions in rabbits infected with metacyclic forms of *Trypanosoma congolense* cultured *in vitro*. *Annals of Tropical Medicine and Parasitology* **75**, 563–564.

Luckins, A.G., Frame, I.A., Gray, M.A., Crowe, J.S. and Ross, C.A. (1986). Analysis of trypanosome variable antigen types in cultures of metacyclic and mammalian forms of *Trypanosoma congolense*. *Parasitology* **93**, 99–109.

Maudlin, I., Turner, M.J., Dukes, P. and Miller, N. (1984). Maintenance of *Glossina morsitans morsitans* on antiserum to procyclic trypanosomes reduces infection rates with homologous and heterologous *Trypanosoma congolense* stocks. *Acta Tropica* **41**, 253–257.

Murray, M., Hirumi, H. and Moloo, S.K. (1985). Suppression of *Trypanosoma congolense*, *T. vivax* and *T. brucei* infection rates in tsetse flies maintained on goats immunized with uncoated forms of trypanosomes grown *in vitro*. *Parasitology* **91**, 53–66.

Nantulya, V.M., Doyle, J.J. and Jenni, L. (1978). Studies on *Trypanosoma (Nannomonas) congolense*. 1. On the morphological appearance of the parasite in the mouse. *Acta Tropica* **35**, 329–337.

Nantulya, V.M., Musoke, A.J., Rurangirwa, F.R., Minja, S. and Saigar, N. (1985). Diagnosis of African trypanosomiasis using monoclonal antibodies. *Proceedings, VII International Congress of Protozoology, 22–29 June, 1985*, Abstract No. 174. Nairobi, Kenya.

Novy, F.G. and MacNeal, W.J. (1903). The cultivation of *Trypanosoma brucei*. A preliminary note. *Journal of the American Medical Association* **41**, 1266–1268.

Nyindo, M. (1985). Specific immunity in rats inoculated with *Trypanosoma brucei brucei* derived from tsetse salivary glands and cultured *in vitro*. *Acta Tropica* **42**, 123–126.

Nyindo, M. and Rurangirwa, F.R. (1981). *Trypanosoma brucei*: continuous cultivation of antigenically stable parasites at 29°C and induction of antigenic variants at 37°C. *Insect Science and Its Application* **2**, 263–266.

Nyindo, M., Darji, N., Kaminjolo, J.S., Chimtawi, M., Patel, N. and Onyango, P. (1978a). *Trypanosoma brucei* — cultivation *in vitro* of infective forms derived from the midgut of *Glossina morsitans*. *Journal of Parasitology* **64**, 469–474.

Nyindo, M., Chimtawi, M., Owor, J., Kaminjolo, J.S., Patel, N. and Darji, N. (1978b). Sleeping sickness: *in vitro* cultivation of *Trypanosoma brucei* from the salivary glands of *Glossina morsitans*. *Journal of Parasitology* **64**, 1039–1043.

Nyindo, M., Patel, N., Darji, N. and Golder, T.K. (1979). *Trypanosoma brucei*: *in vitro* propagation of metacyclic forms derived from the salivary glands of *Glossina morsitans*. *Journal of Parasitology* **65**, 751–755.

Opperdoes, F.R. (1985). Biochemical peculiarities of trypanosomes, African and South American. *British Medical Bulletin* **41**, 130–136.

Opperdoes, F.R. and Steiger, R.F. (1981). Localization of hydrolases in cultured procyclic trypomastigotes of *Trypanosoma brucei*. *Molecular and Biochemical Parasitology* **4**, 311–323.

Opperdoes, F.R. and van Roy, J. (1982). The phospholipases of *Trypanosoma brucei* bloodstream forms and cultured procyclics. *Molecular and Biochemical Parasitology* **5**, 309–319.

Overath, P., Czichos, J., Stock, V. and Nonnengaresser, C. (1983). Repression of glycoprotein synthesis and release of surface coat during transformation of *Trypanosoma brucei*. *EMBO Journal* **2**, 1721–1728.

Pearson, T.W., Liu, M., Gardiner, I.C., Longridge, D., Sayer, P.D., Gould, S.S., Waitumi, J.N. and Njogu, A.R. (1986). Use of procyclic trypanosomes for detection of antibodies in sera from vervet monkeys infected with *Trypanosoma rhodesiense*: an immunodiagnostic test for African sleeping sickness. *Acta Tropica* **43**, 391–399.

Reichenow, E. (1934). Die Züchtung der pathogen Trypanosomen. *Archiv für Schiffs- und Tropen Hygiene* **38**, 292–302.

Richardson, J.P., Jenni, L., Beecroft, R.P. and Pearson, T.W. (1986). Procyclic tsetse fly midgut forms and culture forms of African trypanosomes share stage- and species-specific surface antigens identified by monoclonal antibodies. *Journal of Immunology* **136**, 2259–2264.

Roberts, C.J., Gray, M.A. and Gray, A.R. (1969). Local skin reactions in cattle at the site of infection with *Trypanosoma congolense*. *Transactions of the Royal Society of Tropical Medicine and Hygiene* **63**, 620–624.

Ross, C.A., Gray, M.A., Taylor, A.M. and Luckins, A.G. (1985). *In vitro* cultivation of *Trypanosoma congolense*; establishment of infective mammalian forms in continuous culture after isolation from the blood of infective mice. *Acta Tropica* **42**, 113–122.

Schneider, I. (1979). Tsetse fly tissue culture and its application to the propagation of African trypanosomes *in vitro*. *In* "Practical Tissue Culture Applications" (K.

Maramorosch and H. Hirumi, eds), pp. 373–386. Academic Press, New York and London.

Seebeck, Th. and Kurath, V. (1985). Two simple media for biochemical experimentation with cultured procyclic *Trypanosoma brucei*. *Acta Tropica* **42**, 127–136.

Shapiro, S.Z., Naessens, J., Liesegang, B., Moloo, S.K. and Magondu, J. (1984). Analysis by flow cytometry and DNA synthesis during the life cycle of African trypanosomes. *Acta Tropica* **41**, 313–323.

Steiger, R.F., Steiger, E., Trager, W. and Schneider, I. (1977). *Trypanosoma congolense*: partial cyclic development in a *Glossina* cell system and oxygen consumption. *Journal of Parasitology* **63**, 861–867.

Taylor, A.E.R. and Baker, J.R, (1968). "Cultivation of Parasites *In Vitro*". Blackwell Scientific Publications, Oxford.

Taylor, A.E.R. and Baker, J.R. (eds) (1978). "Methods of Cultivating Parasites *In Vitro*". Academic Press, London and New York.

Tetley, L. and Vickerman, K. (1985). Differentiation in *Trypanosoma brucei*. Host-parasite cell junctions and their persistence during acquisition of the variable antigen coat. *Journal of Cell Science* **74**, 1–19.

Thevenaz, P. and Hecker, H. (1980). Distribution and attachment of *Trypanosoma (Nannomonas) congolense* in the proximal part of the proboscis of *Glossina morsitans morsitans*. *Acta Tropica* **37**, 163–175.

Thomson. J.G. and Sinton, J.A. (1912). The morphology of *Trypanosoma gambiense* and *Trypanosoma rhodesiense* in cultures and a comparison with the developmental forms described in *Glossina palpalis*. *Annals of Tropical Medicine and Parasitology* **6**, 331–356.

Tobie, E.J. (1958). The cultivation of *Trypanosoma congolense in vitro*. *Journal of Parasitology* **44**, 241–242.

Trager, W. (1959). Tsetse fly tissue culture and the development of trypanosomes to the infective stage. *Annals of Tropical Medicine and Parasitology* **53**, 473–491.

Trager, W. (1975). On the cultivation of *Trypanosoma vivax*: a tale of two visits in Nigeria. *Journal of Parasitology* **61**, 3–11.

Vickerman, K. (1971). Morphology and physiological considerations of extracellular blood protozoa. *In* "Ecology and Physiology of Parasites" (A.M. Fallis, ed.), pp. 58–91. University Press, Toronto.

Vickerman, K. (1973). The mode of attachment of *Trypanosoma vivax* in the proboscis of the tsetse fly *Glossina fuscipes*: an ultrastructural study of the epimastigote stage of the trypanosome. *Journal of Protozoology* **20**, 394–404.

Vickerman, K. (1985). Developmental cycles and biology of pathogenic trypanosomes. *British Medical Bulletin* **41**, 105–114.

Chapter 7

Plasmodiidae: Erythrocytic Stages

T. PONNUDURAI

Institute of Medical Parasitology, Catholic University of Nijmegen, The Netherlands

I. INTRODUCTION

Much progress has been made in the biochemistry, immunology and molecular biology of the human malaria parasite *Plasmodium falciparum* during the past 10 years since continuous cultivation of this parasite was first reported (Trager and Jensen, 1976; Haynes *et al.*, 1976). These studies, essential for the development of vaccines against malaria, were possible because relatively large numbers of specific stages of the parasite are available in the laboratory. Even the chromosomes of different isolates have been analysed (Van der Ploeg *et al.*, 1985). Of the four human malaria parasites, *P. falciparum* is the only one that can be maintained in culture for long periods. Some species infecting other mammals can also be continuously cultivated in the laboratory.

II. *PLASMODIUM FALCIPARUM*

The basic culture techniques have been described previously (Trigg, 1978) and some minor modifications have since been reported.

A. GENERAL MATERIALS AND METHODS

1. *Culture medium*

Various additives to the RPMI 1640 culture medium are said to enhance parasite growth. Osisanya *et al*. (1981) used an additional 2.0 g litre^{-1} of glucose together with 40 mM TES buffer. This modification was based on the fact that parasitized erythrocytes utilize glucose very rapidly and produce large amounts of lactic acid (Homewood, 1977). Using this modified medium the authors obtained reasonable parasite multiplication during weekends without the daily medium change that is normally required. Zolg *et al*. (1982) confirmed their findings, and it was later shown that glucose could not be replaced by other sugars (Geary *et al*. 1985). Evidently the key exogenous purine source incorporated by the parasites is hypoxanthine (Sherman, 1979; Geary *et al*., 1985). Zolg *et al*. (1982) reported that added hypoxanthine up to 50 μg ml^{-1} increased the growth rate of the cloned parasite (Tak9/clone 96). Surprisingly, they also reported that hypoxanthine did not stimulate the growth of uncloned parasites. Ifediba and Vanderberg (1981) suggested that the incorporation of an additional 50 μg ml^{-1} of hypoxanthine in the medium was necessary for reproducible maturation of gametocytes infective to mosquitoes: this could not be confirmed by Campbell *et al*. (1982).

Enrichment of RPMI 1640 medium with glutathione (600 μg ml^{-1}) is reported to have a stimulating effect on parasite growth (Zolg *et al*., 1982), but orotic acid failed to have an added effect when hypoxanthine was also included. Cyclic AMP (1 mM) added to the culture medium at a critical period during parasite development enhanced conversion of ring stages to gametocytes (Kaushal *et al*., 1980). This observation has not been confirmed by others. Maswoswe *et al*. (1985) reported that corticosteroids added to cultures stimulated the production of gametocytes in chloroquine-resistant strains but not in chloroquine-sensitive strains. It was not reported whether such gametocytes were able to reach maturity and infect mosquitoes.

2. *Serum*

The serum-free medium used by Iscove and Melchers (1978) for lymphocyte cultures failed to support growth of *P. falciparum* (Zolg *et al*., 1982).

After a period of adaptation lasting 2–3 weeks, growth of four different strains of *P. falciparum* in New Zealand White rabbit serum was no different from that seen in human serum (Sax and Rieckmann,

1980). Ifediba and Vanderberg (1980) reported that commercial calf serum supplemented with 1% of a 15% (w/v) solution of neopeptone (Difco) or proteose-peptone No. 3 (Difco) could replace human serum in the cultivation of the FCR-3 strain. Whether other parasite isolates can grow satisfactorily in this serum has yet to be confirmed. There has not yet been any report of successful serum replacement for gametocyte cultures.

3. *Gas phase*

P. falciparum is said to be a microaerophile (Scheibel *et al.*, 1979). Although a low O_2 tension is essential for the growth of the parasite, it cannot survive anaerobic conditions. Gas mixtures used in different culture systems consist of O_2, 1–10% and CO_2, 3–5%, the balance being N_2. Within these limits the gas composition is not critical. In our laboratory we use 3% O_2, 4% CO_2 and 93% N_2.

4. *Manual cultures*

The size of culture vessels in which medium is changed manually ranges from microtitre wells containing up to 100 microlitres to large tissue culture flasks with 25–50 ml of culture material. The red cell suspension varies from 1–2% to about 8–10% in these systems. Medium is generally changed daily but to avoid changing it at weekends one of two strategies has been adopted — either the method of Osisanya *et al.* (1981) described above, or establishing a low haematocrit value (1–2%) and very low parasitaemia at the start of the weekend.

5. *Automated culture systems*

As mentioned previously, *P. falciparum* cultures normally require medium change at least once or twice daily. Manual medium change is not only laborious but also restricts the volume of culture that can be handled. Scaling up, essential for some of the biochemical and molecular biological work, is possible only when medium change can be effected automatically. An additional advantage is the relative reproducibility in the performance of these cultures.

In addition to the settled layer cultures described by Trager and Jensen (1976) and by Jensen and Trager (1977), parasites can be grown successfully in suspension systems, first described by Butcher (1979, 1981b). Automated systems have been described based on both these types of culture.

B. STATIC CULTURE SYSTEMS

Medium change is achieved either by continuous flow techniques or intermittently.

1. Continuous flow technique

1° Trager's method
(Trager, 1979)

APPARATUS

The flow vessel is illustrated in Fig. 1. It is constructed of 17 × 25 mm rectangular glass tubing with a vertical "chimney" at each end. Tubes for medium inflow and outflow, and for gas delivery, pass through silicone rubber stoppers fitted to the chimneys. A side-arm, closed by a vaccine cap, permits addition and removal of infected erythrocytes. The assembly of culture vessel and tubes is placed in an incubator at 37°C; 12 ml of 8–10% red cell suspension with 1% parasitaemia is introduced into the culture vessel. After the red cells have settled, culture medium is pumped in and out at a rate of 55 ml d^{-1}. The outflow pump speed is set slightly faster than the inflow; the depth of medium is therefore determined by the position of the outflow tube, the tip of which is adjusted so that there is a depth of 3–4 mm of medium. Subculturing is effected by removing 10 ml of the parasitized suspended cells and adding an equal volume of fresh uninfected cells. A battery of such culture vessels can be set up to increase the total volume of the culture.

TECHNIQUE

(i) A 500 ml flask holding 350 ml of sterile RPMI 1640 without bicarbonate, but with HEPES buffer at 25 mM, is provided with 14.7 ml of sterile 5% (w/v) $NaHCO_3$ solution and 40 ml sterile serum. This is the medium reservoir and is held in a small refrigerator at 4°C.

(ii) The flask is provided with a separately sterilized silicone rubber stopper holding two short gas delivery tubes, plugged with cotton wool, and a long glass tube of 4.0 mm outside diameter that reaches to the bottom of the flask.

(iii) At its outer end this tube bears a short piece of silicone rubber tubing capped with aluminium foil.

(iv) Teflon-lined tubing connects the medium reservoir to the culture vessel, through a segment of silicone rubber tubing passing

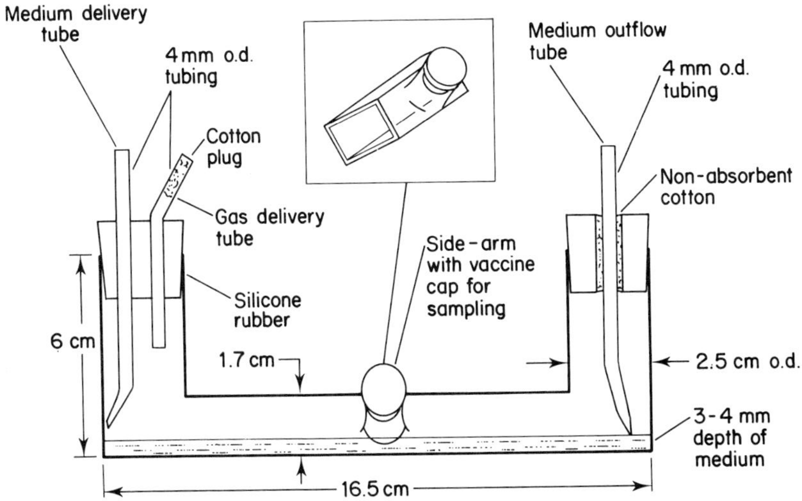

Fig. 1. Cultivation of *P. falciparum*. The flow vessel is made from rectangular 17 × 25 mm tubing with 2 vertical cylindrical necks of 25 mm diameter and a side arm — 25 mm long and 11 mm in diameter, attached at an angle so as to permit access to the bottom by means of a syringe and needle. For sterilization the side arm is capped with aluminium foil, each vertical neck is plugged with cotton, and the whole wrapped in paper. For use, the aluminium foil is removed and a sterile vaccine stopper inserted in the side arm. Infected blood suspension (12 ml of 10% or 8% red cell suspension with an appropriate, usually 1%, parasitaemia) is introduced aseptically through the right-hand vertical neck. This cotton plug is then replaced with a No. 4 silicone rubber stopper bearing the outflow tube. The left-hand cotton plug is similarly replaced with a No. 4 silicone stopper bearing the inflow tube and the gas inlet tube. Both stoppers are individually wrapped and autoclaved separately. Both the inflow and outflow tubes are capped with aluminium foil to protect their sterility until they are connected. (Trager, 1979; reproduced by kind permission of the author and the Society of Protozoologists.)

through a Harvard peristaltic pump, another piece of Teflon-lined tubing and a second short piece of silicone rubber tube. Before this tubing assembly is connected to the culture vessel, it is autoclaved with protective aluminium caps at either end.

(v) The Teflon-lined tubing without the short piece of silicone rubber tubing is connected, through a hole in the refrigerator, to the glass tube of the medium reservoir. The other end, still protected by the aluminium cap, hangs inside the incubator on the left side of the culture vessel.

(vi) The outflow assembly consists of a 250 ml flask bearing a silicone rubber stopper with a cotton-wool outlet to the air and a glass tube running to near the bottom of the flask. A piece of silicone rubber tubing is attached to the glass tubing with its other end protected by aluminium foil. The stopper with the tubing assembly is autoclaved separately and then attached to the sterile flask. The foil-protected tubing end passes into the incubator and hangs down on the right side of the culture flask. A portion of this tubing is inserted into a second peristaltic pump.

(vii) The culture vessel is assembled as shown in Fig. 1. The assembled flow vessel holding the parasitized blood is held in a clamp in the incubator and connected aseptically to the inflow and outflow tubes after removing the aluminium foil caps.

(viii) The gas delivery tube is connected to a tank containing 7% CO_2, 5% O_2 and 88% N_2. Owing to the slow flow of gas and the porosity of the tubing, the actual gas composition was 2% CO_2 and 10% O_2.

(ix) The red cells in the culture vessel are allowed to settle for 1 h. Flow is then begun by setting the peristaltic pump controlling inflow from the reservoir at a high speed until the tubing is full and medium just begins to enter the culture vessel. The pump is then reset to the desired speed. The pump controlling outflow is switched on and set at a speed slightly higher than the inflow.

(x) Once set up, the culture vessel can be kept in operation for about 2 months with periodic addition of fresh cells and replenishment of the medium reservoir flask each week.

2. *Intermittent medium change*

An automated intermittent medium-change culture apparatus was first described by Jensen *et al.* (1979). In this system, culture vessels were tipped periodically (usually once or twice a day); the spent medium was thus removed by gravity, flowing over a retaining lip into a lower chamber. A measured volume of fresh medium was added by the opening of solenoid-controlled clamps. Wickham *et al.* (1980) described a similar system for the cultivation of *P. knowlesi*. Both these systems, however, had the disadvantage of an unacceptable loss of red cells during medium change, particularly for long culture periods such as required for the development of *P. falciparum* gametocytes. The non-tipping system described by Scott-Finnigan *et al.* (1981), and the large-scale apparatus of Palmer *et al.* (1982), though adequate for short-term culture of asexual stages, are unsuitable for prolonged gametocyte cultures of *P. falciparum*, because the medium replacement is never as complete as with tipping systems.

1° "The tipper"
(Ponnudurai *et al.*, 1982b)

APPARATUS

The apparatus consists of a battery of six culture vessels made from rectangular glass tubing of 40 × 20 mm. Each vessel is capable of holding 10 ml of up to 12% red cell suspension (Fig. 2). Medium change occurs automatically, generally twice a day, although it is possible to change medium as often as the red cells have resettled. Fresh medium is pumped (by pump P1) into warming chambers (W) situated on top of the culture chambers (C); P1 is then reversed so that 10 ml of culture medium remains in W. During the tipping process, activated by the cam, a second pump (P2) removes spent medium at a controlled rate. There is a pause of 3 min with the platform at an angle of −30° to ensure that all the spent medium that can be removed safely has been pumped away and to give time for sampling. The platform then tips backwards to an angle of +30°; the red cells stream back in the minimal residual medium and mix with the fresh medium flowing into the lower chamber (C) by gravity. The platform then returns to the horizontal position. Warmed humidified gas flows constantly from the warming chamber over the culture and exits at the front end of the vessels through a port plugged with cotton wool.

TECHNIQUE

(i) The battery of culture vessels, clamped to the removable tipping platform, is attached by silicone rubber tubing via glass manifolds to fresh (R) and waste (S) medium reservoirs. Each tube contains a segment of Technicon pump tubing. The 1.0 litre fresh medium reservoir is fitted with a glass tube reaching to the bottom and is vented through a sterile Millipore "Millex" 0.22 μm filter (F).

(ii) The end of the tubing to the waste bottle is connected to a wide-bore needle and provided with a protective cap during autoclaving. The 10 litre waste bottle, fitted with a vaccine screw cap and vent, is autoclaved separately.

(iii) The gas manifold of the culture apparatus is connected by silicone rubber tubing to an autoclavable hydrophobic 0.3 μm filter. The culture vessels are connected to the manifold by wide-bore silicone rubber tubing.

(iv) The whole assembly of culture vessels, inflow and outflow tubes, and the attached fresh medium reservoir, is autoclaved as one unit. The waste medium reservoir is autoclaved separately and

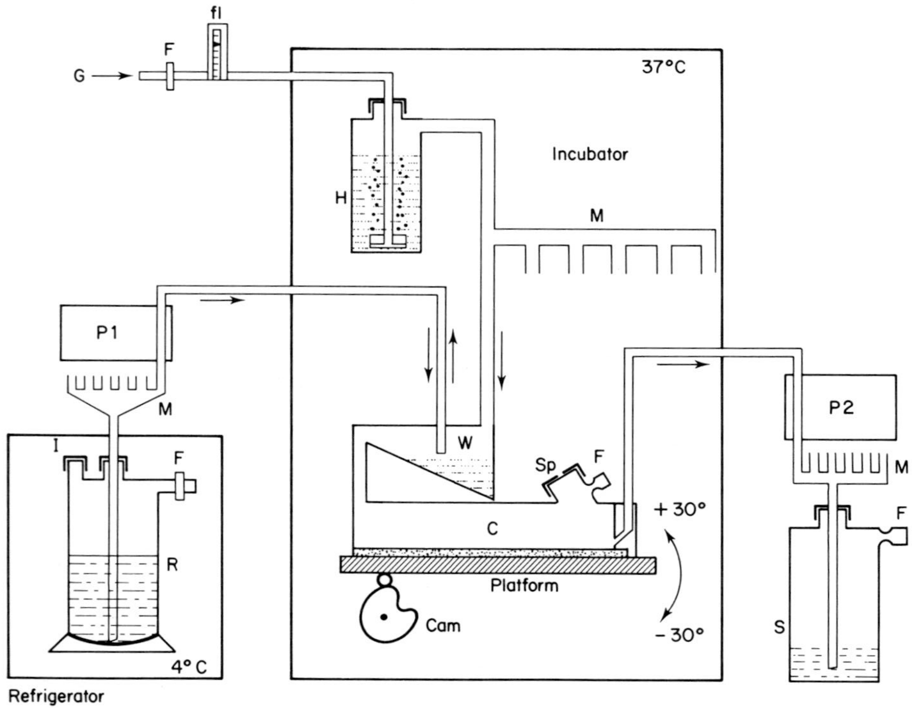

Fig. 2. Cultivation of *P. falciparum*. Diagram showing apparatus for the automatic replacement of the medium whilst maintaining the cultures at the correct temperature and gas conditions. (Ponnudurai *et al*., 1982b; reproduced by kind permission of the authors and the Royal Society of Tropical Medicine and Hygiene.) Key: C, culture chamber; F, filter; fl, flow meter; G, gas (3% O_2; 4% CO_2; 93% N_2); H, humidifier; I, port for filling medium; M, manifold; P1, Watson–Marlow pump; P2, Cenco pump; R, (fresh) medium reservoir; S, (spent) medium reservoir; Sp, sampling port; W, warming chamber.

later attached to the outflow by means of the wide-bore needle.

(v) The tubes pass in and out of the incubator through an aperture cut in the side wall. A similar space is cut in the refrigerator. This makes it possible to sterilize the whole unit without having to make connections aseptically after autoclaving.

(vi) The sterile filter on the end of the gas manifold tubing is connected to a source of warmed humidified gas consisting of 3% O_2, 4% CO_2 and 93% N_2. The flow of gas is adjusted to about 100 ml min^{-1}.

(vii) The fresh medium reservoir is filled with sterile complete medium consisting of 750 ml RPMI 1640 with 25 mM HEPES and 50 mg $litre^{-1}$ hypoxanthine to which is added 32 ml of 5% $NaHCO_3$ and 90 ml serum. This is adequate for 1 week with two medium changes per day.

(viii) When setting up the apparatus for the first time after sterilizing, and before the addition of the culture material, it is usual to run one cycle of medium change in order to test the system and also to wet all tubing and culture vessels with medium.

(ix) Parasitized cells are added and allowed to settle for at least 2 h before medium change is allowed to take place.

(x) The frequency and time of medium change is set on the timer.

(xi) During the forward tipping motion, lasting 6.5 min, red cells accumulate at the front of the vessel in a turbulence-free space whilst the spent medium can be safely pumped away. Sampling and harvesting of parasite culture in a minimum volume can be done during the 3 min pause at the forward tipped position (– 30° in Fig. 2).

2° "The mini-tipper"

(Ponnudurai *et al.*, 1982b)

APPARATUS

A battery of 10 mini-tipper culture vessels, each holding 3.0 ml of culture material, with automatic medium change is used for maintaining different strains or stabilates of *P. falciparum* and also for providing seed material for the larger systems. Medium is changed twice daily.

TECHNIQUE

(i) The mini-tipper vessels are connected with silicone-rubber medium delivery tubing (internal diameter 1.5 mm) and intervening pump tubing (Technicon) to 10 cm long needles via Luer-lock connectors.

(ii) A pair of culture vessels is connected to 60 ml bottles (actual capacity about 70 ml) with vaccine screw caps, by passing the needles through the caps. The needles reach to the bottom of the bottles.

(iii) The outflow silicon rubber tubes (inside diameter 1.5 mm) from the culture vessels are attached to wide-bore needles via a segment of Technicon pump tubing. During sterilization the wide-bore needles are protected by plastic caps.

(iv) After autoclaving the empty fresh medium bottles are replaced by bottles containing 69 ml of fresh sterile medium. A hypodermic needle carrying a Millex 0.22 μm filter serves as the sterile vent. The five bottles containing medium are placed in a refrigerator. Fresh medium bottles are replaced every 4 or 5 d.

(v) The outflow tubes are connected to sterile 1.0 litre waste bottles with vaccine screw caps by means of the wide-bore needles. Two waste bottles are used, each connected to five culture vessels.

(vi) The sterile filter attached to the gas manifold is connected to a source of warmed, humidified gas of (3% O_2, 4% CO_2, 93% N_2).

(vii) Cultures are initiated by withdrawing 0.3 ml of a 5% culture suspension containing parasites into a 1.0 ml syringe with 0.3 ml of 50% uninfected red cell suspension. This constitutes a tenfold dilution of the original culture suspension with fresh red cells.

(viii) This 0.6 ml of diluted red cell suspension is injected into a culture vessel containing 3.0 ml of medium.

(ix) Subculturing occurs every Monday and Friday, irrespective of the parasitaemia. The 3 or 4 d-old culture suspension is withdrawn from the culture vessel using a 5.0 ml syringe and gauge 20 needle. In a laminar flow cabinet the needle is removed, the needle of the 1.0 ml syringe containing fresh red cells is inserted through the nozzle of the 5.0 ml syringe, and the culture suspension is withdrawn into the 1.0 ml syringe as described in (vii).

(x) The contents of two mini-vessels (6.0 ml) are used to initiate cultures for the production of gametocytes in "maxi"-tippers.

C. SUSPENSION CULTURE SYSTEMS

Parasites multiply faster in suspension cultures and reach higher densities (Butcher, 1981a; Zolg *et al.*, 1982; Ponnudurai *et al.*, 1983). The reason lies in the better contact of all parasitized cells with medium and gas compared with static cultures. The cells which form the bottom layer in settled cultures are covered by a layer of other cells through which nutrients, waste matter and gas must diffuse. Suspension cultures are therefore the method of choice for the production of asexual parasites. Very high densities of gametocytes are also produced in suspension cultures. However, here the choice of the parasite isolate is of paramount importance. A relatively slowly growing parasite, such as isolate NF 54, is necessary for success. Even so, many gametocytes are damaged during development and never reach maturity, probably because of the large numbers of asexual parasites present. The quality of mature gametocytes produced in suspension cultures is often poor compared with static cultures, volume for volume. However, this can be more than compensated for by the larger volumes of suspension cultures that can be conveniently handled.

Automated static cultures are therefore necessary for the production of fertile, highly infective gametocytes for mosquito transmission. Two automated suspension culture systems have been described (Butcher, 1981b; Ponnudurai *et al*., 1983; Ponnudurai *et al*., 1986). The first of these systems, described by Butcher (1981b) and modified by Ponnudurai *et al*. (1986), is dependent on the removal of spent medium after allowing red cells to settle for 2 h.

1° Small scale suspension culture
(Ponnudurai *et al*., 1986)

APPARATUS

The culture vessels (CF in Fig. 3) are modified from 100 ml Erlenmeyer flasks. The mouth of the flask is closed and an opening is made on the side of the flask and closed by a vaccine screw cap (SP). The outflow glass tube passes through the roof of the culture vessel and extends down to such a level that, when the supernatant spent medium is removed, none of the red cells of a 7% suspension is lost. The glass inflow tube enters the culture vessel at the side wall close to the top. Warmed, humidified gas passes through a sterile filter (G) and exits through a port (F) plugged with cotton wool. The inflow and outflow glass tubes are connected to silicone rubber tubing (inside diameter 1.5 mm). The medium delivery tubes with Technicon pump tubing are attached to 10 cm long hypodermic needles by Luer-lock connectors. The outflow tubes are connected to wide-bore needles by an intervening segment of Technicon pump tubing.

TECHNIQUE

(i) The needles are protected by plastic caps during autoclaving.
(ii) The culture vessels are fixed at an angle of 15° to the horizontal

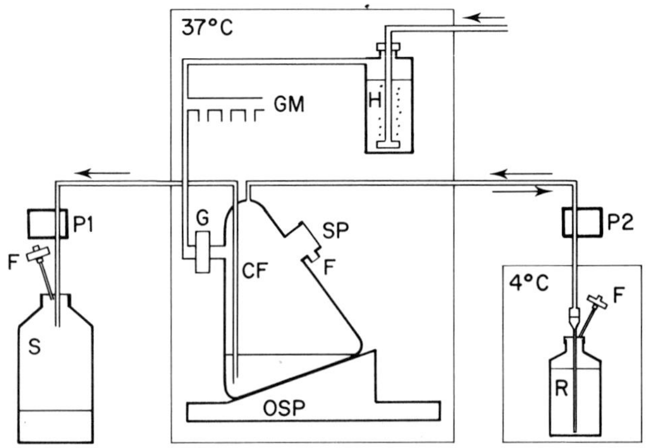

Fig. 3. Diagram of the automated suspension culture system for *P. falciparum*. The culure flask (CF) is placed at 15° to the horizontal on an orbital shaker platform (OSP) and rotated at 100 rev/min. At periodic intervals determined by a 24 h timer, the OSP is stopped for 2 h allowing the red cells to sediment after which pump 1 (P1) is activated to remove the supernatant medium into vessel (S). Once the spent medium has been taken off, the OSP starts rotating again and pump 2 (P2) comes into operation to transfer 10 ml of fresh medium from the reservoir (R) to the CF. H, humidifier: GM, gas manifold; G, hydrophobic gas-inlet filter; F, filtered vents; SP, sampling port. (Ponnudurai *et al.*, 1986; reproduced by kind permission of the authors and Cambridge University Press.)

on an orbital shaking platform (OSP) so that the vessel tops are pointing backwards.

(iii) Twenty such vessels are placed in two rows of 10 each in a New Brunswick G-25 incubator-shaker.

(iv) The inflow and outflow tubes pass through an aperture cut out of the top of the back wall of the incubator.

(v) Having passed the inflow tubing through a Watson-Marlow peristaltic cassette pump (P2), the inflow needles are inserted through vaccine screw caps into 100 ml fresh medium bottles (R) kept in a refrigerator. Each culture vessel is individually connected to its source of fresh medium.

(vi) The outflow needles are connected to sterile 1.0 litre waste bottles (S), five needles per bottle.

(vii) The gas supply tubing is divided into two manifolds (GM) for the two rows of culture vessels, and clamped to the bottom of the platform. Silicone rubber tubing connects the manifolds to the culture vessels.

(viii) To start the cultures 10 ml of 5–7% red cell suspension is introduced into the culture flask through the sampling port and the platform allowed to rotate at 100 rev min^{-1}.

(ix) Generally, two medium changes per day are carried out with the aid of a control unit. The platform is stopped for 2 h to allow the red cells to settle. P1 is then switched on, which gently removes the supernatant spent medium. Once the medium has been removed and P1 is switched off, the platform resumes its rotatory motion and P2 comes into operation to deliver about 10 ml into each culture flask. The cold medium runs down the side of the flask and is rapidly warmed before reaching the cells. P2 then reverses to empty the delivery tubes. Sampling can be carried out at any time by removing a small volume of cell suspension with a long (10 cm) needle, centrifuging in an Eppendorf centrifuge for 5 s, and making a blood film with the deposit.

2° Large scale suspension culture

(Ponnudurai *et al.*, 1983)

APPARATUS

The culture flask (CF in Fig. 4) is the base of a 5.0 litre Erlenmeyer flask, closed on top with ports for removal and replacement of culture and gas entry and exit. The culture vessel is connected to a plasma separator unit (PS) through a set of pre-filters (PF) by means of silicone rubber tubing. When the culture suspension is pumped (by pump P2) through the plasma separator, part of the spent medium is filtered out and collected in a measuring vessel (M). By repeated passage of culture suspension the required volume of spent medium is removed. Once this is accomplished, warmed fresh medium is pumped (by P2) into the silicone rubber tubing circuit so that all cells are flushed back into the culture vessel. This vessel is placed at an angle of 10° to the horizontal on an orbital shaker and cells are kept in suspension by rotation at 60 rev min^{-1}.

TECHNIQUE

(i) A 10 litre vessel, which serves as the fresh medium reservoir (R), is connected to the water jacketed warming vessel (W) by silicone-rubber tubing of 16 mm inside diameter (unit 1). The 500 ml measuring unit (M) is connected to the spent medium collecting vessel (S) by silicone rubber tubing (inside diameter 7 mm) and an overflow tube of 16 mm (unit 2). Unit 3 consists of the culture vessel assembly with silicone rubber tubing (inside diameter 7 mm) and pre-filters. The two open ends of the tubing for connection to the plasma separator unit are covered with autoclave paper. Connection from the culture circuit to the warming vessel (fresh medium) is via a wide-bore needle passing through a vaccine screw cap at the lower end of the latter vessel. The wide-bore needle is protected with a plastic cover during sterilization. Another length of tubing (inside diameter 7 mm) is connected to a wide-bore needle which serves to conduct the spent medium from the plasma separator to the measuring vessel. This is prepared separately for autoclaving.

(ii) The three units described in (i), and the separate tubing, are sterilized by autoclaving. Unit 1 is set up with the medium reservoir in a refrigerator and the warming vessel clamped upright on a retort stand standing on a table at the back of the incubator. The two pumps, a small water bath and control unit are also placed on this table. The silicone rubber tubing passes through a Watson–Marlow peristaltic pump (P1). Unit 2 is placed on the floor with the measuring vessel on top of the collector. Unit 3 is connected, inside a laminar flow cabinet, to a plasma separator unit by removing the protective paper covering the two open ends. The separate length of tubing is connected to the outflow side of the plasma separator. The entire unit 3, together with the attached plasma separator, is mounted inside a "King-size" New Brunswick orbital shaker-incubator (G 25). The loop of tubing which passes through P2, the connecting tubing to the warming vessel, and that from the outflow of the plasma separator all pass through ports in the back wall of the incubator.

(iii) The plastic covers protecting the wide bore needles of the fresh and spent medium tubing are removed and the needles pushed through the rubber caps on the warming and measuring vessels respectively.

(iv) The tubes pass through solenoid clamps as indicated in Fig. 4. The correct sequence of opening of the clamps ensures that, first, culture material passes through the plasma separator and the spent medium is removed; later, when the spent medium has been removed, the closing of the first set of clamps and the opening of others allows fresh medium to enter the circuit and flush the cells back into the culture vessel.

(v) Two culture vessels, of 500 ml each, can be mounted in the same circuit. Medium change is effected alternately from the two vessels with the aid of two double clamps and a switching device.

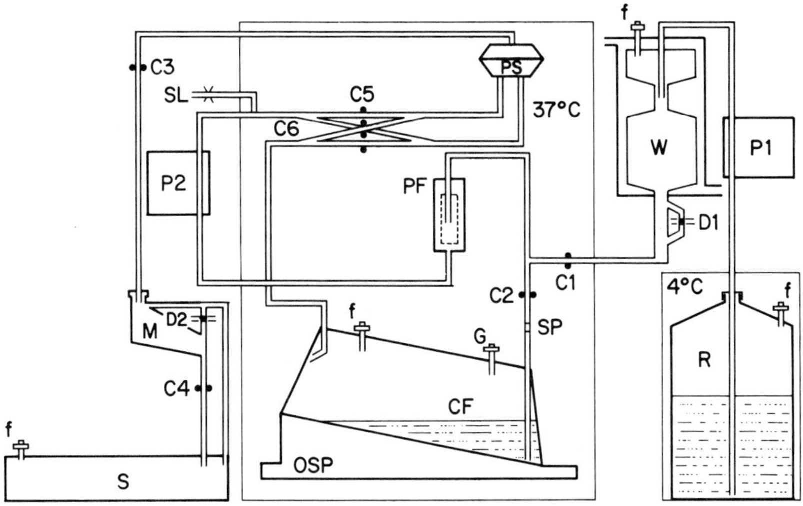

Fig. 4. Diagram of the automated large-scale culture apparatus for *P. falciparum*. The culture flask (CF) is placed at 10° to the horizontal on an orbital shaker platform (OSP) and rotated at 60 rev/min. At periodic intervals determined by a 24 h timer, 500 ml of fresh medium is pumped by pump P1 from the fresh medium reservoir (R) into the warming vessel (W). Twenty minutes later pump P2 comes into operation and simultaneously clamps C2 and C3 are opened. The culture is pumped through the pre-filter (PF) and the plasma separator unit (PS) back into the culture flask. During repeated passage of the culture, spent medium is filtered and collects in a measuring vessel (M). When 500 ml of spent medium has accumulated, the detector (D2) attached to this vessel is triggered, which then stops P2, closes C3 and opens C1. Fresh warmed medium then enters the culture flask via C2. Twenty seconds later C2 closes and C4 opens draining the spent medium into the spent medium collector (S). Escape of fluid from M results in restarting P2 with C1 still open. Fresh medium then flushes the tubing circuit including the pre-filter and the plasma separator. One minute after P2 is restarted C5 and C6 come into operation resulting in the flow of medium through the plasma separator (PS) being reversed. When the fresh medium level in the warming vessel (W) drops to a pre-determined point, detector D1 is triggered which stops P2 and closes C1. Clamps C4, C5 and C6 remain in operation for another 5 min before they are switched off, completing the programme. f, Hydrophobic sterile filter; G, gas inlet; SL, subculture line. (Ponnudurai *et al.*, 1983; reproduced by kind permission of the authors and Cambridge University Press.)

(vi) Cultures are initiated with 500 ml of a 6% suspension of parasitized red cells. Sampling is possible only during medium change, when a small volume is withdrawn through a latex rubber cap over a small side arm on the culture inlet tube. Sterile culture suspension is harvested by pumping the material through an outlet on the circuit, after switching P2 to manual mode and clamping the tubing to the plasma separator.

(vii) Medium change is effected by setting the timer on the control unit. Fresh medium is pumped (by P1) into the warming vessel and, when the pump reverses, 500 ml of the medium is left behind and the delivery tube is emptied.

(viii) About the same time a water-bath heater is switched on and the warm water circulates through the outer jacket of the warming vessel.

(ix) After a warm-up period of 20 min, the culture suspension is pumped (by P2) through the pre-filter and plasma separator. Solenoid clamps automatically open to ensure this. Spent medium accumulates in the measuring vessel until 500 ml has been collected. A photoelectric level detector then shuts off the pump and admits fresh medium into the circuit, clearing the cells in the proximal arm of the circuit by gravity. Later, by the action of P2, the cells in the distal arm are flushed back. The spent medium flows by gravity into the spent medium collector (S) after a clamp is automatically opened, thereby emptying the measuring vessel.

(x) A cross-over circuit ensures that medium flow through the plasma separator unit is reversed, helping to flush out trapped cells.

D. GAMETOCYTE CULTURES AND INFECTION OF MOSQUITOES

Recently the greatest advance in the cultivation of *P. falciparum in vitro* has been made in gametocyte cultures. Although gametocytes were frequently and routinely seen in cultures (Smalley, 1976; Jensen, 1979), and the maturity of these stages was demonstrated by exflagellation (Carter and Beach, 1977), it was not possible routinely to obtain mature parasites for the infection of mosquitoes; only an occasional mosquito infection was obtained (Campbell *et al.*, 1980; Meuwissen *et al.*, 1980). However, routine and reproducible production of infective gametocytes became possible when Ifediba and Vanderberg (1981) reported that hypoxanthine was essential for complete maturation of these stages.

Gametocyte production involves two distinct processes. The first is the induction of gametocytes under the influence of unknown environmental factors (Carter and Miller, 1979; Ponnudurai *et al.*, 1982a). The second is the process of maturation.

The precise circumstances in which gametocyte production is "switched on" in culture are largely unknown. Figure 5 illustrates what

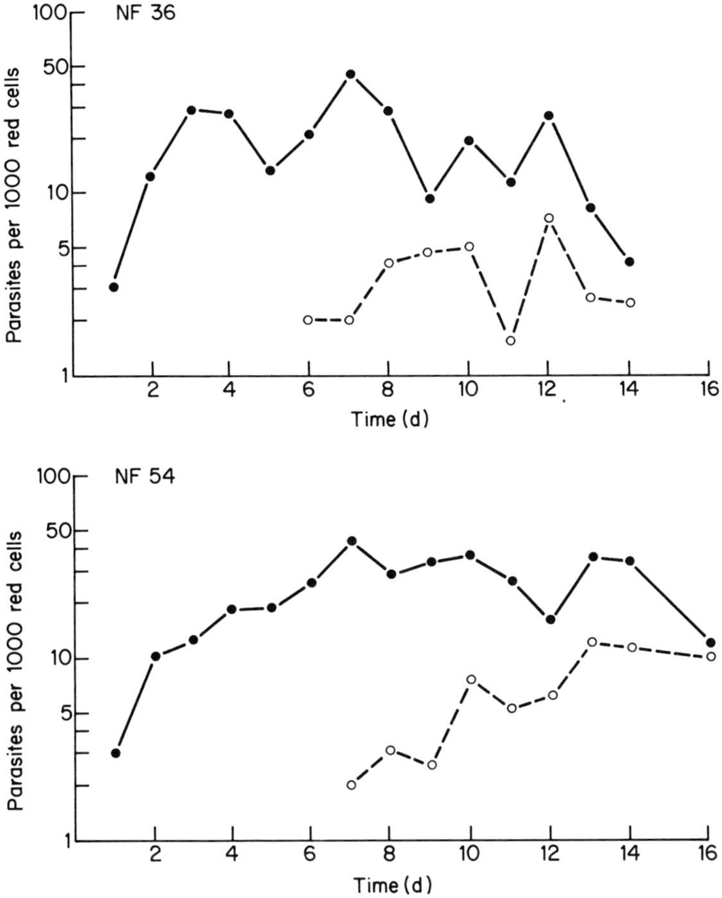

Fig. 5. Asexual parasite and gametocyte counts in cultures of two isolates (NF36 and NF54) of *P. falciparum* when parasite density is not reduced by addition of red cells. • —— • Asexual parasite counts, ○ ---- ○ Gametocyte counts. (Ponnudurai *et al.*, 1982a; reproduced by kind permission of the authors and the Royal Society of Tropical Medicine and Hygiene.)

happens concerning gametocytogenesis when parasites are held in culture for more than 1 week without the addition of red cells. Gametocytogenesis occurs unfailingly so long as the parasite possesses the capacity to produce gametocytes (Fig. 6). At the time of the switch, the growth of asexual parasites slows down. There is a general decline in the quality of the asexual cultures, probably owing to inadequate nutrition to support the growth of the peak parasitaemia and also to the accumulation of

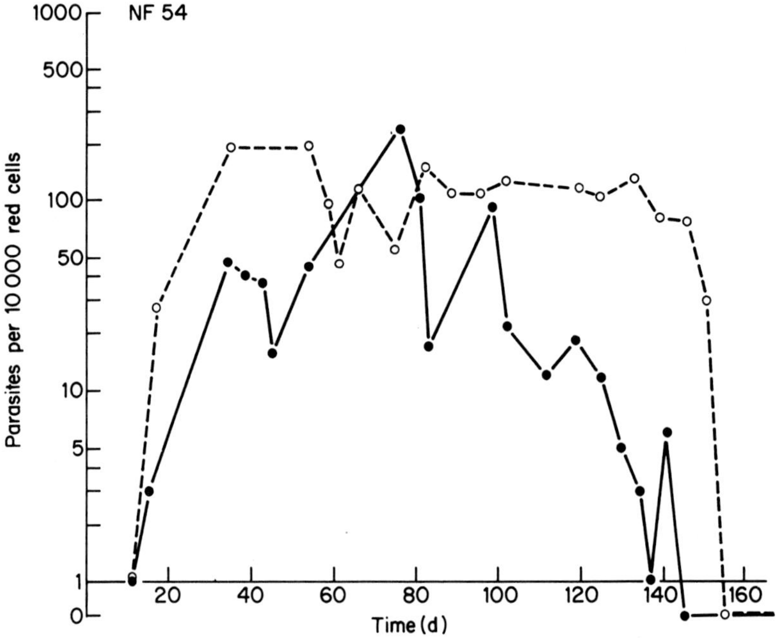

Fig. 6. Gametocyte counts of two stabilates of a single isolate (NF 54) of *P. falciparum* established in continuous *in vitro* culture 5 months apart. Each point on the graph represents a count of all stages of gametocytes from a single plate 10 to 14 days after its commencement. (Ponnudurai *et al.*, 1982a; reproduced by kind permission of the authors and the Royal Society of Tropical Medicine and Hygiene.)

metabolic products. These factors may contribute to the switch to gametocytogenesis (Ponnudurai *et al.*, 1982a).

1. *Strains*

Although gametocytes are seen in all freshly isolated *P. falciparum* strains only some isolates are capable of producing sexual stages that can attain maturity in culture (Ponnudurai *et al.*, 1982a). The correct choice of a strain is important for studies on gametocytes. The isolate NF 54, among others, is a good candidate.

2. *Gametocytogenesis*

Gametocyte production *in vitro* is possible for only a limited period during the "life" of a stabilate. It is therefore necessary to have access to a cryobank where replicates of the chosen isolates can be stored.

3. *Gametocyte maturation*

Once gametocytes are produced, some of them (depending on the type of culture) will mature over a period of 7 d (Ponnudurai *et al.*, 1986). Ifediba and Vanderberg (1981) doubled the volume of the culture medium after 6 d in order to promote maturation. Our unpublished data indicated that, in manual cultures with once-daily medium change, gametocyte maturation was encouraged by low haematocrit values. The presence of too many relatively rapidly growing and actively metabolizing asexual parasites led to damage of the developing sexual stages; this is often a problem in suspension cultures. The total duration of gametocyte cultures is, on average, 14 d.

4. *Gametogenesis*

Mature gametocytes undergo gametogenesis at ambient temperature and atmosphere. Both male and female gametocytes round up — a process called activation — and escape from the red cell. The male gametocyte then undergoes the dramatic process of exflagellation.

1° Preparation of mature gametocytes for infection of mosquitoes

(Modified from Ponnudurai *et al.*, 1982b)

(i) Add 1.5 ml of 14 d old static cultures (e.g., from the "tipper") containing 0.25–0.3 ml packed cell volume (pcv) to a tube prewarmed to 37 °C and containing 0.5 ml pcv of washed fresh red cells overlaid with 3.0 ml of culture medium without serum.
(ii) Centrifuge (560*g*, 2 min) and discard supernatant.
(iii) Withdraw pellet into prewarmed syringe containing 3.25 ml red cells (pcv) in 2.75 ml thawed fresh frozen human serum.
(iv) Introduce into membrane feeders at 37 °C and allow mosquitoes to feed through the stretched parafilm membrane.

Note. Speed of handling and maintenance of temperature at 37 °C throughout processing are essential for good results. The design of the feeder should include as large a feeding surface as possible; the blood in the feeding chamber must have minimal contact with the ambient environment and be adequately warmed by contact with a surface separating it from warm water (Ponnudurai, in preparation).

E. CRYOPRESERVATION

1° Method of Diggs, Aikawa and Haynes
(Modified from Diggs *et al.*, 1977)

FREEZING

(i) Centrifuge parasite suspension (560*g*, 7 min).
(ii) Remove supernatant and add fresh complete culture medium to give a haematocrit value of 50%.
(iii) Add an equal volume of cold 30% glycerol in phosphate-buffered saline (PBS, pH 7.2), a few drops at a time, with gentle agitation.
(iv) Dispense 0.75 ml of the suspension into cryopreservation plastic vials with screw caps.
(v) Place vials upright in a freezing chamber with controlled cooling, using the programme for cryopreservation of leucocytes (Knight, 1980). Here there is rapid cooling to about 0 °C, at which temperature the material is allowed to equilibrate for a while. The temperature is then brought down gradually to − 30 °C and rapidly to − 100 °C. The total duration of the freezing programme is about 45 min. The vials are stored in liquid N_2.

RETRIEVAL

(i) Thaw contents of vial rapidly in a 37 °C water bath and, as soon as thawed, place in melting ice.
(ii) Transfer contents to a calibrated tube and add 2 volumes of cold 27% sorbitol (w/v) in 0.01 M phosphate buffer (pH 7.4); the sorbitol solution should be allowed to run gently down the side of the inclined tube so that the parasitized red cells form a layer above the clear sorbitol layer. Mix by gently rotating the tube in a nearly horizontal position. Allow tube to stand in ice for 12 min.
(iii) Layer 2 volumes of cold 5% sorbitol (w/v) in 0.01 M phosphate buffer (pH 7.4) on top of the mixture in the tube. Mix gently as described above and allow to stand in ice for 10 min.
(iv) Centrifuge (300*g*, 5 min) and replace supernatant with 2 volumes of 5% cold sorbitol. Stir the cells into suspension with the aid of the pipette and allow to stand in ice for 8 min.
(v) Centrifuge (300*g*, 3 min), discard supernatant and wash cells twice in cold complete culture medium (RPMI 1640, HEPES, bicarbonate and serum) by centrifugation (300*g*, 5 min).
(vi) Add 2 volumes of fresh red cells and make up a 2–3% red cell suspension; culture preferably as a suspension culture.

(vii) Add 1% fresh red cells every 2–3 d until a 5% red cell suspension has been obtained. Thereafter subculture as described previously, either as static or suspension cultures.

2° Method of Rowe, Eyster and Kellner
(Modified from Rowe *et al.*, 1968)

FREEZING

(i) Centrifuge parasitized cells (500*g*, 10 min) and add a volume of cryoprotectant equal to the volume of packed cells. The cryoprotectant is prepared by adding 70 ml of glycerol to 180 ml of 4.2% sorbitol in physiological saline, and sterilized by filtration through a 0.45 μm filter.
(ii) The suspension is allowed to equilibrate for 5–10 min at room temperature.
(iii) 0.5 ml aliquots of the suspension are placed in screw-capped vials and frozen rapidly by agitation in an alcohol-dry ice bath or, less preferably, liquid nitrogen.
(iv) Frozen vials are stored in liquid nitrogen.

RETRIEVAL

(i) Thaw contents of vials quickly in a 37 °C water bath and transfer to a calibrated tube.
(ii) Add 0.5 ml of 3.5% NaCl (w/v) and centrifuge (500*g*, 10 min).
(iii) Discard supernatant and add 0.25 ml 3.5% NaCl and centrifuge as before.
(iv) Replace supernatant with 0.5 ml complete culture medium and centrifuge as before.
(v) Add 5 ml complete medium, mix, and centrifuge as before.
(vi) The pellet is then suspended in complete medium to give a haematocrit value of 8% and introduced into culture.

F. CONCENTRATION OF SCHIZONTS AND TROPHOZOITES

1° Method of Pasvol *et al.* (1978)

To obtain the best results, concentration should be carried out when most of the parasites are in the schizont stage.
(i) Centrifuge cells (600*g*, 5 min) and reconstitute in culture medium to give a 40% haematocrit value, in a 15 ml centrifuge tube.

(ii) Add 1 volume of a sterile solution of 3% (w/v) gelatin (Sigma 300 Bloom, G2500) in physiological saline at 37 °C to 2 volumes of the red cell suspension.
(iii) Allow the suspension to sediment for 15 min at 37 °C.
(iv) The supernatant contains 80–90% schizonts, which are washed twice (presumably by centrifugation (600*g*, 5 min) and resuspension) in culture medium.

2° Method of Jensen (1978)

(i) Select the time for concentration when most of the parasites are trophozoites and schizonts.
(ii) Dissolve 1.0 g of gelatin (Sigma 300 Bloom, G2500) in 96 ml RPMI 1640 with 25 mM HEPES by constant stirring at 60 °C. Sterilize by filtration through a 0.45 μm filter and allow to cool. Add 4.0 ml 5% $NaHCO_3$.
(iii) Add an equal volume of the above solution at 37 °C to a parasitized cell suspension (20% haematocrit value) in the complete medium.
(iv) Allow to sediment in a water bath (37 °C, 30 min).
(v) Centrifuge (500*g*, 10 min) supernatant, containing cells infected with trophozoites and schizonts.
(vi) Wash twice with complete medium, by centrifugation as above, before use.

G. SYNCHRONIZATION OF CULTURES

1° Synchronization of asexual parasites using sorbitol

(Based on methods of Jensen, 1978 and Lambros and Vanderberg, 1979)

(i) Concentrate trophozoites and schizonts as described in Section II. F. 1°.
(ii) Place concentrated trophozoites and schizonts in culture with added fresh erythrocytes.
(iii) Four hours later, when freshly released merozoites have reinvaded red cells, wash the cells with 5% aqueous sorbitol (Lambros and Vanderberg, 1979). The sorbitol destroys the cells infected with trophozoites and schizonts, leaving only ring stages which are 4 h old or less.

2° Synchronization of asexual parasites using aphidicolin
(Inselburg and Banyal, 1984)

(i) Trophozoites and schizonts are concentrated (Section II.F.1°).
(ii) The red cells containing the trophozoites and schizonts are attached as a monolayer to 60 mm tissue culture dishes using concanavalin A.
(iii) This layer is then overlaid with a 3% suspension of red cells in medium (see Section II.A.1).
(iv) Incubation of these cultures for 19 h leads to the formation of asynchronous ring-stage parasites in the overlaid cells.
(v) The overlying cells are harvested and cultured until the most mature parasites are early trophozoites, 30 h old.
(vi) The parasites are then incubated with 1.5 μg ml^{-1} of aphidicolin for 15 h. Aphidicolin reversibly blocks DNA synthesis by inhibiting DNA polymerase (alpha). This permits parasites to mature to the early trophozoite stage, at which they appear reversibly blocked.
(vii) 12–15 h after the removal of aphidicolin from the cultures, further maturation of the parasites occurs, with synchronous release of merozoites and reinvasion of red cells.

3° Synchronization of sexual stages
(Ponnudurai *et al.*, 1986)

(i) Trophozoites and schizonts are concentrated as described above.
(ii) Fresh red cells are added to reduce parasitaemia to 5–7%, and the suspension (5% haematocrit value) is introduced into an automated suspension culture system.
(iii) 50 mM N-acetyl glucosamine (GlcNAc) is added to the culture medium (Section II.A.1) in the medium reservoir and the first medium change timed to occur 12 h after the parasites were reintroduced into culture.
(iv) Medium containing GlcNAc is automatically changed twice a day for the next 48 h after which the GlcNAc is withdrawn and normal medium is used. GlcNAc is toxic to schizonts at this concentration (Gupta *et al.*, 1985) and no further reinvasion of red cells occurs in its presence; hence no new gametocyte is formed.

Stage I gametocytes are seen 36 h after parasites were reintroduced into culture, and mature synchronously over a period of 7 d.

III. OTHER MALARIA PARASITES OF MAMMALS

The following parasite species other than *P. falciparum* have been successfully cultivated over long periods: *P. fragile* (Chin *et al.*, 1979), *P. knowlesi* (Butcher, 1979; Wickham *et al.*, 1980), *P. cynomolgi* (Nguyen-Dinh *et al.*, 1981) and *P. berghei* (Janse *et al.*, 1984). The culture conditions for the cultivation of these parasites were similar to those described for *P. falciparum* by Jensen and Trager (1977). Red cells from the appropriate animals were used as host cells. *P. berghei* parasites were grown in a specially developed suspension culture system (Mons *et al.*, 1983); long-term cultivation was possible using daily addition of rat erythrocytes. Medium containing foetal calf serum was changed manually twice a day by centrifugation (200*g* for 5 min).

REFERENCES

Butcher, G.A. (1979). Factors affecting the *in vitro* culture of *Plasmodium falciparum* and *Plasmodium knowlesi*. *Bulletin of the World Health Organization* **57** (Suppl.1), 17–26.

Butcher, G.A. (1981a). A comparison of static thin layer and suspension cultures for the maintenance *in vitro* of *Plasmodium falciparum*. *Annals of Tropical Medicine and Parasitology* **75**, 7–17.

Butcher, G.A. (1981b). A simple automated device for the maintenance of *Plasmodium falciparum* in suspension culture. *Annals of Tropical Medicine and Parasitology* **75**, 111–113.

Campbell, C.C., Chin, W., Collins, W.E. and Moss, D.M. (1980). Infection of *Anopheles freeborni* by gametocytes of cultured *Plasmodium falciparum*. *Transactions of the Royal Society of Tropical Medicine and Hygiene* **74**, 668–669.

Campbell, C.C., Collins, W.E., Nguyen-Dinh, P., Barber, A. and Broderson, J.R. (1982). *Plasmodium falciparum* gametocytes from culture *in vitro* developed to sporozoites that are infectious to primates. *Science* **217**, 1048–1050.

Carter, R. and Beach, R.F. (1977). Gametogenesis in culture by gametocytes of *Plasmodium falciparum*. *Nature* **270**, 240–241.

Carter, R. and Miller, L.H. (1979). Evidence for environmental modulation of gametocytogenesis in *Plasmodium falciparum* in continuous culture. *Bulletin of the World Health Organization* **57** (Suppl.1), 37–52.

Chin, W., Moss, D. and Collins, W.E. (1979). The continuous cultivation of *Plasmodium fragile* by the method of Trager-Jensen. *American Journal of Tropical Medicine and Hygiene* **28**, 591–592.

Diggs, C.L., Aikawa, M. and Haynes, J.D. (1977). Ultrastructure and viability of cryopreserved *Plasmodium falciparum*. *Bulletin of the World Health Organization* **55**, 299–304.

Geary, T.G., Divo, A.A., Bonanni, L.C. and Jensen, J.B. (1985). Nutritional requirements of *Plasmodium falciparum* in culture. III. Further observations on essential nutrients and antimetabolites. *Journal of Protozoology* **32**, 608–613.

Gupta, S.K., Schulman, S. and Vanderberg, J.P. (1985). Stage dependent toxicity of N-acetyl-glucosamine to *Plasmodium falciparum*. *Journal of Protozoology* **32**, 91–95.

Haynes, J.D., Diggs, C.L., Hines, F.A. and Desjardins, R.E. (1976). Culture of human malaria parasites, *Plasmodium falciparum*. *Nature* **263**, 767–769.

Homewood, C.A. (1977). Carbohydrate metabolism of malarial parasites. *Bulletin of the World Health Organization* **55** (Suppl.1), 229–235.

Ifediba, T. and Vanderberg, J.P. (1980). Peptones and calf serum as a replacement for human serum in the cultivation of *Plasmodium falciparum*. *Journal of Parasitology* **66**, 236–239.

Ifediba, T. and Vanderberg, J.P. (1981). Complete *in vitro* maturation of *Plasmodium falciparum* gametocytes. *Nature* **294**, 364–396.

Inselburg, J. and Banyal, H.S. (1984). *Plasmodium falciparum*: synchronization of asexual development with aphidicolin, a DNA synthesis inhibitor. *Experimental Parasitology* **57**, 48–54.

Iscove, N.N. and Melchers, F. (1978). Complete replacement of serum by albumin, transferrin and soya bean lipid in cultures of lipopolysaccharide reactive B lymphocytes. *Journal of Experimental Medicine* **147**, 923–933.

Janse, C.J., Mons, B., Croon, J.J.A.B. and van der Kaay, H.J. (1984). Long-term *in vitro* cultures of *Plasmodium berghei* and preliminary observations on gametocytogenesis. *International Journal for Parasitology* **14**, 317–320.

Jensen, J.B. (1978). Concentration from culture of erythrocytes infected with trophozoites and schizonts of *Plasmodium falciparum*. *American Journal of Tropical Medicine and Hygiene* **27**, 1274–1276.

Jensen, J.B. (1979). Observations on gametogenesis in *Plasmodium falciparum* from continuous cultures. *Journal of Protozoology* **26**, 129–132.

Jensen, J.B. and Trager, W. (1977). *Plasmodium falciparum* in culture: use of outdated erythrocytes and description of the candle-jar method. *Journal of Parasitology* **63**, 883–886.

Jensen, J.B., Trager, W. and Doherty, J. (1979). *Plasmodium falciparum*: continuous cultivation in a semiautomated apparatus. *Experimental Parasitology* **48**, 36–41.

Kaushal, D.C., Carter, R., Miller, L.H. and Krishna, G. (1980). Gametocytogenesis by malaria parasites in continuous culture. *Nature* **286**, 490–492.

Knight, Ç.S. (1980). Freezing of Lymphocytes. *In* "Low Temperature Preservation in Medicine and Biology" (M.J. Ashwood-Smith and J. Farrant, eds), pp. 121–138. Pitman Medical [Churchill Livingstone] Edinburgh.

Lambros, C. and Vanderberg, J.P. (1979). Synchronization of *Plasmodium falciparum* erythrocytic stages in cultures. *Journal of Parasitology* **65**, 418–420.

Maswoswe, S.M., Peters, W. and Warhurst, D.C. (1985). Corticosteroid stimulation of the growth of *Plasmodium falciparum* gametocytes *in vitro*. *Annals of Tropical Medicine and Parasitology* **79**, 607–616.

Meuwissen, J.H.E.Th., Ponnudurai, T. and Leeuwenberg, A.D.E.M. (1980). *In vitro* cultivation of *Plasmodium falciparum* isolates from cases imported into Holland. *In* "Proceedings of the Third International Symposium on the Biochemistry of Parasites and Host-Parasite Relationship: The Host-Invader Interplay (H. Van den Bossche, ed.), Vol. 2, pp. 633–636. Elsevier/North-Holland, New York.

Mons, B., Janse, C.J., Croon, J.J.A.B and van der Kaay, H.J. (1983). *In vitro* culture of *Plasmodium berghei* using a new suspension system. *International Journal for Parasitology* **13**, 213–217.

Nguyen-Dinh, P., Gardner, A.L., Campbell, C.C., Skinner, J.C. and Collins, W.E.

(1981). Cultivation *in vitro* of the vivax-type malaria parasite *Plasmodium cynomolgi*. *Science* **212**, 1146–1148.

Osisanya, J.O.S., Gould, S. and Warhurst, D.C. (1981). A simplified culture technique for *Plasmodium falciparum*. *Annals of Tropical Medicine and Parasitology* **75**, 107–109.

Palmer, K.L., Hui, G.S.N. and Siddiqui, W.A. (1982). A large-scale *in vitro* production system for *Plasmodium falciparum*. *Journal of Parasitology* **68**, 1180–1183.

Pasvol, G., Wilson, R.J.M., Smalley, M.E. and Brown, J. (1978). Separation of viable schizont-infected red cells of *Plasmodium falciparum* from human blood. *Annals of Tropical Medicine and Parasitology* **72**, 87–88.

Ponnudurai, T:, Meuwissen, J.H.E.Th., Leeuwenberg, A.D.E.M., Verhave, J.P. and Lensen, A.H.W. (1982a). The production of mature gametocytes of *Plasmodium falciparum* in continuous cultures of different isolates infective to mosquitoes. *Transactions of the Royal Society of Tropical Medicine and Hygiene* **76**, 242–250.

Ponnudurai, T., Lensen, A.H.W., Leeuwenberg, A.D.E.M. and Meuwissen, J.H.E.Th. (1982b). Cultivation of fertile *Plasmodium falciparum* gametocytes in semi-automated systems. 1. Static cultures. *Transactions of the Royal Society of Tropical Medicine and Hygiene* **76**, 812–818.

Ponnudurai, T., Lensen, A.H.W. and Meuwissen, J.H.E.Th. (1983). An automated large-scale culture system for *Plasmodium falciparum* using tangential flow filtrations for medium change. *Parasitology* **87**, 439–445.

Ponnudurai, T., Lensen, A.H.W., Meis, J.F.G.M. and Meuwissen, J.H.E.Th. (1986). Synchronization of *Plasmodium falciparum* gametocytes using an automated suspension culture system. *Parasitology* **93**, 263–274.

Rowe, A.W., Eyster, E. and Kellner, A. (1968). Liquid nitrogen preservation of red blood cells for transfusion. *Cryobiology* **5**, 119–128.

Sax, L.J. and Rieckmann K.H. (1980). Use of rabbit serum in the cultivation of *Plasmodium falciparum*. *Journal of Parasitology* **66**, 621–624.

Scheibel, L.W., Ashton, S.H. and Trager, W. (1979). *Plasmodium falciparum*: microaerophilic requirements in human red blood cells. *Experimental Parasitology* **47**, 410–418.

Scott-Finnigan, T.J., Wilson, R.J.M. and Williamson, J. (1981). A compact semi-automated system for *Plasmodium falciparum*. *Transactions of the Royal Society of Tropical Medicine and Hygiene* **75**, 292–298.

Sherman, I.W. (1979). Biochemistry of *Plasmodium falciparum* (malarial parasites). *Microbiological Reviews* **43**, 453–495.

Smalley, M.E. (1976). *Plasmodium falciparum* gametocytogenesis *in vitro*. *Nature* **264**, 271–272.

Trager, W. (1979). *Plasmodium falciparum* in culture: improved continuous flow method. *Journal of Protozoology* **26**, 125–129.

Trager, W. and Jensen, J.B. (1976). Human malaria parasites in continuous culture. *Science* **193**, 673–675.

Trager, W. and Jensen, J.B. (1976). Human malaria parasites in continuous culture. *Science* **193**, 673–675.

Trigg, P. (1978). Plasmodiidae. *In* "Methods of Cultivating Parasites *in vitro*" (A.E.R. Taylor and J.R. Baker, eds), pp. 89–110. Academic Press, London and New York.

Van der Ploeg, L.H.T., Smits, M., Ponnudurai, T., Vermeulen, A., Meuwissen, J.H.E.Th. and Langsley, G. (1985). Chromosome-sized DNA molecules of *Plasmodium falciparum*. *Science* **229**, 658–661.

Wickman, J.M., Dennis, E.D. and Mitchell, G.H. (1980). Long term cultivation of a simian malaria parasite (*Plasmodium knowlesi*) in a semi-automated apparatus. *Transactions of the Royal Society of Tropical Medicine and Hygiene* **74**, 789–792.
Zolg, J.W., Macleod, A.J., Dickson, I.H. and Scaife, J.G. (1982). *Plasmodium falciparum*: modifications of the *in vitro* culture conditions improving parasitic yields. *Journal of Parasitology* **68**, 1072–1080.

Chapter 8

Plasmodiidae: Exoerythrocytic Forms

M.R. HOLLINGDALE

Biomedical Research Institute, Rockville, Maryland, USA

I. INTRODUCTION

Although mammalian exoerythrocyctic (EE) malarial parasites were first described in the livers of infected monkeys (Shortt and Garnham, 1948) and subsequently in humans (Shortt *et al.*, 1948), it was not until 1981 that Hollingdale *et al.* reported the first successful *in vitro* culture of EE forms of the rodent malarial parasite, *Plasmodium berghei*. It may be argued, perhaps justifiably, that the focus of malarial research, particularly vaccine development, was primarily directed to the asexual merozoite and sporozoite stages, and that the *in vitro* cultivation of asexual parasites (Trager and Jensen, 1976) and infective gametocytes (Ponnudurai *et al.*, 1982) diverted attention to these parasites at the expense of EE parasites. However, it has become essential to achieve a better understanding of EE development. Candidate malarial vaccines, particularly those based on sporozoite antigens, are likely to include parasite surface antigens involved in parasite invasion of, and intracellular development within, the respective target cells. Thus identification of the circumsporozoite (CS) proteins (Yoshida *et al.*, 1980), and the finding that monoclonal antibodies (Mabs) to the CS proteins neutralize infectivity *in vivo*, require an *in vitro* culture system for the study of sporozoite–hepatic cell interactions. Similarly, antigenic analysis of EE development, particularly that of EE merozoites, is necessary further to delineate targets of anti-merozoite antibody-mediated immunity. With the probability that antimalarial immunity may also involve non-antibody mechanisms (Roberts and Weidanz, 1979), the comparatively long EE development cycle within hepatocytes provides a target for lymphocytes, cytokines or T-cell-mediated mechanisms. Finally, chemotherapy of malaria necessarily demands an understanding of the basic biochemistry of the target parasite. The search for prophylactic drugs, particularly those active against relapsing malarias, is more easily undertaken within suitable *in vitro* culture systems.

In vitro culture of mammalian EE parasites has been hindered by the relative *in vitro* host specificity of sporozoites for hepatocytes. In contrast, avian EE parasites grow in avian fibroblasts (Davis *et al.*, 1966) and rodent fibroblasts (Beaudoin *et al.*, 1974). Although rodent and human hepatocyte cultures can now be maintained *in vitro*, and in fact Doby and Barker had reported parasite-like development within human hepatocytes inoculated with *P. vivax* sporozoites as long ago as 1976, a host cell other than primary hepatocytes, particularly of human origin, is clearly desirable. Strome *et al.* (1979) and Sinden and Smith (1980) reported that *P. berghei* sporozoites would develop into EE schizonts within primary rodent and fibroblast cells, though full maturation of parasites was not observed. In fact, *P. berghei* sporozoites successfully invaded and initiated EE development in a wide variety of rodent, monkey and human cells (Hollingdale *et al.*, 1983b), but the first successful demonstration of complete EE development, with release of infectious EE merozoites, was achieved in primary human embryonic (WI38 cells) cells (Hollingdale *et al.*, 1981). Briefly, WI38 cells were maintained in culture media such as NCTC135 supplemented with 10% foetal bovine serum. Salivary glands of *Anopheles stephensi* mosquitoes, infected with *P. berghei* sporozoites, were dissected into sterile culture medium, teased apart to release sporozoites, and inoculated into the WI38 cell cultures. The cycle of development closely resembled *P. berghei* EE parasites within rat livers. After 24 h, uninucleated EE trophozoites measured 12 μm in diameter and by 48 h EE schizonts had grown to 30 μm. Segmentation began at 60 h, and by 68 h free merozoites were seen, which were infective when injected into mice. Serologically (Hollingdale *et al.*, 1983a), trophozoites more closely resembled sporozoites, whereas schizonts and merozoites closely resembled erythrocytic parasites.

One drawback with WI38 cells was their relatively low susceptibility to *P. berghei* sporozoites; also, sporozoites of the mammalian species *P. falciparum* and *P. vivax* rarely invaded and did not develop further. The alternative approach of using primary hepatocytes has resulted in the complete cycle of development of *P. yoelii* and *P. berghei* in rat and *Thamnomys* hepatocytes (Lambiotte *et al.*, 1981; Mazier *et al.*, 1982;

Pirson, 1982), and *P. vivax* and *P. falciparum* in human hepatocytes (Mazier *et al.*, 1984, 1985; Smith *et al.*, 1984). However, the necessity for repeated isolation of hepatocytes, and the limited availability of human liver, has restricted the use of such culture systems. Many of these restrictions have been overcome by the use of a cloned line of human hepatoma cells (HepG2-A16), which closely resemble primary human hepatocytes (Knowles *et al.*, 1980; Geuze *et al.*, 1984). These cells are highly susceptible to infection with *P. berghei* sporozoites and both invasion and the complete cycle of EE development and merozoite release have been extensively studied (Hollingdale *et al.*, 1983d; Aikawa *et al.*, 1984). Cryopreserved *P. berghei* sporozoites, and sporozoites of both *P. falciparum* and *P. vivax*, also invade HepG2-A16 cells (Hollingdale *et al.*, 1984) and develop into trophozoites, although only *P. vivax* parasites complete schizogony and release merozoites (Hollingdale *et al.*, 1985a, 1986a). Thus, the complete cycle of development of *P. falciparum* EE parasites can as yet be studied only in human hepatocytes. However, the *in vitro* culture of *P. vivax* in HepG2-A16 cells is of considerable interest, as both primary EE schizonts and non-dividing parasites, which persisted after primary schizogony was completed and therefore were similar to the hypnozoite form observed *in vitro* (Krotoski *et al.*, 1980), were seen (Hollingdale *et al.*, 1985a). For strains of south Asian origin, the proportion of primary schizonts to non-dividing parasites (hypnozoites) was approximately equal (Hollingdale *et al.*, 1985a), while for the North Korean strain almost all parasites observed were hypnozoites (Hollingdale *et al.*, 1986). These ratios had been predicted from human volunteer studies (Ungureanu *et al.*, 1976). Ultrastructurally, both primary EE schizonts of *P. vivax* in HepG2-A16 cells, and EE schizonts of *P. falciparum* in human hepatocytes, resembled, but were not identical to, those of *P. berghei* (Aikawa *et al.*, 1984; Meis *et al.*, 1985; Uni *et al.*, 1985).

Inhibition of sporozoite invasion of hepatocytes would completely prevent the development of malaria and block its transmission. Sporozoites are extremely immunogenic, and both live and irradiated sporozoites elicit an antibody response in humans that confers protection (Cochrane *et al.*, 1980). Mabs have identified the species-specific repeat epitope region of the CS protein that confers protection (Cochrane *et al.*, 1982; Zavala *et al.*, 1983), and the genes encoding the CS proteins of *P. knowlesi, P. cynomolgi*, and, now, of *P. falciparum* and *P. vivax* have been cloned into *Escherichia coli* (Godsen *et al.*, 1983; Dame *et al.*, 1984; Enea *et al.*, 1984a, b; McCutchan *et al.*, 1985; Arnot *et al.*, 1985). Immunogenicity studies of recombinant proteins or synthetic peptides common to the repeat epitope region of *P. falciparum* have suggested the rationale for development of an antimalarial vaccine (Young *et al.*, 1985; Zavala *et al.*, 1985; Ballou *et al.*, 1985). Development of an antimalarial vaccine necessarily requires a suitable assay to measure protective antibodies. Since sporozoites of both rodent and human malarias invade HepG2-A16 cells *in vitro*, inhibition of sporozoite invasion (ISI) by specific antisporozoite repeat epitope antibodies might form the basis of an *in vitro* assay. Earlier experiments using *P. berghei* demonstrated that invasion of sporozoites into WI38 or HepG2-A16 cells was specifically blocked when Mabs to the *P. berghei* CS protein (Yoshida *et al.*, 1980) were incorporated into the culture medium (Hollingdale *et al.*, 1982; Leland *et al.*, 1984). Similarly, Mabs against *P. falciparum* and *P. vivax* also blocked invasion of HepG2-A16 cells (Hollingdale *et al.*, 1984). Because these Mabs diminish or abolish sporozoite infectivity to susceptible primate hosts, it was suggested that ISI represents an *in vitro* assay of protective antibodies. This was confirmed by the finding that serum from volunteers immunized with irradiated sporozoites and immune to sporozoite challenge (Clyde *et al.*, 1973) also blocked sporozoite invasion (Hollingdale *et al.*, 1984). The ISI assay was used to demonstrate that candidate *P. falciparum* recombinant protein or synthetic peptide vaccines elicited protective antibodies in immunogenicity studies in mice and rabbits (Young *et al.*, 1985; Zavala *et al.*, 1985; Ballou *et al.*, 1985), and that immunogenicity was similar whether recombinant proteins or peptides were used, and that alum, an adjuvant approved for human use, was as effective as Freund's complete adjuvant (Young *et al.*, 1985; Ballou *et al.*, 1985; Hollingdale *et al.*, 1987). The ISI assay also detected naturally acquired ISI antibodies in areas of the Gambia endemic for malaria (Hollingdale *et al.*, 1984). Further studies, using sera from the Flores region of Indonesia, demonstrated that ISI antibodies were related to antibody levels by enzyme-linked immunosorbent antibody (ELISA) assays, and that ISI activity was completely removed after sera were absorbed with repeat epitope peptides, suggesting that vaccination with these sequences might elicit protection (Hoffman *et al.*, 1986).

II. TISSUE CULTURE METHODS

1° Cultivation of exoerythrocytic forms of *P. berghei* and *P. vivax*

(Hollingdale *et al.*, 1983d, 1985a)

MEDIA

(i) *For HepG2 or HepG2-A16 cells*
MEM (Earle's salts) .. .. 100 ml
Foetal bovine serum .. .. 10 ml
L-Glutamine
(29.2 mg ml^{-1}) .. .. 1.0 ml
Penicillin-streptomycin (5000 iu ml^{-1}, 5000 μg ml^{-1}) .. 1.0 ml

(ii) *For WI38 cells*
NCTC135 .. .. 100 ml
Foetal bovine serum .. .. 10 ml
Penicillin-streptomycin (5000 iu ml^{-1}, 5000 μg ml^{-1}) .. 1.0 ml

(iii) For hepatocyte cultures (human or rat)
Williams' E medium .. .. 100 ml
Foetal bovine serum .. .. 10 ml
Porcine insulin
(20 iu ml^{-1}) .. .. .. 0.1 ml
Dexamethasone (50 mM) .. 0.1 ml
Fungisone (2.5 mg ml^{-1}) .. 0.1 ml
Gentamycine
(50 mg ml^{-1}) .. .. .. 0.1 ml
Vancomycine
(100 mg ml^{-1}) .. .. .. 0.1 ml

Tissue culture reagents from Gibco.

TECHNIQUES

STOCK CULTURES OF HOST CELLS

In our laboratory, the cloned line of human hepatoma (HepG2-A16) cells is routinely used for the *in vitro* culture of *P. berghei* and *P. vivax* EE parasites, and for the ISI assay using *P. falciparum* and *P. vivax* sporozoites. However, both the uncloned parent HepG2 cell line or WI38 cells may also be used, particularly for culture of *P. berghei*. Parent cultures of HepG2 or HepG2-A16 cells are maintained indefinitely in Falcon 25 cm^2 tissue culture flasks in the medium described above at 37 °C in 5% CO_2 in air. The doubling time of these cells is approximately 4 d and to obtain maximal growth care should be taken to seed such flasks with sufficient cells (about 2×10^6) to form an initial culture that is about 25% confluent. Such cultures are subdivided weekly by rinsing with Dulbecco's phosphate buffered saline (PBS), incubating with 0.25% w/v trypsin in Hanks's buffered saline (Gibco), centrifuging cells (1000g, 10 min), and resuspending in complete culture medium.

It is advisable that reserve stocks of cells be cryopreserved (see p. 185).

Cultures of WI38 cells, obtainable from the American Type Culture Collection, or MRC-5 (human embryonic lung fibroblast) cells, are maintained similarly in the culture medium described above at 37 °C in 5% CO_2 in air. However, the restricted longevity of such cultures usually necessitates maintaining frozen stocks, except at high passage number.

In other laboratories, cultures of rodent and human hepatocytes have been established from whole rat livers (Seglen, 1976), and left lobes of human liver (Rijntjes *et al.*, 1986) using a two-step perfusion technique. In general, after perfusion of Ca^{2+}- and Mg^{2+}-free HEPES buffer (pH 7.4), the liver is perfused with HEPES buffer (pH 7.6) containing 5 mM $CaCl_2$ and 0.05% (w/v) collagenase at 37 °C continuously gassed with O_2. Then the hepatocytes are dispensed by tearing the liver tissue apart, purified by filtration (250 μm and 100 μm nylon filters) and centrifugation (50*g*), and incubated at 37 °C with Dulbecco's PBS containing 2% bovine serum albumin. Cells are seeded on to tissue culture dishes pre-coated with homologous extracellular liver matrix (Rojkind *et al.*, 1980) at a density of 1.75×10^5 cells cm^{-2}.

Primary human hepatocyte cultures remain as a viable, non-dividing monolayer for 2 to 4 weeks when incubated at 37 °C in 5% CO_2 in air. Alternatively, hepatocytes may be cryopreserved for future use when required by freezing in programmable freezing racks in a medium containing 10% DMSO and 20% foetal calf serum (Rijntjes *et al.*, 1987).

CRYOPRESERVATION OF HEPG2-A16 CELLS

(i) Remove cells from 25 cm^2 tissue culture flask by trypsinization.
(ii) Centrifuge (1000*g*, 10 min) and resuspend in medium (1° (i) above).
(iii) Centrifuge similarly.
(iv) Resuspend in 1.0 ml medium (1° (i) above) containing 10% v/v dimethyl sulphoxide (DMSO).
(v) Place suspension in slurry of dry ice in ethanol until frozen and subsequently leave to equilibrate for 10 min (do not agitate the tubes); the resultant cooling rate is 70–80 °C min^{-1}.
(vi) Store in liquid nitrogen.
(vii) For retrieval thaw as rapidly as possible in a water bath at 37 °C, centrifuge (1000*g*, 10 min), resuspend in about 4.0 ml medium (1° (i) above), place in a 25 cm^2 flask and allow to grow to confluence.

CULTIVATION OF HEPG2-A16 CELLS FOR INFECTION

(i) Rinse confluent stock cultures, in 25 cm^2 tissue culture flasks, with Dulbecco's PBS, incubate with trypsin, remove cells and centrifuge (1000*g*, 10 min).

(ii) Resuspend cells in 10 ml medium (1° (i) above) and count in a haemocytometer.

(iii) Add about 10^5 cells to microcultures, consisting of 1.0 cm^2 round glass coverslips contained in individual glass vials with aluminium test tube caps. (These have two advantages: firstly, growth of parasites on glass coverslips rather than plastic Petri dishes or microtitre plates allows their removal for study by staining, serology, or processing for electron microscopy; secondly, the separation of individual cultures reduces the occurrence of bacterial or yeast contamination carried by the sporozoite inoculum, and its spread from culture to culture which occurs more readily in microtitre plates.)

(iv) Long term cultures (3 d), for growth of the complete cycle of *P. berghei* EE parasites, are incubated until about 75% confluent before sporozoite inoculation. This allows continuing host cell division to produce a confluent culture after a further 3 d incubation, but without the cell stacking characteristic of HepG2 or HepG2-A16 cells.

(v) For culture of *P. vivax* EE parasites, HepG2 or HepG2-A16 cells must be maintained for a further 9-12 d after sporozoite inoculation (Hollingdale *et al.*, 1985a), otherwise the host cells produce over-confluent cultures making further use for staining or serology impossible. Therefore, cell cultures are grown until almost confluent and then irradiated by a ^{60}Co source with about 10^3–3×10^3 rads (Hollingdale *et al.*, 1985b). Such cultures remain viable for up to 4 weeks without further cell division and are routinely used for *P. vivax*.

SPOROZOITES OF RODENT MALARIAS

(i) Allow 5 d old *A. stephensi* female mosquitoes to feed on mice, infected 3–4 d earlier with blood containing *P. berghei* NK65 or 7 d earlier with *P. berghei* NK65 sporozoites.

(ii) Maintain mosquitoes at 21 °C for 18–21 d, by which time migration into the salivary glands is completed.

(iii) Lightly anaesthetize mosquitoes with chloroform; rinse groups of five in 70% v/v aqueous ethanol, and then with medium 199.

(iv) Place mosquitoes on a glass microscope slide under a dissection microscope.

(v) Add one drop (10 μl) of culture medium, containing 10% foetal bovine serum, to the slide and excise the salivary glands of each mosquito into the culture medium using 1.0 ml tuberculin syringes with 25-gauge needles.

(vi) Transfer each pool of five sets of salivary glands to a glass tissue grinder held in an ice–water mixture. (Pools of salivary glands from up to 100 mosquitoes may be assembled in about 2 h by a trained person.)

(vii) Disrupt the glands by trituration in the grinder and transfer to a sterile plastic tube, rinsing with sterile culture medium, to give a sporozoite suspension usually contained in 250–500 μl.

(viii) Count the number of sporozoites in a haemocytometer.

SPOROZOITES OF HUMAN MALARIAS

P. falciparum sporozoites are isolated from *A. stephensi* mosquitoes infected 14–16 d earlier with *in vitro* cultures of the NF54 strain of *P. falciparum* containing gametocytes (Ponnudurai *et al.*, 1982). *P. vivax* sporozoites from *A. stephensi, A. freeborni* or other species of mosquitoes fed on *P. vivax*-infected chimpanzees have been supplied to this laboratory by Dr W.E. Collins at the Communicable Diseases Center, Atlanta, Georgia, USA (I gratefully acknowledge the continuing interest of Dr Collins and his group in this work). The isolation of *P. falciparum* and *P. vivax* sporozoites is identical to that of *P. berghei*, except that all maintenance of infected mosquitoes and sporozoite isolations must be performed under PIII pathogen containment in the USA (or its equivalent in other countries).

INFECTION AND SUBSEQUENT HANDLING OF CULTURES — GENERAL

Because the invasion rates vary for each species, and depending on the use for the cultures, the numbers of sporozoites added must be initially calculated to obtain maximum use from each sporozoite isolation. The approximate invasion rates (the percentage of sporozoites that invade during 2–3 h) for HepG2-A16 and WI38 cells are:

	HepG2-A16	*WI38*
P. berghei	10.0	1.0
P. vivax	0.01–0.1	0.01
P. falciparum	0.5–2.0	0.01

The rates for *P. vivax* south Asian strains (Collins *et al.*, 1973; 1983) are given for sporozoites obtained from *A. stephensi* mosquitoes fed on infected chimpanzees (Hollingdale *et al.*, 1985a). However, the invasion rate of *P. vivax* sporozoites from *A. stephensi* or *A. freeborni* mosquitoes fed with human volunteers' infected blood through a membrane ranged from 1–10% (Hollingdale *et al.*, 1984), and these differences are not as yet understood.

A further consideration is the use for which the cultures are inoculated.

Principally, two types of cultures are initiated in this laboratory: cultures of complete EE parasite development for serology, drug testing or ultrastructure; and cultures for the ISI assay (reviewed in Section IV).

INFECTION AND SUBSEQUENT HANDLING OF CULTURES — *P. BERGHEI*

Approximately 75% confluent cultures of WI38 or HepG2-A16 cells on 1.0 cm^2 glass coverslips or confluent monolayers of rat hepatocytes on 1.2 cm^2 plastic coverslips are inoculated with *P. berghei* sporozoites diluted such that a known number are added to each culture. In this laboratory, the numbers added usually range from 2×10^4 per culture for serological analysis to 5×10^5 per culture for electron microscopy. After incubation (37°C, 2 h) in 5% CO_2 in air, by which time invasion is completed (Hollingdale *et al.*, 1983c), the culture medium is replaced and the culture further incubated at 37°C. Culture medium is replaced after 24 and 48 h. By 72 h, EE merozoites may be observed, by inverted phase microscopy, contained within detached host HepG2-A16 or WI38 cells floating in the medium, or less frequently within adherent cells.

INFECTION AND SUBSEQUENT HANDLING OF CULTURES — *P. VIVAX* and *P. FALCIPARUM*

Irradiated cultures of HepG2-A16 cells on 1.0 cm^2 glass coverslips, or cultures of human hepatocytes on 1–2 cm^2 plastic coverslips or Petri dishes, are inoculated with known numbers of *P. vivax* or *P. falciparum* sporozoites. In this laboratory, using HepG2-A16 cultures, the numbers of *P. vivax* sporozoites range from 5×10^4 to 1×10^5 for serological studies, or 5×10^5 to 1×10^6 per culture for electron microscopy. After incubation at 37°C for 3 h in 5% CO_2 in air, when invasion is completed (Hollingdale *et al.*, 1984), the culture medium is replaced daily for 9 d by which time EE merozoites have been observed by phase-contrast microscopy as described above for *P. berghei* (Hollingdale *et al.*, 1985a).

For culture of EE parasites of *P. vivax* or *P. falciparum* in primary human hepatocytes, adult human hepatocytes are isolated from liver biopsy specimens by perfusion with collagenase (Sigma type I, EC 3.4. 24.3) and seeded on to "Tissue Tek" chamber slides (Miles) at 8×10^4 cells per 64 mm^2 (Smith *et al.*, 1984) or 5×10^5 cells per 35 mm Falcon Petri dish (Mazier *et al.*, 1985). Culture medium is either Williams' E or MEM supplemented with 10% foetal bovine or human serum, 20 mu ml^{-1} insulin, and 50 nmol $litre^{-1}$ dexamethasone. Groups of pooled salivary glands from 5–10 mosquitoes are added to each culture and release sporozoites directly into the medium. After daily changes of medium, cultures are fixed as described for *P. berghei*, for staining or serology.

III. DETECTION OF EE PARASITES IN CULTURES

Cultures of both rodent and human EE parasites have been examined by Giemsa staining or serological techniques to detect their presence. For either method, cultures, at chosen time intervals, are rinsed twice with Dulbecco's PBS, fixed with methanol at room temperature for 10 min, and rinsed with Dulbecco's PBS.

1° Giemsa staining for examination of cultures

(i) Stain fixed cultures (see above) with 10% Giemsa w/v in PBS at pH 7.2 for 45 min when using WI38 cells; for HepG2-A16 cells, restain cultures for a further 45 min for maximal nuclear and cytoplasmic differentiation.
(ii) Rinse cultures with deionized water and dry.
(iii) View Petri dish cultures directly; view plastic coverslip cultures mounted in water on a glass microscope slide; mount glass coverslip cultures with Permount (Fisher) on a glass microscope slide before viewing.

2° Serological method for examining cultures

Cultures may be examined by immunofluorescent (IFA) and immunoperoxidase (IPA) antibody techniques (Hollingdale and Leland, 1982; Hollingdale *et al.*, 1983a).
(i) React fixed cultures with either Mabs to specific parasite antigens, or hyperimmune mouse or rabbit antiserum to sporozoite or erythrocytic parasite stages.
(ii) Rinse cultures with Dulbecco's PBS.
(iii) For IFA, incubate cultures with the relevant fluorescent isothiocyanate (FITC) conjugated antiserum.
(iv) For IPA, incubate cultures with the relevant horseradish-peroxidase-conjugated antiserum, followed by the 3′,3′-diaminobenzidine substrate.
(v) View IFA-treated cultures mounted in Dulbecco's PBS.
(vi) View IPA-treated cultures after drying and mounting with Permount (Fisher).

IV. INHIBITION OF SPOROZOITE INVASION (ISI) ASSAY

The ISI assay was developed in this laboratory (Hollingdale *et al.*, 1984) as an *in vitro* assay of protective antibodies. The principle is that protective antibodies specifically block sporozoite invasion of HepG2-A16 cells. The assay relies on the observation that *P. falciparum* and *P. vivax* sporozoites both invade HepG2-A16 cells. However, only *P. vivax* sporozoites complete EE development within these cells. In contrast, *P. falciparum* invade and develop the complete EE cycle in primary human hepatocytes, for which the suitability of the ISI assay has been confirmed by Mazier *et al.* (1986), using identical sera in a parallel study with HepG2-A16 cells (Young *et al.*, 1985). However, Mazier also reported that anti-sporozoite antibodies not only blocked sporozoite invasion, but may also inhibit further EE development of those sporozoites that evade antibody neutralization and penetrate the hepatocytes. Similar studies using *P. berghei* sporozoites in HepG2-A16 cells have confirmed this observation (M.R. Hollingdale, unpublished results), and the effect appears to be mediated by bound antibody carried into the hepatocyte by the sporozoite. This intracellular effect enhances the anti-sporozoite activity of immune sera, but does not appear to occur independently of ISI activity. Thus, the ISI assay using HepG2-A16 cells, given its reproducibility and capacity to test relatively large numbers of sera, may be the functional assay of choice.

1° Inhibition of sporozoite invasion assay
(Hollingdale *et al.*, 1984)

(i) Culture HepG2-A16 cells on 1 cm^2 glass coverslips as described for EE parasite development (Section II.1°). However, cell monolayers should be confluent for maximum reproducibility.
(ii) Isolate *P. falciparum* or *P. vivax* sporozoites from mosquito salivary glands as described, pool, and count.
(iii) Dilute test sera, or purified monoclonal or polyclonal antibodies, in complete culture medium. For sera semi-log dilutions are routinely employed: 1/20, 1/100, 1/200, 1/1000, 1/2000, 1/10 000, 1/50 000. For purified antibodies, semi-log dilutions range from 100 $\mu g\ ml^{-1}$ to 0.1$\mu g\ ml^{-1}$.
(iv) Add sera or antibodies to the HepG2-A16 cell cultures contained in 200 μl, followed by 3×10^4 *P. falciparum* sporozoites or 5×10^4 *P. vivax* sporozoites.
(v) Incubate at 37°C in 5% CO_2 in air for 3 h, by which time invasion is complete.
(vi) Rinse cultures with Dulbecco's PBS, fix with methanol, and react in IPA tests usually with a Mab to the relevant sporozoite species CS protein.

(vii) Count intracellular sporozoites at × 200 or × 400 magnification with phase contrast illumination. Intracellular sporozoites can easily be distinguished from attached, non-invaded sporozoites by their morphological appearance (Hollingdale *et al.*, 1983c).

CALCULATION OF RESULTS

(i) The number of sporozoites that invade in the presence of dilutions of sera or purified antibodies is compared with invasion in the presence of pre-immune sera, or similar concentrations of purified antibody specific for a sporozoite species other than the species being assayed. Occasionally pre-immune sera will give a slight (<50%) reduction in invasion at 1/20 dilution. It is essential that all samples be assayed and counted in a double-blind trial, to remove any subjective interpretation by the viewer, and that each sample be tested in duplicate, or preferably triplicate, for statistical analysis.

(ii) Calculate ISI activity as the percentage reduction given by the test sample compared to a suitable control:

$$\text{ISI} = \frac{I_c - I_t}{I_c} \times 100\%$$

where I_c = number invaded in control and I_t = number invaded in test.

From serial semi-log dilutions, three values can be calculated:

ISI_{max} = maximum ISI given by any dilution;
ISI_{100} = lowest concentration giving 100% inhibition;
ISI_{50} = concentration giving 50% inhibition.

(iii) Occasionally some sera and purified antibodies show a prozone effect (see Zavala *et al.*, 1985), in which ISI activity actually increases between serum dilutions of 1/100 and 1/1000, compared to that at 1/20. However, for routine testing of large numbers of sera, particularly from endemic areas, assays at 1/20 have proved successful. Where complete inhibition does not occur at any dilution, ISI_{100} cannot be calculated. ISI_{50} is calculated from linear regression analysis of replicate samples at each concentration with a 95% confidence limit.

V. RELEVANCE FOR VACCINE DEVELOPMENT

A. *IN VITRO* CULTURE OF EE PARASITES

The original rationale (Beaudoin, 1977) for attempting *in vitro* culture of mammalian EE parasites, apart from basic biological studies, was the hope that mammalian EE parasites would grow in continuous culture, as had previously been demonstrated for avian parasites (Davis *et al.*, 1966). However, the realization that mammalian sporozoites invade hepatocytes and undergo only one cycle of EE development before erythrocyctic invasion was entirely corroborated by the similar single cycle of development *in vitro* (Hollingdale *et al.*, 1981). The recent advances in molecular biology have completely transformed the concept of malarial vaccine development to one in which the parasite itself merely represents the complete mosaic of potential protective antigens. Gene cloning and expression of selected antigens can produce the large amounts of purified antigen needed for vaccine development. Nevertheless, vaccine efficacy must eventually be determined using intact parasites. Functional antibody or cell-mediated protective responses require a suitable *in vitro* assay procedure, and development of *in vitro* cultures of EE parasites has led to application of the ISI assay for quantitative measurement of protective anti-sporozoite antibodies.

However, the EE parasite also presents, if indirectly, an immune target suitable for vaccine development. Although the developing EE parasite has been serologically characterized (Hollingdale *et al.*, 1983a, 1985a), whether antigens synthesized by the parasite are inserted into the hepatocyte surface membrane, analogously to parasite-infected erythrocytes, remains to be investigated. It is likely that the invading sporozoite deposits CS protein on the hepatocyte surface (Hollingdale *et al.*, 1983c), in a manner resembling the interaction of glycophorin-binding antigens during merozoite invasion of erythrocytes. The occurrence of hepatocyte-associated antigens, whether derived from sporozoites or EE parasites, may represent an immune target, particularly for cell-mediated immunity generated by sporozoite-based vaccines or, potentially, merozoite-based vaccines.

The rapid progress in development of vaccines based on the repeat epitope region of the sporozoite CS protein has been advanced by the ISI assay, where antibody to these sequences, either as synthetic peptides or recombinant proteins, blocks sporozoite invasion *in vitro*, and may represent, at least in part, the assumed protective mechanism *in vivo*. The potent protective activity of anti-repeat region antibodies or monovalent

Fab fragments (Hollingdale *et al.*, 1982, 1984) suggests that the repeat region itself recognizes the hepatocyte membrane through a specific receptor (or receptors) and mediates sporozoite invasion. Analysis of the nucleotide sequences of the species so far cloned indicates that the repeat region is species-specific, and thus each species would require its own specific hepatocyte receptors. However, the repeat region is flanked by two regions of sequences highly conserved for each species examined (Dame *et al.*, 1984), and results in this laboratory suggest that the CS region that recognizes the hepatocyte surface is the conserved region and not the repeat epitope region. Further analyses using *in vitro* culture of either HepG2-A16 cells or hepatocytes to define the sequence of CS protein-hepatocyte interactions leading to sporozoite invasion may well lead to improved vaccine design.

In vitro cultures provide a system suitable for the assay of lymphokines and other cytokines that may block EE parasite development. For example, administration of γ-interferon to rhesus monkeys completely suppresses *P. cynomolgi* infection (Maheshwari *et al.*, 1986), and delays *P. vivax* infections in chimpanzees (Ferreira *et al.*, 1986). *In vitro* experiments have also demonstrated that interferons will markedly reduce sporozoite invasion of HepG2-A16 cells (Maheshwari *et al.*, 1986) and also block this subsequent EE development in HepG2-A16 cells (Ferreira *et al.*, 1986). As the importance of T-cell-mediated responses becomes more apparent in controlling malaria, T-cell epitopes of sporozoite and merozoite antigens will need to be defined and incorporated into candidate vaccines. *In vitro* EE culture systems will then also be required to assay responses to such vaccination.

B. ISI ASSAY

The ISI assay appears to correlate well with antibody-mediated protective immunity as demonstrated by resistance of experimentally vaccinated human volunteers to live sporozoite challenge (Clyde *et al.*, 1973). However, the relationship of ISI activity to other assays of antibody activity elicited by the sporozoite CS protein remains unclear. In immunogenicity studies with synthetic peptides or recombinant proteins common to the CS repeat epitope region, antibody activity in ELISA assays closely paralleled ISI activity (Ballou *et al.*, 1985; Young *et al.*, 1985; Zavala *et al.*, 1985). Similarly, ELISA activities of sera from persons in malaria endemic regions of Indonesia also correlated well with ISI activities (Hoffman *et al.*, 1986). Until the challenge studies planned for phase 2 of the human volunteer vaccine trials are performed, the levels of ELISA or ISI activity that correlate with protection remain undetermined.

Such challenge studies may well establish the relationship between ELISA and ISI activity and protection. Given the requirement for large numbers of viable *P. falciparum* or *P. vivax* sporozoites to perform the ISI assay, it is likely that the ISI assay will become the "gold standard" assay predictive of protection, against which ELISA or other assays may be standardized.

REFERENCES

Aikawa, M., Schwartz, A.L., Uni, S., Nussenzweig, R.S. and Hollingdale, M.R. (1984). Ultrastructure of *in vitro* cultured exoerythrocytic stage of *Plasmodium berghei* in a hepatoma cell line. *American Journal of Tropical Medicine and Hygiene* **33**, 792–799.

Aley, S.B., Bates, M.D., Tam, J.P. and Hollingdale, M.R. (1986). Synthetic peptides from the circumsporozoite proteins of *Plasmodium falciparum* and *P. knowlesi* recognize the human hepatoma cell line HepG2-A16 *in vitro*. *Journal of Experimental Medicine* **164**, 1915–1922.

Arnot, D.E., Barnwell, J.W., Tam, J.P., Nussenzweig, V., Nussenzweig, R.S. and Enea, V. (1985). Circumsporozoite protein of *Plasmodium vivax*: gene cloning and characterization of the immunodominant epitope. *Science* **230**, 815–818.

Ballou, W.R., Rothbard, J., Wirtz, R.A., Gordon, D.M., Williams, J.S., Gore, R.W., Schneider, I., Hollingdale, M.R., Beaudoin, R.L., Meloy, W.L., Miller, L.H. and Hockmeyer, W.T. (1985). Immunogenicity and biologic activity of antibodies to synthetic peptides of the circumsporozoite protein of *Plasmodium falciparum*. *Science* **228**, 996–999.

Beaudoin, R.L. (1977). Should cultivated exoerythrocytic parasites be considered as a source of antigen for a malaria vaccine? *Bulletin of the World Health Organization* **55**, 371–372.

Beaudoin, R.L., Strome, C.P.A. and Clutter, W.G. (1974). Cultivation of avian malaria parasites in mammalian liver cells. *Experimental Parasitology* **36**, 355–359.

Clyde, D., McCarthy, V., Miller, R. and Hornick, R. (1973). Specificity of protection of man immunized against sporozoite-induced falciparum malaria. *American Journal of Medicine and Science* **266**, 398–403.

Cochrane, A.H., Nussenweig, R.S. and Nardin, E.H. (1980). Immunization against sporozoites. *In* "Malaria" (J.P. Kreier, ed.), pp. 163–202. Academic Press, New York and London.

Cochrane, A.H., Santoro, F., Nussenweig, V. and Nussenzweig, R.S. (1982). Monoclonal antibodies identify the protective antigens of *Plasmodium knowlesi*. *Proceedings of the National Academy of Sciences (USA)* **79**, 5651–5655.

Collins, W.E., Contacos, P.G., Jumper, J., Smith, C.S. and Skinner, J.C. (1973). Studies on human malaria in *Aotus* monkeys. III. Exoerythrocytic stages of the Salvador II strain of *Plasmodium vivax*. *Journal of Parasitology* **59**, 859–866.

Collins, W.E., Warren, M., Skinner, J.C., Huong, A.Y. and Nguyen-Dinh, P. (1983). Observations on the infectivity of two strains of *Plasmodium vivax* from Vietnamese refugees to *Aotus* monkeys and anopheline mosquitoes. *Journal of Parasitology* **69**, 689–695.

Dame, J.B., Williams, J.L., McCutchan, T.F., Weber, J.L., Wirtz, R.A., Hockmeyer, W.T., Meloy, W.T., Haynes, J.D., Schneider, I., Roberts, D., Sanders, A.S.,

Reddy, E.P., Diggs, C.L. and Miller, L.H. (1984). Structure of the gene encoding the immunodominant surface antigen on the sporozoite of the human malaria parasite *Plasmodium falciparum*. *Science* **225**, 593–599.

Davis, A.G., Huff, C.G. and Palmer, T.T. (1966). Procedure for maximum production of exoerythrocytic stages of *Plasmodium fallax* in tissue culture. *Experimental Parasitology* **19**, 1–8.

Doby, J.M. and Barker, R. (1976). Essais d'obtention *in vitro* des formes preérythrocytaires de *Plasmodium vivax* en cultures de cellules hépatiques humaines inoculées par sporozoites. *Comptes Rendus des Séances de la Société de Biologie* **170**, 166–665.

Enea, V., Arnot, D., Schmidt, E.C., Cochrane, A., Gwadz, R. and Nussenzweig, R.S. (1984a). Circumsporozoite gene of *Plasmodium cynomolgi* (Gombak): cDNA cloning and expression of the repetitive circumsporozoite epitope. *Proceedings of the National Academy of Sciences (USA)* **81**, 7520–7524.

Enea, V., Ellis, J., Zavala, F., Arnot, D.E., Asanovich, A., Masuda, A., Quakyi, I. and Nussenzweig, R.S. (1984b). DNA cloning of *Plasmodium falciparum* circumsporozoite gene: amino acid sequence of repetitive epitope. *Science* **225**, 628–630.

Ferreira, A., Schofield, L., Enea, V., Schellekens, H., van der Meide, P., Collins, W.E., Nussenzweig, R.S. and Nussenzweig, V. (1986). Inhibition of development of exoerythrocytic forms of malaria parasites by γ-interferon. *Science* **232**, 881–884.

Geuze, H.J., Slot, J.W., Strous, G.J.A.M., Luzio, J.P. and Schwartz, A.L. (1984). The pathway of the asialoglycoprotein ligand during receptor-mediated endodcytosis: a morphological study with colloidal gold/ligand in the human hepatoma cell line HepG2. *European Journal of Cell Biology* **32**, 38–44.

Godsen, G.N., Ellis, J., Svec P., Schlesinger, D.H. and Nussenzweig, V. (1983). Indentification and chemical synthesis of a tandemly repeated immunogenic region of *Plasmodium knowlesi* circumsporozoite protein. *Nature* **305**, 29–33.

Hoffman, S.L., Wistar, R., Ballou, W.R., Hollingdale, M.R., Wirtz, R.A., Schneider, I., Marwoto, H.A. and Hockmeyer, W.T. (1986). Naturally acquired human antibodies to *Plasmodium falciparum* circumsporozoite protein repeat region mediate biologic reactions associated with protective immunity. *New England Journal of Medicine* **315**, 601–606.

Hollingdale, M.R. and Leland, P. (1982). Detection of exoerythrocytic stages of *Plasmodium berghei* in fixed liver tissue and cultured cells by an immunoperoxidase antibody technique. *Transactions of the Royal Society of Tropical Medicine and Hygiene* **76**, 624–625.

Hollingdale, M.R., Leef, J.L., McCullough, M. and Beaudoin, R.L. (1981). *In vitro* cultivation of the exoerythrocytic stage of *Plasmodium berghei* from sporozoites. *Science* **213**, 1021–1022.

Hollingdale, M.R., Zavala, F., Nussenzweig, R.S. and Nussenzweig, V. (1982). Antibodies to the protective antigen of *Plasmodium berghei* sporozoites prevent entry into cultured cells. *Journal of Immunology* **128**, 1929–1930.

Hollingdale, M.R., Leland, P., Leef, J.L. and Beaudoin, R.L. (1983a). Serological reactivity of cultured exoerythrocytic stages of *Plasmodium berghei* in indirect immunofluorescence or immunoperoxidase antibody tests. *American Journal of Tropical Medicine and Hygiene* **32**, 24–30.

Hollingdale, M.R., Leland, P., Leef, J.L. and Beaudoin, R.L. (1983b). The influence of cell type and culture medium on the *in vitro* cultivation of exoerythrocytic stages of *Plasmodium berghei*. *Journal of Parasitology* **69**, 346–352.

Hollingdale, M.R., Leland, P., Leef, J.L. and Schwartz, A.L. (1983c). Entry of *Plasmodium berghei* sporozoites into cultured cells, and their transformation into

trophozoites. *American Journal of Tropical Medicine and Hygiene* **32**, 685-690.
Hollingdale, M.R., Leland, P. and Schwartz, A.L. (1983d). *In vitro* cultivation of the exoerythrocytic stage of *Plasmodium berghei* in a hepatoma cell line. *American Journal of Tropical Medicine and Hygiene* **32**, 682–684.
Hollingdale, M.R., Nardin, E.H., Tharavanij, S., Schwartz, A.L. and Nussenzweig, R.S. (1984). Inhibition of entry of *Plasmodium falciparum* and *P. vivax* sporozoites into cultured cells; an *in vitro* assay of protective antibodies. *Journal of Immunology* **132**, 909–913.
Hollingdale, M.R., Collins, W.E., Campbell, C.C. and Schwartz, A.L. (1985a). *In vitro* culture of two populations (dividing and nondividing) of exoerythrocytic parasites of *Plasmodium vivax. American Journal of Tropical Medicine and Hygiene* **34**, 216–222.
Hollingdale, M.R., Leland, P. and Sigler, C.I. (1985b). *In vitro* cultivation of the exoerythrocytic stage of *Plasmodium berghei* in irradiated hepatoma cells. *American Journal of Tropical Medicine and Hygiene* **34**, 21–23.
Hollingdale, M.R., Leland, P., Sigler, C.I. and Leef, J.L. (1985c). *In vitro* infectivity of cryopreserved *Plasmodium berghei* sporozoites to cultured cells. *Transactions of the Royal Society of Tropical Medicine and Hygiene* **79**, 206–208.
Hollingdale, M.R., Collins, W.E. and Campbell, C.C. (1986). *In vitro* culture of exoerythrocytic parasites of the North Korean strain of *Plasmodium vivax* in hepatoma cells. *American Journal of Tropical Medicine and Hygiene* **35**, 275–276.
Hollingdale, M.R., Ballou, W.R., Aley, S.B., Pancake, S.J., Miller, L.H. and Hockmeyer, W.T. (1987). *Plasmodium falciparum*: elicitation by peptides and recombinant circumsporozoite proteins of circulating mouse antibodies inhibiting sporozoite invasion of hepatoma cells. *Experimental Parasitology* **63**, 345–351.
Knowles, B.B., Howe, C.C. and Aden, D.P. (1980). Human hepatocellular carcinoma cell lines secrete the major plasma proteins and hepatitis B surface antigen. *Science* **209**, 497–499.
Krotoski, W.A., Krotoski, D.M., Garnham, P.C.C., Bray, R.S., Killick-Kendrick, R., Draper, C.C., Targett, G.A.T. and Guy, M.W. (1980). Relapses in primate malaria: discovery of two populations of exoerythrocytic stages. Preliminary note. *British Medical Journal* **i**, 153–154.
Lambiotte, M., Landau, I., Thierry, N. and Miltgen, F. (1981). Dévelopement de schizontes dans les hépatocytes de rat adulte en culture après infestation *in vitro* par des sporozoites de *Plasmodium yoelii. Compte Rendu hebdomonadaire des Séances de l' Académie des Sciences, Paris* **293**, 431–433.
Leland, P., Sigler, C.I., Danforth, H.D. and Hollingdale, M.R. (1984). Inhibition of *Plasmodium berghei* sporozoite invasion of cultured hepatoma cells. *Transactions of the Royal Society of Tropical Medicine and Hygiene* **78**, 639–640.
Maheshwari, R.K.C.W., Dutta, G.P., Puri, S.K., Dhawan, B.N. and Friedman, R.M. (1986). Recombinant human interferon gamma inhibits simian malaria. *Infection and Immunity* **53**, 628–630.
Mazier, D., Landau, I., Miltgen, F., Druilhe, P., Lambiotte, M., Baccam, D. and Gentilini, M. (1982). Infestation *in vitro* des cellules 'hépatiques de *Thamnomys* adulte par des sporozoites de *Plasmodium yoelii*: schizogonie et libération de merozoites infestans. *Compte Rendu de hebdomonadaire des Séances el Académie des Sciences, Paris* **294**, 963–965.
Mazier D., Landau, I., Druilhe, P., Miltgen, F., Gugen-Guillouzo, C., Baccam, D., Baxter, J., Chigot, J.-P. and Gentilini, M. (1984). Cultivation of the liver forms of *Plasmodium vivax* in human hepatocytes. *Nature* **307**, 367–369.

Mazier, D., Beaudoin, R.L., Mellouk, S., Druilhe, P., Texier, B., Trosper, J., Miltgen, F., Landau, I., Paul, C., Brandicourt, O., Gugen-Gillouzo, C., Langois, P. and Gentilini, M. (1985). Complete development of hepatic stages of *Plasmodium falciparum in vitro. Science* **277**, 440–442.

Mazier, D., Mellouks, S., Beaudoin, R.L., Texier, B., Drulhe, P., Hockmeyer, W., Trosper, J., Parl, C., Charoenvit, Y., Young, J., Miltgen, E., Chedid, L., Chigot, J.P., Galley, B., Brandicourt, O. and Gentilini, M. (1986). Effect of antibodies to recombinant and synthetic peptides on *P. falciparum* sporozoites *in vitro. Science* **231**, 156–159.

McCutchan, T.F., Lal, A.A., de la Cruz, V.F., Miller, L.H., Meloy, W.L., Charoenvit, Y., Beaudoin, R.L., Guerry, P., Wistar, R., Hoffman, S., Hockmeyer, W.T., Collins, W.E. and Wirth, D. (1985). Sequence of the immunodominant epitope for the surface protein on sporozoites of *Plasmodium vivax. Science* **230**, 1381–1383.

Meis, F.G.M., Verhave, J.-P., Ponnudurai, T., Hollingdale, M.R., Rijntjes, J.M., Smith, J.E., Sinden, R.E., Jap, P.H.K., Meuwissen, J.H.E. Th. and Yap, S.H. (1985). Fine structure of the liver forms of the malaria parasite *Plasmodium falciparum* in human hepatocytes *in vitro. Cell and Tissue Research* **244**, 345–350.

Pirson, P. (1982). Culture of the exoerythrocytic liver stages of *Plasmodium berghei* sporozoites in rat hepatocytes. *Transactions of the Royal Society of Tropical Medicine and Hygiene* **76**, 722.

Ponnudurai, T., Meuwissen, J.H.E.T., Leeuwenberg, A.D.E.M., Verhave, J.-P. and Lensen, A.H.W. (1982). The production of mature gametocytes of *Plasmodium falciparum* in continuous culture of different isolates infective to mosquitoes. *Transactions of the Royal Society of Tropical Medicine and Hygiene* **76**, 242–250.

Potocnjak, P., Yoshida, N., Nussenzweig, R.S. and Nussenzweig, V. (1980). Monovalent fragments (Fab) of monoclonal antibodies to a sporozoite surface antigen (Pb44) protect mice against malarial infection. *Journal of Experimental Medicine* **151**, 1504–1513.

Rijntjes, P.J.M., Moshage, H.J., Van Gemert, P.J.L., de Waal, R. and Yap, S.H. (1987). Cryopreservation of adult human hepatocytes: the influence of deep freezing storage on the viability, cell seeding, survival, fine structure and albumin synthesis in primary cultures. *Journal of Hepatology* (in press).

Roberts, D.W. and Weidanz. W.P. (1979). T-cell immunity to malaria in the B-cell deficient mouse. *American Journal of Tropical Medicine and Hygiene* **28**, 1–3.

Rojkind, M., Gatmaitan, Z., Mackensen, S., Giambrone, M., Ponce, P. and Reid, L.M. (1980). Connective tissue biomatrix: its isolation and utilization for long-term cultures of normal rat hepatocytes. *Journal of Cell Biology* **87**, 255–263.

Seglen, P.O. (1976). Preparation of isolated rat liver cells. *In* "Methods of Cellular Biology" (D.M. Prescott, ed.), Vol, 13, pp. 163–202. Academic Press, New York and London.

Shortt, H.E. and Garnham, P.C.C. (1948). Demonstration of a persisting exoerythrocytic cycle in *Plasmodium cynomolgi* and its bearing on the production of relapses. *British Medical Journal* **i**, 1225–1232.

Shortt, H.E., Garnham, P.C.C., Covell, G. and Shute, P.G. (1948). The preerythrocytic stage of human malaria, *Plasmodium vivax. British Medical Journal* **i**, 547.

Sinden, R.E. and Smith, J. (1980). Culture of the liver stages (exoerythrocytic schizonts) of rodent malaria parasites from sporozoites *in vitro. Transactions of the Royal Society of Tropical Medicine and Hygiene* **74**, 134–135.

Smith, J.E., Meis, J.F.G.M., Ponnudurai, T., Verhave, J.P. and Moshage, H.J. (1984). *In vitro* culture of exoerythrocytic form of *Plasmodium falciparum* in adult human hepatocytes. *Lancet* **ii**, 757–758.

Strome, C.P.A., DeSantis, P. and Beaudoin, R.L. (1979). Cultivation of the exoerythrocytic stages of *Plasmodium berghei* from sporozoites. *In Vitro* **15**, 531–536.

Trager, W.T. and Jensen, J.B. (1976). Human malaria parasites in continuous culture. *Science* **193**, 673–675.

Ungureanu, E., Killick-Kendrick, R., Garnham, P.C.C., Branzei, P., Romanescu, C. and Shute, P.G. (1976). Prepatent periods of a tropical strain of *Plasmodium vivax* after inoculation of tenfold dilutions of sporozoites. *Transactions of the Royal Society of Tropical Medicine and Hygiene* **70**, 482–483.

Uni, S., Aikawa, M., Collins, W.E., Campbell, C.C. and Hollingdale, M.R. (1985). Electron microscopy of *Plasmodium vivax* exoerythrocytic stages grown *in vitro* in a hepatoma cell line. *American Journal of Tropical Medicine and Hygiene* **34**, 1017–1021.

Yoshida, M., Nussenzweig, R.S., Potocnjak, P., Nussenzweig, V. and Aikawa, M. (1980). Hybridoma produces protective antibodies directed against the sporozoite stage of malaria parasite. *Science* **207**, 71–73.

Young, J.F., Hockmeyer, W.T., Gross, M., Ballou, W.R., Wirtz, R.A., Trosper, J.H., Beaudoin, R.L., Hollingdale, M.R., Miller, L.H., Diggs, C.L. and Rosenberg, M. (1985). Expression of *P. falciparum* circumsporozoite protein derivatives in *E. coli* for development of a human malaria vaccine. *Science* **228**, 958–962.

Zavala, F., Cochrane, A.H., Nardin, E.H., Nussenzweig, R.S. and Nussenzweig, V. (1983). Circumsporozoite proteins of malarial parasites contain a single immunodominant region with two or more identical epitopes. *Journal of Experimental Medicine* **157**, 1947–1953.

Zavala, F., Tam, J.P., Hollingdale, M.R., Cochrane, A.H., Quakyi, I., Nussenzweig, R.S. and Nussenzweig, V. (1985). Rationale for development of a synthetic vaccine against *Plasmodium falciparum* malaria. *Science* **228**, 1436–1440.

Chapter 9

Babesiidae

ELIZABETH U. CANNING and CAROLINE M. WINGER

Department of Pure and Applied Biology, Imperial College, London, England

I. INTRODUCTION

Babesia spp. are unpigmented intraerythrocytic parasites of mammals and are transmitted by ticks. Together with *Theileria* spp. (Chapter 10) they are commonly known as piroplasms and are assigned to the protozoan phylum Apicomplexa by reason of the apical complex of organelles on the stages (merozoites, kinetes and sporozoites) which enter host cells. Their affinities with other apicomplexan groups are well revealed in a classification recently proposed by Mehlhorn and Schein (see Canning, 1986). In this, the malaria parasites (family Haemosporidae) and piroplasms (family Piroplasmidae) are placed together in the suborder Aconoidina, organisms in which a conoid is absent from the apical complex. The babesias and theilerias are distinguished from one another by several biological features. In particular they have different transmission mechanisms through ticks and pre-erythrocytic schizogony in lymphocytes occurs in the latter only. The subfamily Babesiidae contains a single genus, *Babesia*, with many species parasitic in mammals of several orders.

Babesia spp. have simpler apical complexes than the malaria parasites and can be distinguished further from them by important biological differences. Thus, the babesias have no exoerythrocytic schizogony; intra-erythrocytic division is usually by binary fission, the developing merozoites lying directly in the cytoplasm, not in a parasitophorous vacuole; haemoglobin digestion does not result in the deposition of haemozoin in food vacuoles; merozoites retain invasive ability for several hours after release from erythrocytes; gametocytes are possibly represented by ovoid rather than the usual piriform stages in the blood but are not clearly differentiated as macro- and microgametocytes; transmission is by ticks, in which gametes develop in the form of spiky-ray bodies; fusion of gametes gives motile kinetes which in turn undergo cycles of division giving more kinetes, before invading the ova; infections are acquired only by adult engorged female ticks and infectivity is reached only in the F_1 generation — i.e., transmission is transovarial; the babesias have a narrower host range — mammals only, compared with reptiles, birds and mammals for malaria parasites; and the two groups respond entirely differently to chemotherapy.

It is not surprising, therefore, that the method devised for continuous culture of *Plasmodium falciparum* (Trager and Jensen, 1976) was unsuccessful when applied to *Babesia bovis* (Erp *et al.*, 1978). However, the conditions required for continuous growth of *B. bovis*, first in suspension culture (Erp *et al.*, 1980a) and then in microaerophilous stationary phase (MASP) culture (Levy and Ristic, 1980), were determined and several other species can now be grown in continuous culture. The MASP technique is regarded as the more convenient. Descriptions of these techniques and their modifications for different species form the bulk of this chapter but one section discusses the limited success achieved in culturing the kinete stages in primary cell cultures and cell lines derived from ticks.

Schizogony in lymphoblastoid cells has recently been shown to occur during infections with two piroplasms, *Babesia equi* (Schein *et al.*, 1981) and *Babesia microti* (Mehlhorn and Schein, 1984). These observations, coupled with transstadial transmission by ticks, set these species apart from true *Babesia* spp. and indicate their affinity with *Theileria*. The culture of *B. equi* is described in Chapter 10 (see p. 231).

Four species of *Babesia* parasitizing cattle have been grown in continuous culture. Additionally, *B. canis* has been grown in short-term culture and *B. caballi* of horses has recently been established *in vitro*.

Bovine babesiosis is primarily a disease of adult cattle. Calves generally recover from field infections and become carriers. The mild course of disease in calves is partially attributable to colostral antibodies from immune mothers, but there are important innate blood factors that also determine resistance. Levy *et al.* (1982) demonstrated the inability of calf erythrocytes and serum to support continuous growth of *Babesia bovis in vitro*.

Thus, blood from adult cattle must be used for cultures. Even with adults there are differences between individuals and, when initiating cultures, it is advisable to select suitable donor animals for the regular supply of blood. Erythrocytes can be stored for only about 2 weeks. After adaptation of parasite strains to culture, commercially prepared pooled adult sera may be suitable but, again, batches should be tested for suitability beforehand. Commercial suppliers may be prepared to set aside entire batches, once these have been identified. Sera stored at -20 °C or below can be used for several months.

Other important factors are temperature (best at 37–38 °C), pH, which can vary little either side of neutral (Erp *et al.*, 1980a), and levels of CO_2 and O_2. CO_2 is necessary for invasion of erythrocytes by merozoites (Levy and Ristic, 1980) and O_2 concentration must be low (about 2%) at the level of the erythrocytes. This can be achieved by a low O_2 content gas mixture and shallow culture, or by increasing the depth of the medium overlying the erythrocytes in higher O_2 concentrations (e.g. in air).

Erythrocytes from animals infected with *Babesia*, whether or not they harbour developing parasites, are more fragile than those from healthy animals. Additionally, in *B. divergens* and *B. bigemina* infections, the parasitized erythrocytes are more fragile than the non-parasitized ones in an infected host: the converse is true of *B. bovis*. Care has to be taken during the preparation of blood to avoid disruption of infected cells and consequent loss of developing stages, as mature merozoites may be present in insufficient numbers to establish the cultures. If cultures are prepared from heterologous cells (e.g., parasitized gerbil cells in cultures containing normal bovine cells), there is massive lysis of the donor's cells and the parasitaemia must be high enough to ensure that abundant mature stages are present.

Clotting must be prevented by defibrination of the blood, by shaking with glass beads or by use of heparin as anticoagulant. Some haemolysis is inevitable with both methods, either directly during defibrination or during the extensive washing which must follow the use of heparin. Washing infected and uninfected erythrocytes before addition to cultures is recommended, especially when mixing erythrocytes of different host species (e.g., from gerbil and bovine or human blood), as complement in the donor host's serum will lyse the heterologous cells. When the erythrocytes are sedimented between washes, it is advisable to remove the buffy coat (Timms, 1980).

II. METHODS

A. SUSPENSION CULTURES

1° Suspension culture of *B. bovis*

A method developed by Erp *et al.* (1978) for a short series of subcultures of *B. bovis* when modified only by lowering the pH to 7.0, permitted continuous culture (Erp *et al.*, 1980a). Optimal conditions were evaluated by Erp *et al.* (1980b).

MEDIA

Medium 199, RPMI 1640, or NCTC 135, supplemented with 40% v/v adult bovine serum and buffered with 25 mM *N*-2-hydroxyethylpiperazine-*N*-2-ethane sulphonic acid (HEPES), are all satisfactory.

TECHNIQUE

(i) Blood, collected from infected cattle by jugular venipuncture when parasitaemia reaches 1–5%, is defibrinated by shaking with glass beads and used immediately; it is centrifuged and resuspended in 100 ml volumes of the buffered, serum-supplemented medium to give a packed cell volume (PCV) of 20–30% (see Appendix, pp. 224–225).
(ii) The pH is adjusted to 7.0 with sterile 0.31 N HCl and the cultures are incubated at 38 °C in spinner flasks with loose caps, in a gaseous environment of 5% CO_2 in air and stirred at 100 rev min^{-1}.
(iii) The medium is replaced daily: this involves transferring the blood to sterile centrifuge tubes and centrifuging (1000*g*, 10 min) at room temperature.
(iv) Subculture is necessary at least every 3 d, when cultures are again centrifuged (1000*g*, 10 min) and 50–80% of the erythrocytes are removed. The remaining erythrocytes are returned to the flasks and fresh medium, including serum and erythrocytes prepared from defibrinated blood, is added, keeping the PCV between 20% and 30%. Over a period of 32 d in suspension culture, during which 14 subcultures had been carried out, the numbers of

intraerythrocytic organisms increased by a factor of 5×10^5 (Erp *et al.*, 1980a).

Notes

(i) Growth is markedly affected by pH change: at pH 7.3 or 6.7 there is virtually no growth.

(ii) Heparinized plasma cannot be substituted for serum.

(iii) Erythrocytes stored for more than 6 d at 4°C are not suitable for subculture and serum must be stored at −20°C if it is to be kept more than 1 week.

(iv) Although not specifically stated, the preparations of infected and normal erythrocytes would benefit from washing in medium before being added to the cultures. The cultures support good growth of *B. bovis* but are inconvenient because of the large volumes involved (100 ml per culture) and because medium change and subculture require centrifugation.

2° Suspension culture of *B. canis*

Short term suspension cultures of *B. canis* have been described by Moreau and Soula (1979) and Molinar *et al.* (1982). Further information on the method of Molinar *et al.* (1982) is given by Levy and Ristic (1983). The technique used by both groups is based on that of Erp *et al.* (1978), but there are minor differences between them. The conditions described by Molinar *et al.* (1982) are given below and the differences described by Moreau and Soula (1979) are indicated.

MEDIUM

RPMI 1640 at pH 7.4 supplemented with 24 mM HEPES and 20% normal canine serum (Moreau and Soula (1979) used pH 7.2, 40 mM HEPES and 15% serum from dogs of blood group A −).

TECHNIQUE

(i) Blood, collected from infected splenectomized dogs when the parasitaemia reaches 3–4%, is defibrinated by shaking with glass beads, centrifuged (400*g*, 20 min, 4°C), the buffy coat is removed, and the erythrocytes are washed three times and finally resuspended in four volumes of the serum-supplemented medium.

(ii) Cultures in 12 ml volumes are incubated at 38°C in an atmosphere of 5% CO_2 in air with gentle stirring. The medium is replaced at 48 h.

Notes

(i) Moreau and Soula (1979) collected blood, at 0.5% parasitaemia, into heparin anticoagulant (1000 iu ml^{-1}). The PCV was adjusted to 8% and the cultures were incubated as 100 ml volumes in 200 ml "spinner flasks" with a gas mixture of 6.9% CO_2, 5.1% O_2 and 88% N_2. (No information was given concerning temperature or speed of agitation.)

(ii) Both groups reported increase in parasitaemia up to 24 h but a decline thereafter. The methods as reported might be improved by modification of the conditions, especially of the pH, as suggested by Erp *et al.* (1980a) for *B. bovis.*

3° Suspension culture of *B. divergens*

Pudney (1984) reported 30% parasitaemia of *B. divergens* in human erythrocytes, when maintained in 50 ml volumes in Erlenmeyer flasks on an orbital shaker. No detail was given.

B. MICROAEROPHILOUS STATIONARY PHASE CULTURES

The highly successful microaerophilous stationary phase (MASP) technique devised by Levy and Ristic (1980) for continuous *in vitro* culture of *B. bovis* has been applied, with modifications, to continuous culture of *B. divergens*, *B. major*, *B. bigemina*, *B. caballi*, and, as a short term system, to *B. canis* (procedures 2°–6° below). The principle of the technique was established for cultures initiated from parasitaemias of 0.5% and above. The parasites are allowed to grow in a settled layer of erythrocytes, above which the depth of medium is maintained at a level that does not allow for complete replenishment of O_2 in the settled layer as the gas becomes depleted by parasite metabolism; thus the cultures are "microaerophilous". The gas mixture is high in O_2 (5% CO_2 in air); the replication rate normally gives sufficient parasites to lower the O_2 tension within 4 d and cultures darken as the haemoglobin becomes deoxygenated. Cultures in which parasitaemia did not rise above 1% infected erythrocytes, and which failed to darken within this period, did not become established. The culture depth, using this gas mixture, is critical but, as long as the same surface to volume ratio is maintained, cultures can be grown on any scale from microtitre plates to large flasks.

Rodriguez *et al.* (1983) also found that *B. bovis* will multiply rapidly in cultures maintained at a high O_2 level (5% CO_2 in air), as long as the initial percentage of parasitized erythrocytes is 0.5% or above. However, under these conditions, when the initial percentage is below 0.1%, there

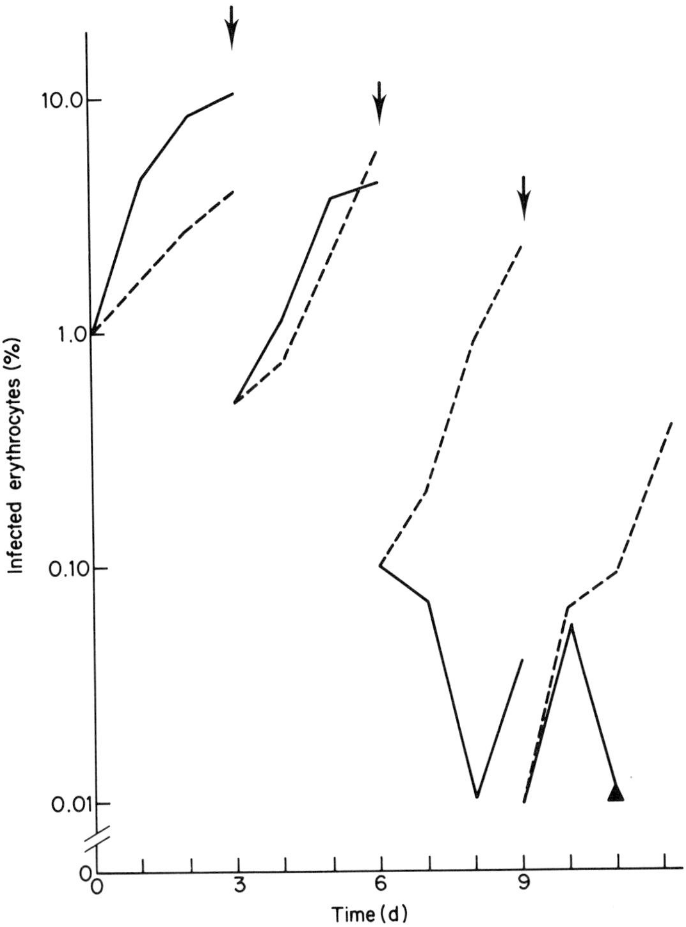

Fig. 1. MASP cultures of *B.-bovis* initiated with 1% parasitaemia at a 6.2 mm fluid level and incubated under 5% CO_2 in air (—) or 2% O_2, 5% CO_2, 93% N_2 (---). The cultures were diluted progressively on days 3, 6 and 9 (arrows) to 0.5%, 0.1% and 0.01%. ▲ = no visible parasites.

is a rapid decline in parasite numbers to extinction. In contrast, under low O_2 conditions (5% CO_2, 2% O_2, 93% N_2), cultures established with 1% parasitized erythrocytes grow slowly and those established at 0.1% grow rapidly (Fig. 1). The level of culture medium is less critical at low O_2 tensions where depths ranging from 2.0 to 6.2 mm can support growth, whereas with high O_2 conditions the culture depth is optimal at 6.2 mm and must not fall below 4.0 mm. Under low O_2 conditions the optimum depth for cultures started at 0.1% parasitaemia, or below, is 4.0–5.0 mm, and the optimum for cultures started at parasitaemias above 0.1% is 2.0–3.0 mm.

1° MASP culture of *B. bovis*

(a) INITIAL PARASITAEMIAS OF 0.1% AND ABOVE

The method is described fully by Levy and Ristic (1980, 1983) and Levy *et al.* (1981).

MEDIUM

M199 in Hanks's balanced salt solution (HBSS), supplemented with 40% adult bovine serum and 15 mM HEPES buffer; 100 mg ml^{-1} streptomycin and 100 iu ml^{-1} penicillin may be added to control bacterial growth.

TECHNIQUE

(i) Blood is collected aseptically from cattle when parasitaemias are between 0.1 and 2.0% and defibrinated by shaking with glass beads, centrifuged (conditions not stated) and the erythrocyte pellet is resuspended in the supplemented medium. A PCV of 9% (see Appendix, pp. 224–225) is usual, although 5–10% PCVs are said to be satisfactory.

(ii) The pH is adjusted to 7.0 with 1 N HC1 and the cultures are dispensed to achieve a depth of 6.2 mm in the culture vessel, i.e, 0.62 ml cm^{-2}.

(iii) Cultures are incubated at 37–38 °C in an atmosphere of 5% CO_2 in humidified air. The medium is sucked off with minimum disturbance of the erythrocyte layer and replaced every 24 h.

(iv) Subcultures are necessary every 48–72 h when, according to the percentage of parasitized erythrocytes, the old culture can be diluted 3–25 times with fresh bovine erythrocytes prepared in the same way as the original infected blood to a PCV of 9% in supplemented medium. Maximum growth rates are achieved when the percentage of infected erythrocytes is reduced to 0.5–1.0% at each subpassage.

Notes

(i) Parasitaemias of about 20% are achieved on the second day and 26% on the third day in the initial cultures, as well as in the passaged cultures, showing that a period of adaptation is not necessary. The maximum parasitaemias are of the order of 38% (Levy and Ristic, 1980).

(ii) Palmer *et al.* (1982) adapted the basic method by using M199 in Earle's balanced salt solution (EBSS), a 10% PCV and TES (2,2-hydroxy-1,1-*bis*(hydroxymethyl)ethylaminoethane sulphonic acid), instead of HEPES buffer. Normal bovine erythrocytes stored at 4°C in Puck's saline G (Gibco; see Puck *et al.*, 1958)[1] containing glucose at 20 g $litre^{-1}$, were used for the subpassages, but the storage time was not given. In this system 15% erythrocytes were parasitized after 6 d of culture.

(b) INITIAL PARASITAEMIAS BELOW 0.1%

The method uses the basic MASP technique, modified by Palmer *et al.* (1982), with adjustment of the culture medium depth and the gas mixture (Rodriguez *et al.*, 1983). Blood with parasitaemias of 0.1% or below is adjusted to a PCV of 9–10% (see Appendix, pp. 224–225) in supplemented medium and cultured in a settled layer of erythrocytes at a depth of 4.0 mm in an atmosphere of 5% CO_2, 2% O_2 and 93% N_2 at 37–38°C. Rodriguez *et al.* (1983) used this method to recover parasites from carrier animals in which no parasite could be seen in blood smears (Section III.C), and to establish clones by limiting dilution (Section III.B).

Notes

(i) Higher parasitaemias were achieved by Levy and Ristic (1980) in MASP cultures than had been achieved in suspension cultures (Erp *et al.*, 1980a). However, the percentage of parasitized erythrocytes was lower in the system used by Palmer *et al.* (1982).

(ii) The success of MASP cultures, whether in wells of microtitre plates or in large volume flasks, the ability of parasites in them to invade erythrocytes quickly so that few organisms are lost with change of medium, and the ease of medium change make MASP culture the system of choice.

(iii) Medium 199 is the only medium reported for use in these cultures but it is likely that others would be equally successful.

(iv) The use of stored erythrocytes (Palmer *et al.*, 1982) is convenient when fresh bovine erythrocytes are not available.

[1]Puck's saline G (g $litre^{-1}$): NaCl, 8.0; KCl, 0.4; $MgSO_4 \cdot 7H_2O$, 0.154; $CaCl_2 \cdot 2H_2O$, 0.016; $Na_2HPO_4 \cdot 7H_20$, 0.29; KH_2PO_4, 0.15; glucose, 1.10; phenol red, 0.0012.

2° MASP culture of *B. divergens*

As well as being a parasite of cattle, *B. divergens* infects the Mongolian gerbil, *Meriones unguiculatus*. It is therefore possible to avoid altogether the use of cattle in setting up MASP cultures of this species, once stabilates have been frozen in liquid N_2. It is difficult to establish cultures directly from frozen stabilates, but stabilates can be used to infect gerbils, from which cultures can be initiated, with immediate transfer to bovine blood (C.M. Winger and E.U. Canning, unpublished results).

The first MASP cultures of *B. divergens* were established using a candle jar to control the gas conditions (Väyrynen and Tuomi, 1982; Konrad *et al.*, 1984). Cultures can equally well be established from bovine or gerbil blood using a gas mixture of 5% CO_2, 2% O_2 and 93% N_2 (Pudney, 1984).

MEDIA

M199 or RPMI 1640 in HBSS, supplemented with 40% v/v adult bovine serum from suitable donors, and 15 mM HEPES or TES.

TECHNIQUE

(i) Stabilates are thawed for 3 min at 38 °C and immediately inoculated intraperitoneally (i.p.) into gerbils. Parasitaemias above 50% should be reached in 4 d but, if they are not, passage i.p. of 2×10^8 parasitized erythrocytes into another gerbil will give the required parasitaemia in 3 d.

(ii) Blood is withdrawn by heart puncture from an anaesthetized gerbil, yielding about 2.0 ml. The blood is defibrinated by means of glass beads in a sterile, glass, 7.0 ml universal container (bijou bottle), then washed three times by centrifugation (200*g*, 10 min, 4 °C) and resuspension in unsupplemented medium to remove complement, which would otherwise lyse the bovine cells in culture. The blood is resuspended to its original volume after removal of the buffy coat; the PCV need not be adjusted because gerbil cells are lysed in the presence of bovine serum.

(iii) Blood from a suitable adult, healthy, bovine donor (see Note (ii)) is defibrinated, washed in unsupplemented medium and resuspended in medium to its original volume after removal of the buffy coat. It is adjusted to a PCV of 18% (see Appendix, pp. 224–225) by dilution with supplemented medium and the pH brought to 7.2 with 1 N HCl.

(iv) The diluted, infected gerbil blood and uninfected bovine blood are mixed in a ratio of 1 : 1, thus bringing the PCV to 9%, when all the gerbil cells are lysed within 24 h. The mixture is dispensed into the wells of 24-well Costar plates to a depth of 0.62 cm.

(v) The plates are incubated at 38 °C in a candle jar (Trager and Jensen, 1976). Owing to lysis of the gerbil cells, developing intra-erythrocytic merozoites are lost but, with high initial parasitaemias, there are enough mature merozoites to invade bovine cells and initiate cultures.

(vi) After 24 h, the medium is changed by removal and replacement of about 0.8 ml of the fluid layer, with minimum disturbance of the settled layer. Thereafter the medium is changed twice daily. When the percentage of parasitized erythrocytes reaches 1%, subculture is carried out at a 1 : 1 ratio of cultured to fresh bovine erythrocytes.

(vii) Subculture in the candle jar system is repeated once, then the cultures can be scaled up to 25 cm^2 flasks and transferred, with loosened caps, to an incubator at 38 °C in an atmosphere of 5% CO_2 in air. Parasitaemias of 10–12% are regularly achieved.

(viii) Further subcultures are necessary every 3–4 d. When parasitaemias reach 10–12%, cultures are diluted 1 : 9 with freshly prepared erythrocytes; if the percentage of parasitized erythrocytes is between 4 and 7%, subcultures should be made at a ratio of 1 : 4.

(ix) When cultures are growing well with 40% serum (10–12% infected cells), the serum requirement can be lowered (see Note (ii)), in 5% steps. First the serum is reduced from 40% to 35%; the cultures take about 1 week (two subcultures) to recover to the 10–12% infection level. Subsequent reductions are from 35% to 30%, and 30% to 25%; the cultures require 2 or more weeks to recover at each step. Cultures do not grow well in bovine cells with less than 25% serum and, when serum levels below 40% are used, fresh serum is essential.

Notes

(i) Pudney (1984) cultured a human strain ('J') of *B. divergens* in bovine cells and in human cells using RPMI 1640 medium with 20% bovine serum or 10% human serum respectively. The gas mixture was 5% CO_2, 2% O_2 and 93% N_2; the low O_2 content was perhaps more suitable than 5% CO_2 in air and may have influenced the serum requirement.

(ii) Serum from cattle varies considerably in its ability to support culture growth. If individual bovine donors are used, they must be carefully selected by preliminary trial; if commercially prepared pooled sera are used, batches must be tested for suitability.

3° MASP culture of *B. major*

B. major may be established from infected bovine erythrocytes using the method for *B. divergens* (Donelly *et al.*, 1984). The rate of multiplication is slightly higher than that of *B. divergens*, but the peak parasitaemias are similar.

4° MASP culture of *B. bigemina*

Difficulties have been encountered in attempts to culture *B. bigemina* by the MASP method which supports growth of *B. bovis*. However, Vega *et al.* (1985a) established continuous MASP cultures of *B. bigemina* by adopting the procedure specified by Rodriguez *et al.* (1983) for initiating cultures from low parasitaemias. Parasitaemias rarely exceeded 3% under these conditions. However, after three subcultures, when the parasitaemia had reached 1%, it was possible to change to the conventional MASP system, using the depth of the culture to control the O_2 supply to the erythrocytes. Cultures then grew to 10% parasitaemia (Vega *et al.*, 1985b).

Additionally, because of fragility of the erythrocytes, great care had to be taken in preparing the blood from infected cattle. Blood was withdrawn into heparinized tubes instead of being defibrinated, and was then washed in a special saline solution (VYM), necessary for the stability of the erythrocytes.

MEDIA

M199 with EBSS and 2.2 g litre^{-1} $NaHCO_3$ or M199 with HBSS and 1.4 g litre^{-1} $NaHCO_3$, supplemented with fresh adult bovine serum at 50% v/v for initiation of cultures and 35–40% v/v when cultures are established.

Two other media, minimal essential medium (MEM-EBSS) plus 2.2 g litre^{-1} $NaHCO_3$ and RPMI 1640 plus 0.35 g litre^{-1} $NaHCO_3$, also supported continuous growth but at a slower rate.

WASHING SOLUTION

VEGA Y MARTINEZ (VYM) WASHING SOLUTION FOR INFECTED AND NORMAL ERYTHROCYTES

The following chemicals are made up in 1 litre of double-distilled deionized water, containing 0.25 mM adenine and 0.5 mM guanosine.

	litre^{-1}
$CaCl_2.2H_2O$	16.0 mg
KCl	400.0 mg
KH_2PO_4	1415.4 mg
$MgSO_4.7H_2O$	154.0 mg
$Na_2HPO_4.7H_2O$	1450.0 mg
NaCl	7077.0 mg
Dextrose	20.5 g

TECHNIQUE

(i) To initiate cultures, blood (Vega *et al.* (1985a) used 35 ml) at low parasitaemia (e.g. 0.2%) is collected from an infected bovine donor into sterile tubes containing 5–10 iu ml^{-1} heparin. The blood is washed three times in VYM solution by centrifugation (1000*g*, 15 min, 4 °C) and resuspension, with removal of the buffy coat and the top third of the erythrocyte layer at each wash.

(ii) Normal blood from an adult bovine donor is defibrinated over glass beads and washed twice in VYM with removal of the buffy coat cells as before. The red cells are resuspended in VYM to a PCV of 30–50% and can be stored at 4 °C for 3–4 d. Serum from this blood can be frozen and used to supplement the culture medium.

(iii) The infected and uninfected red cell suspensions, adjusted to 5% PCV by addition of medium supplemented 1 : 1 with serum (see Appendix, pp. 224–225), are mixed in a ratio of 1 part infected cells to 3 parts of normal cells and the pH is adjusted to 6.8–6.9 with 1 N HCl.

(iv) The final red cell suspension, with a parasitaemia of 0.05%, is distributed into the wells of 24-well Costar plates at a depth of 4.0 mm (805 μl per well) and the plate is incubated at 37 °C in an atmosphere of 5% CO_2, 2% O_2 and 93% N_2.

(v) Medium change is carried out at 24 h intervals by removal of 600 μl from each well and replacement with 600 μl of the 50% serum-supplemented medium; at the 48 h and 168 h changes, however, the replacement medium should contain normal red cells at 2.5% PCV.

(vi) Cultures are subcultured at 4 d and again at 9 d by diluting 1 : 2 with serum-supplemented medium containing normal red blood cells at 7.5% PCV. When the parasitaemia reaches about 1%, further subcultures are performed at about 1 : 10 dilution with supplemented medium containing normal red blood cells at 10% PCV. At this stage, cultures can be scaled up to flasks and the serum content of the medium lowered to about 35% or 40%.

(vii) The blood, infected with the culture-adapted parasites, is adjusted to a PCV of 10% in medium supplemented with 35–40% serum at pH 7.0, added to flasks to a depth of 7.0 mm and incubated at 37 °C in an atmosphere of 5% CO_2 in air. The culture medium is replaced every 24 h and subcultures are performed every 48 h by mixing cultured cells and normal cells, at 10% PCV in serum-supplemented medium, in a 1 : 1 ratio.

5° MASP culture of *B. caballi*

B. caballi has recently been established in culture (Irene Müller and L.P. Phipps, personal communication). Maximum parasitaemias are as yet only 2.7%, indicating that the culture conditions are not ideal, or that the parasite is not fully adapted to *in vitro* growth, or both. However, continuous culture has been achieved for several weeks.

MEDIUM

RPMI 1640 with 1.5 g $litre^{-1}$ $NaHCO_3$ and 0.584 g $litre^{-1}$ L-glutamine, buffered with 20 mM HEPES. Penicillin 150 iu ml^{-1} and streptomycin 150 $\mu g\ ml^{-1}$ are added. The medium is supplemented with 40% v/v pooled horse serum and the pH adjusted to 7.0–7.2 with HCl.

TECHNIQUE

Blood from a donor pony with 1.5% parasitaemia is defibrinated with glass beads and washed twice with medium by centrifugation (details not given), with removal of the buffy coat. The red cells are resuspended to a PCV of 10% (see Appendix, pp. 224–225) in supplemented medium and dispensed into 24-well Costar plates to a depth of 6.2 mm. The cultures are incubated in a candle jar environment (Trager and Jensen, 1976) at 38 °C, medium is changed every 24 h and subcultures are made weekly. At subculture, the cultures are washed in cold supplemented medium, centrifuged (details not given) and diluted with fresh supplemented medium to a parasitaemia of 0.2–0.5%.

6° MASP culture of *B. canis*

The only published account of *B. canis* in stationary phase culture is that of Moreau and Soula (1979); preparation of blood and serum-supplemented medium are the same as for suspension cultures (Section II.A.2°). The cultures are dispensed in 4 ml aliquots into 50 mm diameter Petri dishes and incubated in a desiccator. As with suspension cultures, multiplication is maintained for 24 h only.

Improvements on this performance have been reported by Moreau and Laurent (1984), who showed that cultures can be maintained over several subcultures at 3 d intervals, with medium change every 24 h, but details of the changed conditions were not reported.

C. LOW TEMPERATURE AS AN AID TO CONTINUOUS CULTURE

The ability of *Babesia* spp. to survive at low temperature (about 4°C) for several weeks has been known since Lignières (1900) reported that cattle blood infected with *Babesia argentina* (= *B. bovis*), transported by ship in cold storage from Argentina to France, was infective to cattle on arrival. Other examples of retention of infectivity after long-term storage at low temperatures were given by Nocard and Motas (1902) for *B. canis* and by Kossel *et al.* (1904) for bovine *Babesia*.

Infected blood cells in culture medium, stored at 4°C, also retain infectivity for several weeks and normal growth resumes when the cultures are returned to body temperature. The technique has been used for suspension cultures (Erp *et al.*, 1980a) and MASP cultures (Levy and Ristic, 1983) of *B. bovis* and MASP cultures of *B. divergens* and *B. major* (Konrad *et al.*, 1985), and could probably be applied to other species. The parasitaemia falls off rapidly after the first few days at low temperature, but the decline can be slowed down by weekly changes of medium. Resumption of active growth can be accelerated by subculture when the temperature is raised (Konrad *et al.*, 1985). Erythrocytes remain in the bright red, oxygenated condition throughout the cooled period in contrast to the dark red, deoxygenated condition seen in cultures at body temperature.

Low-temperature incubation of cultures virtually eliminates the need for infection of animals once continuous cultures have been established, as the cultures can be put into "slow growth" between experiments. Also, it is a convenient method of simplifying culture maintenance, by reducing the frequency of subculture to once weekly. The weekly schedule of *B. divergens* employed by C.M. Winger and E.U. Canning (unpublished results) is given below.

Wednesday pm:	subculture to 1% parasitaemia and incubate at 38°C in an atmosphere of 5% CO_2 in air.
Thursday pm:	change medium and transfer to refrigerator at 4°C for the weekend; parasitaemia now about 2%.
Monday am:	transfer to incubator at 38°C and 5% CO_2 in air.
Monday pm:	change medium.
Tuesday am:	change medium.
Tuesday pm:	change medium.
Wednesday am:	change medium; parasitaemia now 10–12%.

D. ISOLATION OF MEROZOITES AND INITIATION OF CULTURES FROM MEROZOITES

In actively growing MASP cultures, *Babesia* merozoites, released from erythrocytes, rapidly reinvade cells and are rarely seen free in the medium. However, if the cultures are deprived of CO_2 by transference to air at 37–38°C, intraerythrocytic stages mature into merozoites which leave the cells but are unable to reinvade and thus accumulate in the medium. For *B. bovis* and *B. bigemina*, it is necessary only to transfer the cultures to air for 4–6 h at 37°C (Levy and Ristic, 1980; Palmer *et al.*, 1982; Vega *et al.*, 1985b). *B. divergens* merozoites require more severe CO_2 deprivation, achieved by transferring them for 6–8 h to a desiccator at 38°C over soda lime to absorb the CO_2 (J. Konrad, personal communication).

Merozoites must be maintained at 4°C throughout the procedure of isolation from the culture supernatant to prevent spontaneous agglutination. The cultures are centrifuged at low speed to remove erythrocytes and the supernatant is recentrifuged at high speed to sediment the merozoites. The following method for *B. bovis* was given by Levy and Ristic (1980).

(i) Cultures, deprived of CO_2 for 4–6 h, are decanted into 50 ml conical centrifuge tubes and centrifuged (400*g*, 5 min).

(ii) The top 30 ml containing the merozoites are removed and centrifuged (750*g*, 10 min).

(iii) The supernatant is again removed and centrifuged at 750*g* for 10 min.

(iv) The final supernatant is centrifuged (1000*g*, 20 min) to produce a pellet of merozoites, contaminated with some erythrocyte stroma.

For *B. divergens* two or three centrifugations are used, with times and speeds as follows: either 300*g* for 15 min followed by 2000*g* for 10 min (L.P. Phipps, M.R. Rankin and J. Donelly, personal communication), or 200*g* for 10 min followed by 400*g* for 10 min and, finally, 1000*g* for 20 min (C.M. Winger and E.U. Canning, unpublished results).

For *B. bigemina* only two centrifugations are used: 200*g* for 10 min and 3000*g* for 10 min (Vega *et al.*, 1985b).

Merozoites can be further purified on a Percoll gradient of 1.02 g ml^{-1} specific density for *B. divergens* (X.-S. Jiang, C.M. Winger and E.U. Canning, unpublished results) or 1.087 g ml^{-1} for *B. bovis* (Rodriguez *et al.*, 1986). Isolated merozoites can be resuspended in medium and added to cultures with adjustment to maintain the correct serum content and depth of medium (see Section II.B).

E. INITIATION OF CULTURES FROM FROZEN STABILATES

Whereas frozen erythrocytes infected with *Babesia* spp. can readily infect new hosts (Dalgleish, 1972), the viability after retrieval from the frozen state is so diminished that it is difficult to use stabilates to establish cultures directly. However, cultures of *B. bovis* and *B. bigemina* have been initiated directly from stabilates, the latter into cultures gassed with mixtures low in O_2, in accordance with the principle established for cultures initiated from low parasitaemias (Section II.B.1°(b)). Cultures of these species and of *B. divergens* have also been initiated from frozen merozoites. Retrieval of infected erythrocytes for culture initiation has been achieved only when polyvinylpyrrolidone 40 has been used as cryoprotectant. Dimethyl suphoxide and glycerine have also been used successfully, but less effectively, with merozoites.

1° Preparation and retrieval of stabilates

MEDIA

The following three media are all satisfactory.

Medium 199 with Earle's balanced salt solution containing 40% v/v adult bovine serum and buffered with 26 mM TES (M199-EBSS-TES).

M199 EBSS with 2.2 g $litre^{-1}$ $NaHCO_3$ and 40% v/v serum (M199-EBSS-$NaHCO_3$).

M199 with Hanks's balanced salt solution containing 15 mM HEPES and 40% v/v serum (M199-HBSS-HEPES).

CRYOPROTECTANTS

Polyvinylpyrrolidone 40 (PVP), 20% or 10% w/v in Puck's saline G containing 20 g litre^{-1} added glucose (PSG +); or PVP, 20% w/v in VYM solution (Section II.B.4°); or dimethyl sulphoxide (DMSO), 10% or 20% v/v, in PSG + .

TECHNIQUE FOR *B. BOVIS*

(Palmer *et al.*, 1982)

(i) To obtain infected erythrocytes, cultures with 15% parasitaemia, prepared as described in Section II.B.1°(a), Note (ii), are centrifuged (200*g*, 10 min, 4°C). To obtain merozoites, cultures are deprived of CO_2 for 4 h (Section II.D) and then centrifuged (as before). The supernatant, containing merozoites, is recentrifuged (3000*g*, 10 min, 4°C) and the pelleted merozoites resuspended in PSG + (final concentration not stated).

(ii) Volumes of the packed erythrocytes or diluted merozoites are mixed 1 : 1 with 20% PVP in PSG + to achieve a final PVP concentration of 10%.

(iii) The mixture is sealed into sterile ampoules in 1.0 ml aliquots and frozen at a cooling rate of about 20°C min^{-1} to − 70°C (see Section 2° below on controlled cooling). Samples are left overnight at − 70°C and then transferred to liquid N_2.

(iv) Retrieval is by rapid thawing with agitation in a water bath at 37°C. Partly thawed samples are washed in 50 ml PSG + at 4°C, centrifuged (3000*g*, 10 min, presumably at 4°C) and 0.5 ml aliquots are resuspended in 1.0 ml aliquots of fresh M199-EBSS-TES containing 40% adult bovine serum and 10% fresh bovine erythrocytes. The suspension is dispensed in 0.2 ml aliquots into the wells of 96-well microtitre plates, maintaining a depth of 6.2 mm, and incubated at 37°C in an atmosphere of 5% CO_2 in air.

TECHNIQUE FOR *B. BIGEMINA*

(Vega *et al.*, 1985b)

(i) Packed erythrocytes obtained by centrifugation of cultures with 6–10% parasitaemia at 200*g* for 10 min at 4°C, or packed merozoites obtained by CO_2 deprivation and centrifugation of the supernatant at 1000*g* for 10 min at 4°C, are mixed 1 : 1 with 20% PVP in VYM to achieve a final concentration of 10% PVP.

(ii) Sterile ampoules containing 1.0 or 2.0 ml aliquots are frozen at a cooling rate of − 20°C min^{-1} to − 80°C (see Section 2° below on controlled cooling), then transferred to liquid N_2.

(iii) For retrieval, the sample is thawed rapidly at 37°C, washed in 50 ml VYM, centrifuged (1000*g*, 10 min, 4°C) and added to M199-EBSS containing 2.2 g litre^{-1} $NaHCO_3$, 40% serum and 10% fresh erythrocytes. The culture depth is 5.0 mm and the atmosphere is 5% CO_2, 2–5% O_2 and 93–90% N_2.

TECHNIQUE FOR *B. DIVERGENS*

(L.P. Phipps, M.R. Rankin and J. Donelly, personal communication)

(i) Merozoites obtained by severe CO_2 deprivation are harvested from the culture supernatant (Section II.D) and resuspended in cryoprotectant consisting of 10% or 20% PVP or DMSO in PSG+.

(ii) 0.5 ml aliquots, sealed in sterile ampoules, are frozen at a cooling rate of 15–20°C min^{-1} to −70°C (see Section 2° below on controlled cooling), then transferred to liquid N_2.

(iii) Retrieval is by rapid thawing in a water bath at 40–45°C. The contents of the ampoule are washed in 20 ml of M199 (pH 7.2) at 4°C, then centrifuged (2000*g*, 10 min, 4°C) and resuspended in 0.5 ml of M199 at pH 7.2. The suspension is added to uninfected cultures (Section II.B.2°) in 24-well Costar plates which have been pregassed in a candle jar (Trager and Jensen, 1976) for at least 2 h. The serum content is adjusted to 40%, and the total volume per well is 1.25 ml.

2° Controlled cooling

Controlled cooling is best done in a programmed cell freezer but, if this is not available, the samples can be immersed in an ethanol–dry ice (solid CO_2) bath (−76°C). L.R. Phipps *et al.* (personal communication) inserted the ampoules into Pasteur pipette teats to insulate them before immersion in the ethanol bath. C.M. Winger and E.U. Canning (unpublished observations) used a −80° freezer, with ampoules packed in a box of expanded polystyrene; after cooling overnight, the ampoules were transferred to liquid N_2.

F. *BABESIA* IN TICK CELL CULTURES

Tick cell culture has not yet been developed to provide a satisfactory environment for the multiplication and differentiation of *Babesia*. Attempts have been made to infect tick cells *in vitro* by adding erythrocytic stages of *B. bigemina* to organ cultures of *Boophilus annulatus* (Hoffman, 1972), and of *B. bovis* to primary cultures or subcultures, or both, of embryonic cells of *Boophilus microplus* (Bhat *et al.*, 1979; Droleskey *et al.*, 1981). Droleskey *et al.* (1981) saw no multiplication of phagocytosed parasites but Bhat *et al.* (1979) obtained a 20-fold increase at 48 h, from the number of parasites at 8 h (when residual free parasites in the inoculum had been removed). Multinucleate stages, seen intracellularly at 16 h, increased in number and size up to 48 h but declined thereafter. Ultrastructurally they resembled blood forms more than tick forms.

Inoculation of tick cell cultures with kinetes has also been only partially successful. Kurtti *et al.* (1983) established primary cell cultures and cell lines from *Anacentor nitans* infected with *B. caballi*, and also added kinetes from the haemolymph of *A. nitans* to cell lines of *Rhipicephalus appendiculatus* and *A. nitans*. Kinetes were seen in the primary cultures but did not persist after 7 d and were not present in subcultures. When kinetes were added to the cell lines, they were seen to attach, enter and leave cells. Intracellular kinetes were often situated close to the host cell nucleus. Again, parasites persisted for up to 5 d but none was detected after 1 week. There was a greater association of kinetes (90% were either attached to cells or were intracellular) when they were added to the vector's(*A. nitans*) cells than when added to *R. appendiculatus* cells, but association was improved when an extract of embryonated eggs of *A. nitans* was added to the medium for *R. appendiculatus* cells.

1° Culture of *Babesia* in tick cell cultures
(Culture of tick cells by Pudney *et al.*, 1973)

MEDIA

(i) 0.5% lactalbumin hydrolysate in HBSS, pH 6.9, supplemented with 10% v/v inactivated foetal calf serum (FCS), 1000 iu ml^{-1} penicillin and 1.0 mg ml^{-1} streptomycin.
(ii) Liebovitz L-15, pH 7.0, supplemented with 15% v/v tryptose phosphate broth, 15% v/v inactivated FCS, and the same concentrations of antibiotics.

TECHNIQUE

(i) *B. microplus*, 1–2 d after engorgement, are surface sterilized by shaking in 1% benzalkonium chloride (Roccal® solution) and soaking in fresh Roccal® for 15–20 min. The ticks are rinsed twice in sterile distilled water, dried on sterile filter paper and incubated in Petri dishes in a moist chamber at 28°C. Oviposition starts 2 d later.

(ii) Halfway through embryonic development (12–13 d) the eggs are crushed in a small quantity of medium (i) with just enough pressure to crack the eggshells.

(iii) The tissue and eggshell suspension is filtered to remove the shells (100–120 μm pore size; two layers of bolting cloth will do).

(iv) The suspension is centrifuged (200*g*, 8 min), washed in PBS at pH 7.2 and containing 1000 iu ml^{-1} penicillin and 1.0 mg ml^{-1} streptomycin, and recentrifuged.

(v) The pellet is resuspended in medium (i); 4.0 ml aliquots are placed in 25 cm^2 plastic flasks and incubated at 28 °C. The ratio of medium to tissue can be varied: a heavy suspension consisting of the embryonic tissue from the eggs of 10 female ticks in 4.0 ml gives a good start. The flasks should be filled with medium one week before use, to condition them. After one week the tissue clumps should be adherent with large granular fibroblast-like cells growing out from them.

(vi) During the second week, and weekly thereafter, the medium is changed.

(vii) At 4 weeks there should be a surface monolayer of epithelial-like cells with relatively few fibroblasts and the cultures can be subcultured. The cells are removed with a rubber "policeman" and suspended in sufficient medium to be dispensed into two or more flasks. After 3–4 subcultures, the cultures can be grown in medium (ii).

(viii) To infect tick cell cultures with intraerythrocytic *Babesia*, infected erythrocytes are collected (with anticoagulant) from splenectomized calves. Erythrocytes infected with *B. bovis* are more resistant to lysis than are uninfected cells, so infected cells can be concentrated by differential lysis (Mahoney, 1967). Suspensions containing 98–100% infected intact erythrocytes are centrifuged, resuspended in L 15 medium, and added to the cultures (Bhat *et al.*, 1979).

(ix) Primary cultures of tick cells infected with kinetes can be established from eggs of *Babesia*-infected ticks. Alternatively, drops of haemolymph, obtained by severing a leg from an infected tick, can be used as a source of kinetes to be added to cell lines.

III. ADVANTAGES OF CULTURES IN BASIC STUDIES ON *BABESIA*

A. REDUCTION IN ANIMAL EXPERIMENTATION

Continuous *in vitro* culture of bovine babesias and the ability to establish cultures directly from frozen stabilates have greatly reduced the need for infection of animals in experimental work. Unfortunately, the economically important *B. canis* can, as yet, be maintained *in vitro* for a few subcultures only and new cultures have to be set up regularly from infected dogs. Determination of the conditions for continuous culture of this species is a laudable target.

Continuous culture over long periods may cause attenuation of strains, which may have potential as vaccines before the era of non-living vaccines begins. Signs of attenuation have been noticed in the Weybridge strain of *B. divergens* grown in continuous culture for 18 months (C.M. Winger and E.U. Canning, unpublished results).

Preliminary experiments in many areas of research can be conducted *in vitro* before trials are conducted in animals. It is particularly useful that gerbils are susceptible to *B. divergens*, since this host can be used for intermediate trials before the expensive use of cattle is contemplated. It must always be kept in mind that *in vitro* conditions do not faithfully reflect those *in vivo*, and failure *in vitro* does not necessarily mean failure *in vivo*.

B. CLONING

The technique evolved by Rodriguez *et al.* (1983) for establishing cultures from parasitaemias below 0.1% (Section II.B.1°(b)) was used by them to derive clones of *B. bovis*.

The number of infected erythrocytes per μl of culture is determined by estimation of the percentage parasitaemia in stained smears and estimation of the total number of erythrocytes by haemocytometer counts. 20 μl of culture are dispensed into wells of 96-well microtitre plates and 180 μl of medium added (200 μl). Serial 10-fold dilutions are made, keeping to the 200 μl volume, until the final dilution is calculated to contain one infected erythrocyte. The contents of the final dilution wells are split between four wells (50 μl each) to make a cloning set. 80 μl of 2.5% erythrocyte suspension in supplemented medium (Section II.B.1°(a)) are added to each well to bring the level to 4 mm, and plates are incubated in 5% CO_2, 2% O_2 and 93% N_2 at 37°C. Medium is changed at 24 h and 48 h, and at 72 h the medium is replaced with 2.5% normal erythrocytes in supplemented medium, to bring the PCV to 5%.

Screening of the cloning wells is performed when dilution wells that theoretically started with 10 infected cells show good multiplication. Cloning wells showing growth are transferred to separate plates and samples are cryopreserved.*

Rodriguez *et al.* (1983) were able to identify fast, intermediate and slow growing clones that, when recloned, retained their growth characteristics. The method clearly has potential in isolating non-virulent strains that can be used as live vaccines.

C. DETERMINATION OF THE CARRIER STATE OF ANIMALS RECOVERED FROM BABESIOSIS

The technique of Rodriguez *et al.* (1983) for establishing cultures from parasitaemias below 0.1% (Section II.B.1°(b)) was used to isolate *B. bovis* from carrier animals. Infected cultures were initiated from five cows that had been recorded as uninfected from the examination of blood films.

D. PREPARATION OF NON-LIVING VACCINES

Work on *B. bovis* has revealed that the major antigenic determinants associated with the induction of protective immunity are localized in the surface coat of the merozoites (Ristic *et al.*, 1981; Ristic and Levy, 1981; James *et al.*, 1981). When merozoites invade fresh erythrocytes, the surface coat is stripped off and remains as soluble glycoprotein antigen in the culture supernatant. Parasitaemias are normally much higher *in vitro* than *in vivo* and free merozoites are not removed from cultures, as they would be by reticulo-endothelial cell activity *in vivo*. Thus higher concentrations of soluble antigens and free merozoites can be obtained from cultures.

Culture supernatants have been tested as non-living vaccines in cattle and have given moderate protection from severe reactions and death after challenge with homologous strains of *B. bovis*. In these experiments, Smith *et al.* (1981) and Kuttler *et al.* (1982) used lyophilized culture supernatant, while Timms *et al.* (1983) used filtered culture supernatant which had been stored at − 20°C. The inoculum was the reconstituted or thawed supernatant, mixed with saponin as adjuvant. The animals were given two injections and were challenged either by exposure to ticks infected with *B. bovis* or by inoculation of infected erythrocytes. A

*Cultures isolated by dilution techniques described here cannot be regarded as definitive clones, since the presence in the cloning wells of only a single parasite is a statistical probability but not a certainty [Eds].

Venezuelan strain gave not only solid homologous protection but also a considerable degree of protection against challenge with heterologous strains (Montenegro-James *et al.*, 1985). Lyophilized vaccines prepared from supernatants from MASP cultures of *B. canis* also gave good protection; in laboratory and field trials they proved to be safe and potent immunogens (Moreau and Laurent, 1984). In contrast, 10 times concentrated culture supernatant from *B. divergens* cultures gave only partial protection in gerbils against homologous challenge, even after three inoculations given with Freund's complete adjuvant (Winger *et al.*, 1987).

Smith and Ristic (1981) showed that, with conventional techniques, it is virtually impossible to prepare supernatant antigens or merozoites entirely free of contaminating erythrocyte stroma from cultures and emphasized that the safety of these non-living vaccines is dependent on freedom from these host proteins. Monoclonal antibody technology provided the means of isolating and characterizing pure babesial antigens. In addition, advances in recombinant DNA technology mean that, once an immunogenic antigen is isolated, there is the possibility of cloning the gene that codes for it and scaling up production of the antigen for commercial development as a vaccine. The potential of these techniques has barely been tapped. To raise monoclonal antibodies, Wright *et al.* (1983) used a lysate of cattle-derived infected erythrocytes after partial purification by affinity chromatography. The degree of contamination with host products was indicated by the facts that 1000 of the hybrids were specific for host antigens and most of the remaining 87 hybrids reacted with both parasite and host components. One monoclonal antibody of class IgM with kappa light chains, which recognized a protective antigen, was used to characterize this antigen as a 29 kDa protein which apparently co-migrated with a haemoglobin subunit (Wright *et al.*, 1985). Contamination of the inoculum used to raise the monoclonal antibodies can be partly overcome by isolating erythrocyte-free merozoites from cultures. Thus, Winger *et al.* (1986) raised monoclonal antibodies to *B. divergens* merozoites isolated from MASP cultures after severe CO_2 deprivation (Section II.D) and used one of these antibodies, of class IgG_1, which strongly inhibited merozoite invasion into erythrocytes, to isolate a 50–60 kDa protein. Gerbils vaccinated with this protein were protected against challenge with the homologous strain.

E. DRUG SENSITIVITY AND BIOCHEMICAL STUDIES

The darkening of MASP cultures which occurs when growth of *Babesia* is established (Section II.B) can be used as a simple index of drug activity and correlates well with parasitaemia (Levy and Ristic, 1980). Drug and correlates well with parasitaemia (Levy and Ristic, 1980). Drug activity can be monitored using serial dilutions of the drug, with replicates, in the wells of microtitre plates. Levy and Ristic (1980) used the system to titrate the action of Berenil on *B. bovis*, because a lethal concentration (3.5×10^{-6} g ml^{-1}) prevents cultures darkening, and Akinboade (1984) similarly investigated the activity of imidocarb on *B. bovis* and determined a lethal concentration of 1.2×10^{-6} mg ml^{-1} by monitoring parasitaemia.

An *in vitro* test using the incorporation of ^{3}H-hypoxanthine as a marker of drug activity was developed by Irvin and Young (1977) using short-term cultures of four species of *Babesia*. Uptake of the tritiated precursor, monitored by liquid scintillation, stopped when the parasites were no longer viable. This test has greater potential as a screen for new drugs now that MASP cultures have been developed.

Biochemical studies on *Babesia* are facilitated in culture. Irvin *et al.* (1978) and Irvin and Young (1979) investigated the uptake of purine and pyrimidine precursors of nucleic acids and their metabolism by *Babesia* spp. and Barry (1982, 1984) used *Babesia*-infected erythrocytes, suspended in basal salts solution or Krebs Ringer phosphate solution, to measure uptake and metabolism of glucose and production of excess free amino acids. The composition of the excess amino acids was in concordance with the composition of host haemoglobin, suggesting that the amino acids were derived from the erythrocyte protein.

F. INVASION OF ERYTHROCYTES BY MEROZOITES

Investigation of the receptors on the surfaces of host erythrocytes and *Babesia* merozoites, which mediate attachment to and invasion of erythrocytes, can conveniently be carried out *in vitro*. Candidate receptors, sugars or lectins, can be added at the time of subculture and their effects can be monitored by parasitaemia counts. Additionally, free merozoites or unparasitized erythrocytes can be pre-incubated with the sugars or lectins, washed and added to the system. This is an area deserving attention, because little is known about the mediators of invasion into erythrocytes by merozoites of *Babesia*. With malaria parasites, however, the three sugars *N*-acetylglucosamine, *N*-acetylgalactosamine and *N*-acetylneuraminic acid have been implicated (Jungery, 1984). The importance of complement in invasion of erythrocytes by *B. rodhaini* had been proposed by Chapman and Ward (1979) and by Jack and Ward (1980a, b). However, heat treatment of bovine serum to inactivate complement before addition to *B. bovis* cultures reduced but did not prevent multiplication of the parasites, demonstrating that invasion of *B. bovis* is not complement mediated (Levy and Ristic, 1983).

G. IMMUNOLOGICAL STUDIES

Immunological studies *in vitro* are restricted by the limited interaction of the various components of the immune system. However, some information can be obtained by examining the components in isolation. Bautista and Kreier (1979) found that immune serum from hamsters inhibited growth of *B. microti* in short-term cultures. When erythrocytes parasitized by *B. microti* were added in culture medium to monolayers of hamster macrophages, there was little inhibition of multiplication when normal macrophages alone were used, but there was increasingly greater inhibition (a) with normal macrophages in the presence of immune serum, (b) with macrophages from immune animals in the absence of serum and (c) with macrophages from immune animals in the presence of immune serum (Bautista and Kreier, 1980).

H. MEROZOITE NEUTRALIZATION ASSAY

A merozoite neutralization assay was developed by Ristic and Levy (1981) to test the effects of antibodies on parasite growth. Uninfected bovine erythrocytes were set up in microtitre plates according to the conditions required for MASP culture of the species under test. These cultures contained normal or immune sera at various dilutions as part of the overall serum content and were preincubated for 1 h at 37–38°C.

Merozoites collected from the culture supernatant after several hours of CO_2 deprivation were suspended in culture medium and 20 μl of the suspension were added to 200μl of the cultured erythrocytes in the wells. After several hours incubation, smears were made and stained to determine the percentage parasitaemia. Alternatively, the merozoites can be pre-incubated for 30 min with the test or control sera, washed, and added to the cultures (Winger *et al.*, 1986).

The merozoite neutralization assay can be used to examine the effects of immune sera or monoclonal antibodies, with or without other components of the immune system, in MASP cultures.

APPENDIX

CALCULATION OF RATIOS OF BLOOD, MEDIUM AND SERUM FOR CULTURES

The original packed cell volume (PCV) of the infected or uninfected blood is divided by the PCV required for use in the cultures to give a dilution factor. For example, when the original PCV = 30 and the required PCV = 9, the dilution factor = 30/9 = 3.3.

The blood should be washed three times (by centrifugation at 200–300*g* for 10 min at 4°C and resuspension in medium) and the buffy coat removed. It can then be resuspended to its original volume in medium–serum mixture and further diluted with the medium–serum mixture according to the dilution factor; e.g., in the above example, 1.0 ml of the resuspended blood would be made up to 3.3 ml.

Alternatively, after washing three times as above, the blood can be resuspended in medium only and further medium and serum added to give the correct dilution. The table below gives the quantities of blood and medium to add to 11 ml of serum to produce 30 ml of culture medium supplemented with 40% serum. The figures must be recalculated if different ratios of medium to serum are used.

PCV	Blood (ml) washed 3× in medium and resuspended to its original volume in medium	Medium (ml)	Serum (ml)
40	6.7	12.4	11.0
39	6.9	12.2	11.0
38	7.1	12.0	11.0
37	7.3	11.8	11.0
36	7.5	11.6	11.0
35	7.7	11.4	11.0
34	8.0	11.1	11.0
33	8.2	11.0	11.0
32	8.4	10.7	11.0
31	8.7	10.4	11.0
30	9.0	10.0	11.0
29	9.3	9.8	11.0
28	9.7	9.4	11.0
27	10.0	9.1	11.0
26	10.4	8.7	11.0
25	10.8	8.3	11.0
24	11.2	7.9	11.0
23	11.7	7.4	11.0
22	12.3	6.8	11.0
21	12.8	6.3	11.0
20	13.5	5.6	11.0
19	14.2	4.9	11.0
18	15.0	4.1	11.0
17	15.9	3.2	11.0
16	16.8	2.3	11.0
15	18.0	1.1	11.0

The volume of liquid dispensed into the culture vessels for MASP cultures is equal to surface area of the vessel multiplied by the specific depth, i.e., 6.2 mm for standard MASP cultures.

ACKNOWLEDGEMENTS

The authors are grateful to the Wellcome Trust for supporting work on culture of *B. divergens* and to the following persons for permission to quote unpublished information: Mr J. Culverhouse on *B. divergens* cultures; Irene Müller and Mr L.P. Phipps on *B. caballi* cultures; Dr J. Konrad on recovery of *B. divergens* merozoites; Mr L.P. Phipps, Mr M.R. Rankin and Mr J. Donelly on cryopreservation of *B. divergens* merozoites for culture initiation.

REFERENCES

Akinboade, O.A. (1984). Effect of imidocarb (Imizol®) on *Babesia* cultures. *Animal Technology* **35**, 33–35.

Barry, D.N. (1982). Metabolism of *Babesia* parasites *in vitro*. Amino acid production by *Babesia rodhaini* compared to *Plasmodium berghei*. *Australian Journal of Experimental Biology and Medical Science* **60**, 175–180.

Barry, D.N. (1984). Metabolism of *Babesia* parasites *in vitro*. Glucose and energy metabolism of *B. bovis*. *Australian Journal of Experimental Biology and Medical Science* **62**, 53–61.

Bautista, C.R. and Kreier, J.P. (1979). Effect of immune serum on the growth of *Babesia microti* in hamster erythrocytes in short term culture. *Infection and Immunity* **25**, 470–472.

Bautista, C.R. and Kreier, J.P. (1980). The action of macrophages and immune serum on growth of *Babesia microti* in short term cultures. *Tropenmedizin und Parasitologie* **31**, 313–324.

Bhat, U.K.M., Mahoney, D.F. and Wright, I.G. (1979). The invasion and growth of *Babesia bovis* in tick tissue culture. *Experientia* **35**, 752–753.

Canning, E.U. (1986). Terminology, taxonomy and life cycles of Apicomplexa. *Insect Science and its Application*, 7, 319–325.

Chapman, W.E. and Ward, P.A. (1977). *Babesia rodhaini*: requirement of complement for penetration of human erythrocytes. *Science* **196**, 67–70.

Dalgleish, R.J. (1972). Effects of low temperature preservation and route of inoculation on infectivity of *Babesia bigemina* in blood diluted with glycerol. *Research in Veterinary Science* **13**, 540–545.

Donelly, J., Phipps, L.P. and Konrad, J. (1984). Infectivity in bovine hosts of *Babesia divergens* and *Babesia major* from *in vitro* culture. *Parasitology* **89**, lxxvi [abstract].

Droleskey, R.E., Holman, P. and Wagner, G.C. (1981). *Babesia bovis* infection of a tick cell line studied by electron microscopy. *In Vitro* **17**, 209 [abstract].

Erp, E.E., Gravely, S.M., Smith, R.D., Ristic, M., Osorno, B.M. and Carson, C.A. (1978). Growth of *Babesia bovis* in bovine erythrocyte cultures. *American Journal of Tropical Medicine and Hygiene* **27**, 1061–1064.

Erp, E.E., Smith, R.D., Ristic, M. and Osorno, B.M. (1980a). Continuous *in vitro* cultivation of *Babesia bovis*. *American Journal of Veterinary Science* **41**, 1141–1142.

Erp, E.E., Smith, R.D., Ristic, M. and Osorno, B.M. (1980b). Optimisation of the suspension culture method for *in vitro* cultivation of *Babesia bovis*. *American Journal of Veterinary Research* **41**, 2059–2062.

Hoffman, G. (1972). Haltung babesieninfizierter Zeckengewebe in kunstlichem Nährmedium. *Zeitschrift für angewandte Entomologie* **71**, 26–34.

Irvin, A.D. and Young, E.R. (1977). Possible *in vitro* test for screening drugs for activity against *Babesia* and other blood protozoa. *Nature* **269**, 407–409.

Irvin, A.D. and Young, E.R. (1979). Further studies on the uptake of tritiated nucleic acid precursors by *Babesia* spp. of cattle and mice. *International Journal for Parasitology* **9**, 109–114.

Irvin, A.D., Young, E.R. and Purnell, R.E. (1978). The *in vitro* uptake of tritiated nucleic acid precursors by *Babesia* spp. of cattle and mice. *International Journal for Parasitology* **8**, 19–24.

Jack, R.M. and Ward, P.A. (1980a). *Babesia rodhaini* interactions with complement: relationship to parasite entry into red cells. *Journal of Immunology* **124**, 1566–1573.

Jack, R.M. and Ward, P.A. (1980b). The role *in vivo* of C3 and the C3b receptor in babesial infection in the rat. *Journal of Immunology* **124**, 1574–1578.

James, M.A., Levy, M.G. and Ristic, M. (1981). Isolation and partial characterisation of culture derived soluble *Babesia bovis* antigens. *Infection and Immunity* **31**, 358–361.

Jungery, M. (1984). *Plasmodium falciparum* invasion of human red cells is partially dependent on parasite lectin-like proteins. *In* "Molecular Biology of Host-Parasite Interactions" (N. Agabian and H. Eisen, eds), pp. 249–259. Alan Liss, New York.

Konrad, J., Phipps, L.P., Canning, E.U. and Donelly, J. (1984). Long term *in vitro* maintenance of *Babesia divergens* in a stationary phase culture. *Parasitology* **89**, lxxvi [abstract].

Konrad, J., Canning, E.U., Phipps, L.P. and Donelly, J. (1985). Maintenance of *in vitro* cultures of *Babesia divergens* and *Babesia major* at low temperatures. *Zeitschrift für Parasitenkunde* **71**, 313–316.

Kossel, H., Schütz, W., Weber, A. and Meissner, H. (1904). Über die Hämoglobinurie der Rinder in Deutschland. *Arbeiten aus dem Kaiserlichen Gesundheitsamte* **22**, 1–77.

Kurtti, T.J., Munderloh, U.G. and Stiller, D. (1983). The interaction of *Babesia caballi* with tick cells. *Journal of Invertebrate Pathology* **42**, 334–343.

Kuttler, K.L., Levy, M.G., James, M.A. and Ristic, M. (1982). Efficacy of a non-viable culture-derived *Babesia bovis* vaccine. *American Journal of Veterinary Research* **43**, 281–284.

Levy, M.G. and Ristic, M. (1980). *Babesia bovis* : continuous cultivation in a microaerophilous stationary phase culture. *Science* **297**, 1218–1220.

Levy, M.G. and Ristic, M. (1983). Cultivation and *in vitro* studies of *Babesia*. In "*In vitro* Cultivation of Protozoan Parasites" (J.B. Jensen, ed.), pp. 221–241. CRC Press, Boca Raton, Florida.

Levy, M.G., Erp, E.E. and Ristic, M. (1981). Cultivation of *Babesia*. *In* "Babesiosis" (M. Ristic and J.P. Kreier, eds), pp. 207–223. Academic Press, New York and London.

Levy, M.G., Clabaugh, G. and Ristic, M. (1982). Age resistance in bovine babesiosis : role of blood factors in resistance to *Babesia bovis*. *Infection and Immunity* **37**, 1127–1131.

Lignières, J. (1900). Sur la 'tristeza'. *13th Congrès International de Médecin. Comptes Rendus de la Section de Bactériologie et Parasitologie*, *Paris*, pp. 108–116.

Mahoney, D.F. (1967). Bovine babesiosis : preparation and assessment of complement fixing antigens. *Experimental Parasitology* **20**, 232–241.

Mehlhorn, H. and Schein, E. (1984). The piroplasms: life cycle and sexual stages. *Advances in Parasitology* **23**, 37–103.

Molinar, E., James, M.A., Kakoma, I., Holland, C. and Ristic, M. (1982). Antigenic and immunogenic studies on cell culture derived *Babesia canis*. *Veterinary Parasitology* **10**, 29–40.

Montenegro-James, S., Ristic, M., Benitez, M.T., Leon, E. and Lopez, R. (1985). Heterologous strain immunity in bovine babesiosis using a culture-derived soluble *Babesia bovis* immunogen. *Veterinary Parasitology* **18**, 321–337.

Moreau, Y. and Laurent, N. (1984). Antibabesial vaccination using antigens from cell culture fluids : industrial requirements. *In* ''Malaria and Babesiosis'' (M. Ristic, P. Ambroise-Thomas and J.P. Kreier, eds), pp. 129–140. Martinus Nijhoff, Dordecht.

Moreau, Y. and Soula, A. (1979). *Babesia canis* : le culture *in vitro* du parasite et son étude ultrastructurale. *Bulletin de la Société des Sciences Vétérinaire et Médicine Comparée, Lyon* **81**, 225–261.

Nocard, E.I.E. and Motas, C.S. (1902). Contribution a l'étude de la piroplasmose canine. *Annales de l'Institut Pasteur* **16**, 257–290.

Palmer, D.A., Buening, G.M. and Carson, C.A. (1982). Cryopreservation of *Babesia bovis* for *in vitro* cultivation. *Parasitology* **84**, 567–572.

Puck, T.H., Cieciura, S.J. and Robinson, A. (1958). Genetics of somatic mammalian cells III. Long-term cultivation of euploid cells from human and animal subjects. *Journal of Experimental Medicine* **108**, 945–955.

Pudney, M. (1984). Cultivation of *Babesia divergens* in bovine and human erythrocytes *in vitro*. *Parasitology* **89**, lxxv–lxxvi [abstract].

Pudney, M., Varma, M.G.R. and Leake, C.J. (1973). Culture of embryonic cells from the tick *Boophilus microplus* (Ixodidae). *Journal of Medical Entomology* **10**, 493–496.

Ristic, M. and Levy, M.G. (1981). A new era of research toward solution of bovine babesiosis. *In* ''Babesiosis'' (M. Ristic and J.P. Kreier, eds), pp. 509–544. Academic Press, New York and London.

Ristic, M., Smith, R.D. and Kakoma, I. (1981). Characterisation of *Babesia* antigens from cell cultures and ticks. *In* ''Babesiosis'' (M. Ristic and J.P. Kreier, eds), pp. 337–380. Academic Press, New York and London.

Rodriguez, S.D., Buening, G.M., Green, T.J. and Carson, C.A. (1983). Cloning of *Babesia bovis* by *in vitro* cultivation. *Infection and Immunity* **42**, 15–18.

Rodriguez, S.D., Buening, G.M., Vega, C.A. and Carson, C.A. (1986). *Babesia bovis* : purification and concentration of merozoites and infected bovine erythrocytes. *Experimental Parasitology* **61**, 236–243.

Schein, E., Rehbein, G., Voigt, W.P. and Zweygarth, E. (1981). *Babesia equi*: 1. Development in horses and in lymphocyte culture. *Tropenmedizin und Parasitologie* **32**, 223–227.

Smith, R.D. and Ristic, M. (1981). Immunisation against bovine babesiosis with culture derived antigens. *In* ''Babesiosis'' (M. Ristic and J.P. Kreier, eds), pp. 485–507. Academic Press, New York and London.

Smith, R.D., James, M.A., Ristic, M., Aikawa, M. and Vega, C.A. (1981). Bovine babesiosis: protection of cattle with culture-derived soluble *Babesia bovis* antigen. *Science* **212**, 335–338.

Timms, P. (1980). Short term cultivation of *Babesia* species. *Research in Veterinary Science* **29**, 102–104.

Timms, P., Dalgleish, R.J., Barry, D.N., Dimmock, C.K. and Rodwell, B.J. (1983). *Babesia bovis* : comparison of culture-derived parasites, non-living antigen and conventional vaccine in the protection of cattle against heterologous challenge. *Australian Veterinary Journal* **60**, 75–77.

Trager, W. and Jensen, J.B. (1976). Human malaria parasites in continuous culture. *Science* **193**, 673–675.

Väyrynen, R. and Tuomi, J. (1982). Continuous *in vitro* cultivation of *Babesia divergens*. *Acta Veterinaria Scandinavica* **23**, 471–472.

Vega, C.A., Buening, G.M., Green, T.J. and Carson, C.A. (1985a). *In vitro* cultivation of *Babesia bigemina*. *American Journal of Veterinary Research* **46**, 416–420.

Vega, C.A., Buening, G.M., Rodriguez, S.D. and Carson, C.A. (1985b). Cryopreservation of *Babesia bigemina* for *in vitro* cultivation. *American Journal of Veterinary Research* **46**, 421–423.

Winger, C.M., Canning, E.U. and Culverhouse, J.D. (1986). A monoclonal antibody to *Babesia divergens* which inhibits merozoite invasion. *Parasitology* **94**, 17–27.

Winger, C.M., Canning, E.U. and Culverhouse, J.D. (1987). Induction of protective immunity to *Babesia divergens* in Mongolian gerbils, *Meriones unguiculatus*, using culture-derived immunogens. *Veterinary Parasitology* **26** (in press).

Wright, I.G., White, M., Tracey-Patte, P.D., Donaldson, R.A., Goodger, B.V., Waltisbuhl, D.J. and Mahoney, D.F. (1983). *Babesia bovis* : isolation of a protective antigen by using monoclonal antibodies. *Infection and Immunity* **41**, 244–250.

Wright, I.G., Mirre, G.B., Rode-Bramanis, K., Chamberlain, M., Goodger, B.V. and Waltisbuhl, D.J. (1985). Protective vaccination against virulent *Babesia bovis* with a low molecular-weight antigen. *Infection and Immunity* **48**, 109–113.

Chapter 10

Theileriidae

C.G.D. BROWN

Centre for Tropical Veterinary Medicine, University of Edinburgh, Easter Bush, Roslin, Scotland

I. INTRODUCTION

Theileria have been cultivated *in vitro* for over 40 years (Tsur-Tchermonoretz, 1945), yet progress has been slow and developments have been rare. Only one stage of the parasite, the macroschizont, has been grown with reproducible techniques that allow widespread application and enable the cultured parasites to be used in the laboratory. Attempts have been made to grow other stages of the life-cycle, but these have proved technically frustrating and yielded only limited success. The piroplasm, or intraerythrocytic schizont (Conrad *et al.*, 1985), undergoes restricted division in culture and an arthropod form, the motile kinete, can be produced from zygotes in tick organ cultures (Bell, 1984).

Some progress has been made in inducing transition of one form of the parasite to another *in vitro*. Limited success has been achieved with the vertebrate cycle so that a number of different theilerial species can be grown from sporozoites harvested from the tick to macroschizont-infected lymphoid cell lines (Brown *et al.*, 1973). However, it has not yet been possible reproducibly to produce merozoites from these schizonts, to infect erythrocytes, or to induce persistent replication of the intraerythrocytic form with invasion of uninfected erythrocytes. Moreover, ticks cannot be infected with cultured erythrocytes. The recent exception has been the report of exoerythrocytic schizogony of *Babesia equi* in cultures of horse lymphocytes by Schein *et al.* (1981) and Rehbein *et al.* (1982), who obtained release of merozoites from schizont-infected lymphocytes with invasion of equine erythrocytes and secondary schizogony into "Maltese cross" forms in the erythrocytes. It is likely that such infected erythrocytes would be infective for ticks, in which case the full vertebrate cycle of this piroplasm has been cultivated *in vitro*, a significant milestone in parasite cultivation technology.

Perhaps equally significant is the implication of this finding in the context of the classification and taxonomy of the piroplasms. Levine (1985) proposed two families of piroplasms, the Babesiidae and the Theileriidae, each with a single genus, *Babesia* and *Theileria*. Thus are lost, at least for the time being, such interesting genera as *Dactylosoma, Cytauxzoon* and *Nuttallia* (Brocklesby, 1978). *Babesia equi* would, until recently, have been *Nuttallia equi*, a small "*Babesia*" exhibiting division into four within the erythrocyte and only trans-stadial, not transovarial, transmission through ticks. Now that exoerythrocytic schizogony has been confirmed, *B. equi* and perhaps other small *Babesia* spp., notably *B. microti*, for which exoerythrocytic schizonts have tentatively been described (Mehlhorn and Schein, 1984), will have to be moved to the family Theileriidae if not to the genus *Theileria*. For the purposes of this chapter, therefore, the techniques described should be considered as being relevant not only to the cultivation of classically accepted *Theileria* spp. but also to *Babesia equi* and possibly other small *Babesia* as well as "*Cytauxzoon*" spp. now recognized as *Theileria* — notably *Theileria taurotragi* and *Theileria felis*.

II. CULTURE METHODS

The methods used for the cultivation of *Theileria in vitro* and their applications have been extensively reviewed in recent years (Brown, 1979, 1980, 1981, 1983). In addition, specific and comprehensive details of techniques for the following species have been provided: *T. annulata* from cattle (Pipano, 1984); *T. parva* from cattle (Kurtti *et al.*, 1981; Stagg *et al.*, 1981); *T. hirci* from sheep (Hooshmand-Rad and Hawa, 1975); *T. lawrencei* from African buffalo, *Syncerus caffer* (Stagg *et al.*, 1974); *T. taurotragi* from eland, *Taurotragus oryx* (Stagg *et al.*, 1976). In all instances, apparently immortalized schizont-infected cell lines were established from the infected host species or by infection of lymphoid cells by sporozoites *in vitro*.

Notable absentees from the list of *Theileria* grown *in vitro* are important species such as *T. mutans* and *T. orientalis* (*T. sergenti*). It is interesting, and perhaps relevant, that these are species in which there is incontrovertible evidence for intraerythrocytic multiplication of the piroplasm, while the exoerythrocytic schizont appears of little importance in the development cycle. Indeed, it is only recently that schizonts of these species have been seen in bovine lymphoid cells (Uilenberg *et al.*, 1974, 1985).

The successful cultivation of *Theileria in vitro* has developed through three distinct phases. In the first of these it was demonstrated that the schizont stage of *T. annulata* could be maintained, and would even multiply *in vitro*, in plasma clots, leading up to early monolayer cultures. Between 1945 and 1963, the pioneering work of Tsur-Tchermonoretz and his colleagues in Israel was made possible by the development of early culture media, basal salt solutions, supplements such as yeast extract and lactalbumin hydrolysate and, perhaps most important, antibiotics (Tsur-Tchermonoretz, 1945; Tsur and Adler, 1962, 1965).

In the second phase, the mode of replication of the theilerial schizont was conclusively demonstrated *in vitro* (Hulliger *et al.*, 1964), while an awareness of the role of feeder layers (Hulliger, 1965; Malmquist and Brown, 1974) and foetal calf serum (FCS) (Malmquist *et al.*, 1970) enabled *T. parva* to be cultured for the first time. Schizont-transformed bovine lymphoid cultures were by then being grown in suspension, developments running hand-in-hand with improvements in techniques for the establishment and maintenance of immortalized human lymphoid cell lines (Moore, 1970), involving the use of complex media, notably Eagle's MEM and RPMI 1640, together with foetal calf serum. Finally, the development of effective fungistats permitted the isolation of sporozoites from ticks and made possible the infection and transformation of lymphoid cells *in vitro* (Brown *et al.*, 1973).

The third, current phase has resulted from interest in immune mechanisms and molecular biology. There has therefore been a need for cell and parasite cloning by techniques that have been made possible by a number of developments. Some of these are common to many tissue culture techniques: the advent of laminar flow hoods to permit open culture handling; the use of mini- and micro-well plates with media volumes of 2.0 ml down to 100 μl; improved pH and humidity control achieved by the use of media such as Leibovitz L15, better buffers such as the zwitterionic HEPES, and finite control of the gas tensions with the use of humidified CO_2 incubators; and, most recently, awareness of the role or growth factors. These, provided in the form of better feeder layers of "filler cells", frequently irradiated, emphasize the importance of cell–cell contact in selecting the concentration and density of cells in a culture and, specifically in relation to lymphocyte growth, involve the use of interleukins or mononuclear cell growth factors. The principles behind these culture techniques have much in common with those used in hybridoma technology developed for the production of monoclonal antibodies.

The outstanding development specific to theileriosis has been an awareness of the necessity for producing heavily infected ticks from which large numbers of sporozoites can be harvested. While ticks of the genus *Hyalomma* are readily infected with *T. annulata* to a high level, even when fed on low parasitaemias (Schein, 1975; Samish and Pipano, 1983; Walker *et al.*, 1985), *T. parva* rarely infects its vector *Rhipicephalus appendiculatus* as easily and great care must be taken to optimize infection rates (Young and Leitch, 1981; Buscher and Otim, 1986).

Techniques used in this laboratory for the cultivation of *T. annulata* are described below, with modifications relevant to other species appended where appropriate. Except when stated, these methods are adaptable from one laboratory to another and tolerate limited changes in medium, culture vessel, serum and gas conditions. Common factors include the following.

(i) *Temperature*. Cultures of the vertebrate stages of the theilerial life cycle are maintained at 37 °C, the arthropod stages at 28 °C.
(ii) *Culture medium*. Media used include RPMI 1640, Eagle's MEM or Leibovitz L15, supplemented, most commonly, with FCS at 15–20% (though horse serum is an excellent substitute) and 2.0 mM L-glutamine (Gibco) freshly added.
(iii) *Buffer/gas tension*. Except when L15 is used, buffering is effected with a CO_2-bicarbonate system involving gassing the cultures with 5% CO_2 in air and sealing them or, if open, holding in a gassed, humidified incubator or sealable plastic box at that level of CO_2 and O_2 tension. RPMI 1640 and MEM are formulated with 25 mM HEPES as a supplementary buffer.
(iv) *Antibiotics/antimycotic*. Penicillin 100 iu ml^{-1}, streptomycin 100 $\mu g\ ml^{-1}$ and, occasionally, gentamycin 50 $\mu g\ ml^{-1}$ are the antibiotics used. The fungistat of choice, used only when tick material is introduced, has been nystatin (Gibco) at 100 $\mu g\ ml^{-1}$.

A. CULTIVATION OF MACROSCHIZONTS

Schizonts are cultivated as *Theileria*-infected lymphoblastoid cell lines, which can be established from infected lymphoid tissue from an animal exhibiting patent parasitosis, either clinically ill or recently dead, or, with greater difficulty, from a convalescent or recovered carrier. Alternatively, cultures can be initiated by inoculating suspensions of sporozoites, harvested from infected ticks, into cultures containing uninfected lymphocytes of a host species susceptible to that particular *Theileria*. In each instance, the critical period of the culture is early, when the infected cell is adapting from its *in vivo* to its *in vitro* form.

Success is achieved by manipulation of a number of interrelated factors: cell concentration, growth factors (from both serum and cells), judicious medium changes aimed at replacing much of the exhausted medium at frequent intervals yet maintaining sufficient "conditioned" medium, pH balance, and gas tension. Monitoring cultures daily via an inverted microscope and, less frequently, with Giemsa-stained smears of samples from the culture, will predict or confirm "transformation" as it occurs, illustrated by the selective, clonal replication of schizont-infected lymphoblastoid cells. Smears can be most readily prepared from 50–100 μl samples using a cytocentrifuge (Cytospin, Shandon).

1° *Theileria* macroschizont cultures from infected tissues

COMPLETE MEDIUM

RPMI 1640, supplemented with 16% FCS, 25 mM HEPES, 2.0 mM L-glutamine, penicillin 100 iu ml^{-1}, streptomycin 100 $\mu g\ ml^{-1}$.

TECHNIQUE

Schizont-infected cell lines can be established from lymphoid tissue obtained from a dead animal up to 2 h or more *post mortem*. Lymph node, spleen, thymus, bone marrow, lung, liver and kidney are all suitable sources of infected lymphocytes. The tissue may be taken aseptically during necropsy, 1.0 cm^3 being quite sufficient, or by needle biopsy. From a living animal the most convenient tissues are lymph node, spleen or liver, from which needle biopsy samples may be obtained using 14–16 gauge needles, careful aseptic skin preparation and routine theilerial diagnostic techniques.

(i) Withdraw the biopsy specimen and transfer to a 30 ml universal bottle containing 10 ml RPMI 1640, antibiotics and 10 iu ml^{-1} preservative-free heparin.

(ii) If the specimen is sufficiently large, transfer in medium to a Petri dish, dissect and discard extraneous tissue and chop with two scalpels into pieces 1–2 mm^3. Further break up tissue clumps using the conical plunger of a disposable syringe in the conical base of a disposable universal container.

(iii) Pipette vigorously to disperse cells further, allow clumps to settle for 2–3 min, then pipette off supernatant cell suspension. (At this stage, if a single-cell suspension is required, pass it through a 60–100 μm mesh stainless steel or nylon sieve; this is not essential.)

(iv) Centrifuge cell suspension (400*g*, 5 min, 5 °C), discard supernate and resuspend in 10 ml complete medium.

(v) Count cells in a Neubauer haemocytometer, or electronic particle counter (Coulter ZB1) calibrated for bovine lymphocytes, and adjust concentration to 2.0×10^6 ml^{-1}.

(vi) Inoculate 10.0 ml into 25 cm^2 flasks, 2.0 ml into 2 cm^2 wells of 24-well plates, 1.0 ml to 1.0 cm^2 wells of 48-well plates or 200 μl to flat-bottomed wells of 96-well microtitre plates (Costar).

Note. For *T. annulata* no feeder layer is necessary but, for more fastidious species such as *T. parva*, pre-established monolayers of bovine embryo spleen (BESP), bovine embryo thymus (BETH), bovine embryo skin (BES), bovine aortic endothelium (BAE) or the buffalo lung cell line IMR31, with their supernatant conditioned medium, assist in the establishment of *Theileria*-infected lines.

2° *Theileria* macroschizont cultures from peripheral blood mononuclear cells — density gradient method

The most readily accessible source of *Theileria*-infected material in the living animal is peripheral blood. Established techniques are available by which mononuclear cells can be separated in large numbers from bovine blood and, with only slight modification, from a wide range of ungulate species. Expected yields are 30–50% of the peripheral blood lymphocytes with variable monocyte levels within the purified mononuclear population; thus 3×10^7 peripheral blood mononuclear cells (PBM) from 10 ml blood or $2–3 \times 10^8$ from 100 ml.

TECHNIQUE

(i) Take blood aseptically into anticoagulants — heparin (20 iu ml^{-1}, preservative-free), Alsever's solution (1 : 2) or acid citrate dextrose (ACD, 1 : 5) — or defibrinate with glass beads or sterile wooden swab sticks. The latter method give a population with a reduced number of adherent cells (less satisfactory for *T. annulata*, perhaps better for *T. parva*) and free of platelets. A convenient method for small volumes is to use sterile 10 ml evacuated containers with lithium heparin (Vacutainer, Becton Dickinson).

(ii) Dilute blood with an equal volume of phosphate buffered saline (PBS, Dulbecco A, calcium- and magnesium-free, pH 7.3), except when taken into Alsever's solution, and layer over a density gradient comprising a high polymer of sucrose used to accelerate erythrocyte (and granulocyte) sedimentation and an iodinated compound of high density. Ficoll 400 with sodium diatrizoate (Ficoll-paque, Pharmacia) or ficoll and sodium metrizoate (Lymphoprep, Nycomed), each formulated to a density of 1.077, are both suitable. It is convenient to layer 20 ml of diluted blood over 8.0 ml density gradient in a disposable 30 ml universal container (Sterilin).

To harvest large numbers of PBM from large volumes of blood, resuspend the buffy coat from 50 ml blood centrifuged (2800*g*, 20 min, 5 °C) in 18 ml PBS and layer the suspension over 8.0 ml density gradient.

(iii) Centrifuge (800*g*, 35 min, 15–20 °C).

(iv) Remove PBM at interface and wash twice in 20 ml PBS (300*g*, 10 min, then 300*g*, 5 min) to remove platelets (not necessary for defibrinated blood).

(v) Resuspend in 10 ml cold complete medium (see Section II.A.1°, above). Keep on ice.

(vi) Count and adjust concentration to 2.0×10^6 ml^{-1} with cold complete medium.

(vii) Dispense in flasks or plates as required, with or without feeder layers as indicated in Section II.A.1°.

3° *Theileria* macroschizont cultures from peripheral blood mononuclear cells — lysis method

An alternative technique, particularly satisfactory for *T. annulata*, is convenient for the rapid isolation of leucocytes from large numbers of blood samples and is an appropriate and sensitive method for monitoring the schizont carrier status of *T. annulata*-infected cattle. The technique can readily be applied to blood volumes of 10 ml or less taken aseptically into sterile heparinized or EDTA Vacutainers.

(i) Centrifuge blood (2000*g*, 20 min, 5 °C).

(ii) Remove and keep plasma.

(iii) Collect buffy coat and resuspend in 4.0 ml PBS in a disposable universal container.

(iv) Add 18 ml ice-cold 0.17 M ammonium chloride, pipette gently and stand in ice for 10 min. This lyses the erythrocytes.

(v) Layer 5.0 ml autologous plasma beneath the suspended buffy coat cells and centrifuge (400g, 7 min, room temperature). The leucocytes are deposited and washed through their own plasma.

(vi) Remove supernatant and resuspend deposited cells in 10 ml cold complete medium (Section II.A.1°). Count, adjust to 2.0 × 10^6 ml^{-1} and proceed as in Section II.A.1°.

4° *In vitro* infection of lymphoid cells with *Theileria* sporozoites

This technique involves the introduction *in vitro* of sporozoites harvested from infective ticks to uninfected lymphocytes from a susceptible host. There are thus two components: the lymphocytes, normally separated from peripheral blood by methods described in Sections II.A.2° and 3°, and sporozoites for which methods of harvesting are given below.

TECHNIQUES

(a) PREPARATION OF *T. ANNULATA* SPOROZOITE SUSPENSION

(i) Feed adult *Hyalomma anatolicum anatolicum* heavily infected with *T. annulata*, whose infection rate has been confirmed by methyl green-pyronin staining of salivary glands (Walker *et al.*, 1979), for 3 d on rabbits to stimulate sporozoite maturation.

(ii) Remove ticks from rabbits' ears, wash in a sieve with running tap water to remove debris and transfer to a disposable universal container.

(iii) Wash once in 1% benzalkonium chloride (Roccal, Winthrop) and three times in 70% ethanol, transferring to a second sterile universal container.

(iv) Wash again with four rinses of warm MEM with Hanks's salts (see Taylor and Baker, 1968, p. 350) and double-strength antibiotics/antimycotic mixture (penicillin 200 iu ml^{-1}, streptomycin 200 μg ml^{-1} and nystatin 100 μg ml^{-1}), leaving for 5–20 min in last wash.

(v) Pipette off wash mixture and transfer ticks to sterile mortar in 5.0 ml ice-cold MEM with 3.5% bovine plasma albumin (BPA; Armour Fraction V, Sigma).

(vi) Grind with pestle with repeated additions of MEM/BPA, pipetting off supernatant suspension from the ground-up ticks (GUTs) on each occasion. Pool the supernates until a concentration of four tick equivalents (t.e.) per ml is reached; e.g. grind 100 3-day-fed ticks in 25 ml MEM/BPA.

(vii) Centrifuge (50–100g, 5 min, 4°C) and remove GUTs.

(viii) Filter through a 25 mm Swinnex filter holder containing a 22 mm AP prefilter and a 25 mm 8.0 μm MF membrane (Millipore).

(ix) Cytocentrifuge 50 μl aliquots of GUTs filtrate, and stain with Giemsa to determine sporozoite concentration. Dilute with MEM/BPA accordingly (usually *T. annulata* from *H. a. anatolicum* is used at 1.0 t.e. ml^{-1}).

Note. Infection levels and times required for sporozoite maturation may vary between species, e.g. *T. parva* in *R. appendiculatus* requires 4–5 d, as does *B. equi* in *H.a. anatolicum*. In addition, where infection levels in the ticks are low, it may be necessary to concentrate the parasites. This can be achieved to some extent by dissecting salivary glands aseptically from pre-fed ticks or centrifuging the GUTs filtrate (2000*g*, 30 min, 4°C) and resuspending the deposited sporozoite pellet in a smaller volume, thus increasing the proportion of sporozoites to extraneous debris and cytotoxic factors derived from the partially fed ticks. Material should be kept at ca 4°C (i.e. on ice) whenever possible.

(b) *IN VITRO* INFECTION OF PBM WITH THEILERIAL SPOROZOITES

This is most conveniently effected in wells of open plates maintained in a gassed (5% CO_2), humidified incubator. Use of the plates facilitates daily monitoring and media changes without the necessity of centrifugation, resuspension and replacement of cells, as has to be done if cultures are initiated and established in flasks. As mentioned in Section II.A.1°, there is no need for feeder layers when setting up *T. annulata* cultures in this way. For *T. parva* and certain other more fastidious species, whose cells have a lower cloning efficiency than *T. annulata*, feeder layers and conditioned medium will prove helpful. In particular the careful adjustment of cell contact, using filler cells comprising irradiated autologous lymphocytes, may prove effective in promoting the transformation and immortalization of *T. parva*-infected cells.

(i) Take equal volumes of PBM suspension at 2.0×10^6 ml^{-1} and sporozoite suspension (GUTs filtrate) at 1.0 t.e. ml^{-1} (it may be useful to use serial 3–10-fold dilutions of the sporozoites to reduce further the toxic effects of tick contaminants).

(ii) Incubate at 37°C for 1 h.

(iii) Dispense replicate aliquots into plates: 2.0 ml into 2.0 cm^2 wells, 1.0 ml into 1.0 cm^2 wells, or 200 μl into wells of 96-well microplates.

(iv) Incubate at 37°C in gassed (5% CO_2), humidified incubator.

(v) Change medium according to pH change and evidence of metabolism, taking 50 μl aliquots for cytocentrifuge preparations at 4 d, 8 d and 12 d to determine infection and transformation rates.

(vi) Once 30% or more cells are infected, necessitating frequent medium changes, transfer cultures to larger wells and, eventually, to 25 cm^2 flasks.

(vii) Subculture or passage, by dilution initially 1 : 3 then 1 : 5, aiming, with an established line, to dilute cells with medium 1 : 10 two or three times weekly. Cell concentration should be maintained between 10^5 to 2.0×10^6 ml^{-1}.

(viii) Cryopreserve 2 to 10 aliquots at intervals after isolation (see Section II.D.1°).

B. CULTIVATION OF PIROPLASMS

Conrad (1983) and Conrad *et al.* (1985) reviewed the evidence for intraerythrocytic multiplication of the piroplasms of *T. annulata* and *T. parva* and reported the progress made in establishing *in vitro* cultures of *T. annulata*. From these cultures they produced evidence to support their claim that *Theileria* undergoes a form of intraerythrocytic schizogony, dividing into four merozoites. This culture system, based on methods developed for the cultivation of *Plasmodium falciparum* by Trager and Jensen (1976) and for *Babesia bovis* by Levy and Ristic (1980), was only partially successful, clearly illustrating intraerythrocytic multiplication of the piroplasm into tetrad forms but failing to achieve transfer of the parasites to uninfected erythrocytes.

The culture system developed by Conrad used media containing high concentrations (40%) of FCS (as opposed to the autologous or homologous serum used in *Plasmodium* and *Babesia* cultures). This was perhaps both the strength and the weakness of the method, which is still being developed, since many permutations of CO_2 and O_2 tensions have yet to be evaluated.

1° Cultivation of intraerythrocytic stages *in vitro*

MEDIA

Two media have proved satisfactory, permitting limited multiplication of theilerial piroplasms over 6–8 d.

(i) MEM alpha medium and 25 mM HEPES (80%) with 20% heat-inactivated FCS and gentamycin 50 μl ml^{-1} supplemented with 2.0 mg glucose, 50 μg G-hydroxypurine and 50 μg reduced glutathione ml^{-1}.

(ii) Medium 199 with Hanks's salts (see Taylor and Baker, 1968, pp. 355–356 and p. 350) and 25 mM HEPES (60%) with 40% heat-inactivated FCS and gentamycin 50 μg ml^{-1}.

(Media, antibiotics and sera are from Gibco, supplements from Sigma.)

TECHNIQUE

Blood from infected cattle in the first few days of patent parasitosis is the optimal source of *T. annulata* piroplasms for culture.

(i) Bleed infected calf between 10 and 14 d after infecting with cryopreserved sporozoites (Section II.D.2°), when exhibiting a parasitaemia of 1–10%.
(ii) Defibrinate blood with glass beads or wooden swab sticks.
(iii) Centrifuge (1000*g*, 10 min, 15°C), remove plasma, buffy coat and upper 20% of erythrocytes.
(iv) Resuspend remaining erythrocytes to a concentration of 3% in medium (i) or (ii).
(v) Dispense into flat-welled microtitre plates (20 μl per well), 2.0 cm^2 wells (1.0 ml) or 25 cm^2 flasks standing vertically (5.0 ml).
(vi) Incubate at 37°C in a humidified atmosphere of 5% CO_2 in air or a candle jar.
(vii) Change medium daily.
(viii) Assess growth by colour change of medium, and examination of cytocentrifuge smears of 50 μl medium stained with Giemsa.

C. CULTIVATION OF ARTHROPOD FORMS

Significant developments in arthropod cell and tissue culture have been recorded in recent years. Those concerning tick tissue culture have been reviewed by Kurtti and Buscher (1979) and detailed technical recipes are given by Pudney *et al.* (1979). Despite this progress, the most promising invertebrate culture system for studying and producing stages in the vector cycle of piroplasms is organ culture of infected ticks. This was predicted by Hadani (1981) from his experience in attempting to produce infective sporozoites of *Nuttallia* (= *Babesia*) *danii* in tick salivary gland cultures (Hadani *et al.*, 1978). The successful production of apicomplexan sporozoites from cell cultures remains as remote as ever.

The system described below is, therefore, all the more interesting (Bell, 1984) as it is a simple technique yielding significant numbers of *T. annulata* kinetes, the migratory form found in the haemolymph of ticks. Using what are literally "backless tick" explants, some 8000 kinetes can be harvested from a 20-tick culture in a single flask over a 5 d period. Thus individual kinetes are available for cloning and large numbers (10^5 or more) of this, the biggest stage in the theilerial life cycle, can be obtained for nucleic acid extraction.

1° Cultivation of *T. annulata* kinetes

MEDIUM

Leibovitz L 15 (70%), 20% FCS and 10% tryptose phosphate broth containing penicillin 100 iu ml^{-1} and streptomycin 100 $\mu g\ ml^{-1}$.

TECHNIQUE

(i) Maintain *H. a. anatolicum* nymphs at 28°C and 85% relative humidity for 7–10 d after detachment from a calf with patent *T. annulata* parasitaemia.

(ii) Wash the nymphs for 2 min in tap water (in a sieve), followed by 5 min in thiomersal 1 : 10000, 5 min in 1% benzalkonium chloride, 30 s in 70% ethanol and two rinses in sterile distilled water.

(iii) Dry the nymphs on sterile filter paper for 10 min and embed them, dorsal aspect uppermost, in sterile wax in a Petri dish by means of a heated spatula and curved iris forceps.

(iv) Cover the embedded nymphs with sterile Hanks's BSS (see Taylor and Baker, 1968, p. 350) plus 0.5% lactalbumin hydrolysate (dissecting medium) at room temperature and carefully dissect them, in a laminar flow hood, using watchmaker's forceps and a scalpel with No. 11 blade.

(v) First remove the dorsal integument, sever tracheae and spiracles and transfer the entire developing adult to a small Petri dish containing sufficient dissecting medium just to cover the explants. (If nymphs are used, 10 d after detachment, it will be necessary also to dissect off the newly formed adult dorsal integument.)

(vi) Pipette 1.5 ml complete L 15 medium in a 25 cm^2 flask and spread over bottom surface. Distribute 20 explants, evenly spaced, centrally in the flask, to adhere.

(vii) Close flask and incubate at 28°C. Examine daily with an inverted microscope ($\times$ 10 objective) for kinetes migrating adjacent to the explants, exhibiting very slow movement.

(viii) Harvest kinetes by removing culture medium and gently rinsing with 1.0 ml fresh medium. This can be repeated daily for 5 d.

D. CRYOPRESERVATION

Freezing parasite stocks or defined strains at regular passage intervals, to provide stabilates of cryopreserved material, is an essential component of both protozoal and cell culture methodology. Two stages of the life cycle of *Theileria* have been cryopreserved to date with reproducible success: sporozoites, harvested from infected ticks, and macroschizont-infected lymphoid cells from cell culture.

The difficulty with freezing intraerythrocytic piroplasms has been assessing their functional integrity on resuscitation by showing that the piroplasms retain their infectivity for ticks. This has been reported only once for cryopreserved piroplasms of *T. parva* by Jongejan *et al.* (1980), who successfully infected *R. appendiculatus* by injecting engorged nymphs percutaneously with frozen and thawed *T. parva* piroplasms from which the cryopreservative had been removed by dialysis. This important development provides a novel method for storage of freshly-isolated field stocks of *Theileria* and a third stage in the theilerial life cycle from which stabilates can be made. Jongejan *et al.* (1980), however, were not able to infect ticks with cultured merozoites of *T. parva* using this method, nor, like most other workers before them, from the first attempt by Hulliger *et al.* (1966) onwards, could they infect erythrocytes *in vitro* with merozoites from transitional stages produced in schizont cultures.

For successful freezing, and preservation in liquid nitrogen, of theilerial sporozoites and schizont-infected lymphocytes, a number of factors must be considered. (i) The high osmolarity of the cryopreservative solutions. (ii) The different behaviour of the two cryoprotectants, dimethyl sulphoxide (DMSO) and glycerol. DMSO passes rapidly across membranes but is toxic at room and body temperature and must, once in contact with cells, be kept ice-cold; glycerol crosses membranes slowly but is relatively non-toxic once osmotic balance is established. Glycerol-treated material can thus be handled at 20–37 °C. (iii) The rate of freezing must be correlated with the rate of thawing. (iv) Thawing and resuscitation requires reinstatement of normal osmolarity, and thus removal of the cryoprotectant by dilution, washing, or dialysis.

The techniques used for cryopreserving stabilates of sporozoites or cell lines are simple, pragmatic and functional methods which effectively preserve enough viable material to serve their purpose. These routine techniques have been described by Brown (1983).

1° Freezing *Theileria*-infected cell lines

A programmable biological freezer (Planer, R204), providing a predetermined wide range of cooling rates between + 25 °C and − 180 °C, has provided the opportunity to study the effects of cooling and thawing rates on the viability of theilerial material. Wathanga *et al.* (1986) described two programmes giving >95% viability of lymphoid cell lines using DMSO concentrations of 5%, 10% or 15%, and the following cooling schedules. (i) Cooling at 1°C min^{-1} from +5°C to −60°C, then rapid cooling (by immersion in liquid nitrogen) to −196°C; or (ii) interrupted cooling, rapidly to −30°C at which temperature cells are held for 30 min, followed by further rapid cooling to −60°C and below.

Cryopreserved cells are thawed rapidly in a 40°C waterbath and then diluted stepwise with culture medium over a 5 min period, adding successively 1.0, 2.0 and 4.0 ml of medium to effect a 2-fold dilution at each step. The cell suspension is then centrifuged, the supernate is discarded and the pellet is resuspended in the required volume of fresh culture medium. 10^7 cells cryopreserved in a 1.0 ml ampoule (Cryotube, Nunc) are resuspended in 10 ml of medium.

Cell cultures handled in this way exhibit a cloning efficiency close to that of the parent cell lines from which they are derived.

The same two cooling rates, with 10% DMSO as cryoprotectant, provide excellent techniques for the storage of bovine lymphocytes and PBM (H.K. Kiara, personal communication). This is of particular value in studies of cell-mediated immunity in theileriosis.

The slow cooling rate (i) can be approximated in the top shelf of a gas-phase liquid nitrogen container (Union Carbide LR40). Interrupted cooling (ii) can be obtained simply by plunging the ampoules into an alcohol bath at −30°C, holding for 30 min, then transferring them directly into liquid nitrogen (Oliver *et al.*, 1984).

2° Freezing theilerial sporozoites

This method is essentially that described by Brown (1983), and probably yields no more than 10% viable, infective sporozoites. This is quite satisfactory for most purposes but results could be improved. Briefly, the method uses 7.5% glycerol in MEM plus 3.5% bovine plasma albumen as cryoprotectant with 40 min equilibration at room temperature before freezing at 1–2°C min^{-1}; thawing is rapid, followed by equilibration at room temperature for 20–40 min before any further dilution. As glycerol has high osmolarity (7.5% is approximately 3 times isotonic) and moves slowly across membranes, if stabilates are diluted suddenly in isotonic medium the sporozoites swell and explode. To avoid this, material for infecting cattle is diluted after thawing with the cryoprotectant mixture (including 7.5% glycerol) and inoculated subcutaneously, *not* intravenously. Material required for infecting cells *in vitro* is allowed to equilibrate, after thawing at room temperature for 30 min before *either* being inoculated into an equal volume of the cell suspension, *or* being serially diluted slowly as described in Section II.D.1°(ii) above, with the doubling dilutions at 20 min intervals.

III. APPLICATIONS AND USES OF THEILERIAL CULTURES

A. ISOENZYME ELECTROPHORESIS

Cell lines infected with schizonts of *T. parva* and *T. annulata* have been used to provide material for detecting genetic differences between stocks based on isoenzyme heterogeneity. Lysates from cell lines have been examined using starch gel electrophoresis (Melrose *et al.*, 1980) and isoelectric focusing (IEF) (Allsopp and Gibson, 1983).

Results of these and later studies (Melrose *et al.*, 1984) show that, using starch gel electrophoresis, there is marked glucose phosphate isomerase (GPI) activity associated with the parasite and distinct from host cell activity, but little evidence of other specific parasite isoenzymes. Surprisingly, considerable GPI polymorphism was detected between stocks of *T. annulata*, yet all *T. parva* stocks examined to date have shown the same GPI enzyme banding pattern (Melrose *et al.*, 1984). However, Allsopp and Gibson (1983) were able to detect *T. parva* lactic dehydrogenase (LDH) using IEF in agarose, and subsequently Allsopp *et al.* (1985) detected up to 12 parasite-associated enzymes in *T. parva* and *T. taurotragi*. They concluded, however, that piroplasms were a better source of material for isoenzyme studies than cultured schizonts.

There is presently no indication that the genetic polymorphism detected using these techniques is necessarily associated with immunological differences between parasite stocks. There is evidence from these studies, however, that many field and laboratory stocks are composed of more than one parasite strain (Irvin *et al.*, 1983). A further application of this technique, which one hopes will be supported by the finding of other *Theileria*-associated isoenzymes, is in confirming clonal selection of parasites in culture.

B. CHEMOTHERAPY

The use of schizont-infected lymphoid cell lines to screen for antitheilerial chemotherapeutic agents has been reviewed by Dolan (1981), McHardy (1982) and Brown (1983). The first breakthrough came with evidence for the activity of the antimalarial naphthoquinone, menoctone (McHardy *et al.*, 1976). This has led to the production of a series of highly effective theilericides, notably parvaquone and buparvaquone, whose activity was first detected when screened against schizonts of *T. parva* and *T. annulata in vitro* (McHardy and Morgan, 1981; McHardy *et al.*, 1985).

Other compounds also showing schizonticidal activity *in vitro* have been halofuginone, a synthetic febrifugine (Uilenberg *et al.*, 1980) and the ionophores, monensin and lasalocid (McHardy and Rae, 1982). These three coccidiostats are all highly toxic, however, and, to date, only halofuginone has shown acceptable *in vivo* activity (Schein and Voigt, 1979). This perhaps illustrates one of the limitations of an otherwise highly sensitive and cost-effective screen, which is, however, the only alternative to testing drugs in cattle.

With reproducible infection and transformation of bovine lymphocytes by theilerial sporozoites *in vitro* now possible, screening of compounds against other stages of the vertebrate cycle of *Theileria* can be undertaken. Vivien Taylor (personal communication), working in this laboratory, has recently tested three anti-theilerials against the sporozoite, the transforming trophozoite-infected lymphocyte and the schizont-infected lymphoblast. Such a screen is potentially able to detect chemoprophylactic, or infection-blocking, activity, and has already proved to be a sensitive bioassay for drug levels in bovine serum.

C. IMMUNOLOGY

Excellent reviews are available of the immunology, immune mechanisms and immunogenesis of theileriosis (Morrison *et al.*, 1986; Irvin and Morrison, 1987). This section is therefore restricted to recent developments in cell culture techniques as they apply to this field of study.

1. *Immunodiagnosis*

Routine serodiagnosis of theileriosis is usually performed using the indirect fluorescent antibody test (IFAT). The methods in general use are based on those developed for *T. parva* using either piroplasms or cultured schizonts as antigen (Burridge and Kimber, 1972). Antigens derived from cell cultures are suitable for detection of antibodies to the species of *Theileria* in which the schizont is the major pathogenic stage in the life-cycle; these include *T. annulata* (see Pipano and Cahana, 1969), *T. p. lawrencei* (see Burridge *et al.*, 1973) and *T. hirci* (see Hawa *et al.*, 1976).

Though other serological tests have been developed, they have not been able to use cultured schizonts as antigen but have instead used piroplasms because of the ease of separation from background host material. The separation of schizonts from lymphocytes has proved particularly intractable (Nyormoi *et al.*, 1981). Thus, when monoclonal antibodies (Mabs) were raised against schizont-infected lymphocytes, the IFAT proved to be the primary screen of choice (Pinder and Hewett, 1980; Minami *et al.*, 1983; Shiels *et al.*, 1986a). Already the few Mabs raised to *T. parva* and *T. annulata* schizonts have proved valuable indicators of strain differences between parasite stocks (Irvin *et al.*, 1983) and heterogeneity within stocks of what, on cloning, have proved to be populations consisting of more than one strain (Shiels *et al.*, 1986a).

2. *Immune mechanisms*

To date two immune mechanisms operating against *Theileria* in cattle have been recognized. In both instances their recognition and characterization have depended on the establishment of a culture system. The most significant development, which enables both humoral and cellular mechanisms to be identified *in vitro*, has been the ability to infect lymphoid cells with sporozoites *in vitro* (Brown *et al.*, 1973). This has enabled antisporozoite antibodies — both monoclonal and polyclonal — to be screened for invasion-blocking activity (Gray and Brown, 1981; Dobbelaere *et al.*, 1984; Musoke *et al.*, 1984; Preston and Brown, 1985). It has also made possible comprehensive investigations of genetically-restricted cell-mediated cytotoxicity in both *T. parva* by Eugui and Emery (1981) and Emery *et al.* (1981) and *T. annulata* by Preston *et al.* (1983). In such studies, infected cell lines established *in vitro* are killed by autologous but not allogeneic lymphocytes from immune cattle.

3. *Vaccines*

Cell culture vaccines have been widely used for the control of tropical theileriosis caused by *T. annulata* (see Pipano, 1981). They are, however, less effective in the control of east coast fever (ECF) caused by parasites of the *T. parva* complex (Brown, 1981). The explanation for this seems to be that there is a histocompatibility barrier to the infection of cattle with *T. parva*-infected cells (Dolan *et al.*, 1984) and that, to date at least, immunity can be induced only by establishing the parasite within cells of the host.

This thesis appears to be confirmed by studies on *T. annulata* infection (Pipano *et al.*, 1977), in which a "vaccine" prepared from killed schizont-infected tissue culture cells failed to protect cattle against a sporozoite challenge. In contrast, Emery *et al.* (1986) reported successful immunization of calves against ECF with membranes from autologous cultured lymphocytes infected with *T. parva*. This latter result offers renewed hope for an inactivated vaccine against theileriosis. Further encouragement has been given by the report of what appears to be an infection-associated antigen on *T. annulata*-transformed cells in culture (Shiels *et al.*, 1986b). The Mab recognizing the epitope on the surface of infected, but not uninfected, lymphoblasts has subsequently been shown to induce complement-mediated lysis of cells infected with *T. annulata* in an *in vitro* screen (Preston *et al.*, 1986). Mabs with such activity may prove to be invaluable probes to screen DNA expression libraries for cloned sequences for the derivation of protective antigens.

D. MOLECULAR BIOLOGY

Apart from providing the tools, in the form of immunofluorescence screens and *in vitro* bioassays, for assessing Mab activity, theilerial cell culture systems play an increasing role in studies of the molecular biology of these parasites. For some time, evidence has accumulated that emphasizes the need to use defined, cloned parasite material for such work. The likelihood that *Theileria* undergoes a sexual cycle has many implications, not least in selecting the stage in the life-cycle from which to clone the organism (Irvin and Boarer, 1980).

Present culture and parasite-handling systems offer a number of sources of material for cloning. There is already evidence that kinetes from tick explant organ cultures (Bell, 1984) could be cloned, with new ticks infected *in vitro* or by percutaneous inoculation of single kinetes. There is still the question of producing infective sporozoites from such material, since these would be required to infect lymphocytes *in vitro* to produce usable quantities of cloned material.

The simplest method to date has been to clone schizont-infected cells *in vitro* (Nelson and Hirumi, 1981). The arguable flaw in this is whether a single lymphocyte necessarily means a single parasite. With parallel development of phenotypic, genotypic and molecular markers, it should soon be possible to resolve that question. Moreover, cloning need not be restricted to cells already infected and transformed. There is now evidence that *Theileria* can be cloned by endpoint titration of sporozoite-infected lymphocytes. This may be done either by inoculating increasing dilutions of sporozoites into a standard number of cells in microtitre wells, or by mixing equal volumes of sporozoite and cell suspension, incubating for 60 min at 37°C, and plating out into 12 to 24 replicate wells at 3- to 4-fold dilutions, to reach a limiting dilution endpoint*. In either method, 2×10^5 PBM is an ideal number for flat-bottomed microtitre wells, or 2×10^4 for round-bottomed wells. *T. annulata* needs no further

*See footnote on p. 220 [Eds].

support but *T. parva* benefits from the presence of irradiated autologous PBM as filler cells and probably, exogenous, T-cell growth factor (TCGF) too. A limit dilution conducted in round-bottomed wells with 2×10^4 filler cells per well has proved optimal, while TCGF can be obtained from the supernatants of bovine lymphoblasts stimulated with concanavalin A, or T-cell lines (Rabin *et al.*, 1981). Mabs raised to *T. parva* cell cultures which identify phenotypically and functionally distinct cell types enable defined populations to be selected according to their surface antigens, using a fluorescence-activated cell sorter (FACS, Becton Dickinson); this can reveal whether they present determinants associated with class I or class II antigens, whether they are monocytes or T-cells and, if T-cells, of which subset (Lalor *et al.*, 1986; Naessens *et al.*, 1985). Current results indicate that sporozoite infectivity and cell transformation can be optimized by presenting the correct cell subset to the respective parasite.

In sum, therefore, tissue culture of *Theileria* helps to elucidate the life-cycle and taxonomy of the piroplasms and the immunology of the diseases they cause, and provides material for biochemical, chemotherapeutic and molecular studies. Perhaps more important than this, however, is the fact that, as schizont-infected cultures, the parasites offer a tool for studying the lymphocyte.

ACKNOWLEDGEMENTS

I am particularly grateful to Lesley Bell for an account of the kinete culture system, and to David Emery, Bruno Godderris, Paul Lalor, Ivan Morrison, Jan Naessens and Alan Teale for discussions on lymphoid culture techniques.

REFERENCES

Allsopp, B.A. and Gibson, W.C. (1983). Isoelectric focusing in agarose: a highly discriminatory method for the detection of enzyme heterogeneity. *Annals of Tropical Medicine and Parasitology* **77**, 169–173.

Allsopp, B.A., Gibson, W.C. and Stagg, D.A. (1985). Characterisation of some East African *Theileria* species isolates by isoenzyme analysis, with particular reference to *T. parva*. *International Journal for Parasitology* **15**, 271–276.

Bell, L.J. (1984). Tick tissue culture techniques in the study of arthropod-borne protozoa: the development of *Theileria annulata* in organ culture of *Hyalomma anatolicum anatolicum*. *In* "Acarology VI" (Proceedings of sixth Acarology Congress), D.A. Griffiths and C.E. Bowman, eds), Vol. 2, pp. 1089–1095. Ellis Horwood, Chichester.

Brocklesby, D.W. (1978). Recent observations on tick-borne protozoa. *In* "Tick-borne Diseases and their Vectors" (J.K.H. Wilde, ed.), pp. 263–286. University of Edinburgh, Centre for Tropical Veterinary Medicine, Edinburgh.

Brown, C.G.D. (1979). Propagation of *Theileria*. *In* "Practical Tissue Culture Applications" (K. Maramorosch and H. Hirumi, eds), pp. 223–254. Academic Press, New York and London.

Brown, C.G.D. (1980). *In vitro* cultivation of *Theileria*. *In* "The *In Vitro* Cultivation of the Pathogens of Tropical Diseases" (D.S. Rowe and H. Hirumi, eds), pp. 127–144. Schwabe, Basel.

Brown, C.G.D (1981). Application of *in vitro* techniques to vaccination against theileriosis. *In* "Advances in the Control of Theileriosis" (A.D. Irvin, M.P. Cunningham and A.S. Young, eds), pp. 104–119. Martinus Nijhoff, The Hague.

Brown, C.G.D. (1983). *Theileria*. *In* "*In Vitro* Cultivation of Protozoan Parasites" (J.B. Jensen, ed.), pp. 243–284. CRC Press, Boca Raton, Florida.

Brown, C.G.D., Stagg, D.A., Purnell, R.E., Kanhai, G.K. and Payne, R.C. (1973). Infection and transformation of bovine lymphoid cells *in vitro* by infective particles of *Theileria parva*. *Nature* **245**, 101–103.

Burridge, M.J. and Kimber, C.D. (1972). The indirect fluorescent antibody test for experimental East Coast fever (*Theileria parva* infection of cattle). Evaluation of a cell culture schizont antigen. *Research in Veterinary Science* **13**, 451–455.

Burridge, M.J., Kimber, C.D. and Young, A.S. (1973). Use of the indirect fluorescent antibody technique in serologic studies of *Theileria lawrencei* infections in cattle. *American Journal of Veterinary Research* **34**, 145–148.

Buscher, G. and Otim, B. (1986). Quantitative studies on *Theileria parva* in the salivary glands of *Rhipicephalus appendiculatus* adults: quantitation and prediction of infection. *International Journal for Parasitology* **16**, 93–100.

Conrad, P.A. (1983). "Studies on the *in vitro* cultivation of the intraerythrocytic stages of *Theileria annulata*, *Theileria parva* and *Babesia bovis*." Ph.D. Thesis, University of Edinburgh.

Conrad, P.A., Kelly, B.G. and Brown, C.G.D. (1985). Intraerythrocytic schizogony of *Theileria annulata*. *Parasitology* **91**, 67–82.

Dobbelaere, D.A.E., Spooner, P.R., Barry, W.C. and Irvin, A.D. (1984). Monoclonal antibody neutralizes the sporozoite stage of different *Theileria parva* stocks. *Parasite Immunology* **6**, 361–370.

Dolan, T.T. (1981). Progress in the chemotherapy of theileriosis. *In* "Advances in the Control of Theileriosis" (A.D. Irvin, M.P. Cunningham and A.S. Young, eds), pp. 186–207. Martinus Nijhoff, The Hague.

Dolan, T.T., Teale, A.J., Stagg, D.A., Kemp, S.J., Cowan, K.M., Young, A.S., Groocock, C.M., Leitch, B.L., Spooner, R.L. and Brown, C.G.D. (1984). A histocompatibility barrier to immunization against East Coast fever using *Theileria parva*-infected lymphoblastoid cell lines. *Parasite Immunology* **6**, 243–250.

Emery, D.L., Eugui, E.M., Nelson, R.T. and Tenywa, T. (1981). Cell-mediated immune responses to *Theileria parva* (East Coast fever) during immunization and lethal infections in cattle. *Immunology* **43**, 323–336.

Emery, D.E., Morrison, W.I. and Jack, R.M. (1986). Induction of immunity against infection with *Theileria parva* (East Coast fever) in cattle using plasma membranes from parasitized lymphoblasts. *Veterinary Parasitology* **19**, 321–327.

Eugui, E.M. and Emery, D.L. (1981). Genetically-restricted cell-mediated cytotoxicity in cattle immune to *Theileria parva* (East Coast fever). *Nature* **290**, 251–254.

Gray, M.A. and Brown, C.G.D. (1981). *In vitro* neutralization of theilerial sporozoite infectivity with immune serum. *In* "Advances in the Control of Theileriosis" (A.D. Irvin, M.P. Cunningham and A.S. Young, eds), pp. 127–131. Martinus Nijhoff, The Hague.

Hadani, A. (1981). *In vitro* development of *Babesia* and *Theileria* parasites in organs and tissues of ticks. *In* "Babesiosis" (M. Ristic and J.P. Kreier, eds), pp. 225–226. Academic Press, New York and London.

Hadani, A., Kaufman, W. and Barnett, S.F. (1978). Development of *Nuttallia danii* (Protozoa: Babesiidae) within tick salivary glands cultured *in vitro. Journal of Parasitology* **64**, 501–503.

Hawa, N., Latif, B.M.A. and Bakir, F.A. (1976). Application of the indirect fluorescent antibody test for diagnosis of *Theileria hirci* infection of sheep, using cell culture schizont antigen. *Tropical Animal Health and Production* **8**, 97–101.

Hooshmand-Rad, P. and Hawa, N.J. (1975). Cultivation of *Theileria hirci* in sheep lymphoid cells. *Tropical Animal Health and Production* **7**, 121–122.

Hulliger, L. (1965). Cultivation of three species of *Theileria* in lymphoid cells *in vitro. Journal of Protozoology* **12**, 649–655.

Hulliger, L., Wilde, J.K.H., Brown, C.G.D. and Turner, L. (1964). Mode of multiplication of *Theileria* in cultures of bovine lymphocytic cells. *Nature* **203**, 728–730.

Hulliger, L., Brown, C.G.D. and Wilde, J.K.H. (1966). Transition of developmental stages of *Theileria parva in vitro* at high temperatures. *Nature* **211**, 328–329.

Irvin, A.D. and Boarer, C.D.H. (1980). Some implications of a sexual cycle in *Theileria. Parasitology* **80**, 571–579.

Irvin, A.D. and Morrison, W.I. (1987). Immunopathology, immunology and immunoprophylaxis of *Theileria* infections. *In* "Immune Responses in Parasitic Infections," vol. 3, "Protozoa" (E.J.L. Soulsby, ed.), pp. 223–274. CRC Press, Boca Raton.

Irvin, A.D., Dobbelaere, D.A.E., Mwamachi, D.M., Minami, T., Spooner, P.R. and Ocama, J.G.R. (1983). Immunization against East Coast fever: correlation between monoclonal antibody profiles of *Theileria parva* stocks and cross immunity *in vitro. Research in Veterinary Science* **35**, 341–346.

Jongejan, F., Perie, N.W., Franssen, F.F.J. and Uilenberg, G. (1980). Artificial infection of *Rhipicephalus appendiculatus* with *Theileria parva* by percutaneous injection. *Research in Veterinary Science* **29**, 320–324.

Kurtti, T.J. and Buscher, G. (1979). Trends in tick cell culture. *In* "Practical Tissue Culture Applications" (K. Maramorosch and H. Hirumi, eds), pp. 351–371. Academic Press, New York and London.

Kurtti, T.J., Munderloh, U.G., Irvin A.D. and Buscher, G. (1981). *Theileria parva*: early events in the development of bovine lymphoblastoid cell lines persistently infected with macroschizonts. *Experimental Parasitology* **52**, 280–290.

Lalor, P.A., Morrison, W.I., Goddeeris, B.M., Jack, R.M. and Black, S.J. (1986). Monoclonal antibodies identify phenotypically and functionally distinct cell types in the bovine lymphoid system. *Veterinary Immunology and Immunopathology* **13**, 121–140.

Levine, N.D. (1985). "Veterinary Protozoology". Iowa State University Press, Ames, Iowa.

Levy, M.G. and Ristic, M. (1980). *Babesia bovis*: continuous cultivation in a microaerophilous stationary phase culture. *Science* **207**, 1218–1220.

Malmquist, W.A and Brown, C.G.D. (1974). Establishment of *Theileria parva*-infected lymphoblastoid cell lines using homologous feeder layers. *Research in Veterinary Science* **16**, 134–136.

Malmquist, W.A., Nyindo, M.B.A. and Brown, C.G.D. (1970). East Coast fever: cultivation *in vitro* of bovine spleen cell lines infected and transformed by *Theileria parva. Tropical Animal Health and Production* **2**, 139–145.

McHardy, N. (1982). The validity of *in vitro* models for the study of chemotherapeutic agents in the theilerioses. *In* "Animal Models in Parasitology" (D.G. Owen, ed.), pp. 205–212. Macmillan, London.

McHardy, N. and Morgan, D.W.T. (1981). Studies *in vivo* and *in vitro* with 993C and halofuginone. *In* "Advances in the Control of Theileriosis" (A.D. Irvin, M.P. Cunningham and A.S. Young, eds), pp. 209–211. Martinus Nijhoff, The Hague.

McHardy, N. and Rae, D.G. (1982). Antitheilerial activity of the coccidiostat monensin. *Tropical Animal Health and Production* **14**, 13–14.

McHardy, N., Haigh, A.J.B. and Dolan, T.T. (1976). Chemotherapy of *Theileria parva* infection. *Nature* **261**, 698–699.

McHardy, N., Wekesa, L.S., Hudson, A.T. and Randall, A.W. (1985). Antitheilerial activity of BW720C (buparvaquone): a comparison with parvaquone. *Research in Veterinary Science* **39**, 29–33.

Mehlhorn, H. and Schein, E. (1984). The piroplasms: life cycle and sexual stages. *Advances in Parasitology* **23**, 37–103.

Melrose, T.R., Brown, C.G.D. and Sharma, R.D. (1980). Glucosephosphate isomerase isoenzyme patterns in bovine lymphoblastoid cell lines infected with *Theileria annulata* and *T. parva*, with an improved enzyme visualisation method. *Research in Veterinary Science* **29**, 298–304.

Melrose, T.R., Brown, C.G.D., Morzaria, S.P., Ocama, J.G.R. and Irvin, A.D. (1984). Glucose phosphate isomerase polymorphism in *Theileria annulata* and *T. parva*. *Tropical Animal Health and Production* **16**, 239–245.

Minami, T., Spooner, P.R., Irvin, A.D., Ocama, J.G.R., Dobbelaere, D.A.E. and Fujinaga, T. (1983). Characterisation of stocks of *Theileria parva* by monoclonal antibody profiles. *Research in Veterinary Science* **35**, 334–340.

Moore, G.E. (1970). The culture of human lymphocytoid cell lines. *In* "Methods in Cancer Research V" (H. Busch, ed.), pp. 423–453. Academic Press, New York and London.

Morrison, W.I., Lalor, P.A., Goddeeris, B.M. and Teale, A.J. (1986). Theileriosis: antigens and host-parasite interactions. *In* "Parasite Antigens: Towards New Strategies for Vaccines" (T.W. Pearson, ed.), pp. 167–212. Marcel Dekker, New York.

Musoke, A.J., Nantulya, V.M., Rurangirwa, F.R. and Buscher, G. (1984). Evidence for a common protective antigenic determinant on sporozoites of several *Theileria parva* strains. *Immunology* **52**, 231–238.

Naessens, J., Newson, J., Bensaid, A., Teale, A.J., Magondu, J.G. and Black, S.J. (1985). *De novo* expression of T cell markers on *Theileria parva*-transformed lymphoblasts in cattle. *Journal of Immunology* **135**, 4183–4187.

Nelson, R.T. and Hirumi, H. (1981). *In vitro* cloning of *Theileria*-infected bovine lymphoblastoid cells: standardization and characterization. *In* "Advances in the Control of Theileriosis" (A.D. Irvin, M.P. Cunningham and A.S. Young, eds), pp. 120–121. Martinus Nijhoff, The Hague.

Nyormoi, O., Bwayo, J.J. and Hirumi, H. (1981). *Theileria parva*: isolation of macroschizonts from *in vitro* propagated lymphoblastoid cells of cattle. *Experimental Parasitology* **52**, 303–311.

Oliver, R.A., Springbett, A.J. and Spooner, R.L. (1984). A method for freezing bovine lymphocytes. *Animal Blood Groups and Biochemical Genetics* **15**, 213–217.

Pinder, M. and Hewett, R.S. (1980). Monoclonal antibodies detect antigenic diversity in *Theileria parva* parasites. *Journal of Immunology* **124**, 1000–1001.

Pipano, E. (1981). Schizonts and tick stages in immunization against *Theileria annulata* infection. *In* "Advances in the Control of Theileriosis" (A.D. Irvin, M.P. Cunningham and A.S. Young, eds), pp. 242–252. Martinus Nijhoff, The Hague.

Pipano, E. (1984). Immunization against *Theileria annulata* infection. *In* "Tick and Tick-borne Disease Control: a Practical Field Manual', II" (P.J. McCosker and R.J.

Tatchell, eds), pp. 508–563. FAO, UN, Rome.

Pipano, E. and Cahana, M. (1969). Fluorescent test for the serodiagnosis of *Theileria annulata*. *Journal of Parasitology* **55**,765.

Pipano, E., Goldman, M., Samish, M. and Friedhoff, K.T. (1977). Immunization of cattle against *Theileria annulata* using killed schizont vaccine. *Veterinary Parasitology* **3**, 11–22.

Preston, P.M. and Brown, C.G.D. (1985). Inhibition of lymphocyte invasion by sporozoites and the transformation of trophozoite infected lymphocytes *in vitro* by serum from *Theileria annulata* immune cattle. *Parasite Immunology* **7**, 301–314.

Preston, P.M., Brown, C.G.D. and Spooner, R.L. (1983). Cell-mediated cytotoxicity in *Theileria annulata* infection of cattle with evidence for BoLA restriction. *Clinical and Experimental Immunology* **53**, 88–100.

Preston, P.M., McDougall, C., Wilkie, G., Shiels, B.R., Tait, A. and Brown, C.G.D. (1986). Specific lysis of *Theileria annulata*-infected lymphoblastoid cells by a monoclonal antibody recognizing an infection-associated antigen. *Parasite Immunology* **8**, 369–380.

Pudney, M., Varma, M.G.R. and Leake, C.J. (1979). Establishment of cell lines from ixodid ticks. *Tissue Culture Association Manual* **4**, 1003–1007.

Rabin, H., Hopkins, R.F., Ruscetti, F.W., Neubauer, R.H., Brown, R.L. and Kawakami, T.G. (1981). Spontaneous release of a factor with properties of T cell growth factor from a continuous line of primate tumor cells. *Journal of Immunology* **127**, 1852–1855.

Rehbein, G., Zweygarth, E., Voigt, W.P. and Schein, E. (1982). Establishment of *Babesia equi*-infected lymphoblastoid cell lines. *Zeitschrift für Parasitenkunde* **67**, 125–127.

Samish, M. and Pipano, E. (1983). Transmission of *Theileria annulata* (Dschunkowsky and Luhs, 1904) by *Hyalomma excavatum* (Koch, 1844). *Parasitology* **86**, 269–274.

Schein, E. (1975). On the life cycle of *Theileria annulata* (Dschunkowsky and Luhs, 1904) in the midgut and haemolymph of *Hyalomma anatolicum excavatum* (Koch, 1844). *Zeitschrift für Parasitenkunde* **47**, 165–167.

Schein, E. and Voigt, W.P. (1979). Chemotherapy of bovine theileriosis with halofuginone. *Acta Tropica* **36**, 391–394.

Schein, E., Rehbein, G., Voigt, W.P. and Zweygarth, E. (1981). *Babesia equi* (Laveran, 1901). 1. Development in horses and in lymphocyte culture. *Tropenmedizin und Parasitologie* **32**, 223–227.

Shiels, B., McDougall, C., Tait, A. and Brown, C.G.D. (1986a). Antigenic diversity of *Theileria annulata* macroschizonts. *Veterinary Parasitology* **21**, 1–10.

Shiels, B.R., McDougall, C., Tait, A. and Brown, C.G.D. (1986b). Identification of infection-associated antigens in *Theileria annulata* transformed cells. *Parasite Immunology* **8**, 69–77.

Stagg, D.A., Brown, C.G.D., Crawford, J.G., Kanhai, G.K. and Young, A.S. (1974). *In vitro* cultivation of *Theileria lawrencei*-infected cell lines derived from a buffalo (*Syncerus caffer*). *Research in Veterinary Science* **16**, 125–127.

Stagg, D.A., Kanhai, G.K., Young, A.S. and Brown, C.G.D. (1976). The establishment of *Theileria*-infected cell lines from an eland (*Taurotragus oryx* Lydekker, 1906). *Research in Veterinary Science* **20**, 122–126.

Stagg, D.A., Dolan, T.T., Leitch, B.L. and Young, A.S. (1981). The initial stages of infection of cattle cells with *Theileria parva* sporozoites *in vitro*. *Parasitology* **83**, 191–197.

Trager, W. and Jensen, J.B. (1976). Human malaria parasites in continuous culture. *Science* **193**, 673–675.

Tsur-Tchernomoretz, I. (1945). Multiplication *in vitro* of Koch bodies of *Theileria annulata*. *Nature* **156**, 391.

Tsur, I. and Adler, S. (1962). Growth of *Theileria annulata* schizonts in monolayer tissue culture. *Refuah Veterinarith* **19**, 225.

Tsur, I. and Adler, S. (1965). The cultivation of lymphoid cells and *Theileria annulata* schizonts from infected blood. *Refuah Veterinarith* **22**, 60–62.

Uilenberg, G., Robson, J. and Pedersen, V. (1974). Some experiments on the transmission of *Theileria mutans* (Theiler, 1906) and *Theileria parva* (Theiler, 1904) by the ticks *Amblyomma variegatum* (Fabricius, 1794) and *Rhipicephalus appendiculatus* (Neumann, 1901) in Uganda. *Tropenmedizin und Parasitologie* **25**, 207–216.

Uilenberg, G., Jongejan, F., Perie, N.M. and Franssen, F.F.J. (1980). Chimiothérapie des theilérioses bovines par un anticoccidien, l'halofuginone. Note préliminaire. *Révue d'Elevage Médecine Vétérinaire des Pays Tropique* **33**, 33–43.

Uilenberg, G., Perie, N.M., Spanjer, A.A.M. and Franssen, F.F.J. (1985). *Theileria orientalis*, a cosmopolitan blood parasite of cattle: demonstration of the schizont stage. *Research in Veterinary Science* **38**, 352–357.

Walker, A.R., McKellar, S.B., Bell, L.J. and Brown, C.G.D. (1979). Rapid quantitative assessment of *Theileria* infection in ticks. *Tropical Animal Health and Production* **11**, 21–26.

Walker, A.R., Fletcher, J.D., McKellar, S.B., Bell, L.J. and Brown, C.G.D. (1985). The maintenance and survival of *Theileria annulata* in colonies of *Hyalomma anatolicum anatolicum*. *Annals of Tropical Medicine and Parasitology* **79**, 199–209.

Wathanga, J.M., Jones, T.W. and Brown, C.G.D. (1986). Cryopreservation of *Theileria*-infected lymphoblastoid cells with functional assessment of viability. *Tropical Animal Health and Production* **18**, 191–197.

Young, A.S. and Leitch, B.L. (1981). Production of *Rhipicephalus appendiculatus* with high infections of *Theileria parva*. *Journal of Parasitology* **67**, 751–752.

Chapter 11

Trematoda

J.A. CLEGG and M.A. SMITH

Wellcome Research Laboratories, Langley Court, Beckenham, Kent, England

I. INTRODUCTION

Cultivation of a trematode through its entire life-cycle remains an elusive goal, indeed the rate of reporting of basic culture techniques has decreased in the last few years. This may reflect difficulties in designing improved culture methods, which are notoriously empirical, or perhaps the shortage of funding for long-term projects. Cultivation of trematodes through a substantial part of the life-cycle will be essential for determining the precise nutritional requirements of the parasite. These methods should also facilitate the study of hormonal interactions between host and parasite, the relatively unexplored area of molecular communication between individual parasites plus many other aspects of parasite behaviour particularly the mechanisms involved in their often complex migration routes within the host.

Much, however, can be achieved with existing short-term culture methods, and in the past decade these have been used extensively to elucidate the immune effector mechanisms against trematodes. Most of this work has concentrated on schistosomes, because of their importance as agents of human disease. A wide range of immune mechanisms capable of killing young schistosomes *in vitro* has been discovered and, although this approach has not yet achieved its main aim — the development of an effective vaccine — it has made several contributions to basic immunology such as a function for the eosinophil and a better understanding of the role of IgE in the control of schistosome infections.

We discuss here the progress made in the development of culture techniques for trematodes, detail the best methods available, outline the results of intensive studies on immune effector mechanisms, and consider the modified culture techniques employed.

II. SCHISTOSOMATIDAE

A. CULTURE OF SNAIL HOST STAGES

Like other trematodes, schistosomes have an asexual phase in a snail intermediate host, involving massive production of invasive cercariae to ensure infection of the vertebrate host, in which development to egg-production occurs. The snail is infected by miracidia which hatch from eggs released from the vertebrate host. Soon after penetration of the snail, a miracidium transforms into a mother sporocyst within which daughter sporocysts develop and migrate to the digestive gland, where at least one further generation of daughter sporocysts is produced before the more complex cercarial stage is formed.

The conversion of the miracidium to a mother sporocyst occurs relatively easily *in vitro*. Di Conza and Basch (1974) employed a defined medium supplemented with 20% human serum to culture *Schistosoma mansoni* (see Taylor and Baker, 1978, pp. 160–162): mother sporocysts increased in length 3–4 times, but no daughter sporocysts were formed. Similar work on this phase of the life cycle, with the bird schistosome *Trichobilharzia ocellata* and a modification of Di Conza and Basch's medium, obtained a greater degree of growth *in vitro* but, as with *S. mansoni*, no daughter sporocysts were produced (Mellink and Van den Bovenkamp, 1985). If cultures are initiated with *S. mansoni* daughter sporocysts recovered from mature mother sporocysts, significant growth can be obtained, microvilli develop on the surface membrane and small embryos form inside the sporocysts, but a second generation of daughter (progeny) sporocysts is not produced (Di Conza and Basch, 1974). *T. ocellata* daughter sporocysts, however, do not show even this degree of development *in vitro* (Mellink and Van den Bovenkamp, 1985).

Hansen, who previously had made major contributions to the development of a continuous culture system for a free-living nematode, recognized the importance of culturing the multiplicative phase of schistosomes and, with her colleagues, devoted considerable effort to the development of a culture system (reviewed by Hansen and Hansen, 1980). Briefly, they found that *S. mansoni* daughter sporocysts could produce progeny-daughters *in vitro* when cultured in association with a mosquito cell line. Medium conditioned by cell cultures supported limited development of daughter sporocysts, but in fresh medium the sporocysts degenerated within a few days. The nature of the requirement for cells in the cultures was resolved when it was discovered that progeny-daughter sporocysts were formed in a cell-free medium if favourable redox conditions were provided by addition of sulphydryl compounds and use of a controlled gas phase (Buecher *et al*., 1974). The effect occurred within a narrow range of concentrations of either dithiothreitol or a combination of cysteine and glutathione, together with a reduced level of oxygen, but not when the reductants or a reduced level of oxygen were used alone. Neither this system, nor a wide range of other media and culture conditions, allowed the production of cercariae *in vitro*. Nevertheless, their attempt is the most successful so far and, since it is likely to form the basis for future work, the technique is given in detail.

1° Culture of *S. mansoni* daughter sporocysts
(Buecher *et al*., 1974)

MEDIUM

30% Schneider's *Drosophila* medium (Grand Island Biological Co., GIBCO) plus 2.0 g litre^{-1} galactose; 10% foetal calf serum (inactivated: 56°C, 30 min); phenol red, 10 mg litre^{-1}.

Osmolality, after dilution with distilled water, is 163 mosmol; pH is 7.0 in gas phase with 0.5% CO_2, 1% (3% or 9%) O_2 in N_2.

SULPHYDRYL COMPOUNDS

Dithiothreitol (Cleland's reagent), stock solution 32 mM (stored frozen) used at 0.1 mM, *or* a combination of cysteine (free base), stock solution 60 mM, used at 1.0 mM and glutathione (reduced) stock solution 30 mM, used at 0.5 mM.

TECHNIQUE

(i) Albino *Biomphalaria glabrata* snails were maintained and infected with *S. mansoni* by standard techniques.
(ii) Five days after exposure to miracidia, snails showing mother

sporocysts in the tentacles are divided into two groups to prolong the availability of sporocysts. One group, maintained at 27.4°C, is used as source of daughter sporocysts at 12–14 d and the other, maintained at 22°C, provides sporocysts at 17–21 d.

(iii) Snails are placed in antibiotic–antimycotic solution (penicillin, 1000 iu ml^{-1}; streptomycin, 1000 μg ml^{-1}; fungizone, GIBCO, 2.5 μg ml^{-1}) for 3 h before gently crushing the shell. Tentacle or foot tissue, containing the mother sporocyst, is excised and placed in culture medium containing 200 iu ml^{-1} penicillin and 200 μg ml^{-1} streptomycin (sterilizing-wash solution).

(iv) The mother sporocyst is teased apart, releasing about 50 daughter sporocysts which are passed through five aliquots of sterilizing-wash solution. *These operations are carried out under a dissecting microscope using aseptic techniques.*

(v) Daughter sporocysts, selected for uniform size (150–200 μm), motility and morphology, are transferred into Nucleopore chambers (13 mm, 8.0 μm pore-size Nucleopore membranes cemented on to 5.0 mm lengths of 10.0 mm diameter glass tubing) partially immersed in 2.0 ml of medium contained in 3.0 cm glass Petri dishes. The culture medium does not contain antibiotics.

(vi) Cultures are held at 27°C in the dark in glass containers (type unspecified). A gas mixture is passed through the container at 200 ml min^{-1} for 60 min before the inlets and outlets are sealed and gassing is discontinued. The most effective gas mixture contains 0.5% CO_2 with approximately 1% O_2 in N_2, although reduction of the O_2 to any level below 9% has a distinct effect. Sulphydryl compounds are effective only within narrow limits (0.1 mM dithiothreitol or 1.0 mM cysteine : 0.5 mM glutathione appear optimal).

(vii) In fresh medium, under air, sporocysts degenerate within 7 d. A controlled gas phase in combination with reductants allows 90% of sporocysts to show marked increase in growth and daughter sporocysts are produced after 7–10 d. Cultures are observed with an inverted microscope. The gas phase is re-established periodically, but replenishment of reductants is detrimental. It is interesting to note that snail haemolymph was toxic to sporocysts. This was also reported for cultures of *T. ocellata* sporocysts (Mellink and Van den Bovenkamp, 1985); apparently toxic substances are released, probably from cellular elements, during preparation of the haemolymph.

B. CULTURE OF VERTEBRATE HOST STAGES

1. *Culture of* Schistosoma mansoni

Successful culture of the mammalian phase of the schistosome life-cycle involves development from the cercaria, which penetrates the hosts' skin, to pairs of adult worms; females produce viable eggs. Until recently the maximum degree of development of *S. mansoni* in culture was from schistosomula, recovered from the skin shortly after infection or from the lungs, to near-adult worms (Clegg, 1965: see Taylor and Baker, 1968, pp. 168–170). Male worms formed testes and spermatozoa, but females stopped short of egg production except in one rare case in which the parasites were held inside cellulose tubing in an intermittently agitated culture system.

Basch (1981a,b) has significantly improved this system, obtaining consistent production of eggs in about 10% of cultured worm pairs after more than 2 months' culture; females produced 15–20 eggs with a few individuals making up to 40 (*in vivo* continuous egg production is initiated during the sixth week). The eggs formed by cultured schistosomes were about half the size of normal eggs and appeared to lack the germinal disk from which the miracidium develops. Electron microscopy of cultured worms showed that the structure of the reproductive tract and the vitelline cells, which secrete egg-shell protein, was essentially similar to that of normal worms grown in an experimental host (Irie *et al.*, 1983). In contrast, oocytes within the ovary tended to degenerate in cultured worms and it seems likely that this is the primary reason for the production of abnormal eggs *in vitro*.

Alteration of the culture conditions in a variety of ways, including variation of the serum concentration between 5 and 50%, use of dialysis tubing to hold the organisms, various flow systems and addition of steroids, had no effect on development.

Extracts of adult male worms, shown by Shaw *et al.* (1977) to stimulate development of female vitellaria *in vitro*, did not improve vitelline cell development or stimulate egg production. A low-oxygen gas mixture (2% CO_2, 3% O_2, 95% N_2), or use of reducing agents to give a low redox potential, also failed to induce normal egg production (Basch, 1981b).

Cultured worm pairs, which had reached the stage of forming abnormal eggs, did not complete development to full maturity when surgically implanted into mouse mesenteric veins (Basch and Humbert, 1981). However, younger schistosomes, cultured *in vitro* for 13 d, were capable of full development on transfer to a host; normal adult worms transferred to cultures for up to 53 d recovered full egg production on being returned to the host. Thus, it seems that schistosomes lack some essential stimulus or growth requirement *in vitro*, acting at some stage between 13 d and adulthood, which cannot be replaced by later contact with the host.

Although viable eggs have not yet been produced by schistosomes grown *in vitro*, adult females continue to lay eggs for short periods; recently interest has centred on the culture conditions required for development of viable miracidia from such eggs (Newport and Weller, 1982a,b). A chemically defined medium containing 0.2% albumin was supplemented with various biological components, of which lactalbumin hydrolysate and casein hydrolysate alone stimulated egg deposition and development of miracidia. Some miracidia, hatched from eggs in medium supplemented with casein hydrolysate and fortified with foetal calf serum, showed normal behaviour and were able to penetrate snails and establish patent infections with the production of cercariae. A minimum essential medium for cultivation of the eggs of *Schistosoma japonicum* has been developed by Kawanaka *et al.* (1983).

These analyses of the requirements for development of schistosome eggs have not been paralleled for the growth of schistosomes, although Basch (1984) has reinvestigated the requirement for feeding on red blood cells (RBC). Early growth and development was more rapid and pairs of near-adult worms formed more readily when schistosomes were fed on "pink ghosts" (which retain about 5% of the original haemoglobin after haemolysis in water) than in cultures containing living, human RBC. Probably the utilization of glucose or the production of lactate by (anaerobically) respiring, living RBC is deleterious to the parasites. Schistosomula remained stunted and underdeveloped when fed on white ghosts, haemolysed in 5.0 mM phosphate buffer at pH 8.0, which contain no detectable haemoglobin, thus demonstrating the importance of this protein for normal development. Addition of free haemoglobin or concentrated erythrocyte lysate to cultures did not successfully replace "pink ghosts", suggesting that the physical, particulate, form of the material is important, probably because schistosomes cannot ingest sufficient haemoglobin in soluble form; c.f. the requirement for a particulate supply of food identified for free-living nematodes cultured *in vitro* by Hansen and Hansen (1978).

1° Culture of *S. mansoni*
(Basch, 1981a, b)

MEDIA

CULTURE MEDIUM 169

BME (Eagle's basal medium GIBCO) 1.0 litre, to which is added 1.0 g lactalbumin hydrolysate and 1.0 g glucose plus the following:

	ml
Hypoxanthine, 10^{-3} M	0.5
Serotonin, 10^{-3} M	1.0
Insulin U-100 (Lilly), 100 iu ml^{-1} ..	2.0
or	
Insulin, crystalline (Bovine origin, Sigma), 8 mg ml^{-1}	1.0
Hydrocortisone, 10^{-3} M	1.0
Triiodothyronine, 2×10^{-4} M	1.0
MEM vitamins (GIbco), 100 times concentrate	5.0
Schneider's medium (GIbco) unconcentrated	50
HEPES buffer	2.4 g
*Serum, unconcentrated	100 ml

*Serum was prepared from frozen, outdated human plasma obtained from a transfusion service. Thawed plasma was aseptically mixed with 5.0 mm glass beads and a large Teflon-coated stirring bar and mixed on a magnetic stirrer. Calcium chloride solution (1.0 M) was added dropwise to a final concentration of 0.2 M and a few drops of reconstituted bovine thrombin to produce the formation of a tight fibrin clot. The serum was decanted, centrifuged (10 000 rpm, 15 min) and stored frozen at – 20°C.

WASH MEDIUM

BME plus 15 mM HEPES buffer with 300 iu ml^{-1} penicillin, 300 $\mu g\ ml^{-1}$ streptomycin and 160 $\mu g\ ml^{-1}$ gentamicin.

The final medium is diluted to 275 mosmol kg^{-1} with triple-distilled water, adjusted to pH 7.4 with NaOH (5.0 N) and sterilized by membrane filtration.

TECHNIQUES

(a) PREPARATION OF SCHISTOSOMULA

A number of techniques for inducing cercariae to transform into schistosomula are detailed by Fried (1978). Basch's (1981a) modification of the mechanical method of Colley and Wikel (1974) is described here and Section II.C.1° gives the skin penetration method that has been used preferentially for the study of immune mechanisms *in vitro*.

(1) *Biomphalaria glabrata* snails, 5–9 weeks after infection, are placed in 20 ml of water in a 100 ml beaker (2–3 h), 8.0 cm below a 75 watt bulb.

(ii) The water containing cercariae is filtered through nylon mesh (Sartorius mesh size 0.1 mm) and transferred into two 50 ml glass screw-top centrifuge tubes, cooled to 10 °C to immobilize cercariae and centrifuged (1000 rpm, 1.0 min) to concentrate them.

(iii) Most of the supernatant is discarded and 20 ml of the washing medium (10 °C) containing antibiotics is added and centrifuged as in (ii). *All subsequent steps are carried out aseptically under a bacteriological hood.*

(iv) The cercariae are resuspended in 25 ml wash medium and spun down as before.

(v) The cercarial pellet is then suspended in 25 ml wash medium in a closed centrifuge tube and rotated on an SI-150 rotator (Scientific Industries) to keep the cercariae suspended for 10 min.

(vi) The cercariae are re-centrifuged (1000 rpm) and transferred to fresh centrifuge tubes containing 25 ml washing medium; the rotation step is repeated, followed by centrifugation as before.

(vii) Steps (v) and (vi) are repeated.

(viii) The pellet of washed cercariae and 2.0 ml of medium are drawn into a 5.0 ml glass hypodermic syringe through a 120 mm length of silastic tubing (Dow Corning Medical Products) attached to a blunted 16-gauge needle.

(ix) The needle and tubing are removed from the syringe containing the cercariae and replaced with a micro-emulsifying syringe needle. The other end of the needle is then attached to a second, empty, glass syringe. The cercariae are forced from one syringe to the other slowly, 12 times, causing shearing of tails from cercarial bodies.

(xi) The syringe containing sheared cercariae is removed and they are expressed into a plastic Petri dish (35 × 10 mm). The dish is swirled with a moderately vigorous rotary motion to concentrate the cercarial bodies centrally, leaving their tails in suspension.

(xii) Under a binocular dissecting microscope, the medium containing tails is removed without disturbing the mound of bodies. Fresh wash medium, containing 5% human serum, is added and the swirling and removal of tails repeated until separation approaches 100%.

(xiii) The cercarial bodies are then transferred to a 15 ml centrifuge tube containing 9.0 ml wash medium plus 5% serum and the rotation, centrifugation and transfer to new medium, steps (v) and (vi), are repeated three times.

(b) **CULTURE OF SCHISTOSOMULA**

(i) Cercarial bodies, now considered to be schistosomula, are transferred into several small Petri dishes (35 × 10 mm) containing 2.0–2.5 ml pre-gassed culture medium 169. Eight small Petri dishes are held in one large dish (150 × 15 mm) and incubated at 36°C in a humidified atmosphere with flowing 5% CO_2 in air.

(ii) Fresh human RBC type O, obtained from a transfusion service, are washed twice by centrifugation in wash medium and stored at 4°C. Samples are checked for sterility. A drop of sterile washed RBC is added to each culture after 24–48 h.

(iii) Culture medium and RBC are changed twice weekly.

(iv) Partially grown male and female worms (five each at stage 18; Basch, 1981a) are transferred to metal-capped Leighton tubes and held at 37–38°C in flowing 5% CO_2 in air.

In this system black intestinal pigment, derived from digestion of haemoglobin, is seen at 5 d in some schistosomula and at 11–12 d some schistosomula show fusion of the gut caecae, a degree of growth closely comparable to development in the host. Pairing of male and female worms first occurs during the seventh week of culture and eggs begin to form in about 10% of paired females after about two months. Vitellaria (stained with fast-red; Johri and Smyth, 1956) remain only partially developed. Males form testes with some indication of development of spermatozoa.

2. *Culture of other species of schistosome*

Basch and O'Toole (1982) cultured *Schistosomatium douthitti*, a schistosome of rodents, in the system described above (Section II.B.1°).

Growth *in vivo* of *Sch. douthitti* is more rapid than that of *S. mansoni* and, *in vitro*, cercariae developed to near-adults with egg production within one month, compared with two or more months for *S. mansoni*. As with *S. mansoni*, the cultured worms did not achieve the same size as in the host, and eggs were not viable, although they appeared more normal in size and shape than those of *S. mansoni* produced *in vitro*. Female *Sch. douthitti* are known to form eggs in the host in the absence of males and, not surprisingly, also produced parthenogenetic eggs *in vitro*. Female *S. mansoni* grown without males in the host do not mature, but Basch and Basch (1984) have found that *Sch. douthitti* males can stimulate *S. mansoni* females to produce eggs, although no transfer of spermatozoa occurs. These parthenogenetic eggs are haploid, but are capable of producing miracidia that develop normally in snails and yield cercariae infective for mice. Apparently, the stimulus for normal egg production is independent of insemination or fertilization, quite unlike the situation in some cestodes in which normal egg production *in vitro* is dependent on physical culture conditions that allow insemination to occur (Smyth and Davies, 1974).

Fried (1978) reviewed the culture of several other species of schistosome including *S. haematobium*, *S. japonicum*, *S. bovis* and the avian schistosome, *Trichobilharzia ocellata*. Although the results obtained with *S. haematobium* (Smith *et al.*, 1976) and *T. ocellata* (Howell and Bourns, 1974) were very encouraging, no further work on these species has been reported. However, Yasuraoka *et al.* (1978) have grown *S. japonicum* in a medium consisting of equal volumes of human or rabbit serum and the defined medium NCTC 109 (see Taylor and Baker, 1968, pp. 351–354) supplemented with washed RBC. Male *S. japonicum* developed to adults, with spermatozoa in the anterior testes, and paired with female worms, but eggs were not produced. *S. mansoni* developed to a similar stage in this medium.

Yasuraoka *et al.* (1978) initiated their cultures with cercariae, using the method of Eveland and Morse (1975) detailed by Fried (1978). After preliminary washing to remove contaminating microorganisms, 60–70% of *S. japonicum* and about 20% of *S. mansoni* transformed into schistosomula capable of growth when transferred into the above culture medium containing human or rabbit serum, but the conversion of some organisms took as long as 72 h. Nearly all *S. mansoni* cercariae transformed within 3 h on contact with 50% rat serum (Eveland and Morse, 1975). Many of the schistosomula produced thus, without mechanical removal of the cercarial tail or penetration of skin, grew to a late stage of development in a suitable medium containing rabbit serum (Eveland *et al.*, 1979). Apparently, rat serum is more effective at stimulating cercarial transformation than human serum and analysis of the component involved could be very interesting.

C. *IN VITRO* DISSECTION OF IMMUNE MECHANISMS

Experimental observations of schistosome infections in laboratory hosts have shown that adult worms of a primary infection are often resistant to the immunity they induce; the main target of the immune response seems to be the young schistosomula of the second and later infections. This phenomenon, acquired resistance to reinfection in the presence of adult worms from the primary infection, has been called concomitant immunity by analogy with a similar immune situation seen with some tumours (Smithers and Terry, 1969). Early *in vitro* work on the nature of the immune mechanisms acting against *S. mansoni* schistosomula established that the young organism could be killed by antibody and complement, but that protection against attack developed very rapidly both *in vitro* and in the host (Clegg and Smithers, 1972). The mechanism of protection may involve the uptake of host macromolecules, which has been shown to occur at the surface membrane, but there is now evidence for a number of other protective mechanisms (see Capron *et al.*, 1982a). Development of the *in vitro* approach has generated much information on the immune mechanisms that can kill young schistosomula outside the host and has been particularly advantageous in opening up the study of human effector mechanisms. The field has recently been extensively reviewed (McLaren, 1980; Butterworth *et al.*, 1982; Capron *et al.*, 1982a,b; Butterworth, 1984) and it is appropriate here to summarize only the main findings without reference to individual papers.

Although schistosomula are vulnerable to direct attack by antibodies and complement in immune rhesus monkey serum, the rate of damage is relatively slow and death of some organisms may take as long as 4 d. With human immune serum, much less extensive destruction is mediated by antibodies, especially with homologous complement. Butterworth and his colleagues showed that normal eosinophils can destroy schistosomula, labelled with IgG antibodies in the absence of complement, although most other workers have found that complement can enhance the effectiveness of antibody/eosinophil attack (Butterworth *et al.*, 1982; Butterworth, 1984). A role for the eosinophil as an effector cell in immunity to schistosomes and other helminths has generated much interest, partly because the function of this cell type was previously obscure and because it offered an explanation for the well-known association of eosinophilia with helminth infections (Kay, 1979; Butterworth, 1984). Eosinophils derived from normal or eosinophilic donors are less effective *in vitro* than those from slightly eosinophilic ones, suggesting that these cells are activated; indeed, a considerable amount of evidence on the mechanism of activation is now available (Butterworth, 1984). Activation of eosinophils has also been studied very effectively in the rat model in which antibody-dependent eosinophil-mediated killing of schistosomula has been shown to require an accessory signal from mast cells (Capron *et al.*, 1982a). The mast cells bind to schistosomula but do not damage them directly; they release eosinophil chemotactic factors (ECF-A) which appear to enhance the destructive capacity of the eosinophils as well as attracting more of them to the schistosomulum. The mast cell is triggered to release ECF-A tetrapeptides by anaphylactic antibody and antibodies of this type also bind the eosinophils to the schistosomula. In the rat system IgG_{2a} anaphylactic antibodies are involved during the first 6 weeks of a schistosome infection; subsequently IgE is more important (Capron *et al.*, 1982a). Human eosinophils are also activated by ECF-A tetrapeptides and have receptors for IgE, but the role of IgE in the human system is less clear: IgG_1, rather than IgE, has been implicated in attachment of eosinophils to schistosomula, but there is evidence that the IgE/mast cell/eosinophil complex probably plays a significant part in the immune mechanism (Butterworth, 1984).

The role of the neutrophil granulocyte in cell-mediated damage to *S. mansoni* schistosomula has generated considerable, as yet unresolved, controversy. Several groups have described antibody- and/or complement-dependent effects of human or rat neutrophils on schistosomula, whereas others have found neutrophils much less effective (McLaren, 1980; Butterworth, 1984). These inconsistent results from different laboratories emphasize the variability which plagues many *in vitro* systems and underline the importance of controlling as many technical variables as possible.

Mechanisms of damage to *S. mansoni* schistosomula demonstrated *in vitro* are not restricted to eosinophil and neutrophil granulocyte-mediated effects. Extensive work by Capron and his colleagues has shown that rat macrophages can bind IgE antibodies complexed with schistosome antigen and mediate specific destruction of schistosomula. Human monocytes from the blood and baboon peritoneal macrophages also develop the capacity to kill schistosomula *in vitro* after binding IgE/antigen complexes. Blood platelets, of rat and human origin, are capable of killing schistosomula, and again the antibody involved is of the IgE class. Evidently, IgE may have an important role in immunity in addition to its involvement in mast cell/eosinophil-mediated damage (Capron *et al.*, 1982a,b). In addition, macrophages, stimulated by conventional T-lymphocyte mechanisms, have been shown to kill schistosomula *in vitro* in the absence of antibody. Lymphoid cells have not been found to mediate specific destruction, but activated T-cells can show non-specific toxicity for schistosomula (see Butterworth, 1984).

The great variety of immune effector mechanisms shown to be effective against schistosomula *in vitro* emphasizes the general point that *in vitro* results define the range of possible mechanisms but cannot demonstrate which mechanisms actually operate in the immune host. It is difficult to show unequivocally that any particular mechanism occurs in a laboratory host, but this is an important and active area of experimental study and substantial evidence for the involvement of antibody-dependent cell-mediated mechanisms in the host has been accumulated (Capron *et al.*, 1982a; Butterworth, 1984). The difficulty is still greater in man, where direct experiment is precluded, but correlation of epidemiological studies with the results of *in vitro* assays is yielding valuable information (Butterworth, 1984). The ultimate objective of such studies is the identification of the target antigens of the effective immune mechanism leading to development of a vaccine.

Study of the effects of various cell types on schistosomula *in vitro* has required some compromise between ideal conditions for maintenance of cell function and for culture of the parasite. The medium used by Butterworth's group (Eagle's minimal essential medium buffered with HEPES and containing 10% heat-inactivated foetal calf serum) gives adequate survival of effector cells and allows survival of schistosomula for 48 h, after which death occurs progressively (Butterworth and Vadas, 1979; Section II.C.1°). This period of survival is quite long enough for assay of the destructive effect of cells, which reaches a maximum in about 16–18 h, but, since schistosomula are not able to grow in this medium, there is some doubt whether they can be considered completely normal during the assay. Research towards a medium that would allow growth of the parasite as well as normal function of effector cells would be well worthwhile. Such a medium will be essential for study of immune mechanisms that may act against older schistosomes in culture. The concept that only young schistosomula are susceptible to immune attack may be an oversimplification. Recent *in vivo* evidence strongly suggests that there is a second phase of destruction in immune animals acting against developing juvenile schistosomes (see Butterworth, 1984).

Initial studies towards a defined medium for maintenance of *S. mansoni* schistosomula have shown that addition of selenite and insulin to a chemically defined cell culture medium allowed survival for 13–22 d. A low-molecular-weight fraction from human serum increased survival to 30 d, but no growth was observed (Levi-Schaffer and Smolarsky, 1981).

1. *Detailed Techniques*

1° Preparation of *S. mansoni* schistosomula
(Butterworth and Vadas, 1979)

MEDIA

(i) Earle's balanced salt solution (see Taylor and Baker, 1968, p. 350), containing lactalbumin hydrolysate, 0.5%, HEPES, 10 mM; penicillin, 100 iu ml^{-1}; streptomycin, 100 μg ml^{-1}, pH 7.2 (E/LAC).
(ii) Hanks's balanced salt solution (see Taylor and Baker, 1968, p. 350), containing the same antibiotics (HBSS).
(iii) Foetal calf serum (FCS), inactivated (56°C, 1 h): each batch must be tested separately for direct toxicity to schistosomula sera from Flow Laboratories gave consistently good results).

TECHNIQUES

(a) PENETRATION ASSEMBLIES

Each assembly is made up from two glass O-ring joints of internal diameter 20 mm, an O-ring and a metal clamp (Laboratory Glass Inc.). One joint is cut off to yield an open-ended tube 10 cm high. The other joint is cut similarly and then heat-fused to give a closed-end tube about 8.0 cm long and 20–30 ml in capacity with a conical base. Four such assemblies, sterilized by autoclaving, are used for each preparation of schistosomula.

(b) SKIN PENETRATION TECHNIQUE

Young male rats are killed by asphyxiation using solid CO_2 or chloroform. The hair is removed from the abdomen and surface of the thorax with an electric animal clipper, care being taken not to the nick the skin. The surface is thoroughly washed with tap water and a large strip of skin extending over the entire abdomen and thorax is excised carefully so as to remove as little subcutaneous fat as possible. The skin is placed in a plastic Petri dish containing HBSS and its under surface is rubbed vigorously with sterile gauze pads until the subcutaneous tissue and part of the dermis have been removed. The skin, which now adheres to the Petri dish, is washed three times with HBSS and checked for the absence of holes. The skin from one rat is sufficient for two penetration assemblies. Skin from male mice, hamsters and other laboratory mammals can be used, but it is generally more delicate than rat skin.

The closed half of each glass assembly is filled with E/LAC containing 5% FCS. The skin is mounted, inner side down, over the medium, care being taken to avoid trapping air bubbles between the medium and the skin: this is best achieved by applying the piece of skin while it is still adhering to the Petri dish. The O-ring is then positioned between the epidermal surface of the skin and the upper joint and the two joints are clamped together. [See Clegg and Smithers (1972) for a diagram of a completed assembly]. The upper tube and skin surface are then rinsed three times with aquarium water and the assembly is checked for leaks by inversion. Assemblies are mounted in a water bath at 37 °C so that only the lower joint is immersed.

Cercariae of *S. mansoni*, collected not more than 3 h after shedding, are added to the upper tube which is then covered with aluminium foil to exclude light. *Rubber gloves should be worn and great care should be exercised when handling cercariae*. The assembly is incubated for 3 h at 37 °C, allowing the cercariae to penetrate the skin, lose their tails and fall as transformed schistosomula to the bottom of the lower tube. Up to 50 000 cercariae may be added to each assembly before contamination with untransformed cercariae rises above the base level (not more than 5%). After incubation the water in the upper tube is removed, the apparatus is disassembled and the skin is discarded. Two-thirds of the medium in the lower tube is then removed and the remainder containing the schistosomula transferred to a 50 ml centrifuge tube. The lower tube is rinsed twice with E/LAC/5% FCS and the washings are also transferred. The schistosomula are washed twice by centrifugation in E/LAC/5% FCS and finally resuspended in E/LAC/10% FCS. Damage to the organisms is minimized if the centrifuge is brought rapidly to 200*g*, held at this speed for not more than 30 s and allowed to decelerate slowly.

A sample of the schistosomula is examined and counted under a dissecting microscope. A yield of 20–30% of added cercariae is usual and viability should be greater than 95%. A preparation made late in the day can be stored at 4 °C overnight for use in cytotoxicity assays on the following day, but better results are obtained with fresh organisms.

2° Antibody-dependent cell-mediated assays for *S. mansoni*

(Butterworth and Vadas, 1979)

MEDIA USED FOR PREPARATION OF EFFECTOR CELLS

(i) Eagle's minimal essential medium (see Taylor and Baker, 1968, p. 355) containing glutamine, 2.0 mM; HEPES, 20 mM; penicillin, 100 μg ml^{-1}; streptomycin, 100 μg ml^{-1}; pH 7.2 (MEM).
(ii) MEM containing 10% heat-inactivated (56°C,1 h) FCS (MEM/10% FCS).
(iii) E/LAC (Section II.C. 1°).
(iv) Phosphate buffered saline, pH 7.2 (PBS).
(v) Heparin (preservative-free), 100 iu ml^{-1} in PBS.
(vi) Metrizamide (Nyegaard, Oslo).
(vii) Tyrode's balanced salt solution supplemented with 0.1% gelatin and 30 mg litre^{-1} deoxyribonuclease (TGD).

TECHNIQUES

(a) PREPARATION OF EFFECTOR CELLS

Antibody-dependent cell-mediated cytotoxicity can be demonstrated with unpurified human peripheral blood leukocytes; this is the most convenient method for clinical studies. Venous blood is drawn into a syringe containing 100 iu of heparin. After mixing by inversion the syringe is held vertically for 30 min at 37°C during which time the RBC sediment. The leukocyte-rich supernatant is gently expelled through a needle into a fresh tube and the cells are washed twice in MEM/FCS by centrifugation (250*g*, 5 min) to remove platelets. The unpurified leukocytes are resuspended in MEM/FCS and counted (Coulter counter or haemocytometer). Appropriate dilutions are made in MEM/FCS; effector to target ratios of 2000 : 1 or 4000 : 1 (100 schistosomula and 4 $\times$ 10^5 cells per tube) are usually sufficient.

Purified eosinophils or neutrophils can be prepared as follows (for full details see Vadas *et al.*, 1979). Discontinuous gradients of hypertonic Metrizamide are prepared from a fresh stock solution of 30% in TGD diluted to 18, 20, 22, 23, 24 and 25% in TGD by layering 2.0 ml aliquots of each concentration on top of each other in a 15 ml conical centrifuge tube, with the most dense at the bottom. Unpurified leucocytes (not more than 70 $\times$ 10^6) are applied in 1.0–2.0 ml and the tubes are centrifuged (1200*g*, 45 min, 20°C). Discrete bands form at each interface; these can be carefully withdrawn with a Pasteur pipette, washed twice with MEM/FCS and counted. Smears are made and stained with Giemsa. Preparations contain >95% eosinophils in the 23/24 or 24/25 interface.

(b) PREPARATION OF SERA

Antisera are prepared from patients with chronic *S. mansoni* infections. Venous blood is defibrinated by rotation at room temperature in Universal bottles containing glass beads until clotting occurs. Serum is then withdrawn and freed from contaminating RBC by two cycles of centrifugation (2000*g*, 5 min). Aliquots of serum are stored at −20°C and inactivated at 56°C for 1 h immediately before use. Appropriate dilutions, at ½ $\log_{10}$ intervals, are made in MEM/FCS; a range of 1/10 to 1/1000 (final dilution) is usually adequate. Serum from immune mice, rats, baboons or rhesus monkeys may also be used.

Complement can be prepared from guinea-pig blood, obtained by cardiac puncture, by defibrination and centrifugation as above. Aliquots must be used immediately or stored at −70°C and thawed only once. The serum should be used at a final dilution of 1/10 to 1/40 for cell- and complement-dependent assays. Some guinea-pig sera show direct toxicity to schistosomula, and freshly prepared human or baboon serum may be preferable.

MICROSCOPICAL ASSAYS FOR CELL ADHERENCE AND DAMAGE TO SCHISTOSOMULA

The assay is carried out in a total volume of 0.4 ml in 38 × 7 mm round-bottomed plastic tubes. There replicates of each sample are set up. 0.1 ml of E/LAC/10% FCS (Section II.C.1°) containing 100 schistosomula is dispensed by micropipette into the required number of tubes. Aliquots of 0.1 ml of diluted antiserum and 0.2 ml effector cells are added. Control tubes receive an equivalent volume of MEM/FCS. Tubes containing schistosomula, antibody and effector cells are incubated (37°C) in humidified air-tight plastic boxes. Studies involving effector cells, with or without complement, can be scored at 2–4 h for cell adherence or at 18–48 h for damage. At the end of the incubation period, most of the supernatant is withdrawn from the assay tube and discarded. The pellet is gently resuspended in 0.3–0.4 ml MEM/FCS and allowed to sediment (at 1*g*) for 5 min to eliminate unbound cells which may obscure scoring. The supernatant is again removed and the organisms transferred, in about 50 μl of medium, to a slide that has previously been coated with toluidine blue in methanol and allowed to dry. A coverslip is applied and may be sealed with nail varnish. Separate preparations without toluidine blue can be examined by phase-contrast or interference-contrast microscopy.

Effector cell adherence is scored by counting the number of cells bound to each organism. A minimum of 50–100 schistosomula in each of two or three replicates is scored. Recording of results is arbitary and depends on the nature and number of effector cells. *In all cases direct comparisons are made between test and control preparations.*

Damage to schistosomula is characterized by loss of motility, swelling or shrinking and change of shape, apparent internal granulation and dense uptake of toluidine blue, especially into nuclei. Some of these changes (especially slight uptake of toluidine blue into the surface of the organisms) may occur as the preparation dries; *it is therefore essential to prepare and examine each slide within 10 min.*

A cytotoxicity assay using ^{51}Cr-labelled schistosomula has been used extensively, particularly in clinical studies involving the testing of large numbers of sera. The advantage of this isotopic assay is that it is objective, sensitive and highly reproducible, but the release of ^{51}Cr does not necessarily reflect irreversible damage to the organisms and the microscopic assay is essential for accurate estimation of the extent of damage to schistosomula. For details of the isotopic assay and methods of assessing results, see Butterworth and Vadas (1979).

III. FASCIOLIDAE

A. CULTURE OF *FASCIOLA HEPATICA*

Although *Fasciola hepatica* is responsible for causing disease in the livers of sheep and cattle that is of significant economic importance, relatively few attempts have been made to culture either the stages found in the snail intermediate host or those from the vertebrate host (see Fried, 1978). In the last few years, there has been only one report of an attempt to culture the snail stages (Davies, 1982), but some progress has been made with culture of the metacercarial stage that hatches from a resistant cyst in the duodenum of the vertebrate host.

Davies and Smyth (1978) cultured metacercariae in the defined cell culture medium NCTC 135 supplemented with a high concentration of chicken serum and a small volume of sheep RBC. In one experiment, some organisms reached a length of 1.2 mm with development of early genital primordia. Smith and Clegg (1981) tried to culture metacercariae in a medium (Earle's balanced saline supplemented with lactalbumin hydrolysate (E/LAC, Section II.C. 1°), 50% human serum and 2% human RBC, that was known to support considerable *in vitro* development of *Schistosoma haematobium* and *S. mansoni* (Smith *et al.*, 1976). Growth in this medium was compared with that in cultures in which the E/LAC component had been replaced by defined cell culture media (NCTC 135 or RPMI 1640). Significantly better growth and development were obtained in the medium based on RPMI 1640 than in either of the other two media. Growth was essentially linear during a 14-week period of culture at a rate of about one-quarter of the rate of growth in an experimental host. The worms reached a mean length of 3.1 mm and showed partial development of the uterus, cirrus and testes. A few individual worms (7 in 4 experiments) began to grow more rapidly after 6 weeks in culture and reached a length of 6–7 mm, comparable with the size of sexually mature *Fasciola* grown in mice. These larger worms all showed extensive development of the uterus and a morphologically normal cirrus and cirrus-sac. Testes formed in all the specimens and in two, fully developed spermatozoa were detected in squash preparations. Vitellaria developed extensively, but the ovary was rudimentary and there was no sign of egg formation. The authors comment that the individual worms that suddenly developed more rapidly after 6 weeks slower growth *in vitro* seemed to have adapted in some way to the nutritional or physical culture conditions.

1° Excystment of *Fasciola hepatica* metacercaria
(Dixon, 1966; Smith and Clegg, 1981)

MEDIA

(i) *Activation medium*. Pure CO_2 is bubbled for 30 s through 10 ml of distilled water at room temperature in a screw-top plastic Universal tube. Dry sodium dithionite is added to give 0.02 M final concentration and the tube is closed with a cap and brought to 37 °C. A fine cloudy precipitate forms.

(ii) *Emergence medium*. Hanks's balanced salt solution (HBSS: see Taylor and Baker, 1968, p. 350), buffered to pH 7.4 with 30 mM HEPES, with 10% v/v rabbit bile (this is aspirated aseptically from freshly removed gall bladders and stored in small volumes at − 20 °C after snap-freezing).

TECHNIQUE

This simple two-step method is based on the stimuli for excystation determined by Dixon (1966). The original paper should be consulted for the experimental basis for the stimuli chosen.

(i) Cercarial emergence from infected *Lymnaea truncatula* is enhanced by placing the snails in a glass crystallizing dish of tap water and reducing the temperature to 14 °C with a beaker of ice.

As the temperature returns to ambient (22 °C), cercariae emerge in large numbers and encyst on the glass surface. After 24 h they can be scraped off gently with a scalpel and stored at 4 °C in distilled water containing 100 iu ml^{-1} penicillin and 100 μg ml^{-1} streptomycin (cysts can be used after storage for up to 10 months).

(ii) Incubate cysts (1 h, 37 °C) in 10 ml of *activation medium* by pouring pre-warmed medium on to the cysts in a plastic Universal screw-top tube and closing it immediately. This supplies three conditions essential for activation of the metacercariae within the cysts: elevated temperature, a high pCO_2 and reducing conditions.

(iii) Wash cysts free of activation medium (which is toxic to hatched metacercariae) by allowing the cysts to sediment twice through distilled water; transfer to a fresh plastic tube.

(iv) Add 5.0 ml of pre-warmed *emergence medium* to the cysts and incubate (37 °C, 2–2.5 h). Emergence of metacercariae begins approximately 40 min after transfer of the cysts to the emergence medium and reaches high levels (70–85 %) after about 2 h.

A similar excystment method has been described (Tielens *et al.*, 1981) in which the reductant was omitted from the activation medium and sheep bile or taurocholic acid was added, after 1 h incubation, to stimulate emergence. An experimental apparatus was used to allow large numbers of emerging metacercariae to escape into a medium with a low bile content, since prolonged exposure to bile was found to be toxic for metacercariae. Fried and Stromberg (1985) have reported an excystment method for the related species *Fascioloides magna*.

2° Culture of *Fasciola hepatica* metacercariae
(Smith and Clegg, 1981)

MEDIA

(a) CULTURE MEDIUM

(i) 50% v/v RPMI 1640 (a chemically defined cell-culture medium devised at the Rothwell Park Memorial Institute, Buffalo USA, obtained from Gibco) plus 50% v/v human serum (prepared from type A donated blood allowed to clot at room temperature). The serum was separated by centrifugation (10 000*g*, 30 min) and then filtered through a Sartorius positive-pressure filter using a glass-fibre prefilter and a cellulose nitrate membrane (0.22 μm pore size). Enough serum was obtained from a single donor to complete a single culture experiment.

(ii) Type A human RBC [washed twice by centrifugation (2000*g*, 5 min), stored (4 °C) in RPMI 1640 (pH 7.4) at a concentration of 20% v/v and used at a concentration of 2% v/v]; 50 iu ml^{-1} penicillin and 50 $\mu g\ ml^{-1}$ streptomycin are added immediately before use.

(b) WASHING MEDIUM

Hanks's balanced salt solution (HBSS: see Taylor and Baker, 1968, p. 350), buffered to pH 7.4 with 30 mM HEPES, plus 30% v/v human serum (the high serum content is necessary to prevent metacercariae sticking to Pasteur pipettes).

TECHNIQUE

All culture procedures should be carried out in a laminar-flow hood using standard aseptic technique.

(i) Newly excysted metacercariae are freed from contaminating micro-organisms by sedimentation through 10 ml of washing medium five times.
(ii) Washed metacercariae (15–20 per tube) are transferred to plastic Leighton type culture tubes (Nuclon, Gibco) and 2.0 ml of culture medium added.
(iii) 2% v/v washed human RBC is added (from 20% stock) and each culture tube is gassed (30 s, 8% CO_2 in air) through a sterile cotton-plugged Pasteur pipette. The screw-tops of the culture tubes are additionally sealed with Parafilm and incubated at 37 °C.
(iv) Culture medium and RBC are replaced at 3–4 d intervals. Metacercariae are observed through the optical base of the culture tube with an inverted microscope and their growth measured with a calibrated eye-piece micrometer.
(v) The growth rate is expressed on the basis of the mean length, ± standard deviation, of the six most rapidly growing parasites in each culture tube.
(vi) At termination of culture organisms can be fixed in ethanol and stained in Gower's carmine for general morphological examination.

Aceto-orcein squash preparations can be made from small pieces of the worm to detect developing spermatozoa (Smyth, 1956).

B. IMMUNE MECHANISMS *IN VITRO*

Butterworth (1984) has recently reviewed the relatively limited amount of work that has been carried out on the nature of immune mechanisms that may act on. *F. hepatica*. Acquired immunity to reinfection with *Fasciola* has been demonstrated in cattle and laboratory rats but not in sheep, the major host.

Antibodies from infected cattle can bind to the surface of recently excysted metacercariae but do not mediate damage in the presence of bovine eosinophils or neutrophils. Similarly, rat peritoneal exudate cells are able to bind to the surface of metacercariae *in vitro*, in the presence of immune serum, but they are apparently undamaged since the yield of adult worms was not reduced on transfer to the peritoneal cavity of normal rats. However, metacercariae injected into the peritoneal cavity of immune rats and subsequently examined by electron microscopy showed severe damage to the surface, associated with adherence of eosinophils. Apparently, the damage that occurs *in vivo* is not reproduced in the *in vitro* systems used so far. This may be because, under the conditions of the assay, rapid turnover of the worms' surface (which has been observed *in vitro*) was sufficient to prevent damage, or because the cells used *in vitro* were not appropriately activated (see Butterworth, 1984).

C. *IN VITRO* SCREENING OF DRUGS

Screening assays are often used in the search for new anti-nematode agents (Jenkins, 1982) but the discovery of new drugs active against trematodes of economic importance, such as the schistosomes and *F. hepatica*, still depends on *in vivo* assays in laboratory hosts. *In vitro* assays require far lower concentrations of new compounds than *in vivo* systems, thus allowing preliminary testing of a wider range of potential drugs. Using the above culture system for *F. hepatica* (Section III.A.1°, 2°) Ibarra and Jenkins (1984) examined the susceptibility of newly excysted, 6-week and 12-week-old worms to a range of known fasciolicides *in vitro*. Newly excysted organisms were susceptible only to diamphenithide, the free amine of diamphenethide, emetine hydrochloride and albendazole. The older worms were affected by a much wider range of drugs, including the chlorinated hydrocarbons, the substituted phenols and the salicylanilides, but their susceptibility to diamphenethide and its active metabolite, the free amine, was significantly lower. The activity of the known fasciolicides *in vitro* is thus closely similar to their activity *in vivo*. Several broad-spectrum anti-nematode drugs showed activity *in vitro* against newly excysted worms, but 12 potent anti-protozoal agents were inactive. The authors conclude that the *in vitro* assay system is likely to be useful for detecting anti-*Fasciola* activity in new compounds.

IV. CULTURE OF OTHER TREMATODES

Fried (1978) reviewed cultivation techniques for trematodes belonging to 10 families. The greatest degree of development was obtained with *Cotylurus lutzi* (Strigeidae) by Basch *et al.* (1973), who cultured the tetracotyle (an encysted metacercaria with characteristic muscular pseudosuckers) to the adult stage with production of normal eggs. Miracidia from eggs laid *in vitro* were infective for snails and developed to form tetracotyles (which encyst in the snail tissue). These tetracotyles developed normally into adult worms when fed to finches, the definitive host (see Taylor and Baker, 1978, pp. 172–173). Similarly excellent growth *in vitro* was also achieved for *Sphaeridiotrema globulus* (Psilostomatidae) by Berntzen and Macy (1969). Chemically excysted metacercariae were cultured to adult worms which produced eggs capable of developing hatchable miracidia (see Taylor and Baker, 1978, pp. 178–179).

Surprisingly few additional reports of cultivation have appeared. Mitchell *et al.* (1978) cultured *C. erraticus* (Strigeidae) in tissue culture media enriched with high concentrations of chicken serum. Worms grown in medium with the greatest amount of serum (80% v/v) showed the fastest rate of development, measured by the time taken for eggs to appear in the uterus. The testes, ovaries and vitellaria of these organisms were comparable in structure and histochemistry with those of worms grown in host gulls, but the eggs had abnormal shells and failed to embryonate.

Schnier and Fried (1980) cultured *Amblosoma suwaense* (Brachylaimidae) from metacercariae to egg-producing adults in NCTC 135 supplemented with 20% hen's egg yolk. Eggs were formed with 4 d and some contained developing embryos. The mean body area of the adult worm grown *in vitro* was 20% less than that of the metacercaria but the gonads and vitellaria increased in size. As in other trematodes with well-developed metacercariae it is difficult to determine whether there is real growth *in vitro* (increase in total number of cells) or a reorganization of existing tissue. A similar decrease in body area accompanies increase in gonad and vitellaria size in metacercariae of *Amblosoma* cultured in living chick embryos (Fried *et al.*, 1981). Fried and Pentz (1983) have also grown metacercariae of *Echinostoma revolutum* (Echinostomatidae) in chick embryos using whole eggs or embryos maintained *in vitro*. In both systems, worms developed to a stage with a coiled uterus in 6–8 d, comparable with worms grown in the chick host for 7 d, although somatic growth was less than *in vivo* and no eggs formed.

Fried and his colleagues have examined the factors responsible for attracting metacercariae of *A. suwaense* to each other, an interesting demonstration of the kind of behavioural problem that can be approached *in vitro*. The organisms apparently respond to lipophilic worm products and thin-layer chromatographic separation (TLC) of lipids released by worms maintained *in vitro* showed that chemo-attraction was stimulated by sterol esters but not by free sterols (Fried and Robinson, 1981). Similar work on *E. revolutum* (Fried *et al.*, 1980) showed that metacercariae were not stimulated in this way, but that adult worms were strongly attracted towards each other. Again, the lipophilic fraction of material secreted by the worms *in vitro* was attractive for other adults and in this case the active substance was found in the free sterol fraction which consisted mainly of cholesterol.

Initial work towards *in vitro* culture of *Orthocoelium scoliocoelium* (Amphistomatidae) has been carried out by Sharma and Sharma (1982). Trematodes of this family have not previously been used for cultivation studies. Adult amphistomes recovered from the rumen of buffalos were maintained *in vitro* with apparently normal egg production for 15 d in a synthetic medium based on the contents of rumen fluid. It is not clear from this report whether any new eggs were produced or whether the eggs laid *in vitro* were already present in the adult worms.

Several studies of *in vitro* excystment of trematode metacercariae have been made that would be of value to workers intending to culture these species: *Himasthla quissetensis* (Echinostomatidae; Kirschner and Bacha, 1980); *Himasthla rhigedana* (Leflore and Bass, 1982); *Bucephaloides grascilescens* (Bucephalidae; Johnston and Halton, 1981); *Cloacitrema michiganensis* (Philopthalmidae; Leflore and Bass, 1983) and *Cotylurus variegatus* (Strigeidae; Thomson and Halton, 1982).

V. CULTURE OF TREMATODE CELLS *IN VITRO*

Media able to support proliferation of helminth cell lines *in vitro* have not yet been developed. Clearly, the ability to grow parasite cells *in vitro* is a crucial step in the study of the cell biology of parasites. At present this is virtually an untouched field, although attempts to culture trematode cells have been going on for some time and many unsuccessful results have not been reported.

Capron and Dupas (1972) found that cells migrated out of cut pieces of tissue taken from juvenile *S. mansoni* and remained alive in mammalian cell culture media while in contact with the explanted tissue. No independent proliferation occurred. On the other hand, Weller and Wheeldon (1970, 1982) showed that mammalian cell culture media were unsuitable for the maintenance of isolated *S. mansoni* cells and carried out exhaustive trials to find a suitable medium. One capable of maintaining several cell types was developed but multiplication of cells remained an elusive goal. An alternative approach, the fusion of trematode cells with a line of mammalian cells able to proliferate easily *in vitro*, has been described by Howell (1981). Hybrid cells that multiplied *in vitro* were formed by fusion of *F. hepatica* cells with cells of a rat fibroblast line using the polyethylene glycol method. Hybridization may well provide a method for studying characteristics of helminth cells *in vitro*, and wider application of this technique is likely.

REFERENCES

Basch, P.F. (1981a). Cultivation of *Schistosoma mansoni in vitro*. I. Establishment of cultures from cercariae and development until pairing. *Journal of Parasitology* **67**, 179–185.

Basch, P.F. (1981b). Cultivation of *Schistosoma mansoni in vitro*. II. Production of infertile eggs by worm pairs cultured from cercariae. *Journal of Parasitology* **67**, 186–190.

Basch, P.F. (1984). Development and behaviour of cultured *Schistosoma mansoni* fed on human erythrocyte ghosts. *American Journal of Tropical Medicine and Hygiene* **33**, 911–917.

Basch, P.F. and Basch, N. (1984). Intergeneric reproductive stimulation and parthenogenesis in *Schistosoma mansoni*. *Parasitology* **89**, 369–376.

Basch, P.F. and Humbert, R. (1981). Cultivation of *Schistosoma mansoni in vitro*. III. Implantation of cultured worms into mouse mesenteric veins. *Journal of Parasitology* **67**, 191–195.

Basch, P.F. and O'Toole, M.L. (1982). Cultivation *in vitro* of *Schistosomatium douthitti* (Trematoda: Schistosomatidae). *International Journal for Parasitology* **12**, 541–545.

Basch, P.F., Di Conza, J.J. and Johnson, B.E. (1973). Strigeid trematodes (*Cotylurus lutzi*) cultured *in vitro*: production of normal eggs with continuance of life cycle. *Journal of Parasitology* **59**, 319–322.

Berntzen, A.K. and Macy, R.W. (1969). *In vitro* cultivation of the digenetic trematode *Sphaeridiotrema globulus* (Rudolphi) from the metacercarial stage to egg production. *Journal of Parasitology* **55**, 136–139.

Buecher, E.J., Perez-Mendez, G., Hansen, E.L. and Yarwood, E. (1974). Sulphydryl compounds under controlled gas in culture of *Schistosoma mansoni* sporocysts. *Proceedings of the Society for Experimental Biology and Medicine* **146**, 1101–1105.

Butterworth, A.E. (1984). Cell-mediated damage to helminths. *Advances in Parasitology* **23**, 143–235.

Butterworth, A.E. and Vadas, M.A. (1979). Immunological studies on schistosomes cultured *in vitro*. *In* "Practical Tissue Culture Applications" (K. Maramorosch and H. Hirumi, eds). pp. 287–307. Academic Press, London and New York.

Butterworth, A.E., Taylor, D.W., Veith, M.C., Vadas, M.A., Dessein, A., Sturrock, R.F. and Wells, E. (1982). Studies on the mechanisms of immunity in human schistosomiasis. *Immunological Reviews* **61**, 5–39.

Capron, A. and Dupas, H. (1972). Attempts to obtain cell cultures from explants of

Schistosoma mansoni. Comptes rendus des Séances hebdomadaire de l' Académie des Sciences de Paris **275**, 1243–1246.

Capron, A., Dessaint, J-P., Capron, M., Joseph, M. and Torpier, G. (1982a). Effector mechanisms of immunity to schistosomes and their regulation. *Immunological Reviews* **6**, 41–66.

Capron, A., Dessaint, J-P., Haque, A. and Capron, M. (1982b). Antibody dependent cell-mediated cytotoxicity against parasites. *Progress in Allergy* **31**, 234–267.

Clegg, J.A. (1965). *In vitro* cultivation of *Schistosoma mansoni. Experimental Parasitology* **16**, 133–147.

Clegg, J.A. and Smithers, S.R. (1972). The effects of immune rhesus monkey serum on schistosomula of *Schistosoma mansoni* during cultivation *in vitro. International Journal for Parasitology* **2**, 79–98.

Colley, D.G. and Wikel, S.K. (1974). *Schistosoma mansoni*: simplified method for the production of schistosomules. *Experimental Parasitology* **35**, 44–51.

Davies, C. (1982). *In vitro* transformation and development of miracidia of *Fasciola hepatica. Parasitology* **84**, LXI [abstract].

Davies, C. and Smyth, J.D. (1978). *In vitro* cultivation of *Fasciola hepatica* metacercariae and of partially developed flukes recovered from mice. *International Journal for Parasitology* **8**, 125–131.

Di Conza, J.J. and Basch, P.F. (1974). Axenic cultivation of *Schistosoma mansoni* daughter sporocysts. *Journal of Parasitology* **60**, 757–763.

Dixon, K.E. (1966). The physiology of excystment of the metacercaria of *Fasciola hepatica* L. *Parasitology* **56**, 431–456.

Eveland, L.K. and Morse, S.I. (1975). *Schistosoma mansoni*: *in vitro* conversion of cercariae to schistosomula. *Parasitology* **71**, 327–335.

Eveland, L.K., Ludwikowska, A. and Madden, K.G. (1979). *Schistosoma mansoni*: long-term culture of *in vitro* derived schistosomula. *Experimental Parasitology* **48**, 375–383.

Fried, B. (1978). Trematoda. *In* "Methods of Cultivating Parasites *In Vitro*" (A.E.R. Taylor and J.R. Baker, eds), pp. 151–192. Academic Press, London and New York.

Fried, B. and Pentz, L. (1983). Cultivation of excysted metacercariae of *Echinostoma revolutum* (Trematoda) in chick embryos. *International Journal for Parasitology* **13**, 219–224.

Fried, B. and Robinson, G.A. (1981). Pairing and aggregation of *Amblosoma suwaense* (Trematoda; Brachylaemidae) metacercariae *in vitro* and partial characterization of lipids involved in chemo-attraction. *Parasitology* **82**, 225–230.

Fried, B. and Stromberg, B.E. (1985). *In vitro* excystment of metacercariae of *Fascioloides magna* (Trematoda). *Journal of Parasitology* **71**, 263–264.

Fried, B., Tancer, R.B. and Fleming, S.J. (1980). *In vitro* pairing of *Echinostoma revolutum* (Trematoda), metacercariae and adults and characterisation of worm products involved in chemo-attraction. *Journal of Parasitology* **66**, 1014–1018.

Fried, B., Heyer, B.L. and Pinski, A.K. (1981). Cultivation of *Amblosoma suwaense* (Trematoda; Brachylaemidae) in chick embryos. *Journal of Parasitology* **67**, 50–52.

Hansen, E.L. and Hansen, J.W. (1978). Nematoda parasitic in animals and plants. *In* "Methods of Cultivating Parasites *In Vitro*" (A.E.R. Taylor and J.R. Baker, eds), pp. 231. Academic Press, London and New York.

Hansen, E.L. and Hansen, J.W. (1980). Present status of culture of schistosomes. *In* "The *in vitro* Cultivation of the Pathogens of Tropical Diseases", Proceedings of workshop, Nairobi, Kenya, 4–9 February, 1979, pp. 1179–1186. Schwabe, Basel.

Howell, M.J. (1981). An approach to the production of helminth antigens *in vitro*, the formation of hybrid cells between *Fasciola hepatica* and a rat fibroblast cell line. *International Journal for Parasitology* **11**, 235–242.

Howell, M.J. and Bourns, T.K.R. (1974). *In vitro* culture of *Trichobilharzia ocellata*. *International Journal for Parasitology* **4**, 471–476.

Ibarra, O.F. and Jenkins, D.C. (1984). An *in vitro* screen for new fasciolicidal agents. *Zeitschrift für Parasitenkunde* **70**, 655–661.

Irie, Y., Basch, P.F. and Beach, N. (1983). Reproductive ultrastructure of adult *Schistosoma mansoni* grown *in vitro*. *Journal of Parasitology* **69**, 559–566.

Jenkins, D.C. (1982). *In vitro* screening tests for anthelmintics. *In* "Animal Models in Parasitology" (D. Owen, ed.), pp. 173–186. Macmillan, London.

Johnston, B.R. and Halton, D.W. (1981). Excystation *in vitro* of *Bucephaloides gracilescens* metacercaria (Trematoda; Bucephalidae). *Zeitschrift für Parasitenkunde* **65**, 71–78.

Johri, L.H. and Smyth, J.D. (1956). A histochemical approach to the study of helminth morphology. *Parasitology* **46**, 107–116.

Kawanaka, M., Hayashi, S. and Ohtomo, H. (1983). A minimum essential medium for cultivation of *Schistosoma japonicum* eggs. *Journal of Parasitology* **69**, 991–992.

Kay, A.B. (1979). The role of the eosinophil. *Journal of Allergy and Clinical Immunology* **64**, 90–104.

Kirschner, K. and Bacha, W.J. (1980). Excystment of *Himasthla quissetensis* (Trematoda; Echinostomatidae) metacercariae *in vitro*. *Journal of Parasitology* **66**, 263–267.

Leflore, W.B and Bass, H.S. (1982). *In vitro* excystment of metacercariae of *Himasthla rhigedana* (Trematoda; Echinostomatidae). *Journal of Parasitology* **68**, 1177–1178.

Leflore, W.B. and Bass, H.S. (1983). *In vitro* excystment of the metacercariae of *Cloacitrema michiganensis* (Trematoda; Philopthalmidae). *Journal of Parasitology* **69**, 200–204.

Levi-Schaffer, F. and Smolarsky, M. (1981). *Schistosoma mansoni*: effect of insulin and a low-molecular-weight fraction of serum on schistosomula in chemically defined media. *Experimental Parasitology* **52**, 378–385.

McLaren, D.J. (1980). "*Schistosoma mansoni* : The Parasite Surface in Relation to Host Immunity". Research Studies Press, John Wiley, New York.

Mellink, J.J. and Van den Bovenkamp, W. (1985). *In vitro* culture of intramolluscan stages of the avian schistosome *Trichobilharzia ocellata*. *Zeitschrift für Parasitenkunde* **71**, 337–352.

Mitchell, J.S., Halton, D.W. and Smyth, J.D. (1978). Observations on the *in vitro* culture of *Cotylurus erraticus* (Trematoda; Strigeidae). *International Journal for Parasitology* **8**, 389–398.

Newport, G.R. and Weller, T.H. (1982a). Deposition and maturation of eggs of *Schistosoma mansoni in vitro*: importance of fatty acids in serum-free media. *American Journal of Tropical Medicine and Hygiene* **31**, 349–357.

Newport, G.R. and Weller, T.H. (1982b). Miracidia infective for snails derived from eggs laid by adult *Schistosoma mansoni in vitro*. *Parasitology* **84**, 481–490.

Schnier, M.S. and Fried, B. (1980). *In vitro* cultivation of *Amblosoma suwaense* (Trematoda; Brachylaemidae) from the metacercaria to the ovigerous adult. *International Journal for Parasitology* **10**, 391–396.

Sharma, A.N. and Sharma, P.N. (1982). Preliminary experiments on the *in vitro* cultivation of an amphistome *Orthocoelium scoliocoelium* (Trematoda; Digenea). *Journal of Helminthology* **56**, 169–176.

Shaw, J.R., Marshall, I. and Erasmus, D.A. (1977). *Schistosoma mansoni*: *in vitro* stimulation of vitelline cell development by extracts of male worms. *Experimental Parasitology* **42**, 14–20.

Smith, M.A. and Clegg, J.A. (1981). Improved culture of *Fasciola hepatica in vitro*. *Zeitschrift für Parasitenkunde* **66**, 9–15.

Smith, M., Clegg, J.A. and Webbe, G. (1976). Culture of *Schistosoma haematobium in vivo* and *in vitro*. *Annals of Tropical Medicine and Parasitology* **70**, 101–107.

Smithers, S.R. and Terry, R.J. (1969). The immunology of schistosomiasis. *Advances in Parasitology* 7, 41–93.

Smyth, J.D. (1956). Studies on tapeworm physiology. VIII. Occurrence of somatic mitosis in *Dyphyllobothrium* spp. and its use as a criterion for assessing growth *in vitro*. *Experimental Parasitology* **5**, 260–270.

Smyth, J.D. and Davies C. (1974). *In vitro* culture of the strobilar stage of *Echinococcus granulosus* (sheep strain) a review of the basic problems and results. *International Journal for Parasitology* **4**, 443–445.

Taylor, A.E.R. and Baker, J.R. (1968). "The Cultivation of Parasites *In Vitro*". Blackwell Scientific Publications, Oxford and Edinburgh.

Taylor, A.E.R. and Baker, J.R. (eds) (1978). "Methods of Cultivating Parasites *In Vitro*". Academic Press, London and New York.

Thomson, M. and Halton, D.W. (1982). Observations on excystment *in vitro* of *Cotylurus variegatus* metacercariae (Trematoda; Strigeidae). *Zeitschrift für Parasitenkunde* **68**, 201–210.

Tielens, A.G.M., Van den Meer, P. and Van den Bergh, S.G. (1981). *Fasciola hepatica*; simple large-scale *in vitro* excystment of metacercariae and subsequent isolation of juvenile liver flukes. *Experimental Parasitology* **51**, 8–12.

Vadas, M.A., David, J.R., Butterworth, A.E., Pisani, N.T. and Siongok, T.A. (1979). A new method for the purification of human eosinophils and neutrophils and a comparison of the ability of these cells to damage schistosomula of *Schistosoma mansoni*. *Journal of Immunology* **122**, 1228–1236.

Weller, T.H. and Wheeldon, S.K. (1970). Cell cultures from helminth tissues. *Journal of Parasitology* **56**, 365.

Weller, T.H. and Wheeldon, S.K. (1982). The cultivation *in vitro* of cells derived from *Schistosoma mansoni*. I. Methodology — criteria for evaluation of cultures and development of media. *American Journal of Tropical Medicine* **31**, 335–348.

Yasuraoka, K., Irie, Y. and Hata, H. (1978). Conversion of schistosome cercariae to schistosomula in serum-supplemented media and subsequent culture *in vitro*. *Japanese Journal of Experimental Medicine* **48**, 53–60.

Chapter 12

Cestoda[1]

C. ARME

Parasitology Research Laboratory,
Department of Biological Sciences,
University of Keele,
Keele,
Staffordshire, England

I. INTRODUCTION

In the previous editions of this text Taylor and Baker (1968) and Voge (1978) reviewed work, largely of the past half-century, representing the golden age of *in vitro* culture of cestodes. However, in the eight years since the publication of the second book (Taylor and Baker, 1978), a number of significant, new developments have taken place. Little of what was included in the previous two books is repeated here, nor has an attempt been made to include all relevant references. The selection is perhaps idiosyncratic, emphasizing those areas that seem to me to be the most important, and attention has been drawn only to accounts, published in refereed journals, in which substantial technical details are given. In the field of *in vitro* culture, perhaps above all others, conference reports and short communications have little value, and these have been omitted, despite an abundance of such references in the literature.

[1]Dedicated to the late Marietta Voge, the rightful author of this chapter, and Wynne Owen, whose gift to me was a love of parasites.

The lines of demarcation between *in vitro* experiments, *in vitro* maintenance and *in vitro* culture are sometimes blurred, and the view has been taken that certain observations of considerable biological significance, even if obtained from short-term *in vitro* studies, are worthy of mention. Most references to *in vitro* observations on cestode nutrition have been omitted, since this area has been reviewed on several occasions during the last 10 years (e.g. by Pappas, 1983).

II. GYROCOTYLIDEA

Many features of these animals, not least their evolutionary relationships, remain controversial and enigmatic. By including them in this chapter it is not my intention further to fuel these fires, but studies on their life-cycles demand attention. Such observations may be facilitated by the ease with which eggs may be hatched, releasing lycophore larvae. Malmberg (1974) maintained eggs from *Gyrocotyle urna* in natural sea water (60–65 ml) containing penicillin (type not stated) 100 iu ml^{-1} and streptomycin (100 μg $litre^{-1}$). Hatching occurred over a range of temperatures (6°–12°C), and accounts of egg hatching in other species of *Gyrocotyle* have been provided by Simmons (1974) and Allison and Oakley (1973).

III. TETRARHYNCHIDEA AND TETRAPHYLLIDEA

Voge (1978) could cite only one relevant paper on *in vitro* culture of a tetrarhynch, and little more can be added here. Mudry and Dailey (1971) collected eggs from *Acanthobothrium olseni*, *Parachristianella monomegacantha* and *Lacistorhynchus tenuis* and maintained them in sea water (natural?) at 16°C, before attempting to infect a variety of invertebrate intermediate hosts. *A. olseni* eggs were embryonated upon release; those of *P. monomegacantha* were at the 2- to 4-cell stage but, after 3 d in sea water, contained well-developed oncospheres. *P. monomegacantha* did not require a free-swimming coracidial stage to be infective to copepods, although release of coracidia was observed in *L. tenuis* (see also Voge and Edmonds, 1969).

Sakanari and Moser (1985) studied the effects of salinity and temperature on *Lacistorhynchus tenuis* egg viability, coracidial hatching times and survival. Proglottides were placed in sea water (25%, 50%, 75% and 100%) and released eggs were maintained in the same seawater concentrations at 11, 15 and 19°C in a 12 h light–dark cycle. Tanning of eggs occurred only in 75% and 100% sea water. No coracidia emerged in 25% sea water, and 15°C and 75% sea water were optimal with respect to hatching times and coracidial survival.

Hamilton and Byram (1974) studied aspects of the development of a tetraphyllidean plerocercoid, recovered from the gut of the gastropod *Fasciolaria tulipa* and assigned to the genus *Acanthobothrium*. A variety of culture methods and media (both diphasic and monophasic) were tried. Differentiation of the scolex (as judged by hook production, loss of apical sucker, modification of each quadriloculate bothridium into an anterior accessory sucker and triloculate bothridium) readily occurred in monophasic media, but this was not accompanied by metacestode growth or strobilization. The authors properly distinguished between the morphogenetic response, described above, and activation. This latter term should be reserved for broader aspects of host–parasite interactions, leading to parasite establishment, as discussed by Read (1970).

1. *Detailed technique*

1° *Acanthobothrium* sp.: differentiation of the plerocercoid scolex

(Hamilton and Byram, 1974)

MEDIA

(i) BASIC SALINE

Basic saline is used for washing and in certain pilot experiments to investigate the effects of urea and "bile alcohol".

NaCl	250 mM
KCl	4.4 mM
$MgCl_2$	5.1 mM
Tris(hydroxymethyl) amino methane maleate buffer at pH 7.2	10 mM

(ii) MONOPHASIC MEDIUM

This was based on the minimal essential medium of Eagle (MEM; see Taylor and Baker, 1968, p. 355) with Hanks's salts (See Taylor and Baker, 1968, p. 350), L-glutamine and non-essential amino acids (Travenol). A 10 times concentrated stock solution of MEM was prepared using sterile distilled water and adjusted to pH 7.2 by adding sodium bicarbonate, 0.35 g litre^{-1}.

Supplemented minimal essential medium (sMEM) was prepared thus:

MEM (10 times concentrate; pH 7.2)	1 part
300 mM urea (10 times concentrate)	1 part
0.25 mM Tris maleate buffer (10 times concentrate; pH 7.2)	1 part

Sterilization was by Millipore filter and the solution was diluted to working strength before use. The addition of 1 part of plasma (inactivated at 56 °C for 30 min) from outdated human blood, antibiotics (100 iu penicillin ml^{-1} and 100 μg streptomycin ml^{-1}), glucose (1 mg ml^{-1}), or 0.25% elasmobranch bile obtained from the gall bladder of rays and stored in 5 times its own volume of ethanol did not enhance development when compared with sMEM, although survival of the worms was enhanced in the presence of glucose; shaking of the culture vessels was also ineffective.

TECHNIQUE

Plerocercoids were removed from *F. tulipa* and washed twice in sMEM; "uninjured" worms were selected using a dissecting microscope and incubated under air at 22 °C in 50 ml Micro-Fernbach culture flasks containing 10 ml of medium and 10 plerocercoids. Dead parasites were removed daily. Usually the medium was changed daily but, although this was stated not to be necessary, the minimum frequency of change was not given. Changes in pH during culture were stated to be "minimal". Using this method, scolex differentiation occurred after 48–72 h, although no plerocercoids survived for more than 120 h.

IV. PSEUDOPHYLLIDEA

The techniques of J.D. Smyth for the *in vitro* cultivation of *Ligula intestinalis* and *Schistocephalus solidus* (see Taylor and Baker, 1968, pp. 198, 201) have not been superseded.

Hoole and Arme (1985) studied the interactions between host leucocytes and plerocercoids of *L. intestinalis*. Parasite slices were cultivated in several media, the relative merits of which were evaluated using ultrastructural and physiological (glucose uptake) criteria. Plerocercoids were incubated at 15 °C under air in 50 ml conical flasks containing 25 ml medium, which was changed every 5 h. Media tested were teleost Ringer (Lockwood, 1963; pH 7.3, 290 mosmol), Eagle's medium with Earle's salts (see Taylor and Baker, 1968, p. 355) but minus L-glutamine (pH 7.5, 309 mosmol; Gibco), M199 (pH 7.4, 291 mosmol, Gibco: see Taylor and Baker, 1968, pp. 355–356) and Leibovitz' medium with L-glutamine (pH 7.4, 291 mosmol; Gibco). The latter was found to be the most suitable and, in studies on tegumental dynamics, a turnover rate of surface components of 12–24 h was calculated. Hoole and Arme (1986) also demonstrated that serum was involved in the adherence of leucocytes and damage to plerocercoids of *L. intestinalis in vitro*.

Ligula intestinalis was found to secrete proteolytic enzyme(s) and protease inhibitors into culture media (Matskasi and Juhasz, 1977). Plerocercoids were incubated for 48 h at 37 °C in Tyrode, (see Taylor and Baker, 1968, p. 350), Krebs Ringer Tris (KRT of Read *et al.*, 1963; pH 7.4), Tris-sucrose (pH 7.5), TC 135 (Difco) or TC 199 (Difco) with a medium volume : body weight ratio of 10 : 1. Each medium also contained penicillin, 2000 iu ml^{-1}, streptomycin, 1.0 $\mu g\ ml^{-1}$, and 100 $\mu g\ ml^{-1}$ nystatin. The properties of enzymes and inhibitors were further investigated by Matskasi and Nemeth (1979). Sterry and McManus (1982) and McManus and Sterry (1982) studied the biochemical composition, carbohydrate utilization, oxygen consumption and intermediary carbohydrate metabolism of plerocercoids and adults of *L. intestinalis*. Adult worms were produced *in vitro* using essentially the method of Smyth (1947). Plerocercoids, weighing 1–4 g fresh weight, were cultured at 40 °C in 50 ml TC 199 (Difco: See Taylor and Baker, 1968, pp. 355–356), 20 ml newborn calf serum (Flow Laboratories) and 1 ml antibiotic mixture (5000 iu ml^{-1} penicillin and 5000 $\mu g\ ml^{-1}$ streptomycin; Flow Laboratories). Culture vessels were shaken (50 cycles min^{-1}) and, after every 24 h period, 50 ml of culture medium was removed and replaced with 30 ml fresh TC 199 and 20 ml newborn calf serum, to remove the bulk of toxic metabolic end-products. Arme *et al.* (1982) cultured *L. intestinalis* plerocercoids in Tyrode (see Taylor and Baker, 1968, p. 350) + 1% glucose (1 g parasite : 50 ml medium) for 7 d, changing medium daily. The medium was then analysed for the presence of antigonadotrophins, in an attempt to explain the suppressive effect of this parasite on host pituitary function.

Freze *et al.* (1985) utilized a coracidial agglutination technique to distinguish between species of the genus *Diphyllobothrium*. A ready supply of coracidia was obtained from eggs stored in distilled water in the light at "room temperature". The water was renewed daily and the eggs were stirred. Under these conditions coracidia hatched within 5–7 d, usually in the morning.

The plerocercoid of *Spirometra mansonoides* produces a growth factor (PGF) that, in many respects, behaves as an analogue of human growth hormone. Phares and Ruegamer (1973) and Chang *et al.* (1973) demonstrated that PGF was secreted into culture media *in vitro*. It proved difficult, however, to obtain sufficient material to effect purification, since concentration of large volumes of media by lyophilization and flash evaporation resulted in the loss of 75–100% of PGF activity. Phares (1984) devised a system partially to overcome this difficulty. The anterior 0.5 cm was removed from parasites grown *in vivo* and introduced into other mice to regrow, and thus produce more parasite material. The remaining posterior strobila were cultured *in vitro* in a range of media and temperatures. The greatest amounts of PGF were produced in solutions of 0.1 mM $NaHCO_3$ in 0.12 M NaCl (pH 7.6) (1755 units PGF g^{-1} per worm during 24 h at 43°C) and Krebs–Ringer bicarbonate (pH 7.4) (1450 units PGF g^{-1} per worm during 48 h at 43°C). Techniques were also described to obtain PGF by solubilization of strobila in Triton X-100 in 0.025 M Tris-HCl, pH 8.5).

Culture methods for *Spirometra mansonoides*, and for studying the absorption of vitamin B_{12} by plerocercoids *in vitro*, were reviewed by Voge (1978). Friedman *et al.* (1983) further characterized the B_{12}-binding sites on isolated brush-border membranes of the plerocercoid, and studies on the internal transport and metabolism of the vitamin were described by Friedman *et al.* (1982). Each plerocercoid (80–100 mm in length) was incubated in 3.0 ml of M 199 (Gibco: see Taylor and Baker, 1968, pp. 355–356) containing 25 mM HEPES (pH 7.4), Earle's salts, 100 iu ml^{-1} potassium penicillin b, 100 μg ml^{-1} streptomycin sulphate plus radio-labelled vitamin B_{12}. After 1 h at 37°C, radioactivity in different regions of the worm was determined. Most vitamin B_{12} was found in the anterior 2.0 mm of the plerocercoid and any absorbed in the posterior region was transported anteriorly along the strobila. The mechanisms involved in this anterior movement are unknown, but the authors proposed that high concentrations of vitamin B_{12} in the anterior strobila were associated with the increased metabolic demands of this region, especially during differentiation of the metacestode into the adult worm.

Spirometra erinacei plerocercoids were shown to produce factors chemotactic for guinea-pig eosinophils and neutrophils, *in vitro* (Horii *et al.*, 1984). Fifty intact plerocercoids were washed twice with "saline" and cultured in 10 ml of RPMI 1640 (Gibco) at 37°C for up to 180 min. Significant activity in the culture medium was detected after 5–15 min *in vitro*.

Studies on the effects *in vitro* of anthelmintics on Pseudophyllidea are few. Some references are given by Andrews *et al.* (1983), and Pool (1985) investigated the effects of praziquantel on *Bothriocephalus acheilognathi*. Worms were not killed after maintenance for 60 min in 0.9%

saline at 18 °C containing 10^{-5} g $litre^{-1}$ of drug, and no ovicidal effect of praziquantel was observed.

V. CYCLOPHYLLIDEA

A. HYMENOLEPIS

Voge (1978, pp. 198–209) gave a comprehensive account of the available methods for *in vitro* cultivation of various stages of the life cycle of *H. citelli*, *H. diminuta*, *H. microstoma* and *H. nana*. This was followed by an excellent review by Evans (1980), which contained only a few additional references, and it is evident that the past eight years have produced more in the way of modifications of existing techniques than in original approaches. Most of the newer work has been on strobilate stages of *H. nana* and *H. microstoma* although Valkounova (1979) gave brief details of excystation techniques for nine species of Hymenolepididae, the cysticercoids of which are found in Copepoda and Ostracoda.

Ferretti and Gabriele (1973, 1978, 1979) described the culture *in vitro* of *H. nana*, harvested 14 d after infection, from mice. A complex apparatus was devised, allowing microscopical examination of a large number of culture chambers (up to 60). Many permutations of media and antibiotics were employed, but it was concluded that M 199 (see Taylor and Baker, 1968, pp. 355–356) plus 20% horse serum was best able to support parasite growth for 24 h. Bortoletti and Gabriele (1979) used this technique to evaluate the action of the anthelmintics "Atebrina" and "Yomesan" against *H. nana*. Osuna-Carrillo *et al.* (1982) used the basic culture medium of Sinha and Hopkins (1967) (see Taylor and Baker 1968, pp. 236–237) to study the effects on the growth of *H. nana* of serum and liver extracts from normal males, castrated males and animals treated with testosterone propionate. Only serum and liver extracts from intact males supported normal development. Neither the above techniques nor that described by Guevara-Benitez *et al.* (1978) appear to offer advantages over methods previously published (e.g., by Sinha and Hopkins, 1967), so technical details have been omitted.

The basic techniques for the culture of *H. microstoma* are those of Sinha and Hopkins (1967), subsequently modified by Evans (1970) and described by Voge (1978, p. 206). Khan and De Rycke (1975) found that the basal medium of Eagle (Flow Laboratories) could be substituted for the modified Eagle's medium of Evans and, in a number of subsequent studies, De Rycke, and co-workers have further investigated the culture requirements of this cestode. For example Khan and De Rycke (1976a) demonstrated that haemoglobin and haemin were important growth-promoting substances and Khan and De Rycke (1976b) confirmed the importance of yeast extracts. Chowdhury (1978) demonstrated that growth was inversely proportional to the concentration of a mixture of vitamins (A, D_3 and E). Chowdhury and De Rycke (1979a,b) investigated the effects of calf and horse serum (including inter-batch variations) and pancreatin additives respectively. Serum, with particular reference to lipid and cholesterol, was further studied by Chowdhury and De Rycke (1981) and the effect of ox bile on calcareous corpuscle formation investigated by Chowdhury and De Rycke (1985).

The techniques for culture of *H. diminuta* are basically as described by Voge (1978). Survival characteristics of oncospheres were found to be similar in Tyrode's saline (see Taylor and Baker, 1968, p. 350) with or without glucose supplements (Anderson and Lethbridge, 1975). The system of Schiller (1965), for adult worms, was improved by Fioravanti and MacInnis (1976a), who replaced whole blood and serum additives with defined media (Triple Eagle's and NCTC-135). An important feature of their work was the use of both observational criteria and metabolic indices to assess development. Although *H. diminuta* can synthesize farnesol, the presence of farnesol (a juvenile hormone precursor) or farnesal in culture media did not promote growth (Fioravanti and MacInnes, 1976b).

Hymenolepis diminuta remains the doyen for experimental studies on cestodes. The literature in this area is voluminous and the selection below is offered with some diffidence. Many of the data have been derived from short-term maintenance of worms *in vitro* rather than long-term culture *in vitro*, and the attention of the reader is drawn to the concluding paragraph of this review.

Almost all studies *in vitro* have utilized 10-day-old parasites, and recent knowledge concerning the metacestode is fragmentary, being largely restricted to work on protein synthesis and nutrition by Jeffs and Arme (1984, 1985a,b). New views on adult nutrition were put forward by Podesta (1977a,b) and Podesta *et al.* (1977a,b) and certain of these were reappraised by Murphy and Lumsden (1984). The possible involvement of microtubules in intracellular metabolite translocation was studied by Etges and Bogitsh (1985). Mettrick (1982) reviewed the behavioural cues that were associated with the circadian migration of *H. diminuta*, and the effects *in vitro* of 5-hydroxytryptamine, acetylcholine, histamine and somatostatin were studied by Sukhdeo *et al.* (1984). Andreassen and co-workers (see Bøgh *et al.*, 1986) have undertaken a series of studies of the effect of complement on various *Hymenolepis* spp.

There have been several important studies on the activity of substances released by cestodes into culture media (usually balanced salt solutions). Uglem and Just (1983) reported the production of trypsin inhibitors that, together with inhibition of host trypsin by worm-induced acidification of the gut, protected the parasite from digestion. *H. diminuta* exhibits a crowding effect in the rat, so that an increase in parasite density leads to a decrease in worm size. Insler and Roberts (1980) demonstrated that worm-conditioned saline, prepared with crowded worms, could inhibit the incorporation of tritiated thymidine into *H. diminuta* DNA. Zavras and Roberts (1984, 1985) extended these studies, and suggested that the compounds responsible for inhibition of DNA synthesis were succinate, acetate, D-glucosaminic acid and cGMP.

Hymenolepis diminuta has been used in numerous studies on cestode biochemistry and as a model for determining the mode of action of anthelmintics (for examples, see Andrews and Thomas, 1979 and Branford-White and Hipkiss, 1985). Recent reviews can be found in Arme and Pappas (1983).

B. TAENIA

Although *Taenia pisiformis* has been successfully grown *in vitro* from oncosphere to metacestode (see Voge, 1978), culture of the cysticercus to adult has been achieved only relatively recently (Osuna-Carrillo and Mascaro-Lazcano, 1982) in a diphasic medium similar to that employed for other canine taeniids (Section V.B.1°). A comparison *in vitro* and *in vivo* of viability of *T. pisiformis* eggs, aged under controlled conditions, was made by Coman and Rickard (1977). They concluded that *in vitro* techniques were not reliable for predicting infectivity of eggs to rabbits. Protease inhibitors are produced by metacestodes of *T. pisiformis* and these can be recovered from *in vitro* culture media (Nemeth and Juhasz, 1980, 1981; Nemeth *et al.*, 1979).

Brandt and Sewell (1981) investigated the hatching and activation of *T. taeniaeformis* oncospheres. They noted that methods previously described (Voge, 1978) gave inconsistent results; in particular hatched and activated oncospheres showed poor survival in hatching fluid, and variations were encountered with different "strains" of the parasite. A method was developed that overcame this problem (Section V.B.2°) and which, with minor modification, yielded good results with *T. saginata*.

T. taeniaeformis metacestodes have been successfully maintained *in vitro* to determine the nature of excretory–secretory substances liberated into the media (e.g., Hammerberg and Williams (1978) isolated complement-interacting factors). More recently Rikihisa *et al.* (1984, 1985) found that worm-conditioned media contained a factor(s) that stimulated the growth of rat and dog gastric cells and inhibited the production of testosterone by Leydig cells that were under stimulation by human chorionic gonadotrophin. The culture system was simple, with washed metacestodes being maintained for 24 h in Eagles's basal medium with Hanks's salts (Gibco), 1.0 ml of medium for each metacestode.

The culture of eggs of taeniids has featured in many studies on antigen production (see Chapter 15). However, Lawrence *et al.* (1980) developed a technique of more general interest for the inducement of oncospheres of *T. ovis* to enter post-oncospheral development. Previous studies (see Taylor and Baker, 1978) resulted in the development of only a small proportion of oncospheres. Because *in vivo* the developing parasite is surrounded by fibroblastic tissue, the effect of co-culture with a fibroblast cell line was examined. Seven developmental stages were identified — phase 1, the motile oncosphere, through to phase 7, when scolex anlagen began to form. In a small proportion of phase 7 parasites, invagination of the anlagen occurred. At all stages, secretions or tegumental material were shed from the parasite, although the significance of these is not known. At 7 d, the largest of the cultured parasites was approximately equal in size to those grown *in vivo* but, thereafter, growth *in vitro* was retarded.

Stevenson (1983) achieved a mean activation of 82.7% with *T. saginata* eggs and Owen (1985) described a method, using tetrazolium salts, to determine the viability of *T. saginata* embryos *in vitro*. Prokopic and Jelenova (1980) studied the effect of seven fertilizers on the survival of *T. saginata* eggs *in vitro*. They demonstrated that all fertilizers tested reduced egg viability (*c.f.* Karimova, 1972, who stated that fertilizers were without effect!) and that lime-based materials had the greatest ovicidal activity.

T. saginata metacestodes were found to undergo some strobilization in both monophasic and diphasic media (Brandt and Sewell, 1980), and Rémond *et al.* (1980) achieved survival of cysticerci for up to 5 weeks in NCTC 135 containing 20% foetal calf serum (FCS). Evagination occurred in phosphate buffered saline (PBS) plus glucose (400 mg dl^{-1}). A number of culture systems for *T. saginata* have been utilized for antigen production, and these are described in Chapter 15, but it is clear that successful culture of *T. saginata* and *T. solium* (see below) remains a distant goal.

The hatching *in vitro* of eggs of *T. solium* has been reported by Salazar *et al.* (1984). Ten gravid proglottides were obtained from a patient 8 h after treatment with the anthelmintic niclosamide. They were washed in 0.85% NaCl and stored on saline-moistened filter paper for 1 week at room temperature. (This appeared to be only a convenience and not necessary for hatching.) Eggs were dissected from proglottides and placed in 10 ml of Solution 1 (pH 2.0) in a Petri dish for 15 min at 41 °C (Solution 1: 120 mg pepsin, 1 : 10 000 (Difco); 2.0 ml 1 N HCl; 20 ml distilled water). They were then washed three times in 0.85 % NaCl and pipetted into a 10 mm diameter screw-top tube containing Solution 2 (pH 7.2–7.4: 10 ml 0.85 % NaCl; 35 mg trypsin type XII, Sigma, 0.5 ml whole bile from pig, kept frozen (time unspecified) before use; 1.0 ml 10% $NaHCO_3$ in distilled water). The tube was then tightly capped (to conserve CO_2) and incubated, first at 41 °C for 15 min, and then at room temperature (21–22 °C) for a further 15 min. The embryophores had disintegrated after 30 min in Solution 2, but most oncospheres showed little motility. However, after a further 15 min at room temperature, 98% of oncospheres were active. The following safety procedures were recommended: all contaminated glassware was fully submerged in 95 % ethanol and work areas periodically swabbed with 70% ethanol; hands were rinsed in 70% ethanol before using soap and water. Observations indicated that ethanol concentrations of >50% were lethal to eggs, but in ethanol concentrations of <35%, eggs remained normal over a 15 min period of observation under the light microscope.

Morales and Canedo (1982) attempted to culture *T. solium* from metacestode to adult. The medium used was minimal essential medium (Gibco) at pH 7.4, supplemented with (ml^{-1}) 4.0 mg glucose, 50 iu penicillin G, 50 μg streptomycin sulphate and 30 μg gentamycin. Parasites from pig muscle were selected on the basis of an armed rostellum being present. They were initially incubated in Erlenmeyer flasks for 3 h at 37 °C under 95 % air plus 5 % CO_2 and 100% relative humidity. After incubation, metacestodes were evaginated in the culture medium plus 0.1% trypsin, washed four times in culture medium and transferred to fresh medium supplemented with 10% FCS. Metacestodes grew to a maximum length of 39 mm after 31 d, no growth occurring thereafter.

Bruckner (1979) described the culture of pieces of racemose cysticercus of *T. solium* in RPMI 1640 plus 10% FCS. Survival up to 3 months was achieved and damaged cysticerci reformed a bladder which became filled with fluid.

Laclette *et al.* (1981) incubated cysticerci of *T. solium* in RPMI (pH 7.3) containing FCS (1% w/v) and (ml^{-1}) 50 iu penicillin G, 50 μg streptomycin sulphate and 30 μg gentamycin. Pairs of metacestodes were maintained in 2.0 ml of medium at 37 °C under 95% air plus 5% CO_2 and 100% humidity. Some cultures contained 0.01% w/v mebendazole and, after 16 h incubation, unusual paracrystalline bundles appeared in the secretory cells of the bladder wall. These were stable in colchicine and at low temperature, and their presence was related to disturbances in tubulin assembly associated with the anthelmintic.

For *T. hydatigena*, evagination of cysticerci has been successfully achieved (Voge, 1978), and observations on the roles of O_2, pH, enzymes and bile have been made by Osuna-Carrillo *et al.* (1977, 1978b, 1980). Non-fertile adults of *T. hydatigena* from cysticerci have been produced *in vitro* by Osuna-Carrillo *et al.* (1978a), using a technique similar to that described for *T. pisiformis* (Osuna-Carrillo and Mascaro-Lazcano, 1982) (see Section V.B.1°).

No additional methods for culture of *T. crassiceps* have appeared since those reported by Voge (1978). However, studies *in vitro* on the physiology of this cestode have been of interest and special attention is drawn to the work of Dunn and Threadgold (1984) on endocytosis. *T. crassiceps* may prove an ideal model for the study of the uptake of macromolecules and surface membrane recycling.

1. *Detailed techniques*

1° *Taenia pisiformis*: cysticercus to adult
(Osuna-Carrillo and Mascaro-Lazcano, 1982)

LIQUID MEDIUM

(See Taylor and Baker, 1968, p. 356)

CMRL 1066	100 ml
Foetal calf serum	25 ml
2% KCl	0.5 ml
30% glucose	2.2 ml
Penicillin	100 iu ml^{-1}
Streptomycin	100 μg ml^{-1}

TECHNIQUE

Metacestodes were removed from infected rabbits and stored for 12 h at 4°C. They were then washed in benzalconium chloride (1 : 10 000), followed by further washes in sterile Hanks's saline (see Taylor and Baker 1968, p. 350). After removal of the pericystic membrane, the cysticerci were washed three times in Hanks's and exposed for 35 min to a 0.5% pepsin solution (1 : 10 000) in Hanks's saline at pH 1.5. Two washes in Hanks's saline (pH 7.2) were followed by treatment with evaginating solution (0.5% trypsin, 0.3% pancreatin and 0.05% dog bile in Hanks's saline; pH 7.2) for 8 h at 38°C in a shaking water bath. Evaginated parasites were then placed in a Roux flask (Falcon) in which there was a solid base of 5.0 ml coagulated bovine serum and 10 ml of a liquid medium comprising 5.0 ml of evaginating solution and 5.0 ml culture medium, as described above. The gas phase was 10% O_2 and 5% CO_2 in N_2 and Parafilm-covered flasks were shaken in a water bath at 38 °C for 24 h. Five scoleces were than transferred to a further culture flask with 5 ml coagulated bovine serum and (presumably) 10 ml of liquid medium. All solutions were sterilized by filtration and FCS was inactivated at 56 °C for 30 min. The final pH was adjusted to 6.8 with 0.1 M HCl and the liquid medium was changed and gassed every 72 h.

Segmentation occurred after 7–8 d and, by 22 d, the genital pore was open, although in some proglottides malformations occurred, producing double genital pores. The genitalia of the parasites appeared normal but, although the cirrus was frequently observed inserted into the vagina, no oncospheres were produced — a problem commonly encountered in the cultivation of Taeniidae.

2° *Taenia taeniaeformis:* hatching and activation of oncospheres
(Brandt and Sewell, 1981)

MEDIA

(i) *Hypochlorite solution* (to replace artificial gastric solutions of previous workers). Deosan (Diversey) a proprietary hypochlorite solution containing a minimum available chlorine concentration of 10%. Diluted 1 : 7.5 in distilled water (pH 11.7).

(ii) *Artificial intestinal solution (AIS)*. The basic solution is RPMI 1640 (Gibco) modified with 25 mM HEPES buffer plus L-glutamine (Gibco), to which was added sodium pyruvate (5.0 mM), penicillin (100 iu ml^{-1}) and streptomycin (100 μg ml^{-1}).

To this modified RPMI 1640 is added trypsin (10 mg ml^{-1}), 10% v/v ox bile and 10% v/v heat-inactivated FCS. (The solution was sterilized by passing it through a 0.22 μm membrane filter before adding the sterile FCS).

TECHNIQUE

Eggs were collected in antibiotic saline (145 mM NaCl with 100 iu ml^{-1} penicillin, 100 μg ml^{-1} streptomycin and 2.0 μg ml^{-1} amphotericin B. The egg suspension was then passed through a 150 μm pore sieve on to a micropore plate (20 μm pore size) on which the eggs were retained. Washed eggs were stored in antibiotic saline at 4 °C for up to 14 d before use. Before experimentation, 500–5000 eggs in antibiotic saline were centrifuged at 1500*g* for 10 min and the supernatant was removed, leaving 1–2.5 ml fluid and sedimented eggs. An equal volume of the dilute hypochlorite solution was added, and the suspension was left for 4 min at room temperature. Two volumes of distilled water were added and the tube was centrifuged (1000*g*, 5 min). Two further distilled water washes were each followed by centrifugation (1000*g*, 10 min). Tubes were then filled with AIS and immediately centrifuged (1000*g*, 10 min). The supernatant was drawn off down to 2.0 ml before stationary incubation in a water bath (38°C), during which the proportion of hatched and activated oncospheres was determined after 30 min and 3 h.

The authors recognized that the use of hypochlorite to disaggregate embryophore blocks is open to criticism. They noted, however, that the method is simple, effective and yields a sterile suspension of oncospheres and that hypochlorite treatments have also been used in other species (e.g., see Osborn *et al.*, 1982). The authors also discussed the possible effect of serum in protecting activated oncospheres.

C. MESOCESTOIDES

The research potential of *Mesocestoides corti* and methods for culture from oncosphere to tetrathyridium have been discussed by Voge (1978). Recently, however, cultivation *in vitro* of the tetrathyridium to sexually mature adults has been achieved independently by Thompson *et al.* (1982) (see Section V.C.1°) and Barrett *et al.* (1982) (see Section V.C.2°).

Despite the fact that self-copulation occurred in the cultures of Thompson *et al.* (1982) there was no sperm transfer into the copulation canal; oncospheres were thus not produced. Growth and strobilization were comparable both *in vitro* and *in vivo* (in dogs) during the first 18 d; thereafter, total worm length and proglottis number decreased *in vivo* but continued to increase *in vitro* past 18 d. Some adults were removed from dogs 12 and 18 d after infection and transferred to *in vitro* culture (see Kumaratilake and Thompson, 1981). However, no further maturation occurred *in vitro* and shelled eggs that were present were resorbed within 14 d.

Barrett *et al.* (1982) reported the results of experiments in which there occurred the spontaneous sexual differentiation of a single *M. corti* tetrathyridium in a culture 50 d old, the remaining parasites of which were undergoing asexual multiplication. A series of additional experiments was then designed to determine the conditions favouring segmentation in which the following were varied: type of liquid medium; amount of medium per culture; solid : liquid medium ratio; gas phase; pretreatment (pepsin/trypsin); shaken, still or roller cultures; and intervals between changing medium. Although segmentation proved to be sporadic, rather than consistent as described by Thompson *et al.* (1982), numerous sexually mature parasites were produced using the protocol described below (Section V.C.2°). In one parasite a few shelled eggs containing hooked oncospheres were observed. Although this might indicate that insemination had occurred, it could represent a "haploid oncosphere" as reported by Smyth (1950) in *Schistocephalus solidus* cultures.

It is not known why fertilization did not readily occur in either the system of Thompson *et al.* (1982) or that of Barrett *et al.* (1982). Perhaps adequate nutritional factors were not provided by the culture medium or, alternatively, appropriate physical features were missing (see Smyth and Davies, 1974, for relevant discussion).

Ong and Smyth (1986) investigated the effects of various physicochemical factors on the sexual differentiation of *M. corti in vitro*. To produce sexually mature adults it was necessary to pretreat tetrathyridea in Hanks's saline (see Taylor and Baker, 1968, p. 350) overnight, before culture. Twenty-five tetrathyridea were then transferred to 10 ml of monophasic S10E-H (see Section V.C.2°) but with 20 mM HEPPSO (*N*-hydroxyethyl-piperazine-*N* -2-hydroxypropane sulphonic acid) as buffer instead of HEPES. Using a gas phase of 5% CO_2 and 95% N_2 and a pH of 7.4–7.8, development *in vitro* occurred at a rate comparable with that observed in dogs (Thompson *et al.*, 1982). Segmentation also occurred in diphasic media but at a lower rate. Although sexually mature adults were consistently obtained, eggs containing motile oncospheres were produced only sporadically and possible reasons for this are discussed.

Baldwin *et al.* (1978) studied the incorporation of a variety of cations into the calcareous corpuscles of tetrathyridia of *M. corti*, using culture methods similar to those described by Voge and Coulombe (1966). To M155 (developed by Berntzen and Mueller, 1964) was added 20% Ficoll (Sigma), to replace the horse serum used in previous studies by Kegley *et al.* (1970). Care was required in the introduction of certain cations because of their relative insolubility.

1. *Detailed techniques*

1° *Mesocestoides corti* : tetrathyridium to adult
(Thompson *et al.*, 1982)

MEDIUM

A diphasic system was recommended, the solid base being prepared by coagulating bovine serum in an oven at 76 °C for 75 min.

A liquid phase using CMRL 1066 was preferable to one using M 199 (see Taylor and Baker, 1968, pp. 355–356). Its composition was as follows:

CMRL 1066 with Earle's salts and L-glutamine (Flow Laboratories)	325 ml
Foetal calf serum (Commonwealth Serum Laboratories)	125 ml
30% D-glucose (BDH)	7.25 ml
5% dog bile in Hanks's saline ..	2.1 ml
5% yeast extract in Hanks's saline (Difco)	45 ml
Sodium bicarbonate	2.1 g
HEPES	20 mM
Neomycin–penicillin (Commonwealth Serum Laboratories)	0.4 μg ml^{-1}
Gentamycin (Schering) ..	100 μg ml^{-1}
Amphotericin (Squibb).. ..	2 μg ml^{-1}

TECHNIQUE

Quackenbush mice were killed with chloroform, and 1% iodine in 95% ethanol was applied to the shaved abdomen. Hanks's balanced salt solution (see Taylor and Baker, 1968, p. 350) (2.5 ml, pre-warmed to 38 °C) was injected into the body cavity and the abdomen was palpated. A 1 ml serological pipette was then used to withdraw tetrathyridia, which were placed in a sterile solution of 0.05% pepsin (Sigma) in HCl (pH 2.0) for 5 min at 38 °C. Parasites were then washed in three changes of Hanks's saline, containing 0.4 μg ml^{-1} each of penicillin and neomycin before transfer to 30 ml sterile culture vessels at 38 °C. Approximately 150 tetrathyridia in 10 ml of culture medium were gassed every third day with 10% O_2 and 5% CO_2 in N_2; culture fluid was changed every 3 d. Results are discussed in Section V.C.

2° *Mesocestoides corti* : tetrathyridium to adult
(Barrett *et al.*, 1982)

MEDIUM

A diphasic system was recommended in which the solid base was newborn calf serum (Flow Laboratories) coagulated at 76°C until "set". The liquid medium was S10E-H:

	ml
Foetal calf serum (Gibco)	100.0
CMRL 1066 (Flow Laboratories; see Taylor and Baker, 1968, p. 356) including 26 000 iu penicillin and 26 000 μg streptomycin (Flow Laboratories)	260.0
5% yeast extract (Gibco) in Hanks's saline	36.0
30% glucose (Hopkins and Williams) in distilled water	5.6
5% dog bile in Hanks's saline (or 0.2% sodium taurochlolate (Sigma) in distilled water)	1.4

TECHNIQUE

Tetrathyridea, recovered from the peritoneal cavity of mice, were washed at least twice with Hanks's saline (see Taylor and Baker, 1968, p. 350) at 38°C, rotating the vial during rinsing. The use of pepsin or trypsin pre-treatments was unnecessary, since washing stimulated the evagination of most protoscoleces, the remainder evaginating in culture. Best development occurred when 100–200 metacestodes were cultured in 5–10 ml S10E-H in Leighton tissue culture tubes (19 × 105 mm) slowly rotated at 38°C, under a gas phase of 10% CO_2 and 5% O_2 in N_2. Medium was changed every 2–3 d. Results are discussed in Section V.C.

D. ECHINOCOCCUS GRANULOSUS

Methods for cultivating this species have recently been reviewed by Howell (1986). Viability of *E. granulosus* and *E. multilocularis* protoscoleces during storage has been discussed by Andersen and Loveless (1978), Saad and El Abin (1982) and Jensen *et al.* (1984). Harris and Heath (1983) described a rapid procedure for processing small metacestodes, cultured *in vitro*, for ultrastructural examination.

1. *Culture of oncospheres*

There have been few attempts to culture oncospheres, because of the obvious dangers associated with the handling of *E. granulosus* eggs. Heath and Lawrence (1976) described procedures for the maintenance of infected dogs, the recovery of adult parasites from the intestine and the isolation of eggs. Using NCTC 135, with calf serum and rabbit erythrocytes, cysts grew to a mean diameter of 16 mm after 120 d (maximum diameter, 20 mm), a growth rate at least equalling that observed *in vivo* (Section V.D.1°). Heath and Lawrence (1981) observed the development of oncospheres in McCoy's 5A medium (Gibco), to which was added serum from sheep infected with, or immunized against, *E. granulosus*. They demonstrated that lysis of oncospheres occurred, apparently dependent upon antibody and complement. Of greater interest, from the point of view of *in vitro* culture, was the fact that rabbit erythrocytes could be omitted from the culture system if heat-inactivated normal sheep serum was substituted for FCS (see Section V.D.1°).

2. *Culture of protoscolex*

The protoscolex is able to develop either into a hydatid cyst or a strobilate worm. It thus provides a useful model for studies on control of morphogenesis, a problem elegantly addressed on a number of occasions by J.D. Smyth (see Howell, 1986).

(a) Protoscolex to cystic form. Previous work has shown that, in culture, protoscoleces undergo two principal patterns of cystic development. Some swell or vesiculate, others develop a posterior bladder. Smyth (1962, 1967) found that both of these routes could give rise to a hydatid cyst, which eventually developed a laminated layer. Heath and Osborn (1976) cultured protoscoleces from sheep in a defined medium containing no macromolecules (NCTC 135; Gibco). Protoscoleces were washed twice in culture medium and any attached brood capsule wall was removed by digestion at 37 °C in an artificial gastric fluid (AGF) for 10 or 60 min (AGF = 1% of 1 : 2500 pepsin (BDH) and 1% 10 M HCl in 0.85% NaCl). AGF was removed by three washes in NCTC 135 and parasites were incubated in 250 ml Falcon flasks, completely filled with medium, for 8 weeks, with medium changes every 3–4 d. For the first 7 d, penicillin G (Sigma, 100 iu ml^{-1}) and streptomycin sulphate (Glaxo; 100 μg ml^{-1}) were added to cultures but, thereafter, they were omitted. Treatment with AGF for 60 min yielded a culture containing, after 14 d, approximately 50% protoscoleces with posterior bladders and

40% vesiculated. By 28 d, the beginnings of a laminated layer could be seen around vesiculated parasites and, after 8 weeks, the rate of development of vesiculated protoscoleces was similar to that found in mice. Treatment with AGF for 10 min produced almost 100% vesiculated protoscoleces and, in these cultures, laminations were obvious after 17 d. This study was especially valuable because it demonstrated that (a) protoscoleces could develop *in vitro* to cysts in a defined medium, (b) the laminated layer is of parasite origin (although host material may be eventually associated with it *in vivo*) and (c) the length of AGF treatment (and possible damage resulting therefrom) could influence the future development of protoscoleces. Matters relating to the cultivation of protoscoleces in the cystic direction have generated some controversy. For example, Bénex (1968) claimed that posterior bladders were formed only from protoscoleces that retained no portion of germinal membrane and that hydatid cysts, complete with laminated layer, developed only from vesiculated protoscoleces. Heath and Osborn (1976), however, showed that the posterior bladders of several protoscoleces could coalesce and that these coalesced bladders could develop a laminated layer if they were derived from as many as 10–40 protoscoleces.

Success has been achieved in culturing the horse strain of *E. granulosus* to produce cysts, complete with laminated layers, using a monophasic culture system (NCTC 135 plus 20% v/v heat-inactivated FCS plus antibiotics). The cysts were formed by the vesiculation of protoscoleces and by two additional pathways involving posterior bladders and ruptured everted brood capsules (Rogan and Richards, 1986). The posterior bladders, which were shown to be portions of protoscolex attachment stalk, could detach from protoscoleces and produce small cysts with laminated layers. Similarly, portions of brood capsule wall were seen to form a vesicle during culture, which produced a laminated layer when devoid of protoscoleces (see Smyth, 1979a,b, and Smyth and Davies, 1979, for discussion of *in vitro* culture of *E. granulosus* strains). No cyst cultured *in vitro* in any system has proved fertile, and possible reasons for this are discussed by Howell (1986).

(b) Protoscolex to strobilate form. The culture method of choice is basically that of Smyth and Davies (1974), described by Voge (1978). However because of the importance of this technique, and to incorporate certain modifications described by Smyth (1979a), an abbreviated account of the method is included here (Section V.D.2°). An important consideration is the source of protoscoleces. In Britain, organisms from sheep readily mature to produce worms with three proglottides and a uterus filled with ova, but insemination *in vitro* has never been achieved.

In contrast, protoscoleces of horse origin have never been grown to maturity and these isolates are now regarded as a distinct strain. Using a diphasic system, protoscoleces develop only as far as pre-segmentation stage 3 (Smyth, 1979a) and, in the monophasic system of Rogan and Richards (1986), a small proportion (<5%) of elongated protoscoleces were produced with spined microtriches over their entire surface (Richards and Rogan, 1986). Such protoscoleces had clearly begun to develop in a strobilar direction, although development did not progress as far as proglottis formation. Questions relating to physical and chemical requirements for culture, strain differences, etc., have been considered by Smyth and Davies (1974, 1979), Smyth (1979a,b, 1982), Kumaratilake *et al.* (1983), Macpherson and Smyth (1985) and Howell (1986).

3. In vitro *maturation of worms partially matured* in vivo

An alternative method of producing *E. granulosus* eggs *in vitro* is to allow fertilization to occur *in vivo* and then to transfer inseminated worms to *in vitro* culture where egg development is completed. Smyth and Howkins (1966) and Thompson and Smyth (1976) showed that viable eggs could be produced by this method, using as criteria for viability oncospheral motility and *in vitro* hatching and activation; Kumaratilake and Thompson (1981) extended these studies to include an additional assessment of viability, *viz.* infectivity of eggs to mice. The subsequent development of protoscoleces in fertile hydatid cysts enabled these workers to complete the entire life-cycle of *E. granulosus* in the laboratory (Section V.D.3°).

The development of *in vitro* techniques has been accompanied by significant advances in our understanding of the basic biology of *E. granulosus*. Many of these, including strain differences, morphogenesis, biochemistry, physiology, immunology and chemotherapy, have been reviewed recently by Thompson (1986) and will not be repeated here. Because of the problems associated with chemotherapy, Jeffs and Arme (1986) have investigated the permeability properties of the parasite with the objective of seeking parasite-specific receptors that may possibly be used in drug targeting. An unusual feature of their work was the demonstration in the everted protoscolex that D-amino acids had a higher affinity for amino acid membrane transport systems than L-forms. Does the developing parasite have a specific requirement for this enantiomer? The biological reason for such a dependence, if indeed it exists, is obscure, but it is interesting to note that Kumaratilake and Thompson (1981) included a D–L mixture of β-phenylalanine in their culture medium. It may be of value to investigate whether the presence of

D-amino acids in the culture system of Smyth (1979a) would provide conditions suitable for the production of fertile eggs.

4. Detailed techniques

1° *Echinococcus granulosus* : oncosphere to cyst
(Heath and Lawrence, 1976)

MEDIA AND TECHNIQUE

The dangers of growing adult *E. granulosus in vivo* and handling gravid worms and eggs are strongly emphasized. This should not be attempted unless by trained personnel working with appropriate containment facilities.

Worms were dissected from dog intestine in 0.85% w/v NaCl (saline) at 37°C and chopped into 1.0 mm pieces in 10 ml of saline. The eggs in the supernatant were washed through a 0.05 mm mesh sieve, concentrated by centrifugation, washed twice in distilled water and exposed to a 1 : 5000 solution of chlorhexidine for 1 h. After a distilled water wash, eggs were stored at 4°C in Krebs–Ringer phosphate buffered saline (KRB). Egg hatching and oncosphere activation were as described by Heath and Smyth (1970; see Voge, 1978, pp. 218 and 213 for details). Activated oncospheres are counted as per Heath (1982): a dilution of eggs is determined that will result in a concentration of *mature eggs* of 10–50 per 30 μl. After agitation, eggs in 30 μl aliquots are counted under the microscope. It is recommended that two dilutions are made, from each of which five 30 μl aliquots are counted. The mean number of eggs from 10 samples is then used to calculate the total number of eggs in the original sample. Five thousand activated larvae were placed in a 50 ml Falcon screw-top culture vessel, completely filled with medium. Three media were tested (NCTC-135; RPMI 1640; Modified McCoy's 5A (Gibco)). To these was added 20% FCS (serum from 10-week-old rabbits was less satisfactory) and 1% v/v erythrocytes from 10-week-old rabbits. (Blood from the ear artery was washed in sterile KRB and erythrocytes stored at a concentration of 50% in KRB at 4°C.)

Media were changed every 3–4 d. If the pH dropped in the 50 ml culture tubes (indicator goes yellow) developing parasites were transferred to 250 ml vessels and/or the number of cultured parasites was halved (for discussion of results see Section V.D.1.).

2° *Echinococcus granulosus* : protoscolex to strobilate adult
(Smyth, 1979a)

MEDIA

Before culturing, it is necessary to treat protoscoleces with pepsin to free them from brood capsules and remove dead organisms, and then with bile salts to induce evagination.

(i) Peptic digestion utilized a 1 : 10 000 pepsin solution (2 mg ml^{-1}) in Hanks's saline (see Taylor and Baker, 1968, p. 350) at 38°C (made pH 2.0 with HCl) sterilized by membrane filtration.

(ii) Evagination occurred in Parker's medium 858 containing dog bile (1 ml dl^{-1} of 5% dog bile or 2 mg dl^{-1} sodium taurocholate. The latter should be as pure as possible since crude preparations of "tauroglycocholate" often contain impurities that lyse protoscoleces (see Smyth and Haslewood, 1963).

(iii) *Diphasic culture medium*
Solid base: to milk dilution bottles (Kimax, USA) or 300 ml Falcon flasks is added 20 ml newborn calf serum which is then coagulated by heating in an oven at 76–80°C for 30–60 min. After cooling, 20 ml of Medium S-10 is added (various other glass and plastic culture vessels have been used successfully).
Medium S-10

Parker 858 (or NCTC 135 or CMRL 1066 (see Taylor and Baker, 1968, p. 356))	130 ml
30% glucose (aqueous)	2.8 ml
Foetal calf serum (inactivated at 56°C)	50 ml
5% yeast extract (in Parker 858) ..	18ml
5% dog bile or 0.2% sodium taurocholate in Hanks's saline	0.7 ml
Antibiotics : penicillin ..	100 iu ml^{-1}
streptomycin	100 iu ml^{-1}

TECHNIQUE

Techniques for removal of protoscoleces and viability checks have been described by Smyth and Davies (1974; see Voge, 1978, p. 220) and Smyth (1979a). It is strongly recommended that those removing protoscoleces from hydatid cysts, and observers, should wear protective glasses. Pepsin digestion is aided by rolling or shaking vessels. The process of digestion was complete when inspection revealed that all the protoscoleces were free from brood capsules (15–45 min). Parasites should then be washed four times in Hanks's saline at 38 °C, each rinse lasting 15 min, and protoscoleces transferred to 50–100 ml of evagination solution for 18–24 h at 38 °C. Removal of most of the evaginating solution allowed the protoscoleces to settle in the small volume of fluid remaining and 0.1 ml of settled protoscoleces (about 10 000 parasites) were then transferred to each culture vessel. The gas phase was not explicitly stated by Smyth (1979a), but Smyth and Davies (1974) used 10% O_2, 5% CO_2 and 85% N_2. Medium was changed every 48 h by Smyth (1967). A wide range of developmental stages ensued with growth *in vitro* being, in general, some 50% slower than that *in vivo* (see text for results and discussion: Section V.D.2(b)).

3° *Echinococcus granulosus* : *in vivo* and *in vitro* technique for completing the life-cycle in the laboratory

(Kumaratilake and Thompson, 1981, media and technique modified from Smyth and Howkins, 1966)

MEDIUM

(for *in vitro* cultivation of partially developed worms from dogs)

Medium 199 (modified) with Earle's salts and L-glutamine (Flow Laboratories)	400 ml
Foetal calf serum (Commonwealth Serum Laboratories)	100 ml
2% potassium chloride	5.0 ml
30% D-glucose (BDH)	7.5 ml
L-cystine (Sigma)	1.0 mM
DL-β-phenylalanine (BDH) ..	0.5 mM
Neomycin-penicillin (Commonwealth Serum Laboratories)	0.4 μg ml^{-1}
Gentamycin (Schering) ..	100 μg ml^{-1}
Amphotericin (Squibb) ..	2.0 μg ml^{-1}

TECHNIQUE

In vivo, fertilization of *E. granulosus* takes place 25–30 d after infection (a.i.) and shelled eggs containing fully developed oncospheres first appear in the uterus after 40–48 d. Therefore, at 35 d a.i., parasites were transferred from dogs to *in vitro* culture. Collection of worms was as previously described (Thompson, 1977). The small intestine was opened longitudinally and incubated for 30 min in 1.0 litre of Hanks's at 37 °C. Detached parasites were washed six times (time, temperature not stated) in Hanks's saline (see Taylor and Baker, 1968, p. 350) containing antibiotics in the concentrations noted above. They were then washed a further three times in culture medium before transfer of approximately 75 parasites to 10 ml of monophasic medium in 30 ml Falcon flasks. Cultures were maintained at 38 °C, gassed daily with 10% oxygen and 5% carbon dioxide in nitrogen, and medium was changed every 2 d. After an average of 6 d, each worm produced a mean of 200 (range 18 to over 600) thick-shelled eggs containing morphologically normal oncospheres. (This is approximately 13% of the number produced *in vivo*.) Gravid worms were used to provide a suspension of eggs and their infectivity assessed by infecting male Quakenbush mice. After 300 d, fertile cysts were recovered, and protoscoleces from these were morphologically normal, exhibited flame cell activity and were infective to dogs.

E. *ECHINOCOCCUS MULTILOCULARIS*

Smyth and Davies (1975), Smyth (1979a) and Smyth and Barrett (1979) attempted to culture protoscoleces to the strobilate form. Development of this organism is strikingly different from *E. granulosus* since, although sexually mature forms can be produced, segmentation is suppressed and they are essentially monozoic. In longer cultures (50–80 d), a second set of genitalia has been observed to develop and, more rarely, a second scolex. Growth occurred in monophasic as well as in diphasic media (c.f. *E. granulosus* — Smyth, 1979a, Richards and Rogan, 1986) but no fertile eggs were obtained. (Section V.E.1°).

Thompson and Eckert (1982) adapted the method of Kumaratilake and Thompson (1981) for the *in vitro* cultivation of *E. multilocularis* (partially grown *in vivo*). No morphological abnormalities were observed and the onset of egg production *in vitro* occurred after the same time interval as that *in vivo*. The number of eggs produced *in vitro* was only 11% of those produced *in vivo* and, although the morphological appearance of the eggs was normal, it was not possible adequately to assess their viability.

As with *E. granulosus*, relevant information regarding the use of *E. multilocularis* in experimental systems has been recently summarized by Thompson (1986), and will not be duplicated here.

1. Detailed technique

1° *Echinococcus multilocularis* : protoscolex to adult
Smyth (1979a)

(A fuller account of the development of these procedures is given by Smyth and Davies, 1975.)

MEDIA AND TECHNIQUE

The removal of *E. multilocularis* protoscoleces is complicated by the spongy nature of the cysts and their frequent calcification. Eye protection should be worn and cyst dissection should be carried out in a safety cabinet.

The body wall of a shaved animal was painted twice with iodine (1% in 95% ethanol) and the body cavity was opened. Approximately 10 g of cyst material was transferred into 200 ml of prewarmed sterile pepsin (0.5%, Sigma, 1 : 10 000) in Hanks's saline (see Taylor and Baker, 1968, p. 350) adjusted to pH 2 with 1 N HCl in a sterile Waring Blender. Phenol red was added as indicator. Homogenization for 10–15 sec released calcareous corpuscles from the cyst which made the solution alkaline. The mixture was transferred to a 500 ml beaker on a magnetic stirrer at 38°C and HCl was added frequently to readjust the pH to 2.0. After the pH had stabilized at 2.0 (usually after 15 min) the mixture was left for a further 30 min to 1 h. After digestion, the solution was sieved and then, by repeated washings in Hanks's followed by sedimentation of protoscoleces and removal of the supernatant, an almost pure suspension of protoscoleces was obtained. Culture Medium S10 (Section V.D.2°) was used and 0.05 ml settled and evaginated protoscoleces were used per Falcon flask. Worm growth occurred well in both monophasic (S10 only) and diphasic media (c.f. *E. granulosus*), but was some 50% slower *in vitro* than *in vivo*, and no fertilization was achieved.

VI. CONCLUSION

In vitro culture, perhaps above all other disciplines, demands that a detailed account of materials and methods be given and, regrettably, serious omissions of these occur with some frequency. In particular, details regarding storage of media, gassing procedures, serum (e.g. inactivated or not), osmolarity, solvent used for additives to the basic medium and type of antibiotics are often omitted. Few investigators have made thorough studies of the optimum concentration of certain reagents used, even though it is well established that this can be of great importance for some species (for example the effect of antibiotic concentration on growth and differentiation of *Hymenolepis microstoma*; Evans, 1978). The purity of chemicals (Analar, Reagent grade, etc.) also is frequently not stated, although, in the case of bile used in evagination media for Cyclophyllidea, this is of critical importance (Smyth, 1979a).

The majority of workers in the field of *in vitro* culture use morphological development of parasites as a criterion for assessing the suitability, or otherwise, of their experimental system. Positive evidence of morphogenesis is of course unequivocal, but it could be of value to use additional approaches to explore the deficiencies of some techniques. For example, it may be appropriate to examine parasites for ultrastructural integrity and to investigate whether certain physiological parameters, for example, nutrient absorption, protein or glycogen synthesis, are supported by the medium (see Fioravanti and MacInnis, 1976a; Hoole and Arme, 1985). Methods for assessing parasite viability also require further investigation. Death of worms during an experiment is often obvious, but the viability of the parasite population initially selected for culture is sometimes difficult to assess. For example, in *Echinococcus*, flame cell motility and eosin exclusion are often used to determine the viability of protoscoleces. False negatives can sometimes be obtained with the former, unless prolonged observations are undertaken, and false positives can be obtained with eosin (Robinson and Arme, 1985).

The production of defined media remains a priority (see Heath and Osborn, 1976). Progress in this area may be facilitated by elucidation of the role of the serum in culture media. Some cestode species absorb macromolecules by endocytosis (see Dunn and Threadgold, 1984, for references). Whether this is necessary for normal development is not known, but there appears to be no correlation between the ability of a species to endocytose and the ease of its culture *in vitro*. Nevertheless, the membrane recycling associated with endocytotic mechanisms may be important for some cestodes, and studies on induction and mechanisms of endocytosis may contribute to solving some of the problems encountered in *in vitro* culture.

Techniques for the culture of the entire life-cycles of Taeniidae of medical and veterinary importance are wanting, and urgently required. Some progress has been made since the previous book (Taylor and Baker, 1978) and the importance of incorporating a cell line in cultures, as reported for *Taenia ovis* by Lawrence *et al.* (1980), may repay investigation for other species.

In vitro culture is a powerful tool for studies on morphogenesis. The elegant studies of J.D. Smyth have highlighted the value of *E. granulosus* as a model for studying the environmental triggers that determine whether protoscoleces develop in a cystic or strobilate direction. However, knowledge of morphogenesis in all species of tapeworm is woefully inadequate and even such obvious questions as the assembly of microtriches still require elucidation.

The study of movement of molecules across the tegument of cestodes has in the past largely been dominated by studies on nutrition. However, in recent years transport has been recognized to be a two-way process, with a variety of materials, some of great pharmacological potential, being secreted/excreted into the environment. *In vitro* culture represents a relatively simple procedure for collecting such compounds and important advances in our understanding in this field have been made (see *S.mansonoides, L. intestinalis, T. taeniaeformis* and *T. pisiformis*, above).

The contributions of *in vitro* culture to taxonomy have been discussed by Smyth (1979b). Emphasis was placed on the identification of unknown, immature stages of the life cycle by growing them to adults, the identification of strains and the importance of these (especially *E. granulosus*) in epidemiological studies, and the identification of taeniid eggs. Undoubtedly these approaches to taxonomic problems will have an enduring value, although the use of modern techniques in molecular biology will play an increasingly important role (see, for example, Craig *et al.*, 1986; McManus and Simpson, 1985).

There are many examples in the literature of the effects of chemotherapeutic agents on cestodes *in vitro* (see, for example, Eckert, 1986, for *Echinococcus* spp. and Andrews *et al.*, 1983, for praziquantel). *In vitro* culture will continue to be of value in the future, both in screening potential anthelmintics and elucidating their mode of action. It would, however, be incautious to assume that deleterious effects on parasites observed *in vitro* also occur *in vivo*, and there will be continuing requirement for animal models to study drug metabolism, pharmacokinetics, etc.

A much underdeveloped area is the culture *in vitro* of cestode cells. Sakamoto (1978) observed a rapid proliferation of cells from the germinal layer of *E. multilocularis*, together with the production of syncytia. These cultured cells could develop into hydatid cysts when enclosed in diffusion chambers implanted into rodents, and the potential of this approach for morphogenetic studies is great and merits further study. Rubino *et al.* (1983) studied the cytoskeleton of cells cultured from *E. granulosus* and described the properties of microtubules and microfilaments, using immunological techniques and specific anti-microtubule drugs.

Voge (1978) concluded her review with a note of caution. It was her opinion that studies *in vitro* of the physiology and biochemistry of cestodes (using short-term incubations rather than extended culture) give rise to conclusions "of dubious validity . . . misinformation rather than meaningful facts". It is undoubtedly true that care should be exercised in the extrapolation of conclusions drawn from data derived from *in vitro* studies to the intact symbiosis. It is my view, however, that short-term *in vitro* maintenance, can, in ways not possible using other techniques, provide powerful insights into the physiology and biochemistry of tapeworms. In support of this can be cited the great body of information on the functional diversity of the cestode tegument, derived largely from studies *in vitro* of *H. diminuta*. In those few instances in which comparable studies have been undertaken *in vivo*, qualitatively and quantitatively similar data have been obtained (see Arme, 1975). I consider, therefore, that studies *in vitro*, of the type criticized by Voge will, properly interpreted, continue to contribute to our understanding of the biology of tapeworms.

ACKNOWLEDGEMENT

I am grateful to Professor J.D. Smyth for helpful comments on the manuscript. The errors that remain, however, are my responsibility.

REFERENCES

Allison, F.R. and Oakley, A. (1973). The two species of *Gyrocotyle* in the elephant fish, *Callorhynchus milii* (Bory). *Journal of the Royal Society of New Zealand* **3**, 381–392.

Andersen, F.L. and Loveless, R.M. (1978). Survival of protoscolices of *Echinococcus granulosus* at constant temperatures. *Journal of Parasitology* **64**, 78–82.

Anderson, R.M. and Lethbridge, R.C. (1975). An experimental study of the survival characteristics, activity and energy reserves of the hexacanths of *Hymenolepis diminuta*. *Parasitology* **71**, 137–151.

Andrews, P. and Thomas, H. (1979). The effect of praziquantel on *Hymenolepis diminuta in vitro*. *Tropenmedizin und Parasitologie* **30**, 391–400.

Andrews, P., Thomas, H., Pohlke, R. and Seubert, J. (1983). Praziquantel. *Medicinal Research Review* **3**, 147–200.

Arme, C. (1975). Tapeworm-host interactions. *In* "Symbiosis". Symposium of the Society for Experimental Biology No. 29 (D. Jennings and D.L. Lee, eds) pp. 505–532. Cambridge University Press.

Arme, C. and Pappas, P.W. (1983) (eds). "Biology of the Eucestoda", Vols I and II. Academic Press, London and New York.

Arme, C., Griffiths, D.V. and Sumpter, J.P. (1982). Evidence against the hypothesis that the plerocercoid larva of *Ligula intestinalis* (Cestoda : Pseudophyllidea) produces a sex steroid that interferes with host reproduction. *Journal of Parasitology* **68**, 169–171.

Baldwin, J.L., Berntzen, A.K. and Brown, B.W. (1978). *Mesocestoides corti* : cation concentration in calcareous corpuscles of tetrathyridia grown *in vitro*. *Experimental Parasitology* **44**, 190–196.

Barrett, N.J., Smyth, J.D. and Ong, S.J. (1982). Spontaneous sexual differentiation of *Mesocestoides corti* tetrathyridea *in vitro*. *International Journal for Parasitology* **12**, 315–322.

Bénex, J. (1968). Considerations experimentales nouvelles sur l'evolution *in vitro* en milieu les larves d'*Echinococcus granulosus*. *Annales de Parasitologie Humaine et Comparée* **43**, 561–572.

Berntzen, A.K. and Mueller, J.F. (1964). *In vitro* cultivation of *Spirometra mansonoides* (Cestoda) from the procercoid to the early adult. *Journal of Parasitology* **50**, 705–711.

Bøgh, H.O., Christensen, J.P.B. and Andreassen, J. (1986). Complement mediated lysis *in vitro* of newly excysted tapeworms : *Hymenolepis diminuta, Hymenolepis microstoma, Hymenolepis nana* and *Hymenolepis citelli*. *International Journal for Parasitology* **16**, 157–162.

Bortoletti, G. and Gabriele, F. (1979). Azione di antielmintici su *Hymenolepis nana* coltivata *in vitro*. *Rivista di Parassitologia* **40**, 97–104.

Brandt, J.R.A. and Sewell, M.M.H. (1980). Preliminary observations on the *in vitro* culture of metacestodes of *Taenia saginata*. *Veterinary Research Communications* **3**, 317–324.

Brandt, J.R.A. and Sewell, M.M.H. (1981). *In vitro* hatching and activation of *Taenia taeniaeformis* oncospheres. *Veterinary Research Communications* **5**, 193–199.

Branford White, C.J. and Hipkiss, J.B. (1985). The effect of trifluoperazine (Stellazine) on *Hymenolepis diminuta in vitro*. *Zeitschrift für Parasitenkunde* **71**, 365–372.

Brucker, D.A. (1979). *In vitro* growth of racemose cysticercus of *Taenia solium* from human brain. *Journal of Parasitology* **65**, 450.

Chang, T.W., Raben, M.S., Mueller, J.F. and Weinstein, L. (1973). Cultivation of the sparganum of *Spirometra mansonoides in vitro*, with prolonged production of sparganum growth factor. *Proceedings of the Society for Experimental Biology and Medicine* **143**, 457–459.

Chowdhury, N. (1978). Effects of fat soluble vitamins (vitamin A, D_3 and E) on axenically *in vitro* growth of *Hymenolepis microstoma*. *Zeitschrift für Parasitenkunde* **56**, 29–38.

Chowdhury, N. and De Rycke, P.H. (1979a). Comparative studies on the development of *Hymenolepis microstoma in vitro* in different types of sera. *Japanese Journal of Parasitology* **28**, 273–284.

Chowdhury, N. and De Rycke, P.H. (1979b). Effects of pancreatin on axenically *in vitro* growth of *Hymenolepis microstoma*. *Japanese Journal of Parasitology* **28**, 261–269.

Chowdhury, N. and De Rycke, P.H. (1981). Further observations on the *in vitro* development of *Hymenolepis microstoma* in relation to the lipid and cholesterol content of the serum. *Japanese Journal of Parasitology* **30**, 225–234.

Chowdhury, N. and De Rycke, P.H. (1985). Effect of ox bile on the formation of calcareous corpuscles during *in vitro* growth of *Hymenolepis diminuta*. *Current Science, India* **54**, 240–242.

Coman, B.J. and Rickard, M.D. (1977). A comparison of *in vitro* and *in vivo* estimates of the viability of *Taenia pisiformis* eggs aged under controlled conditions, and their ability to immunise against a challenge infection. *International Journal for Parasitology* **7**, 15–20.

Craig, P.S., Macpherson, C.N.L. and Nelson, G.S. (1986). The identification of eggs of

Echinococcus by immunofluorescence using a specific anti-oncospheral monoclonal antibody. *American Journal of Tropical Medicine and Hygiene* **35**, 152–158.

Dunn, J. and Threadgold, L.T. (1984). *Taenia crassiceps* : temperature, polycations, polyanions and cysticercal endocytosis. *Experimental Parasitology* **58**, 110–124.

Eckert, J. (1986). Prospects for treatment of the metacestode stage of *Echinococcus. In* "The Biology of *Echinococcus* and Hydatid Disease" (R.C.A Thompson, ed.), pp. 250–284. George Allen and Unwin, London.

Etges, D.J. and Bogitsh, B.J. (1985). The effect of colchicine on translocation of incorporated [^{3}H] proline in *Hymenolepis diminuta. Journal of Parasitology* **71**, 290–296.

Evans, W.S. (1970). The *in vitro* cultivation of *Hymenolepis microstoma* from cysticercoid to egg producing adult. *Canadian Journal of Zoology* **48**, 1135–1137.

Evans, W.S. (1978). The effect of streptomycin and penicillin on the growth and differentiation *in vitro* of *Hymenolepis microstoma. Canadian Journal of Parasitology* **56**, 1210–1211.

Evans, W.S. (1980). The cultivation of *Hymenolepis in vitro. In* "Biology of the tapeworm *Hymenolepis diminuta*" (H.P. Arai, ed.), pp. 425–448. Academic Press, New York and London.

Ferretti, G. and Gabriele, F. (1973). *In vitro* cestode cultures. I. Culture apparatus. *Rivista di Parassitologia* **34**, 235–237.

Ferretti, G. and Gabriele, F. (1978). *In vitro* cestode cultures. II. The action of antibiotics on parasite growth. *Rivista di Parassitologia* **39**, 63–70.

Ferretti, G. and Gabriele, F. (1979). *In vitro* cestode cultures. III. Relation of parasite growth to culture medium with particular regard to the non-synthetic component. *Rivista di Parassitologia* **40**, 105–114.

Fioravanti, C.F. and MacInnis, A.J. (1976a). Metabolic indices for evaluating the *in vitro* maintenance of *Hymenolepis diminuta* in the presence and absence of various additives. *Journal of Parasitology* **62**, 741–748.

Fioravanti, C.F. and MacInnis, A.J. (1976b). The *in vitro* effects of farnesol and derivatives on *Hymenolepis diminuta. Journal of Parasitology* **62**, 749–755.

Freze, V.I., Dalin, M.V. and Sergeeva, E.G. (1985). Coracidia agglutination reaction of pseudophyllid cestodes and its application in cestode studies. *Angewandte Parasitologie* **26**, 85–90.

Friedman, P.A., Weinstein, P.P., Mueller, J.F. and Allen, R.H. (1982). Internal transport and metabolism of cobalamin in *Spirometra mansonoides* (Cestoda) spargana. *Journal of Parasitology* **68**,983–992.

Friedman, P.A., Weinstein, P.P., Mueller, J.F. and Allen, R.H. (1983). Characterization of cobalamin receptor sites in brush-border plasma membranes of the tapeworm *Spirometra mansonoides. Journal of Biological Chemistry* **258**, 4261–4265.

Guevara-Benitez, D.C., Osuna-Carrillo, A. and Suarez-Carrillo, L.E. (1978). Cultivo "*in vitro*" de *Hymenolepis nana. Revista Iberica de Parasitologia* **38**, 783–791.

Hamilton, K.A. and Byram, J.E. (1974). Tapeworm development : the effects of urea on a larval tetraphyllidean. *Journal of Parasitology* **60**, 20–28.

Hammerberg, B. and Williams, J.F. (1978). Physicochemical characterisation of complement-interacting factors from *Taenia taeniaeformis. Journal of Immunology* **120**, 1039–1045.

Harris, A. and Heath, D.D. (1983). Rapid processing of *in vitro* cultured microscopical metacestodes for ultrastructural studies. *Journal of Parasitology* **69**, 998–1000.

Heath, D.D. (1982). *In vitro* culture of cysticerci: an aid to investigations of morphological development and host-parasite relationships. *In* "Cysticercosis. Present state of knowledge and perspectives'' (A. Flisser, *et al.*, eds), pp. 477–493. Academic Press, New York and London.

Heath, D.D. and Lawrence, S.B. (1976). *Echinococcus granulosus*: development *in vitro* from oncosphere to immature hydatid cyst. *Parasitology* **73**, 417–423.

Heath, D.D. and Lawrence, S.B. (1981). *Echinococcus granulosus* cysts: early development *in vitro* in the presence of serum from infected sheep. *International Journal for Parasitology* **11**, 261–266.

Heath, D.D. and Osborn, P.J. (1976). Formation of *Echinococcus granulosus* laminated membrane in a defined medium. *International Journal for Parasitology* **6**, 467–471.

Heath, D.D. and Smyth, J.D. (1970). *In vitro* cultivation of *Echinococcus granulosus, Taenia hydatigena, T. ovis, T. pisiformis* and *T. serialis* from oncosphere to cystic larva. *Parasitology* **61**, 329–343.

Hoole, D. and Arme, C. (1985). The *in vitro* culture and tegumental dynamics of the plerocercoid of *Ligula intestinalis* (Cestoda: Pseudophyllidea). *International Journal for Parasitology* **15**, 609–615.

Hoole, D. and Arme, C. (1986). The role of serum in leucocyte adherence to the plerocercoid of *Ligula intestinalis* (Cestoda: Pseudophyllidea). *Parasitology* **92**, 413–424.

Horii, Y., Owhashi, M., Ishii, A., Bandou, K. and Usui, M. (1984). Leukocyte accumulation in sparganosis: demonstration of eosinophil and neutrophil chemotactic factors from the plerocercoid of *Spirometra erinacei in vivo* and *in vitro*. *American Journal of Tropical Medicine and Hygiene* **33**, 138–143.

Howell, M.J. (1986). Cultivation of *Echinococcus* species *in vitro*. *In* ''The Biology of *Echinococcus* and Hydatid Disease'' (R.C.A. Thompson, ed.), pp. 143–163. George Allen and Unwin, London.

Insler, G.D. and Roberts, L.S. (1980). Developmental physiology of cestodes. XVI. Effects of certain excretory products on incorporation of [^{3}H]thymidine into DNA of *Hymenolepis diminuta*. *Journal of Experimental Biology* **211**, 55–61.

Jeffs, S.A. and Arme, C. (1984). *Hymenolepis diminuta*: protein synthesis in cysticercoids. *Parasitology* **88**, 351–357.

Jeffs, S.A. and Arme, C. (1985a). *Hymenolepis diminuta*: characterisation of the neutral amino acid transport loci of the metacestode. *Comparative Biochemistry and Physiology* **81A**, 387–391.

Jeffs, S.A. and Arme, C. (1985b). *Hymenolepis diminuta* (Cestoda): uptake of cycloleucine by metacestodes. *Comparative Biochemistry and Physiology* **81A**, 495–499.

Jeffs, S.A. and Arme, C. (1986). *Echinococcus granulosus*: absorption of cycloleucine and α-aminoisobutyric acid by protoscoleces. *Parasitology* **92**, 153–163.

Jensen, L.A., Short, J.A. and Andersen, F.L. (1984). The survival of protoscolices of *Echinococcus multilocularis* (Cestoda) at constant temperatures. *Proceedings of the Helminthological Society of Washington* **51**, 343–345.

Karimova, K.K. (1972). Dissemination of *Taenia saginata* oncospheres in the environment and means of their destruction. *Sbornik nauchnykhtrudov po teniarinkhosu, Dushanbe, USSR*, pp. 63–68 [quoted from Prokopic and Jelenova, 1980].

Kegley, L.M., Brown, B.W. and Berntzen, A.K. (1970). *Mesocestoides corti*: environmental cation concentration in calcareous corpuscles. *Experimental Parasitology* **27**, 88–94.

Khan, Z.I. and De Rycke, P.H. (1975). The *in vitro* cultivation of *Hymenolepis microstoma* with particular reference to the role of serum for strobilization and gametogenesis. *Biologisches Jahrbuch Dodonaea* **43**, 151–172.

Khan, Z.I. and De Rycke, P.H. (1976a). Studies on *Hymenolepis microstoma in vitro*. I.

Effect of haeme compounds on growth and reproduction. *Zeitschrift für Parasitenkunde* **49**, 253–261.

Khan, Z.I. and De Rycke, P.H. (1976b). Studies on *Hymenolepis microstoma in vitro*. II. Effect of yeast extract on development and maturation. *Zeitschrift für Parasitenkunde* **50**, 73–79.

Kumaratilake, L.M. and Thompson, R.C.A. (1981). Maintenance of the life cycle of *Echinococcus granulosus* in the laboratory following *in vivo* and *in vitro* development. *Zeitschrift für Parasitenkunde* **65**, 103–106.

Kumaratilake, L.M., Thompson, R.C.A. and Dunsmore, J.D. (1983). Comparative strobilar development of *Echinococcus granulosus* of sheep origin from different geographical areas of Australia *in vivo* and *in vitro*. *International Journal for Parasitology* **13**, 151–156.

Laclette, J.P., Merchant, M.T., Willms, K. and Cañedo, L. (1981). Paracrystalline bundles of large tubules, induced *in vitro* by mebendazole in *Cysticercus cellulosae*. *Parasitology* **83**, 513–518.

Lawrence, S.B., Heath, D.D., Parmeter, S.N. and Osborn, P.J. (1980). Development of early larval stages of *Taenia ovis in vitro* using a cell monolayer. *Parasitology* **81**, 35–40.

Lockwood, A.P.M. (1963). "Animal Body Fluids and Their Regulation". Heinemann, London.

McManus, D.P. and Simpson, A.J. (1985). Identification of the *Echinococcus* (hydatid disease) organisms using cloned DNA markers. *Molecular and Biochemical Parasitology* **17**, 171–178.

McManus, D.P. and Sterry, P.R. (1982). *Ligula intestinalis*: intermediary carbohydrate metabolism in plerocercoids and adults. *Zeitschrift für Parasitenkunde,* **67**, 73–85.

Macpherson, C.N.L. and Smyth, J.D. (1985). *In vitro* culture of the strobilar stage of *Echinococcus granulosus* from protoscoleces of human, camel, cattle, sheep and goat origin from Kenya and buffalo origin from India. *International Journal for Parasitology* **15**, 137–140.

Malmberg, G. (1974). On the larval protonephridial system of *Gyrocotyle* and the evolution of Cercomeromorphae (Platyhelminthes). *Zoologica Scripta* **3**, 65–81.

Matskasi, I. and Juhasz, S. (1977). *Ligula intestinalis* (L. 1785): Investigation of plerocercoids and adults for protease and protease inhibitor activities. *Parasitologia Hungarica* **10**, 51–60.

Matskasi, I. and Nemeth, I. (1979). *Ligula intestinalis* (Cestoda: Pseudophyllidea): studies on the properties of proteolytic and protease inhibitor activities of plerocercoid larvae. *International Journal for Parasitology* **9**, 221–227.

Mettrick, D.F. (1982). Behavioural and physiological cues of cestodes with particular reference to 5–HT. *In* "Cues that Influence Behaviour of Internal Parasites" (S.W. Bailey, ed.), pp. 85–109. U.S. Department of Agriculture, New Orleans.

Morales, E. and Canedo, L. (1982). *In vitro* study of the early transition of *Taenia solium* from metacestode to adult. *In* "Cysticercosis. Present state of knowledge and perspectives" (A. Flisser, *et al.*, eds), pp. 495–498. Academic Press, New York and London.

Mudry, D.R. and Dailey, M.D. (1971). Postembryonic development of certain tetraphyllidean and trypanorhynchan cestodes with a possible alternative life cycle for the order Trypanorhyncha. *Canadian Journal of Zoology* **49**, 1249–1253.

Murphy, W.A. and Lumsden, R.D. (1984). Reappraisal of the mucosal epithelial space associated with the surface of *Hymenolepis diminuta* and its effect on transport parameters. *Journal of Parasitology* **70**, 516–521.

Nemeth, I. and Juhasz, S. (1980). A trypsin and chymotrypsin inhibitor from the metacestodes of *Taenia pisiformis*. *Parasitology* **80**, 433–466.

Nemeth, I. and Juhasz, S. (1981). Properties of a trypsin and chymotrypsin inhibitor secreted by larval *Taenia pisiformis*. *International Journal for Parasitology* **11**, 137–144.

Nemeth, I., Juhasz, S. and Bainter, K. (1979). A trypsin and chymotrypsin inhibitor from *Taenia pisiformis*. *International Journal for Parasitology* **9**, 515–522.

Ong, S.J. and Smyth, J.D (1986). Effects of some culture factors on sexual differentiation of *Mesocestoides corti* from tetrathyridia *in vitro*. *International Journal for Parasitology* **16**, 361–368.

Osborn, P.J., Lawrence, S.B. and Heath, D.D. (1982). The chemical removal of embryophoric blocks from eggs of *Taenia ovis* and *Taenia hydatigena* prior to *in vitro* cultivation. *Journal of Parasitology* **68**, 339–341.

Osuna-Carrillo, A. and Mascaro-Lazcano, M.C. (1982). The *in vitro* cultivation of *Taenia pisiformis* to sexually mature adults. *Zeitschrift für Parasitenkunde* **67**, 67–71.

Osuna-Carrillo, A., Mascaro-Lazcano, M.C., Geuvara-Pozo, D. and Guevara-Benitez, D. (1977). Efecto del pH y de la tension de O_2 disuelto en la desenvaginacion "*in vitro*" de *Cysticercus tenuicollis*. *Revista Ibérica de Parasitologia* **37**, 365–374.

Osuna-Carrillo, A., Mascaro-Lazcano, M.C., Geuvara-Poro, D. and Guevara-Benitez, D. (1978a). Cultivo "*in vitro*" de *Taenia hydatigena*. *Revista Ibérica de Parasitologia* **38**, 289–299.

Osuna-Carrillo, A., Mascaro-Lazcano, M.C., Guevara-Pozo, D. and Guevara-Benitez, D. (1978b). Efecto de la tripsina, pancreatina y bilis canina en la desenvaginacion "*in vitro*" de *Cysticercus tenuicollis*. *Revista Ibérica de Parasitologia* **38**, 569–577.

Osuna-Carrillo, A., Mascaro-Lazcano, M.C., Guevara-Pozo, D. and Guevara-Benitez, D.C. (1980). Desenvaginacion "*in vitro*" de *Cysticercus tenuicollis* sin tratamiento pepsico. *Revista Ibérica de Parasitologia* **40**, 295–300.

Osuna-Carrillo, A., Suarez-Carrillo, L.E., Guevara-Benitez, D.C. and Gamarro-Conde, F. (1982). Cultivo "*in vitro*" de *Hymenolepis nana* (Cestoda): influencia en su desarrollo de los niveles hormonales del suero y extracto hepatico empleados en el cultivo. *Revista Ibérica de Parasitologia* **42**, 399–409.

Owen, R.R. (1985). Improved *in vitro* determination of the viability of *Taenia saginata* embryos. *Annals of Tropical Medicine and Parasitology* **79**, 655–656.

Pappas, P.W. (1983). Host-parasite interface. *In* "Biology of the Eucestoda" (C. Arme and P.W. Pappas, eds), Vol. 2, pp. 297–334. Academic Press, London and New York.

Phares, C.K. (1984). A method for solubilisation of a human growth hormone analogue from plerocercoids of *Spirometra mansonoides*. *Journal of Parasitology* **70**, 840–842.

Phares, C.K. and Ruegamer, W.R. (1973). *In vitro* preparation of a growth factor from the plerocercoids of the tapeworm *Spirometra mansonoides*. *Proceedings of the Society for Experimental Biology and Medicine* **142**, 374–377.

Podesta, R.B. (1977a). *Hymenolepis diminuta*: marker distribution volumes of tissues and mucosal extracellular spaces. *Experimental Parasitology* **42**, 289–299.

Podesta, R.B. (1977b). *Hymenolepis diminuta*: unstirred layer thickness and effects on active and passive transport kinetics. *Experimental Parasitology* **43**, 12–21.

Podesta, R.B., Evans, W. and Stallard, E. (1977a). *Hymenolepis diminuta* and *Hymenolepis microstoma*: effect of ouabain on active nonelectrolyte uptake across the "epithelial" syncytium. *Experimental Parasitology* **43**, 25–38.

Podesta, R.B., Stallard, H., Evans, W., Lussier, P., Jackson, D. and Mettrick, D.F.

(1977b). *Hymenolepis diminuta*: determination of unidirectional uptake rates for nonelectrolytes across the surface "epithelial" membrane. *Experimental Parasitology* **42**, 300–317.

Pool, D.W. (1985). The effect of praziquantel on the pseudophyllidean cestode *Bothriocephalus acheilognathi in vitro. Zeitschrift für Parasitenkunde* **71**, 603–608.

Prokopic, J. and Jelenova, I. (1980). Effect of fertilisers on *Taenia saginata*. Goeze, 1782, on egg viability *in vitro. Folia Parasitologica (Praha)* **27**, 343–347.

Read, C.P. (1970). "Parasitism and Symbiology". The Ronald Press Co., New York.

Read, C.P., Rothman, A.H. and Simmons, J.E., Jnr. (1963). Studies on membrane transport, with special reference to parasite-host integration. *Annals of the New York Academy of Sciences* **113**, 154–205.

Rémond, M., Perret, C. and Soulé, C. (1980). Conditions de survie "*in vitro*" *de "Cysticercus bovis". Bulletin Académie Véterinaire de France* **53**, 175–180.

Richards, K.S. and Rogan, M.T. (1986). Surface ultrastructure of a strobilate form of the horse strain of *Echinococcus granulosus* cultured in monophasic medium. *Annals of Tropical Medicine and Parasitology* **80**, 267–268.

Rikihisa, Y., Letonja, T., Pratt, N. and Lin, Y.C. (1984). *Taenia taeniaeformis*: characterisation of larval metabolic products and growth of host gastric cells *in vitro. Experimental Parasitology* **58**, 230–238.

Rikihisa, Y., Lin, Y.C. and Fukaya, T. (1985). *Taenia taeniaeformis*: inhibition of rat testosterone production by excretory-secretory product of the cultured metacestode. *Experimental Parasitology* **59**, 390–397.

Robinson, R.D. and Arme, C. (1985). *Echinococcus granulosus*: failure of the eosin-exclusion test to demonstrate death of protoscoleces. *Annals of Tropical Medicine and Parasitology* **79**, 117.

Rogan, M.T. and Richards, K.S. (1986). *In vitro* development of hydatid cysts from posterior bladders and ruptured brood capsules of equine *Echinococcus granulosus. Parasitology* **92**, 379–390.

Rubino S., Fiori, P.L., Lubinu, G., Monaco, G. and Cappuccinelli, P. (1983). The cytoskeleton of hydatid cyst cultured cells and its sensitivity to inhibitors. *European Journal of Cell Biology* **30**, 182–190.

Saad, M.B. and El Abin, E.A.Z. (1982). The survival rate of protoscolices from hydatid fluid kept at constant temperatures. *Sudan Journal of Veterinary Research* **4**, 157.

Sakamoto, T. (1978). Development of echinococcal tissue cultured *in vitro* and *in vivo. Memoirs of the Faculty of Agriculture, Kagoshima University* **14**, 109–115.

Sakanari, J. and Moser, M. (1985). Salinity and temperature effects on the eggs, coracidia and procercoids of *Lacistorhynchus tenuis* (Cestoda: Trypanorhyncha) and induced mortality in a first intermediate host. *Journal of Parasitology* **71**, 583–587.

Salazar, P.M., de Haro, I. and Voge, M. (1984). Hatching *in vitro* of oncospheres on *Taenia solium. Journal of Parasitology* **70**, 161–162.

Schiller, E.L. (1965). A simplified method for the *in vitro* cultivation of the rat tapeworm *Hymenolepis diminuta. Journal of Parasitology* **59**, 122–129.

Simmons, J.E. (1974). *Gyrocotyle*: a century-old enigma. *In* "Symbiosis in the Sea" (W.B. Vernberg, ed.), pp. 195–218. University of South Carolina Press.

Sinha, D.P. and Hopkins, C.A. (1967). *In vitro* cultivation of the tapeworm *Hymenolepis nana* from the larva to adult. *Nature (London)* **215**, 1275–1276.

Smyth, J.D. (1947). Studies on tapeworm physiology II. Cultivation and development of *Ligula intestinalis in vitro. Parasitology* **38**, 173–181.

Smyth, J.D. (1950). Studies on tapeworm physiology. IV. Further observations on the maturation of *Schistocephalus solidus* under sterile conditions *in vitro. Journal of Parasitology* **36**, 371–381.

Smyth, J.D. (1962). Studies on tapeworm physiology. X. Axenic cultivation of the hydatid organism, *Echinococcus granulosus*; establishment of a basic technique. *Parasitology* **52**, 441–457.

Smyth, J.D. (1967). Studies on tapeworm physiology. XI. *In vitro* cultivation of *Echinococcus granulosus* from the protoscolex to the strobilate stage. *Parasitology* 57, 111–133.

Smyth, J.D. (1979a). *Echinococcus granulosus* and *E. multilocularis*: *in vitro* culture of the strobilar stages from protoscoleces. *Angewandte Parasitologie* **20**, 137–147.

Smyth, J.D. (1979b). An *in vitro* approach to taxonomic problems in trematodes and cestodes especially *Echinococcus. Symposia of the British Society for Parasitology,* **17**, 75–102.

Smyth, J.D. (1982). The insemination-fertilization problem in cestodes cultured *in vitro. In* "Aspects of Parasitology" (E. Meerovitch, ed.), pp. 392–406. Institute of Parasitology, McGill University, Montreal, Canada.

Smyth, J.D. and Barrett, N.J. (1979). *Echinococcus multilocularis*: further observations on strobilar differentiation *in vitro. Revista Ibérica de Parasitologia* **39**, 39–53.

Smyth, J.D. and Davies, Z. (1974). *In vitro* culture of the strobilar stage of *Echinococcus granulosus* (sheep strain): a review of basic problems and results. *International Journal for Parasitology* **4**, 631–644.

Smyth, J.D. and Davies, Z. (1975). *In vitro* suppression of segmentation in *Echinococcus multilocularis* with morphological transformation of protoscoleces into monozoic adults. *Parasitology* **71**, 125–135.

Smyth, J.D. and Davies, C.J. (1979). *In vitro* differentiation of *Echinococcus* as a taxonomic tool, with special reference to its use in "strain" identification. *Helminthologia* **16**, 5–12.

Smyth, J.D. and Haslewood, G.A.D. (1963). The biochemistry of bile as a factor determining host specificity in intestinal parasites with particular reference to *Echinococcus granulosus. Annals of the New York Academy of Sciences* **113**, 234–260.

Smyth, J.D. and Howkins, A.B. (1966). Studies on tapeworm physiology. XII. A technique for the cultivation of partly developed strobila of *Echinococcus granulosus* from dogs to maturity *in vitro. Parasitology* **56**, 763–766.

Sterry, P.R. and McManus, D.P. (1982). *Ligula intestinalis*: biochemical composition, carbohydrate utilisation and oxygen consumption of plerocercoids and adults. *Zeitschrift für Parasitenkunde* **67**, 87–98.

Stevenson, P. (1983). Observations on the hatching and activation of fresh *Taenia saginata* eggs. *Annals of Tropical Medicine and Parasitology* **77**, 399–404.

Sukhdeo, M.V.K., Hsu, S.C., Thompson, C.S. and Mettrick, D.F. (1984). *Hymenolepis diminuta*: behavioural effects of 5-hydroxytryptamine, acetylcholine, histamine and somatostatin. *Journal of Parasitology* **70**, 682–688.

Taylor, A.E.R. and Baker, J.R. (1968). "The Cultivation of Parasites *In vitro*". Blackwell Scientific Publications, Oxford.

Taylor, A.E.R. and Baker, J.R. (eds). (1978). "Methods of Cultivating Parasites *In vitro*". Academic Press, London and New York.

Thompson, R.C.A. (1977). Growth, segmentation and maturation of the British horse and sheep strains of *Echinococcus granulosus* in dogs. *International Journal for Parasitology* **7**, 281–285.

Thompson, R.C.A. (ed.) (1986). "The Biology of *Echinococcus* and Hydatid Disease". George Allen and Unwin, London.

Thompson, R.C.A. and Eckert, J. (1982). The production of eggs by *Echinococcus multilocularis* in the laboratory following *in vivo* and *in vitro* development. *Zeitschrift für Parasitenkunde* **68**, 227–234.
Thompson, R.C.A., and Smyth, J.D. (1976). Attempted infection of the rhesus monkey (*Macaca mulatta*) with the British horse strain of *Echinococcus granulosus*. *Journal of Helminthology* **50**, 175–177.
Thompson, R.C.A., Jue Sue, L.P. and Buckley, S.J. (1982). *In vitro* development of the strobilar stage of *Mesocestoides corti*. *International Journal for Parasitology* **12** 303–314.
Uglem, G.L. and Just, J.J. (1983). Trypsin inhibition by tapeworms : antienzyme secretion or pH adjustment? *Science* **220**, 79–81.
Valkounova, J. (1979). Excystation of scoleces of cestodes of the family Hymenolepididae from the cysts *in vitro*. *Folia Parasitologica* (Praha) **26**, 50.
Voge, M. (1978). Cestoda. *In* "Methods of Cultivating Parasites *In Vitro*" (A.E.R. Taylor and J.R. Baker, eds), pp. 193–225. Academic Press, London and New York.
Voge, M. and Coulombe, L.S. (1966). Growth and asexual multiplication *in vitro* of *Mesocestoides* tetrathyridia. *American Journal of Tropical Medicine and Hygiene* **15**, 902–907.
Voge, M. and Edmonds, H. (1969). Hatching *in vitro* of oncospheres from coracidia of *Lacistorhynchus tenuis* (Cestoda: Tetrarhynchidea). *Journal of Parasitology* **55**, 571–573.
Zavras, E.T. and Roberts, L.S. (1984). Developmental physiology of cestodes: characterization of putative crowding factors in *Hymenolepis diminuta*. *Journal of Parasitology* **70**, 937–944.
Zavras, E.T. and Roberts, L.S. (1985). Developmental physiology of cestodes: cyclic nucleotides and the identity of putative crowding factors in *Hymenolepis diminuta*. *Journal of Parasitology* **71**, 96–105.

Chapter 13

Nematoda Except Parasites of Insects

FRANK W. DOUVRES and JOSEPH F. URBAN Jr.[1]

Helminthic Diseases Laboratory, Animal Parasitology Institute, Agricultural Research Service, US Department of Agriculture, Maryland, USA

I. INTRODUCTION

An updated review of studies on the *in vitro* cultivation of nematode parasites of animals (except insects) and plants since 1977 has been compiled on the assumption that the reader is familiar with the basic techniques and concepts of nematode cultivation presented in the earlier volumes (Taylor and Baker, 1968; Hansen and Hansen, 1978). Here, material is organised in sections according to an alphabetized classification of nematode superfamilies (Anderson *et al.*, 1983). A summary of salient details for the cultivation of nematode parasites of vertebrates is presented in Table I, with additional general information and some selected techniques being presented in the text in Section II. Sections III and IV briefly review the cultivation of nematode parasites of invertebrates and plants, respectively.

[1] Author to whom correspondence should be addressed.

II. NEMATODE PARASITES OF VERTEBRATES

A. SUPERFAMILY: ANCYLOSTOMOIDEA

No major advance has been reported in the cultivation of these nematodes since 1977 (Hansen and Hansen, 1978). Various undefined media were used by Fenglin *et al.* (1979) to maintain *in vivo* derived *Ancylostoma caninum* and *A. duodenale* young adults (YA) and mature adults (MA) for over 1 month: both species grew and oviposited *in vitro*, and copulating pairs of *A. duodenale* were observed. A semi-defined medium was used by Xainxiang and Jiajun (1985) to maintain *Necator americanus* adults which also copulated and oviposited *in vitro*.

B. SUPERFAMILY: ASCARIDOIDEA

Carvajal *et al.* (1981) cultured infective third-stage larvae (L3) of *Anisakis simplex* and *Phocanema decipiens*, derived from Chilean hake, to YA in an undefined medium prepared as described by Van Banning (1971; also see Hansen and Hansen, 1978, p. 242). Development of *P. decipiens* L3 to YA *in vitro* proceeded through two moults: one after 4–7 d and a second after 20–25 d. This sequence was similar to the *in vivo* development of larvae in seals (McClelland, 1980), but slightly longer.

Difficulties in the specific identification of infective anisakine larvae are illustrated in the work of Oshima *et al.* (1982). They cultured "so-called *Anisakis* Type I and Type II larvae . . .", isolated from mackerel and pollacks, in a two-step culture system in which defined medium supported the development of L3 to fourth-stage (L4) (Sommerville and Davey, 1976), and selected L4 were then cultured in an undefined medium modified from that described by Van Banning (1971; also see Hansen and Hansen, 1978, p. 242). Type I larvae developed to YA identified as *An. simplex*, whereas the development of Type II larvae was limited to early L4. Bier and Douvres (unpublished) cultured *An. simplex* L3 (isolated from rockfish) to MA, including ovipositing females, in a roller culture system using Medium API-l plus bovine haemin (see Section II.B.1.1°).

Considerable progress has been made in the cultivation of *Ascaris suum*. Douvres and Urban (1983) designed a three-step roller culture system to grow *As. suum* second-stage larvae (L2), hatched from eggs, to MA, including ovipositing females. Media used were KW-2 (Douvres, 1970), API-l (Douvres and Malakatis, 1977, and Section II.B.1.1°.) and API-18 (Douvres and Urban, 1983). Optimal growth *in vitro* required medium supplements L-cysteine and bovine haemin, and the precise application of two different gas phases during cultivation (see Section II.B.1.2°). Large numbers of L2 developed to L4 after 14–15 d, and a few developed to YA after 31–34 d. One MA male and two MA females developed in separate cultures after 67–80 d; unfertilized eggs were detected in the culture vessels. Comparative development of *As. suum* L4, YA, and MA *in vivo* is 10 d, 22–25 d, and 40–42 d after inoculation of swine, respectively.

Third-stage larvae of *As. suum*, obtained from the lungs of pigs, developed to L4, YA, and MA in a two-step roller culture system after 14, 20, and 50 d, respectively (Douvres and Urban, 1986). Media API-18 and API-l were used with a single gas phase; bovine haemin was an essential supplement for optimal development (see Section II.B.1.3°). Third-stage larvae also developed to YA in a one-step culture system with medium API-1, but yields of YA were marginally lower than in the two-step system (a supplement of 10 mM reduced glutathione was required for the first 7 d). Worm copulation was observed *in vitro* and fertilized eggs were subsequently detected in the culture medium. Medium API-1 was modified (see Section II.B.1.1°) to yield: (1) a semi-defined medium (API-22) lacking the complex mammalian tissue extracts and peptide digests of API-1, and (2) an undefined medium (medium API-23) containing all the components of API-1 except for its chemically defined amino acids, vitamins, and nucleic acids; L3 developed to MA in medium API-23, but not API-22 (Douvres and Urban, 1986).

Ascaris suum larvae have also been cultured in defined media. In general all of these provided parasite-derived excretory-secretory products (ES) free from the complex media components present in undefined and semidefined media. Stromberg *et al.* (1977) developed a roller culture system with a defined medium to grow L3, obtained from the lungs of rabbits, to L4. ES products were isolated from the medium which, when injected into guinea pigs, induced protective immunity to a challenge infection (Stromberg, 1979), and modulated a reaginic antibody response to a non-parasite antigen (Stromberg, 1980). Urban and Douvres (1981) designed a simple stationary culture system (multi-well plates) for the development of *As. suum* L3 to L4 in defined media. This versatile basic culture technique was used subsequently to evaluate systematically the nutritional requirements of larval growth *in vitro* (Urban *et al.*, 1984); for the production of ES capable of inducing protective immunity in pigs (Urban and Romanowski, 1985); as an anthelmintic screening system (Rew *et al.*, 1986); as a bioassay for steroid effects on larval growth and moulting *in vitro* (Fleming, 1985a, b); and as a test for the effects of proline analogs on cuticular synthesis and larval growth *in vitro* (Fetterer and Urban, 1985); and to characterize aerobic–anaerobic transition of L3–L4 by measurement of branch-chained fatty acid production (Komuniecki and Vanover, 1986). A similar culture system was used to grow L2 to L3 in a defined medium (Urban and Tromba, 1982).

1. *Detailed techniques*

1° Preparation and storage of Animal Parasitology Institute (API) media

TABLE I. Ingredients of API media

Media stock components	Stock concentration	API media 1	6	16	22	23	27	28
		Volumes (ml) added to give 1.0 litre of medium						
1. Bovine calf serum	—	250	200	250	250	250	250	250
2. EMEM ess. AA	50×	20	20	20	20	—	20	20
3. EMEM non-ess. AA	100×	10	10	10	10	—	10	10
4. BME vitamins	100×	20	20	20	20	—	20	20
5. Vitamin B_{12}	1%	1	1	1	1	—	1	1
6. Vitamin A	0.025%	0.4	0.4	0.4	0.4	—	0.4	0.4
7. Vitamin K	0.025%	0.04	0.04	0.04	0.04	—	0.04	0.04
8. Vitamin E	0.025%	0.04	0.04	0.04	0.04	—	0.04	0.04
9. Vitamin D_3	1%	0.1	0.1	0.1	0.1	—	0.1	0.1
10. PABA	0.5%	0.75	0.75	0.75	0.75	—	0.75	0.75
11. NAG	5.0%	1.5	1.5	1.5	1.5	—	1.5	1.5
12. L-glutamine	2.9%	10	10	10	10	—	10	10
13. TCA mixture	5×	20	20	20	20	20	—	20
14. Nucleic acid mix.	1×	50	50	50	50	—	50	50
15. Fatty acid mix.	10×	10	10	10	10	10	—	—
16. Ascorbic acid	1%	5	—	5	5	—	5	5
17. Cysteine-HCl	2.5%	5	—	5	5	—	5	5
18. Maltose	8.5%	10	—	10	10	10	10	10
19. Dextrose	10%	7	20	2	7	7	7	7
20. Trehalose	7%	5	—	5	5	5	5	5
21. Fructose	3.6%	5	—	5	5	5	5	5
22. EBSS	10×	55.5	0.5	0.5	55.5	55.5	55.5	55.5
23. $NaH_2PO_4 \cdot H_2O$	8.1%	0.1	—	4.1	0.1	0.1	0.1	0.1
24. KCl	8.95%	—	—	10	—	—	—	—
25. $NaHCO_3$	8.4%	20	3.6	20	20	20	20	20
26. REE	20%	75	250	75	—	75	—	—
27. CLE	20%	70	100	70	—	70	—	—
28. Bacto-peptone	25.8%	20	—	20	—	20	—	—
29. Trypticase	20%	25	50	25	—	25	—	—
30. Yeast extract	30%	10	16.7	10	—	10	—	—
31. Casein	15%	10	—	10	—	10	—	—
32. Lipid mix.	—	—	—	12	—	—	—	—
33. tH_2O	—	283.6	215.4	317.6	493.6	407.4	523.6	503.6

(ii) *Preparation*. The components of these media are prepared and stored as follows (all aqueous stocks are filter-sterilized through 0.22 μm filters and stored at – 20°C unless contra-indicated).

Stock solution 1. Blood (pooled from 2–6 helminth-free, 3- to 6-month-old Holstein-Friesian steer calves) is allowed to clot overnight at 4°C and the serum then separated by centrifugation and stored in 250–500 ml aliquots and heat-inactivated (56°C, 30 min) before use.

Stock solutions 2, 3, and 4. These are, respectively, sterile solutions of 50 times concentrated essential amino acids used for the preparation of Eagle's minimal essential medium (EMEM), 100 times concentrated non-essential amino acids for EMEM, and 100 times concentrated basal medium Eagle's (BME) vitamin solution (Gibco).

Solution 5. This is an aqueous 1% stock of Vitamin B_{12}.

Solutions 6, 7, and 8. These are 0.025% stocks of fat soluble vitamins A, K and E (Sigma), prepared by dissolving 25 mg of each vitamin in 5.2 ml of 10% triethanolamine; and made up to 100 ml in triple-distilled water (tH_2O) with slight stirring and heating.

Solution 9. 1% stock of vitamin D_3 (Sigma) prepared by adding 1.0 g to 1.0 ml of 5% Tween 80 and made up to 100 ml in tH_2O.

Solution 10. 0.5% stock solution of paraaminobenzoic acid (PABA) prepared by dissolving 500 mg of PABA in 5.2 ml of 10% triethanolamine at pH 10 or more and made up to 100 ml with tH_2O.

Solution 11. 5% *N*-acetyl-D-glucosamine (NAG) (Gibco).

Solution 12. 2.92% L-glutamine (Gibco).

Solution 13. tricarboxylic acid (TCA) mixture (sodium pyruvate, 2.5 g; malate, 1.5 g; α-ketoglutarate, 1.0 g; fumarate, 1.8 mg dissolved in 100 ml tH_2O).

Solution 14. Nucleic acid mixture (Sigma): adenosine 5′-triphosphate-2-Na, 95 mg; yeast adenylic acid, 87 mg; cytidine 2′-3′ monophosphoric acid, 77 mg; yeast guanylic acid (Na salt), 101 mg; thymine, 24.5 mg; dissolved in tH_2O to 50 ml. This mixture is freshly prepared before use.

Solution 15. A fatty acid mixture (sodium acetate, 3.5 g; propionic acid, 1.0 g; butyric acid, 0.5 g; dissolved in tH_2O to 100 ml).

Solutions 16, 17, 18, 19, 20, 21. These are aqueous stock solutions of ascorbic acid (1%), L-cysteine-HCl (2.5%), maltose (8.5%), dextrose (10%), trehalose (7%), and fructose (3.6%), respectively.

Solution 22. Sterile stock of 10 times concentrated Earle's balanced salt solution (EBSS) without glucose, $NaHCO_3$ and phenol red.

Solutions 23, 24, 25. These are aqueous salt solutions of $NaH_2PO_4{\cdot}H_2O$ (8.1%), KCl (8.95%) and $NaHCO_3$ (8.4%) respectively.

Solution 26. Rabbit embryo extract (REE) prepared from fresh embryos or embryos frozen for up to 1 year at − 20°C (Diamond and Douvres, 1962; see Taylor and Baker, 1968, p. 299). Briefly, they are rinsed in tH_2O, cut into pieces and placed in an Erlenmeyer flask with 100 ml of 0.02 N HCl per 20 g of tissue (4–8 litre preparations can be made at one time). The tissue is extracted at 4°C for 4 d with periodic shaking. The suspension is passed through cheese cloth, centrifuged (2000*g*, 30 min, 4°C) and then clarified by passage through a series of 25, 8, 4.5, 1.2 and 0.22 μm filters. Aliquots of 100 ml can be stored at − 20°C for up to 1 year.

Solution 27. Bovine calf liver extract (CLE) prepared from fresh livers after 24 h at 4°C or from livers frozen (− 20°C) for less than 6 months. The procedure is the same as that for REE.

Solutions 28, 29, 30, 31. These are bacto-peptone (25.8%) (Difco), trypticase (= casein digest of peptone #97023) (20%), yeast extract (30%) (BBL) and salt- and vitamin-free casein (15%) (ICN Nutritional Biochemicals), respectively. They are dissolved in tH_2O (may require heating), filtered, and added fresh to the medium.

Solution 32. Lipid mixture of stock solutions of 0.25% palmitic acid (ethyl ester; Sigma), 0.5% lecithin (ICN), 0.5% cholic acid (Na salt; Sigma Chemical Co.) and 0.25% cholesterol (ash-free; ICN), prepared separately by dissolving each of the lipids in 100 ml of 5% Tween 80 (cholesterol is first dissolved in 1–5 ml of 95% methanol). The solutions are autoclaved for 15–20 min at 120°C under 15 PSI and a stock mixture is prepared by adding the lipids in the following proportion, 3 : 6 : 3 : 6, respectively.

Solution 33. Triple-glass-distilled water (tH_2O).

After mixing all the stock components, antibiotics are added to give a final concentration of penicillin 1000 iu ml^{-1}, streptomycin 1.0 mg ml^{-1} and amphotericin B 10 μg ml^{-1}; the pH is then adjusted with 1 N HCl or 1 N NaOH as required.

2° *Ascaris suum* L2 to MA
(Douvres and Urban, 1983)

MEDIA

(i) Preparation of medium API-1 is described above (Section II.B.1.1°).

(ii) Media KW-2 and API-18 are prepared from combinations of the following fractions: (A) 500 ml of bovine calf serum, derived from several helminth-free animals; (B) 500 ml of defined Medium 135 supplemented with yeast extract, bactopeptone and dextrose to final concentrations of 2.25 mg ml^{-1}, 2.81 mg ml^{-1} and 3.81 mg ml^{-1} respectively; (C) 60 ml of 50 times concentrated Eagle's Minimal Essential Medium (EMEM) essential amino acids (without glutamine) (Gibco); (D) 60 ml of 100 times concentrated EMEM non-essential amino acids (Flow Laboratories).

Medium KW-2 is prepared by combining fractions A and B, while medium API-18 is prepared by combining fractions A, B, C, and D. All media are adjusted to pH 6.8, filter-sterilized with Millipore filters (0.22 μm), and streptomycin sulphate, 1.0 mg ml^{-1}, penicillin G potassium, 1000 iu ml^{-1} and amphotericin B, 10 μg ml^{-1}, are then added. Media can be used immediately or after storage at −20 °C for < 6 months.

(iii) A 1.0 M L-cysteine·HCl·H_2O stock solution is filter-sterilized and kept at −20 °C for less than 2 weeks before use.

(iv) A 0.24% bovine haemin stock solution is prepared and sterilized as described in Section II.I.1.1°(i)(b).

(v) Washing solution with antibiotics (EBSSA): Earle's balanced salt solution without phenol red (EBSS) containing 1000 iu ml^{-1} penicillin, 1.0 mg ml^{-1} streptomycin and 10 μg ml^{-1} amphotericin B.

TECHNIQUE

Fertilized *As. suum* eggs are freed from the anterior 3–6 cm of excised uteri of gravid females by preparing a suspension in 0.5 N NaOH using a glass Dounce tissue grinder. The suspension is washed with distilled water by centrifugation (200*g*) several times (until the supernatant is clear). Eggs are suspended in 0.1% formalin, placed in a separating funnel, aerated with a small pump, and maintained at approximately 25 °C for 28–32 d to produce infective eggs (the larvae appear as first-molt or L2 when pressed free from eggs between a microscope slide and coverslip). Infective eggs can be stored for over 1 year at 4 °C in 0.1% formalin. The hatching procedure is initiated by washing (centrifugation at 200*g* for 1 min at room temperature with no braking of the rotor) the eggs with saline (0.85% NaCl) five times in 40 ml glass tubes (a packed egg volume of 1.0 ml will yield approximately 1 million viable larvae). The pellet of eggs is resuspended in 20 ml of sodium hypochlorite (5.25%) for 3 min followed by 10 washes in saline. The eggs are then suspended aseptically in 10 ml of EBSS and transferred to a flat-bottom 125 ml flask containing a layer of glass beads (4–6 mm in diameter) and a spinner bar. The flask is maintained in an incubator at 37 °C over a magnetic stirrer which slowly moves the beads. Hatching is monitored by microscopical examination of samples and generally occurs within 15–30 min. The hatched larval suspension is then washed three times in EBSS; larvae are layered over a pad of cotton gauze (dry weight 2.5 g) that is immersed in EBSS in a 10 cm covered funnel and maintained at 37°C for 2–6 h in an atmosphere of 5% CO_2 in air. Larvae, which settle in the neck of the funnel, are collected aseptically and washed 10 times with Dulbecco's phosphate buffered saline (DPBS), followed by five washes in EBSSA. Suspensions of 80 000 to 640 000 L2 in 1.0 to 2.0 ml of EBSSA are used to inoculate cultures.

Larvae develop to MA in a three-step culture system (culture vessels are 1.0 or 1.8 litre glass bottles containing 80 ml of medium and rotated at 1 rev every 1.5 min at 39°C). Step 1 cultures contain medium KW-2 with 10 mM L-cysteine in an atmosphere of 95% N_2 : 5% CO_2 for 4 d, thereafter the gas phase is 85% N_2 : 10% CO_2 : 5% O_2; from days 4–11 medium KW-2 with 5.0 mM L-cysteine is used. Step 2 cultures contain medium API-18 from days 11–18. Step 3 cultures contain medium API-1 supplemented with haemin (24 μg ml^{-1}) from d 18 until termination. Inocula are transferred to fresh media and culture vessels at 4 d, and at weekly intervals thereafter. To avoid damage to large larval stages, worm suspensions are poured into a 40 ml tube and maintained upright for 10–15 min; supernatants are aspirated (including many dead larvae) and the settled-worms are poured into fresh culture vessels; adults are transferred using a tooth pick. Inocula of less than 640 000 L2 produce comparable, or slightly higher survival rates and yields of L4 and YA than those obtained in cultures of 640 000 L2 up to 32 d; thereafter, starting inocula of 640 000 L2 develop better.

3° *Ascaris suum* L3 to MA
(Douvres and Urban, 1986)

MEDIA

(i) Medium API-18 (see Section II.B.1.2°) and medium API-1 (see Section II.B.1.1°) at pH 6.8.
(ii) Glutathione (reduced) is prepared as a 1.0 M solution, filter-sterilized, and stored at −20°C until used.
(iii) A 0.24% bovine haemin stock solution is prepared as described in Section II.I.1.1°(i)(b).

TECHNIQUE

Helminth-free pigs 1–3 months old or 2–3 kg New Zealand White rabbits are inoculated orally with 250 000 or 75 000 infective *As. suum* eggs respectively. Lungs are removed from the animals 7 d later and homogenized briefly (4–8 s) in saline (0.85% NaCl) to prepare a slurry which is placed on a Baermann apparatus (a glass funnel, 20 cm diameter, filled with warm saline at 37 °C and containing a wire screen insert of 60 mesh below which a triple layer of fine cheesecloth is secured). 50 ml of the larval suspension is passed from the neck of the funnel into a large glass settling cone after successive incubation periods of 90 and 60 min at 37 °C. The cone is then filled with 250 ml of saline and 15 min later the supernatant is decanted from above the settled larvae; this process is repeated until the supernatant is clear. Larvae are then transferred aseptically to a 40 ml glass tube and washed five times at room temperature with saline (centrifugation at 200*g* for 5 s with no braking of the rotor). The larvae are then incubated for 1 h at 37 °C in Dulbecco's phosphate buffered saline (DPBS) containing 100 iu ml^{-1} of penicillin, 100 μg ml^{-1} of streptomycin and 10 μg ml^{-1} of amphotericin B, followed by five additional washes in DPBS. Suspensions of 2500 to 10 000 L3 in 1.0 ml of DPBS are used to inoculate cultures.

Larvae develop to YA and MA best in a two-step system (1.0 litre glass bottles, rotating at 1 rev per 1.5 min, containing 40 ml of medium in a gas phase of 5% O_2, 10% CO_2, 85% N_2, for the first 7 d and 80 ml thereafter). Inocula are transferred to fresh media and culture bottles at weekly intervals starting at 7 d, as in Section II.B.1.2°. Larvae are first cultured in medium API-18 for 7 d and thereafter in medium API-1 plus haemin (24 μg ml^{-1}). A one-step system consisting of medium API-1 containing 10 mM reduced glutathione for the first 7 d, followed by medium API-1 supplemented with haemin (24 μg ml^{-1}) thereafter, was marginally less supportive of L3 growth. Medium API-23 with haemin (24 μg ml^{-1}) could successfully be substituted for medium API-1 with haemin in the two-step system, indicating that the complex mammalian tissue extracts and peptide digest of API-1 are essential for development of L3 to MA. An inoculum of 10 000 L3 was generally less viable than one containing fewer than 10 000 L3 from 7–28 d, but the yields of advanced stages were marginally greater. A genital girdle or body constriction at the level of the vulva was present on MA females from cultures lacking MA males, but such females laid only infertile eggs. Females from a culture containing MA males and where copulation was observed, lacked a genital girdle and laid both infertile and fertile eggs.

C. SUPERFAMILY: FILARIOIDEA

Although *in vitro* cultivation of the Filarioidea has received increasing attention in the last decade, it still remains "one of the pressing needs in filariasis research'' (Weinstein, 1986). As with other systems designed to grow nematodes of veterinary or medical importance *in vitro*, the cultivation of filariids has focused on the regulation and control of development, the isolation of stage-specific antigens for immunological studies, the screening of drugs and elucidation of drug action.

Development of the exsheathed *Brugia malayi* and *B. pahangi* microfilaria phase (mf) of first-stage larvae (L1) was examined in undefined medium with one of four mosquito cell lines derived from insect larval tissues (Devaney and Howells, 1979). Although some development to a late L1 or ''sausage form'' (sf) larvae did occur with both species, not all cell lines supported growth equally. Ogura *et al.* (1981) obtained L2 of *B. pahangi* from pieces of mosquito thorax that were floated on a semi-defined medium. Larvae developed to the second moult (M2) and ecdysed to L3 in this system; insect hormones such as the juvenile hormone analogue, ecdysone, and extract of insect neurohaemal organs had no effect on moulting *in vitro*. Wong *et al.* (1982), and later Mak *et al.* (1983), demonstrated that *B. pahangi* L3, derived from mosquitoes, could develop to L4 in semi-defined medium conditioned with mammalian cell lines. Interestingly, when *B. malayi* L3 were cultured similarly they developed from L4 to YA (Mak *et al.*, 1983). Mature adult females of *B. pahangi*, isolated from the peritoneum of jirds, have been maintained metabolically active *in vitro* for up to 3 d and released mf into the culture medium (Furman and Ash, 1983). These were morphologically identical to *in vivo*-derived mf, but could be differentiated from them by a characteristic fluorescein-conjugated lectin-binding pattern on their surface.

Weinstein (1986) has urged a more systematic approach to the cultivation of nematodes through the careful analysis of growth factors, nutrients, hormones, trace elements and gas phases required for optimal *in vitro* development, so as to provide more reproducible culture systems. The application of some of these principles to the cultivation of *Dip. viteae* led to the development of YA from L3 derived from the arthropod host, in a semi-defined medium with carefully controlled gas phase, pH (Franke and Weinstein, 1983; 1984a), and medium composition (Franke and Weinstein, 1984b); a pO_2 of <20 mm Hg or >50 mm Hg was detrimental to larval survival and development. Optimal development in this system depended strictly on the concentration, source, and batch of the serum supplement. Tanner (1981) found that *Dip. viteae* L3, implanted subcutaneously in micropore chambers into jirds, were ''triggered'' *in vivo* for subsequent development to L4 on transferrence to semi-defined media *in vitro*. L3 derived from ticks failed to develop under these conditions but would develop to L4 when the medium was conditioned with a hamster kidney cell line. Miegeville *et al.* (1981) developed a semi-defined maintenance system for *Dip. viteae* MA females that supported survival and release of mf; L2 dissected from ticks survived only briefly in this system without development.

Several workers have used defined or semi-defined media, or media conditioned by mosquito cell lines, to culture L1 of *Dirofilaria immitis* derived from dog blood. In general, the sf of L1 was the greatest development attained. Ando *et al.* (1980) could maintain viable sf for 24 d using semi-defined media and *Aedes* cells; the number of developing larvae increased as the culture temperature decreased from 36 °C to 27 °C. Subsequently, Ando and Kitamura (1982) showed that the osmotic pressure was important for larval development *in vitro*. Sneller and Weinstein (1982) grew *D. immitis* L1 to sf in a defined medium with *A. albopictus* cells and a gas mixture of 95% N_2, 5% CO_2; here serum and O_2 were inhibitory to development, while the addition of blood enhanced the percentage of developing worms. Importantly Abraham *et al.* (1986) maintained *D. immitis* L1 *in vitro* in a defined medium for the collection of ES, and mf were still infective after 2 weeks *in vitro*. Devaney (1981) attempted to improve the development of *D. immitis* L1 in organ culture (see also Taylor and Baker, 1968, p. 302) by maintaining them in Malpighian tubules, dissected from infected mosquitoes, kept in a variety of semi-defined insect and mammalian culture media with a mosquito feeder cell layer, but she obtained development to only L2.

The L3 of *D. immitis*, derived from mosquitoes, have so far been cultured only to L4. Thus Wong *et al.* (1982), with a semi-defined medium and a variety of mammalian cell lines, obtained optimum growth and survival (23 d) with gerbil skeletal muscle; Devaney (1985) used a feeder layer of dog sarcoma cells with a variety of semi-defined media and demonstrated ultrastructurally that the L4 cuticle of *D. immitis* was synthesized *in vitro*. However, Lok *et al.* (1984a) and Kumar *et al.* (1984) obtained development of *D. immitis* L3 to L4 in semi-defined media without feeder cells. Lok *et al.* (1984a) found that both motility and moulting were enhanced with increasing concentrations of foetal bovine serum (FBS) up to 20% (v/v) and that the commercial medium supplement, NuSerum (Collaborative Research), gave comparable results. Abraham *et al.* (1987) observed that *D. immitis* L3 cultured in semi-defined medium for 48 h could be transferred to defined medium and moulting and growth to L4 proceeded; temperature and albumin in the medium were key growth factors. Survival of MA female *D. immitis*, obtained from dogs, in a semi-defined medium for 25–36 d has been reported recently by Tamashiro and Palumbo (1985):

females produced mf *in vitro*, and the system was used as an anthelmintic screen.

The L3 of *Litomosoides carinii*, obtained from multimammate rats ("*in vivo* triggered larvae"), developed to L4 in a defined medium; development was enhanced with a supplement of either FBS or agammaglobulinaemic horse serum (Nelson *et al.*, 1982). Mossinger and Wenk (1986) have maintained female *L. carinii in vitro* in various defined media which were assessed by the degree of *in utero* embryonic development.

Culture systems that support the growth of *Onchocerca* spp. *in vitro* have, in general, been as limited as those for *D. immitis* and *L. carinii*. The best development of *O. gutturosa*, *O. lienalis* and *O. volvulus* mf *in vitro* was to sf in semi-defined media containing cell lines of invertebrate hosts (Pudney and Varma, 1980; Devaney and Howells, 1983). However, Schiller *et al.* (1979) cultured mf from human skin snips and nodules in a diphasic system (see also Fried, 1978, p. 164) and "the length of . . . juveniles . . . [increased to] . . . almost 12 times that of the microfilariae from which they had originated and about 6 times that of a third stage larva." They also described cryogenic techniques for the maintenance of larvae in frozen human cutaneous tissues or in the invertebrate vector, *Simulium*: metabolic activity and morphogenesis were resumed *in vitro* after thawing. Lok *et al.* (1984b) grew both *O. lienalis* and *O. volvulus* L3 to L4 in the semi-defined media and culture conditions described by Franke and Weinstein (1983; 1984a) for the development of *D. viteae*. Subsequently this system was used to test the anthelmintic activity of ivermectin on the development of *O.lienalis* L3 to L4 (Lok *et al.*, 1986).

Maintenance systems for MA females have been described for *Onchocerca* spp. (Townson and Connelly, 1986; Engelbrecht and Schultz-Key, 1984). The enzymatic isolation procedure used to obtain *O. volvulus* adults from skin nodules by Engelbrecht and Schultz-Key (1984) produced MA that survived 1 month *in vitro*.

Development of *Wuchereria bancrofti* mf to sf has been reported by Kharat *et al.* (1980) in semi-defined media. Recently, Kharat and Harinath (1985) reported the development of *W. bancrofti* mf to late L1 in cultures containing *Culex fatigans* cells.

1. Detailed technique

1° *Brugia malayi* L3 to YA; *B. pahangi* L3 to L4
(Mak *et al*., 1983)

MEDIUM

(i) RPMI 1640 (with L-glutamine, Gibco) buffered with sodium bicarbonate (2.0 g litre^{-1}) and HEPES (5.94 g litre^{-1}) to pH 7.2–7.4; penicillin (100 iu per ml^{-1}) and streptomycin (100 μg ml^{-1}) are included.

(ii) *Complete culture medium*. RPMI 1640 medium, as described above, supplemented with 10% inactivated(?) human AB serum(?) and LLC-MK_2 rhesus monkey cell line (Virus Research Division of the Institute of Medical Research, Kuala Lumpur, Malaysia) as a feeder layer.

TECHNIQUE

Mosquitoes (*Aedes togoi*), fed 10–14 d earlier on microfilaraemic cats, are crushed between a test tube and glass plate and then transferred to a Baermann apparatus, containing medium (i), to recover L3. The L3 are subsequently washed (details unspecified) with medium (i) three times, and 40–50 L3 are inoculated into 30 ml screw-capped tissue culture flasks (Costar, 3025) containing 5.0 ml of complete medium (ii); cultures are incubated at 37 °C in air; medium is changed every other day; the cell line is allowed to overgrow on the bottom and sides of the flask. A pH as low as 6.8 is tolerated by the larvae. After 7 d, 13% of the *B. pahangi* were L4, and by 14 d 17.7% of *B. malayi* were L4; 100% of both species were L4 by 28 d. Only *B. malayi* larvae developed to YA (moulting at 44 d); larvae were viable at times in excess of 7 weeks in culture.

2° *Dipetalonema viteae* L3 to YA
(Franke and Weinstein, 1983, 1984a, b)

MEDIA

(i) NCTC 135 and Iscove's modified Dulbecco's medium (IMDM) are obtained in powdered form (Gibco) and prepared separately in 1.0 litre volumes; sodium bicarbonate is added to each medium at 2.2 g litre^{-1} and 3.024 g litre^{-1} respectively. *N*′-2-hydroxyethylpiperazine-*N*′-ethanesulfonic acid (HEPES) is added to NCTC 135 at 5.96 g litre^{-1} (24 mM final concentration). The medium is brought to 1.0 litre with glass-distilled water, filter-sterilized (0.22 μm filters, Millipore) under 5% CO_2 in N_2 and stored separately at 4 °C in the dark for less than 3 weeks. Complete medium is prepared just before use by mixing a 1 : 1 ratio (v/v) of NCTC 135 and IMDM supplemented with 10% or 20% non-heat-inactivated FBS (M.A. Bioproducts). NB. Optimal development is dependent on the serum batch and source; no antibiotics are used in the culture medium.

(ii) *Larval washing media.* (a) NCTC 135 containing 10% FBS plus 1000 iu ml^{-1} penicillin and 100 μg ml^{-1} streptomycin; (b) NCTC 135 containing 10% FBS plus 100 iu ml^{-1} penicillin and 100 μg ml^{-1} streptomycin; (c) NCTC 135 and IMDM [1 : 1 (v/v), antibiotic-free].

TECHNIQUE

Dipetalonema viteae are maintained in jirds (*Meriones unguiculatus*). Infected argasid ticks (*Ornithodorus tartakovskyi*) are surface-sterilized in 70% ethanol and rinsed in sterile 0.85% NaCl; L3 are dissected from the ticks into washing medium (ii.a), then washed twice (details unspecified) in medium (ii. b), followed by two additional washes in medium (ii. c). Culture vessels are glass tubes (16 × 125 mm) containing 2.0 ml of medium. A gas phase of 5% CO_2 in N_2 is passed through a 0.22 μm filter and a 5.0 ml cotton-plugged pipette over the surface of the medium at a flow rate of 65 ml s^{-1} for 60 s; the tubes are plugged with non-toxic rubber stoppers (Bellco Glass) and maintained for 20 min at room temperature. Cultures are then inoculated aseptically with 8–20 L3, regassed, stoppered, incubated at 36.5 °C in the dark and held vertically. The culture medium is changed every 4 d by removing all but 0.1–0.2 ml of the supernatant with a Pasteur pipette and adding 2.0 ml of fresh medium, which is subsequently regassed. Larvae moulted to L4 within 8–20 d and to YA after 42 d.

D. SUPERFAMILY: GNATHOSTOMATOIDEA

Human gnathostomiasis is prevalent in Thailand and a parasite-derived antigen(s) is needed for rapid and specific serodiagnosis. Recently, Morakote *et al.* (1985) described a simple *in vitro* system for the maintenance of *Gnathostoma* larvae in a defined medium which allowed the collection of ES for over a month.

E. SUPERFAMILY: HELIGMOSOIDEA

Studies of the *in vitro* development of *Nematospiroides dubius* (= *Heligmosomoides polygyrum*) (Durette-Desset, 1968) have included analysis of the effects of steroids on development of the free-living stages (Mauro and Weinstein, 1979) and the inhibitory effects of azasteroids (Bottjer *et al.*, 1984). Jenkins and Ibarra (1984) recently described the culture of L4 (from mouse intestines) to YA in multi-well plates and a semi-defined medium: this system was used as an anthelmintic screen.

F. SUPERFAMILY: METASTRONGYLOIDEA

Uga and colleagues (Uga and Matsumura, 1982; Uga *et al.*, 1984) obtained adult *Angiostrongylus cantonensis* from rats infected with L3 derived from aquatic snails, *Biomphalaria glabrata*. Eggs passed from adult worms *in vitro* became embryonated in a semi-defined medium, yielding L1 that were infective to snails.

Hata and colleagues (Hata *et al.*, 1980, 1981, 1983) obtained infected L3 of *Metastrongylus apri*, from earthworms fed embryonated eggs, and described procedures for the exsheathment of L3 and development to YA in a semi-defined medium: the effects of several antibiotics on the development of L3 to L4 were also tested.

G. SUPERFAMILY: STRONGYLOIDEA

The importance of medium API-1 (Section II. B. 1. 1°) for the complete growth of several gastrointestinal worms *in vitro* and of exogenous haemin for their moulting was demonstrated initially with infective larvae [L3 (M2)] of *Oesophagostomum radiatum* which developed to YA in roller culture (Douvres, 1983). This system has been used to analyse the abnormal cuticular development of L4 (Douvres and Wergin, 1984), to trace the end-product fatty acid production of *in vitro* and *in vivo* derived worms using radiolabelled glucose (Rew *et al.*, 1983) and to acquire ES that strongly inhibit bovine immune responses *in vitro* and protect calves from parasite-induced weight loss (Gasbarre *et al.*, 1985; Gasbarre and Douvres, 1986). The rapid development of L3 to late L4 in medium API-1, followed by transfer of the larvae to a defined medium to obtain ES free of medium-derived macromolecules, shows how an *in vitro* system can be used to obtain parasitic stages of nematodes that are difficult or expensive to obtain from the natural host.

Recently, Farrar and Klei (1985) reported the first successful cultivation of *Strongylus edentatus*, *S. vulgaris*, and *S. equinus*. The L3 developed to

L4 in a variety of defined and semi-defined media, with and without organ explants. The limited availability of developing L4 of these species makes such *in vitro* culture techniques particularly appealing to workers interested in the large strongyles of horses.

1. Detailed technique

1° *Oesophagostomum radiatum* infective L3 (M2) to YA
(Douvres, 1983)

MEDIUM

(i) Medium API-1 (see Section II.B.1.1°).

(ii) A 0.24% bovine haemin stock solution is used (see Section II.I.1.1°(i)) as a source of "unbound" or soluble haemin (S-haemin). A modified Fildes' enrichment medium supplement (Blair *et al.*, 1970), called Fildes' reagent, is prepared from calf blood and used as a source of "bound" haemin (F-haemin). Briefly, 50 ml of defibrinated calf blood (from 4- to 6-month-old helminth-free Holstein-Friesian steer calves) is aseptically placed in a glass vessel with 150 ml of 0.85% NaCl (saline), 6.0 ml of 12 N HCl, and 1.0 g of pepsin (P-53, N.F., Fisher), and mixed in a waterbath–shaker at 56°C for 18–24 h. After cooling, the pH is adjusted to 7.5 with 5 N NaOH, then to pH 7.0 with 1 N HCl; stocks are stored at −20°C. (Stocks typically contain 2.4–3.0 mg F-haemin ml^{-1}, and 25–36 mg protein ml^{-1}.)

TECHNIQUE

Oesophagostomum radiatum is maintained by passage through helminth-free calves. Infective L3 (M2) are recovered from 14-day-old faeces-sphagnum moss cultures using a Baermann apparatus and cleaned with multiple washes in sterile 0.85% NaCl (saline) and EBSSA (see Section II.B.1, 2°(v)); larvae can be stored for 5–50 d in tap water at 4°C. Clean larvae are exsheathed by treatment with a 1.25% sodium hypochlorite solution in saline (Diamond and Douvres, 1962; see Taylor and Baker, 1968, p. 300), then washed five times with saline and EBSSA. Suspensions of 80 000 L3 in 1.0 ml of EBSSA are used as the inocula.

Larvae develop best in a one-step roller culture system. Culture vessels are 1.8 litre screw-topped glass bottles containing 40 ml of medium (up to 14 d) or 80 ml of medium (after 14 d), in a gas phase of 5% O_2, 10%

CO_2, 85% N2; screw-tops are tightened and secured with tape after gassing, and bottles are rotated at 1 rev every 1.5 min; medium API-1 supplemented with 20 mM reduced glutathione is used from 0–7 d; medium API-1 alone is used from 7–14 d; thereafter medium API-1 supplemented with haemin is used. Development of L3 to YA is optimal and viability and yields are maximal when 48 μg ml^{-1} of haemin, in the form of F-haemin, is provided in the medium; L3 develop to YA more slowly with S-haemin at 24, 48, and 96 μg ml^{-1} or F-haemin at 24 μg ml^{-1}. Development from L3 to YA *in vivo* (in calves) requires a minimum of 19 d.

2° *Strongylus edentatus, S. vulgaris, S. equinus* L3 (M2) to L4
(Farrar and Klei, 1985)

MEDIUM

(i) *Larval culture medium*. RPMI 1640 (Gibco) supplemented with 10% FBS (Gibco) and antibiotics: potassium penicillin G (500 iu ml^{-1}); dihydrostreptomycin sulphate (500 μg ml^{-1}); amphotericin B (5 μg ml^{-1}) at pH 7.2

(ii) *Organ culture solutions*. (1) Washing solutions: (a) Hanks's balanced salt solution (HBSS) containing 10 times concentrated antibiotics (1000 iu penicillin, 1000 μg ml^{-1} streptomycin and 10 μg ml^{-1} amphotericin B); (b) EMEM containing 10 times concentrated antibiotics. (2) Organ culture medium: EMEM supplemented with 10% FBS, HEPES (25 mM) and unconcentrated antibiotics.

TECHNIQUE

Ponies, infected monospecifically with *S. vulgaris*, *S. edentatus* or *S. equinus*, are used as a source of parasites; contaminated faeces are mixed loosely with vermiculite and water and incubated for 10–14 d at room temperature. Larvae are isolated from cultures in a Baermann apparatus and stored in tap water at 4 °C. Inocula of infective L3 (M2) are sterilized and exsheathed in a 1.3% sodium hypochlorite solution, washed five times with culture medium, then suspended in culture medium. Larvae develop optimally in bicarbonate buffered RPMI 1640 medium containing 10% FBS and gerbil (*Meriones unguiculatus*) caecum explant cultures. Initially, a gerbil caecum is split longitudinally in washing solution (a) and the ingesta removed by gentle swirling and dipping of the tissue into fresh washing solution (a) (three washes, 15 min). The caecum is cut into 2.0 mm squares and transferred to washing solution (b) for 45 min followed by transfer on to a wire grid in 60 × 15 mm organ culture dishes (Falcon) containing organ culture medium. Cultures are incubated at 37 °C for 48 h to check for explant viability and the absence of microbial contamination. Subsequently, tissues are transferred to 16 mm multi-well culture plates (Linbro) containing 2.5 ml of larval culture medium; approximately 100 L3 are inoculated into each culture well at this time. Cultures are maintained at 37 °C in 5% CO_2 in air; medium is changed at 3 d intervals by removing 1.0–1.5 ml of spent medium and replacing it with an equal volume of fresh medium. Cultures in which out-growths from the organ explant form a cellular monolayer produce maximal growth and differentiation of L3. Note: HEPES buffer (25mM) severely limits L3 viability and development. The first moulting of *S. vulgaris*, *S. edentatus* and *S. equinus* L3 to L4 *in vitro* was observed at 6, 5, and 7 d, respectively and maximum viability was maintained through 10, 78 and 48 d, respectively. In a cell-free culture system, larvae moult to L4 in RPMI 1640 medium at relatively low levels; moulting increases in RPMI 1640 medium supplemented with serum.

H. SUPERFAMILY: TRICHINELLOIDEA

Although Berntzen (1962, 1965, 1966) reported the development of muscle-stage L1 of *Trichinella spiralis* to MA over 20 years ago, this area of cultivation has remained controversial. Sakamoto (1979) duplicated parts of Berntzen's work using enzymatic and mechanical procedures to enhance the development of *T. spiralis* L1 *in vitro*: larvae showed loose sheath separation from the cuticle by 8 h in culture and worms that exsheathed showed considerable differentiation of their genital region by 48 h, but no further development. However, larvae (stage not given) cultured for 3 d would develop to MA after further enzymatic and mechanical treatment, including the use of Berntzen's de-sheathing apparatus (see Taylor and Baker, 1968, pp. 253–255). These MA grown *in vitro* shed newborn larvae (NBL); NBL obtained from freshly explanted adult *T. spiralis* could penetrate a monolayer of mouse embryonic cells *in vitro* and develop stichocytes (Sakamoto, 1979).

Jenkins and Carrington (1981) maintained *T. spiralis* L1 in multi-well plates and an undefined medium: viability and change from a "coiled" to a "serpentine" form *in vitro* were used as the basic criteria for the screening of anthelmintics.

Gamble and his colleagues (Gamble *et al.*, 1983; Gamble, 1985; Murrell *et al.*, 1986) have used a stationary culture system with a defined medium to obtain ES, from muscle-stage L1, which proved to be a source of highly specific and sensitive serodiagnostic antigens, capable of inducing some protective immunity in mice and swine.

The maintenance of *T. spiralis* MA (from normal infected mice or previously immunized mice) in a stationary culture system with a semi-defined medium allowed Despommier *et al.* (1977) to assess the immune effects *in vivo* on worm fecundity *in vitro*.

I. SUPERFAMILY: TRICHOSTRONGYLOIDEA

Recent advances in the cultivation of species from this superfamily include the growth of infective L3 (M2) of *Cooperia oncophora* to MA, with ovipositing females, in a roller culture system and semi-defined medium (Douvres, unpublished), and development of infective L3 (M2) of *Dictyocaulus viviparus* to L4 in a stationary system with a defined medium (Bos and Panhuijzen, 1982). Stringfellow (1984) obtained the development of *Haemonchus contortus* L3 (M2) to YA in medium API-1 (Section II.B.1.1°) supplemented with Fildes' reagent (Section II, G, 1, 1°); addition of a filtrate of ovine gastric contents induced development of the L4 to MA, with females ovipositing *in vitro* (Stringfellow, 1986).

Simplified semi-defined media derived from API-1 (API-27, API-28; see Section II.B.1.1°) were designed by Douvres and Fetterer (1986) for maintaining adult *Hae. contortus* for several days *in vitro*, thus replacing the less favourable short-term culture systems that others had used when examining adult biochemical pathways and nutritional requirements *in vitro* (Kaur and Sood, 1982; Kapur and Sood, 1984; Rew *et al.*, 1982; Ward and Huskisson, 1978). Boisvenue *et al.* (1977) have tested the effects of compounds with juvenile hormone activity on the *in vitro* development of infective larvae and Folz *et al.* (1986) recently described a highly automated anthelmintic assay system in which the effects of test compounds on larval motility *in vitro* (measured with a sensitive micromotility meter whereby signal analysis is rapidly recorded using software designed for a personal computer) was correlated with anthelmintic activity. There is also the novel *in vitro* demonstration of *Hae. contortus* larval development *in utero* (Ayalew and Murphy, 1986).

The early successful cultivation of *Nippostrongylus brasiliensis* from egg to adult *in vitro* (see Weinstein and Jones, 1956 and in Taylor and Baker, 1968, pp. 288–289) made it available as a laboratory model for studies of nematode parasitism of vertebrates, so, not surprisingly, this parasite has been studied extensively *in vitro*. Thus, Jenkins and Carrington (1982) cultivated *Nip. brasiliensis* L4 (derived from rodent intestines) to YA in a multi-well stationary culture system using a variety of defined and undefined media, a system that has been used primarily for anthelmintic screening. Mauro and Weinstein (1979) compared the effects of steroids on the development of free-living stages of *Nip. brasiliensis* with *Nem. dubius*, and Bonner (1979) showed that elevation of the temperature and certain nutritional components of defined and semi-defined media stimulated transcription and subsequent developmental changes *in vitro*. Bolla and Weinstein (1980) found that L3, cultured in a medium containing proteins or peptides, had increased acid protease activity that was not detectable in L3 cultured in a medium consisting of only amino acids and vitamins; increased protease activity was blocked by pretreatment of L3 with actinomycin D. Bone and colleagues (Bone *et al.*, 1977; Ward and Bone, 1983; Swanson and Bone, 1984; Ward *et al.*, 1984; Bottjer and Bone, 1985) have maintained *Nip. brasiliensis* adults in Tyrode's solution as a bioassay for feeding behaviour and pheromone responsiveness *in vitro*.

Cultivation of other trichostrongyloid worms of veterinary importance has focused primarily on the production of ES from stages thought to be either immunogenic or pathogenic *in vivo*. Rose (1978) cultured *Ostertagia circumcincta* L3 in an undefined medium to late L4 and YA, then transferred the worms to a defined medium to obtain ES free from medium-derived macromolecules. The injection of such ES into 3-month-old lambs resulted in reduced egg production and lower worm burdens than in the controls. Douvres and Malakatis (1977) cultured infective L3 (M2) of *Os. ostertagi* to MA, including ovipositing females, in a two-step semi-defined system. However, Herlich and Douvres (1979) were unsuccessful in immunizing calves, against challenge, with cultured *Os. ostertagi* larvae (third moult, L4 and fourth moult) or with ES derived from a defined medium in which these stages had been maintained. This two-step system also showed that insect growth-disrupting amines and amides exerted lethal effects against *Os. ostertagi in vitro* at the time of moulting (Douvres *et al.*, 1980). Douvres (unpublished) recently obtained development of *Os. ostertagi* L3 (M2) to MA and ovipositing females in a one-step roller culture system using medium API-1 (see Table I). Douvres (1979) cultured infective L3 (M2) of *Trichostrongylus axei* to YA, by modifying medium API-1 to API-6 (see Section II.B.1.1°). Previously, the cultivation of *Tri. colubriformis* from L3 to YA had required a two-step roller culture system with a semi-defined medium (Douvres, 1980), but recently Douvres (unpublished results) obtained comparable development in a one-step system (see Table I). In related studies *Tri. colubriformis* L4 and YA, obtained from the intestines of infected guinea pigs or sheep, were cultured in a defined medium: the L4 developed to YA with the production of ES (Nelson and

Douvres, 1984). Rapson *et al.* (1985), using an undefined medium in a multi-well plate stationary culture system, obtained development of *in vivo* derived L4 of *Tri. colubriformis* to YA and used this as an anthelmintic assay. Also *Tri. colubriformis* MA have been maintained in a defined medium thus allowing assessment of their feeding and pheromone responses (Bone and Bottjer, 1984, 1985; Bottjer *et al.*, 1985).

The cultivation of free-living stages of *Tri. colubriformis* from L1 to infective L3 was first described by Khan and Dorsman (1978) in an undefined medium containing chick embryo extract, FBS, and yeast extract. Recently, Dorsman and Bijl (1985) developed another undefined medium, free of bacteria, animal tissue and serum, that was capable of supporting development of L1 to L3 *in vitro*.

1. *Detailed technique*

1° *Ostertagia ostertagi* L3 (M2) to MA
(Douvres and Malakatis, 1977)

MEDIUM

(i) Medium RFN (an artificial rumen fluid medium) is prepared as a 2 times concentrated stock solution by sequential addition of the following 5 fractions (a-e): (a) 1.0 g each of L-arabinose, fructose, glucose, maltose, trehalose, sodium pyruvate, L-cysteine·HCl·H_2O; 0.5 g each of bovine liver glycogen and soluble starch; polypeptone (5.626 g); yeast extract (5.0 g) in double-distilled (d) H_2O (700 ml) (requires agitation and heating to approximately 60°C). (b) 20 mg of haemin (Type I: Bovine; Sigma) (dissolved in a solution of 56 mg KOH in 5.0 ml of 95% ethanol) brought to 20 ml in d H_2O. (c) 480 mg of K_2HPO_4 in 80 ml d H_2O. (d) 960 mg NaCl; 960 mg $(NH_4)_2SO_4$; 480 mg KH_2PO_4; 97.4 mg $MgSO_4$, respectively in 40 ml of d H_2O; plus 128 mg $CaCl_2.2H_2O$ in 40 ml of d H_2O. (e) Volatile fatty acid mixture of acetic, propionic, n-butyric, n-valeric, isovaleric and DL-α-methylbutyric acid in volumes of 3.41, 1.2, 0.8, 0.2, 0.2, and 0.2 ml, respectively. The final pH is adjusted to 6.5 with 10% NaOH, and d H_2O is added to a final volume of 950 ml. The solution is autoclaved (120°C, 15 min), cooled, and used freshly prepared or after storage at −20°C for less than 6 months. Working medium (RFN) is prepared by adding RFN (2 times concentrated) 1 : 1 (v/v) with d H_2O; $NaHCO_3$ (50 mM) and the following antibiotics: penicillin (1000 iu ml^{-1}); streptomycin (1 mg ml^{-1}); and amphotericin B (10 ug ml^{-1}); pH is adjusted to 7.5.

(ii) Medium API-1 (Section II.B.1.1°).

(iii) Medium API-1 plus pepsin: the pepsin [from a 10% sterile stock solution of pepsin (twice crystallized; ICN Pharmaceutical) in d H_2O which is prepared and used freshly or after storage at −20°C for less than 1 week] is added to API-1 to give a final concentration of 2.0 mg ml^{-1} (0.2%); pH is adjusted to 4.5 or 6.0 with 1 N HCl. NB. API-1 plus pepsin at pH 6.0 is a clear fluid; at pH 4.5 there is some precipitation and turbidity.

TECHNIQUE

O. ostertagi are maintained by continuous passage through helminth-free calves. The infective L3 (M2) are recovered from 14-day-old faeces-sphagnum moss cultures by passage through a single layer of 25 μm pore nylon cloth in a Baermann apparatus at room temperature and used directly or after storage in tap water at 4°C for less than14 d. Inocula of 100 000 larvae are cleaned by centrifugation (700*g*, room temperature) with 0.85% NaCl (saline) and EBSSA (see Section II.B.1.2°), five washes each; they are exsheathed by resuspending in 10 ml of RFN medium in a 50 ml Erlenmeyer flask (supplied with a Teflon-lined screw cap) and incubated in an oscillatory water bath shaker (120 strokes min^{-1}, 37°C) under the following gas phases (delivered at 200 ml min^{-1}): (1) 40% CO_2, 60% N_2 for 10 min: (2) a non-gassing phase in a sealed flask for 30 min; (3) compressed air for 5 min; (4) a non-gassing phase in a sealed flask for 25 min. The resulting larval suspension serves as the inoculum for the culture vessel.

Exsheathed larvae develop optimally in a two-step roller culture system (1.8–2.8 litre glass bottles containing a final volume of 100 ml of medium, rotated at 1 rev every 1.5 min, 39°C). Step 1 cultures contain medium RFN (pH 7.3, gas phase 5% CO_2 in air) for the first 3 d. Step 2 cultures contain medium API-1 plus pepsin in air at pH 4.5 from 3 to 9 d, and thereafter at pH 6.0. Inocula are transferred to fresh media and new culture vessels every 6 or 7 d. Larvae develop to MA males and females in 28 and 29 d, respectively, and to females laying fertile eggs in 34 to 41 d. Comparable development times *in vivo* (in cattle) are 15 d for MA and 18 d for egg-laying females.

2° *Haemonchus contortus* L3 (M2) to MA
(Stringfellow, 1986)

MEDIA

(i) Fildes' reagent (see Section II.G.1.1°).
(ii) Ovine gastric contents (OGC) prepared as a medium supplement: cannulae are surgically implanted into the abomasum of two sheep (5-month-old, neutered, Polled Dorset males) raised helminth-free, except for minimal infection with *Strongyloides papillosus*; after 2 weeks the cannula plug is removed and the contents from the abomasa drained into a bucket. The OGC are filtered through gauze, centrifuged, and the clear supernatant filter-sterilized (0.2 μm, Millipore), and stored at – 20°C.
(iii) *Culture medium*. Medium API-1 (see Section II.B.1.1°) supplemented with Fildes' reagent (representing 52 $\mu g\ ml^{-1}$ of F-haemin) and OGC (16%).

TECHNIQUE

Haemonchus contortus infective L3 (M2) are isolated from faecal cultures derived from infected sheep, cleaned three times by centrifugation (700*g*, room temperature) with 0.85% NaCl (saline) and exsheathed by treatment with 1.25% sodium hypochlorite in saline, followed by washes in both saline and EBSSA (see Section II.B.1.2°). Suspensions of 40 000 L3 in 1.0 ml of EBSSA are used to inoculate culture vessels (1.8 litre glass bottles containing 40 ml of medium, rotating at 1 rev every 1.5 min, 39°C, gas phase of 5% O_2, 10% CO_2, 85% N_2.

Larvae develop optimally to MA in a one-step system in which the culture medium is at pH 6.4 for the first 7 d and at pH 6.8 thereafter; inocula are transferred to fresh medium and culture vessels weekly. The L3 develop to YA and egg-laying females at 17 and 36 d, respectively: without OGC the L3 develop only to YA. Comparable development times to YA and patent females *in vivo* (sheep) are 9–13 d, and 18–20 d, respectively.

3° *Trichostrongylus axei* L3 (M2) to YA
(Douvres, 1979)

MEDIA

(i) Medium API-6 (see Section II.B.1.1°) is used immediately after preparation or after storage at – 20°C for less than 6 months.

(ii) Pepsin stock solution (10%) (see Section II.I.1.1°(iii)) is added to medium API-6 to provide a final concentration of 0.2%.

(iii) Stock "reducing solution" (Berntzen and Mueller, 1964; also see Taylor and Baker, 1968, p. 209) contains reduced glutathione (1.0 g) and L-cysteineHCl·H_2O (1.0 g) in 100 ml H_2O at pH 9.0; this solution is used immediately after filter sterilization or after freezing at −20°C for less than 1 month; 1.0 ml is added to 50 ml of medium to provide final concentrations of 0.66 mM and 1.3 mM, respectively.

TECHNIQUE

Trichostrongylus axei are maintained by alternate passage through helminth-free calves and sheep. Infective L3 (M2) are recovered from 14-day-old faeces-sphagnum moss cultures with a Baermann apparatus, cleaned with 0.85% saline and EBSSA (see Section II.B.1.2°) and stored for 8–54 d in tap water at 4°C. Inocula are prepared by first exsheathing the larvae with 1.25% hypochlorite solution (Diamond and Douvres, 1962, see Taylor and Baker, 1968, p. 300) and then washing the L3 in saline and EBSSA. Suspensions of 100 000 L3 in 0.5–1.0 ml of EBSSA are used to inoculate culture vessels (1.8 litre glass bottles) containing 50–52 ml of medium gassed with 5% CO_2 in air, rotated at 1 rev every 1.5 min at 39°C.

Larvae develop optimally in a one-step culture system containing API-6 with periods of exposure to reducing agents, pepsin and changes in pH using 12 N NaOH: medium API-6 at pH 5.5 supplemented with L-cysteine and reduced glutathione and pepsin for 0–8 d; then medium API-6 supplemented with pepsin at pH 6.0 for 8–15 d, and pH 6.6 for 15–21 d; thereafter with medium API-6 unsupplemented. Inocula are placed in fresh culture vessels at each medium change. Larvae develop from L3 to L4 after 9 d, and from L4 to YA after 18 d; comparable development times *in vivo* are 7 and 15 d, respectively.

4° *Trichostrongylus colubriformis* L3 (M2) to YA
(Douvres, 1980)

MEDIA

(i) Medium RFN-2 is prepared by combining the following 6 solutions: (a) 475 ml of stock (2 times concentrated) medium RFN (see Section II.I.1.1°); (b) 10 ml of 12 mM KCl; (c) 50 ml of 50 mM $NaHCO_3$; (d) 40 ml of 20 mM reduced glutathione; (e) 25 ml of sodium taurocholate (10 mg ml^{-1}; Difco); (f) 25 ml of trypsin

(10 mg ml^{-1}: 1:300; Gibco). The volume is adjusted to 1.0 litre with triple-distilled (t)H_2O. Antibiotics are added to give the following concentrations: penicillin, 1000 iu ml^{-1}; streptomycin, 1.0 mg ml^{-1}; amphotericin B, 10 μg ml^{-1}; pH is adjusted to 6.8 with 1.0 N HCl.

(ii) Medium API-16 (see Section II.B.1.1°), adjusted to pH 6.4, is used immediately after preparation or after storage at −20°C for less than 6 months.

(iii) Pepsin stock solution (10%) (see Section II.I.1.1°(iii)) is added to medium API-16 to give a final concentration of 0.2%

(iv) Stock solution of 15.36% reduced glutathione (filter-sterilized and stored at −20°C for less than 1 month) is added to medium API-16 during its preparation by replacing an equal volume of t H_2O to give a final concentration of 10 mM.

TECHNIQUE

Tri. colubriformis is maintained by alternate passage through helminth-free calves and sheep. Infective larvae are recovered from 14-day-old faeces-sphagnum cultures in a Baermann apparatus and stored from 13 to 65 d in tap water at 4°C. Larvae are exsheathed (see Section II.I.1.1°) in 10 ml of medium RFN-2, and this suspension is used to inoculate culture vessels (1.8 litre glass bottles containing 40 ml of medium, rotating at 1 rev every 1.5 min, at 39°C).

Larvae develop optimally in a two-step culture system at 39°C: step 1 cultures contain medium RFN-2 (pH 6.8) with a gas phase of 5% CO_2 in air for the first 2 d; followed by transfer to step 2 cultures containing medium API-16 (Section II.B.1.1°) supplemented with pepsin and reduced glutathione at pH 6.8 with 5% CO_2 in air from 2 to 9 d, and thereafter in medium API-16 supplemented with pepsin at pH 6.8 in 5% O_2, 10% CO_2, 85% N_2. Inocula are placed in fresh culture vessels at the time of each medium change. Larvae develop from L3 to L4 after 7 d, and to YA after 16 d; comparable development times *in vivo* (in cattle) are 5 and 10 d, respectively.

III. NEMATODE PARASITES OF INVERTEBRATES

A. SUPERFAMILY: RHABDITOIDEA

Rhabditis maupasi is commonly found in the mantle cavity of North African food snails. Brockelman and Jackson (1978) cultured its "stage-3 larvae", derived from stock cultures grown on rabbit kidney over nutrient agar, in a defined medium (composed of salts, trace metals, purines, and pyrimidines, Krebs-cycle intermediates, butyric acid, vitamins, urea, and dextrose, to which were added 18 amino acids (Jackson, 1973; and see Platzer, 1977) at 21 °C. By systematically examining larval growth and reproduction in this "complete medium", or media deficient in one of the 18 amino acids, they demonstrated the "nematode essential" amino acids necessary for maintenance or development *in vitro*. In the "complete medium", or that lacking the "nematode non-essential" amino acids, larvae developed to MA and produced an F1 generation. However, continuous cultivation of this nematode *in vitro*, with the production of multiple generations, required haemin or another iron porphyrin as well as a sterol such as cholesterol.

IV. NEMATODE PARASITES OF PLANTS

A. SUPERFAMILY: APHELENCHOIDEA

Monoxenic cultivation of plant parasitic nematodes has been described using plant roots, callus tissues or fungal mats (Platzer, 1977). Recent attempts to cultivate axenically a nematode parasite of plants have been limited to a report by Bolla and Jordan (1982) on the development of *Bursaphelenchus xylophilus*, the pine wilt or pine wood nematode. This parasite was maintained on fungal mats in the laboratory and a population of worms was sterilized by washing in a solution of antibiotics. These workers cultured the worms either in an undefined medium of soy peptone/yeast extract supplemented with cholesterol and bovine haemoglobin or in a defined one (CbMM; Gibco) supplemented with cholesterol and bovine haemoglobin: growth, development and reproduction were best in the former undefined medium, but adequate increases in worm populations were obtained in the latter. Cultured worms maintained their pathogenicity after axenic cultivation and were able to induce wilting of 45-day-old *Pinus sylvestris* seedlings.

V. CONCLUSIONS

Much of the successful *in vitro* cultivation of nematode parasites over the last decade has benefited from knowledge of the basic biochemistry and physiology of the worm. For example, the discovery of a general lack of *de novo* cholesterol synthesis in nematodes supported the use of exogenous cholesterol as a medium supplement. Defined media and supplements are now replacing those complex media thought previously to be essential for optimal growth *in vitro*. Expanded information on trace elements, hormones, growth factors and serum supplements used currently to support mammalian cell growth *in vitro* will surely be applied to the cultivation of parasites. The rapid advance in techniques and concepts of molecular biology are lessening the need for large numbers of worms for basic studies. Also, the increased sensitivity of current identification techniques for antigens and the potential for their large-scale production by genetic engineering mean that unique, parasite-derived products, originating from worms cultured *in vitro*, are rapidly becoming available for serodiagnostic and immunoprotective studies. Thus, *in vitro* cultivation of nematode parasites over the next decade will certainly continue to be a fruitful area of research.

REFERENCES

Abraham, D., Lauria, S., Mika-Grieve, M., Lok, J.B. and Grieve, R.B. (1986). Survival and viability of *Dirofilaria immitis* microfilariae in defined and undefined culture media. *Journal of Parasitology* **72**, 776–777.

Abraham, D., Mok, M., Mika-Grieve, M. and Grieve, R.B. (1987). *In Vitro* culture of *Dirofilaria immitis* third- and fourth-stage larvae under defined conditions. *Journal of Parasitology* **73**, 377–383.

Anderson, R.C., Chabaud, A.C. and Willmott, S. (1983). "CIH Keys to the nematode parasites of vertebrates." (Nos. 1–10, 1974–1983). Commonwealth Agricultural Bureaux, Royal Bucks, England.

Ando, K. and Kiamura, S. (1982). Osmotic pressure dependent development of microfilariae of *Dirofilaria immitis in vitro*. *Japanese Journal of Parasitology* **31**, 219–222.

Ando, K., Chinzei, Y. and Kitamura, S. (1980). Conditions of *in vitro* culture for development of microfilariae of *Dirofilaria immitis*. *Japanese Journal of Parasitology* **29**, 483–488.

Ayalew, L. and Murphy, B.E.P. (1986). *In vitro* demonstration of *in utero* larval development in an oviparous parasitic nematode: *Haemonchus contortus*. *Parasitology* **93**, 371–381.

Berntzen, A.K. (1962). *In vitro* cultivation of *Trichinella spiralis*. *Journal of Parasitology* **48** (section 2, supplement 2), 48.

Berntzen, A.K. (1965). Comparative growth and development of *Trichinella spiralis in vitro* and *in vivo*, with redescription of the life cycle. *Experimental Parasitology* **16**, 74–106.

Berntzen, A.K. (1966). A controlled culture environment for axenic growth of parasites. *Annals of the New York Academy of Science* **139**, 176–189.

Berntzen, A.K. and Mueller, J.F. (1964). *In vitro* cultivation of *Spirometra mansonoides* (Cestoda) from the procercoid to the early adult. *Journal of Parasitology* **50**, 705–711.

Blair, J.E., Lennette, E.H. and Truant, J.P. (1970). "Manual of Clinical Microbiology, American Society for Microbiologists," p. 615. The Williams and Wilkins Company, Baltimore, MD.

Boisvenue, R.J., Emmick T.L. and Galloway, R.B. (1977). *Haemonchus contortus* : effects of compounds with juvenile hormone activity on the *in vitro* development of infective larvae. *Experimental Parasitology* **42**, 67–72.

Bolla, R.I. and Jordan, W. (1982). Cultivation of the pine wilt nematode, *Bursaphelenchus xylophilus*, in axenic culture media. *Journal of Nematology* **14**, 377–381.

Bolla, R. and Weinstein, P.P. (1980). Acid protease activity during development and aging of *Nippostrongylus brasiliensis*. *Comparative Biochemistry and Physiology* **66B**, 475–481.

Bone, L.W. and Bottjer, K.P. (1984). Characterization of male adaption to pheromone of female *Trichostrongylus colubriformis* (Nematoda). *Journal of Chemical Ecology* **10**, 1749–1758.

Bone, L.W. and Bottjer, K.P. (1985). Cuticular carbohydrates of three nematode species and chemoreception by *Trichostrongylus colubriformis*. *Journal of Parasitology* **71**, 235–238.

Bone, L.W., Shorley, H.H. and Gaston, L.K. (1977). Sexual attraction and pheromonal dosage response of *Nippostrongylus brasiliensis*. *Journal of Parasitology* **63**, 845–848.

Bonner, T.P. (1979). Initiation of development *in vitro* of third-stage *Nippostrongylus brasiliensis*. *Journal of Parasitology* **65**, 74–78.

Bos, H.J. and Panhuijzen, J. (1982). Development of larval stages of *Dictyocaulus viviparus in vitro*. "Proceedings of the 5th International Congress of Parasitology". *Journal of Molecular and Biochemical Parasitology* (Suppl.), p. 159 [abstract].

Bottjer, K.P. and Bone, L.W. (1985). *Nippostrongylus brasiliensis*: *In vitro* feeding activity. *International Journal for Parasitology* **15**, 15–19.

Bottjer, K.P., Weinstein, P.P. and Thompson, M.J. (1984). Effects of azasteroids on growth and development of the free-living stages of *Nippostrongylus brasiliensis* and *Nematospiroides dubius*. *Comparative Biochemistry and Physiology* **78B**, 805–811.

Bottjer, K.P., Klesius, P.H. and Bone, L.W. (1985). Effects of host serum on feeding by *Trichostrongylus colubriformis* (nematode). *Parasite Immunology* **7**, 1–9.

Brockleman, C.R. and Jackson, G.J. (1978). Amino acid, heme and sterol requirements of the nematode *Rhabditis maupasi*. *Journal of Parasitology* **64**, 803–809.

Carvajal, J., Barros, C., Santander, G. and Alcalde, C. (1981). *In vitro* culture of larval anisakid parasites of the Chilean hake *Merluccius gayi*. *Journal of Parasitology* **67**, 958–959.

Chen, S.N. and Howells, R.E. (1979). The *in vitro* cultivation of the infective larvae and early mammalian stages of the filarial worm, *Brugia pahangi*. *Annals of Tropical Medicine and Parasitology* **73**, 473–486.

Court, J.P. and Lees, G.M. (1983). Improvement of *in vitro* culture conditions of *Brugia pahangi* four-day-old developing larvae for use in antifilarial drug assay. *Tropenmedizin ünd Parasitologie* **34**, 162–168.

Court, J.P., Bianco, A.E., Townson, S., Ham, P.J. and Friedheim, E. (1985). Study on the activity of antiparasitic agents against *Onchocerca lienalis* third stage larvae *in vitro*. *Tropenmedizin ünd Parasitologie* **36**, 117–119.

Dennis, D.T., Despommier, D.D. and Davis, N. (1970). Infectivity of the new born larva of *Trichinella spiralis* in the rat. *Journal of Parasitology* **56**, 974–977.

Despommier, D.D. Campbell, W.C. and Blair, L.S. (1977). The *in vivo* and *in vitro* analysis of immunity to *Trichinella spiralis* in mice and rats. *Parasitology* **74**, 109–119.

Devaney, E. (1981). The development of *Dirofilaria immitis* in cultured malpighian tubules. *Acta Tropica* **38**, 251–260.

Devaney, E. (1985). *Dirofilaria immitis*: the moulting of the infective larva *in vitro*. *Journal of Helminthology* **59**, 47–50.

Devaney, E. and Howells, R.E. (1979). The development of exsheathed microfilariae of

Brugia pahangi and *Brugia malayi* in mosquito cell lines. *Annals of Tropical Medicine and Parasitology* **73**, 387–392.

Devaney, E. and Howells, R.E. (1983). The differentiation of microfilariae of *Onchocerca lienalis in vitro. Annals of Tropical Medicine and Parasitology* **77**, 103–105.

Diamond, L.S. and Douvres, F.W. (1962). Bacteria-free cultivation of some parasitic stages of the swine nematode *Hyostrongylus rubidus* and *Oesophagostomum quadrispinulatum* (*O. longicaudum*). *Journal of Parasitology* **48**, 39–42.

Dorsman, W. and Bijl, A.C. (1985). Cultivation of free-living stages of *Trichostrongylus colubriformis* in media without bacteria, animal tissue extract, or serum. *Journal of Parasitology* **71**, 200–203.

Douvres, F.W. (1970). Cultivation of *Oesophagostomum radiatum* developmental stages in cell culture systems. *Journal of Parasitology* **56** (Section II, Part I), 83–84.

Douvres, F.W. (1979). *In vitro* development of *Trichostrongylus axei*, from infective larvae to young adults. *Journal of Parasitology* **65**, 79–84.

Douvres, F.W. (1980). *In vitro* development of *Trichostrongylus colubriformis*, from infective larvae to young adults. *Journal of Parasitology* **66**, 466–471.

Douvres, F.W. (1983). The *in vitro* cultivation of *Oesophagostomum radiatum*, the nodular worm of cattle. III. Effects of bovine heme on development to adults. *Journal of Parasitology* **69**, 570–576.

Douvres, F.W. and Fetterer, R.H. (1986). Survival of *Haemonchus contortus* adults in defined, semi-defined and complex cell-free media. *Proceedings of the Helminthological Society of Washington* **53**, 263–269.

Douvres, F.W. and Malakatis, G.M. (1977). *In vitro* cultivation of *Ostertagia ostertagi*, the medium stomach worm of cattle. I. Development from infective larvae to egg-laying adults. *Journal of Parasitology* **63**, 520–527.

Douvres, F.W. and Urban, Jr., J.F. (1983). Factors contributing to the *in vitro* development of *Ascaris suum* from second-stage larvae to mature adults. *Journal of Parasitology* **69**, 549–558.

Douvres, F.W. and Urban, Jr., J.F. (1986). Development of *Ascaris suum* from *in vivo* derived third-stage larvae to egg-laying adults *in vitro. Proceedings of the Helminthological Society of Washington* **53**, 256–262.

Douvres, F.W. and Wergin, W.P. (1984). Cuticular abnormality of *Oesophagostomum radiatum* (Nematoda) fourth-stage larvae grown *in vitro. Proceedings of the Helminthological Society of Washington* **51**, 334–337.

Douvres, F.W., Thompson, M.J. and Robbins, W.E. (1980). *In vitro* cultivation of *Ostertagia ostertagi*, the medium stomach worm of cattle. II. Effect of insect-growth-disrupting amines and amides on development. *Veterinary Parasitology* **7**, 195–205.

Durette-Desset, M.C. (1968). Identification des strongyles des mulots et campagnols décrits par Dujardin. *Annales de Parasitologie* **43**, 387–404.

Englebrecht, F. and Schulz-Key, H. (1984). Observations on adult *Onchocerca volvulus* maintained *in vitro*. *Transactions of the Royal Society of Tropical Medicine and Hygiene* **78**, 212–215.

Farrar, R.G. and Klei, T.R. (1985). *In vitro* development of *Strongylus edentatus* to the fourth larval stage with notes on *Strongylus vulgaris* and *Strongylus equinus*. *Journal of Parasitology* **71**, 489–499.

Fenglin, W., Rongjun, H., Xianxiang, H. and Xihui, Z. (1979). *In vitro* cultivation of *Ancylostoma duodenale* and *Ancylostoma caninum*. *Chinese Medical Journal (English Edition)* **92**, 839–846.

Fetterer, R.H. and Urban, Jr., J.F. (1985). The inhibition of growth and development of *Ascaris suum* larvae *in vitro* by proline analogs. *Comparative Biochemistry and Physiology* **82C**, 331–335

Fleming, M.W. (1985a). *Ascaris suum*: Role of ecdysteroids in molting. *Experimental Parasitology* **60**, 207–210.

Fleming, M.W. (1985b). Steroidal enhancement of growth in parasitic larvae of *Ascaris suum*: validation of a bioassay. *Journal of Experimental Zoology* **233**, 229–233.

Folz, S.D., Pax, R.A., Thomas, E.M., Bennett, J.L., Lee, B.L., Conder, G.A. and Tesar, F.B. (1986). Detecting *in vitro* anthelmintic effects using a micromotility meter. *Proceedings of the American Association of Veterinary Parasitologists* **31**, 41.

Franke, E.D. and Weinstein, P.P. (1983). *Dipetalonema viteae* (Nematoda: Filarioidea): culture of third-stage larvae to young adults *in vitro*. *Science* **221**, 161–163.

Franke, E.D. and Weinstein, P.P. (1984a). *In vitro* cultivation of *Dipetalonema viteae* third stage larvae: effect of the gas phase. *Journal of Parasitology* **70**, 493–498.

Franke, E.D. and Weinstein, P.P. (1984b). *In vitro* cultivation of *Dipetalonema viteae* third stage larvae: evaluation of culture media, serum, and other supplements. *Journal of Parasitology* **70**, 618–628.

Fried, B. (1978). Trematoda. *In* "Methods of Cultivating Parasites *In Vitro*" (A.E.R. Taylor and J.R. Baker, eds.), pp. 152–192. Academic Press, London and New York.

Furman, A. and Ash, L.R. (1983). Characterization of the exposed carbohydrates on the sheath surface of *in vitro* derived *Brugia pahangi* microfilariae by analysis of lectin binding. *Journal of Parasitology* **69**, 1043–1047.

Gamble, H.R. (1985). *Trichinella spiralis*: Immunization of mice using monoclonal antibody affinity-isolated antigens. *Experimental Parasitology* **59**, 398–404.

Gamble, H.R., Anderson, W.R., Graham, C.E. and Murrell, K.D. (1983). Diagnosis of swine trichinosis by enzyme-linked immunosorbent assay (ELISA) using an excretory-secretory antigen. *Veterinary Parasitology* **13**, 349–361.

Gasbarre, L.C. and Douvres, F.W. (1987). Protection from parasite induced weight loss by the vaccination of calves with excretory-secretory products of larval *Oesophagostomum radiatum*. *Veterinary Parasitology* (in press).

Gasbarre, L.C., Romanowski, R.D. and Douvres, F.W. (1985). Suppression of antigen- and mitogen-induced proliferation of bovine lymphocytes by excretory-secretory products of *Oesophagostomum radiatum*. *Infection and Immunity* **48**, 540–545.

Hansen, E.L. and Hansen, J.W. (1978). Nematoda parasitic in animals and plants. *In* "Methods of Cultivating Parasites *In vitro*" (A.E.R. Taylor and J.R. Baker, eds.), pp. 227–277. Academic Press, London and New York.

Hansen, E.L., Silverman, P.H. and Buecher, E.J. 1966. Development of *Haemonchus contortus* in media designed for studies on *Caenorhabditis briggsae*. *Journal of Parasitology* **52**, 137–140.

Hata, H., Fujita, J. and Yasuraoka, K. (1980). *In vitro* cultivation of *Metastrongylus apri*. *Japanese Journal of Experimental Medicine* **50**, 365–374.

Hata, H., Oki, Y. and Fujita, J. (1981). *In vitro* cultivation of *Metastrongylus apri*. III. Sex specific antigen for evaluating development of cultured worms. *The Bulletin of the Nippon Veterinary and Zootechnical College* **29**, 101–105.

Hata, H., Sakamoto, Y., Fujita, J. and Yasuraoka, K. (1983). Effects of antibiotics on the development of *Metastrongylus apri in vitro*. *Japanese Journal of Parasitology* **32**, 485–487.

Herlich, H. and Douvres, F.W. (1979). Gastrointestinal nematode immunization trials in cattle. *American Journal of Veterinary Research* **40**, 1781–1782.

Illgen, B. (1982). Experiments *in vitro* with *Litomosoides carinii*, Nematoda Filarioidea. 2. Influence of different sera on the release of microfilariae and experiments on the readiness for copulation of the worms. *Zeitschrift für Parasitenkunde* **67**, 227–236.

Jackson, G.J. (1973). *Neoaplectana glaseri*: Essential amino acids. *Experimental Parasitology* **34**, 111–114.

Jenkins, D.C. and Carrington, T.S. (1981). An *in vitro* screening test for compounds active against the parenteral stages of *Trichinella spiralis. Tropenmedizin ünd Parasitologie* **32**, 31–34.

Jenkins, D.C. and Carrington, T.S. (1982). An *in vitro* screen for anthelmintics employing *Nippostrongylus brasiliensis* in a defined medium. *Veterinary Parasitology* **11**, 223–230

Jenkins, D.C. and Ibarra, O.F. (1984). *Nematospiroides dubius*: response of the late fourth-stage larva to anthelmintics *in vitro. Zeitschrift für Parasitenkunde* **70**, 395–402.

Jenkins, D.C., Armitage, R. and Carrington, T.S. (1980). A new primary screening test for anthelmintics utilizing the parasitic stages of *Nippostrongylus brasiliensis*, *in vitro. Zeitschrift für Parasitenkunde* **63**, 261–269.

Kaur, R. and Sood, M.L. (1982). *Haemonchus contortus*: The *in vitro* effects of DL-tetramisole and rafoxanide on glycolytic enzymes. *International Journal for Parasitology* **12**, 585–588.

Kapur, J. and Sood, M.L. (1984). Amino acid biosynthesis in *Haemonchus contortus* from ^{14}C-labeled precursors *in vitro. Veterinary Parasitology* **15**, 293–299.

Khan, Z.I. and Dorsman, W. (1978). *Trichostrongylus colubriformis*: cultivation of free-living stages. *Journal of Parasitology* **64**, 1024–1027.

Kharat, I. and Harinath, B.C. (1985). *In vivo* and *in vitro* development of *Wuchereria bancrofti* microfilariae. *Indian Journal of Medical Research* **82**, 127–132.

Kharat, I. Satyanarayana, U., Ghirnikar, S.N. and Harinath, B.C. (1980). *In vitro* cultivation of *Wuchereria bancrofti* microfilariae. *Indian Journal of Experimental Biology* **18**, 1245–1247.

Komuniecki, P.R. and Vanover, L. (1987). Biochemical changes during the aerobic-anaerobic transition in *Ascaris suum* larvae. *Molecular and Biochemical Parasitology* **22**, 241–248.

Kumar, H., Sahai, R. and Rao, C.K. (1984). *In vitro* cultivation of infective larvae of *Dirofilaria immitis. Journal of Communicable Diseases* **16**, 77–81.

Lok, J.B., Mika-Grieve, M. and Grieve, R.B. (1983). Cryopreservation of *Dirofilaria immitis* microfilariae and third-stage larvae. *Journal of Helminthology* **57**, 319–324.

Lok, J.B., Mika-Grieve, M., Grieve, R.B. and Chin, T.K. (1984a). *In vitro* development of 3rd and 4th stage larvae of *Dirofilaria immitis*: comparison of basal culture media, serum levels and possible serum substitutes. *Acta Tropica* **41**, 145–154.

Lok, J.B., Pollack, R.J., Cupp, E.W., Bernardo, M.J., Donnelly, J.J. and Albiez, E.J. (1984b). Development of *Onchocerca lienalis* and *O. volvulus* from the third to fourth larval stage *in vitro. Tropenmedizin ünd Parasitologie* **35**, 209–211.

Lok, J.B., Pollack, R.J. and Donnelly, J.J. (1987). Studies of the growth-regulating effects of ivermectin on larval *Onchocerca lienalis in vitro. Journal of Parasitology* **73**, 80–84.

McClelland, G. (1980). *Phocanema decipiens*: molting in seals. *Experimental Parasitology* **49**, 128–136.

McDairmid, S.S. and Matthews, B.E. (1982). Survival and behavior of *Onchocerca gutturosa* microfilariae in vitro. *Journal of Parasitology* **68**, 470–477.

Mak, J.W., Lim, P.K.C., Sim, B.K.L. and Liew, L.M. (1983). *Brugia malayi* and *B. pahangi*: cultivation *in vitro* of infective larvae to the fourth and fifth stages. *Experimental Parasitology* **55**, 243–248.

Matthews, B.E. (1982). Behaviour and enzyme release by *Anisakis* sp. larvae (Nematoda: Ascaridida). *Journal of Helminthology* **56**, 177–183.

Mauro, N.A. and Weinstein, P.P. (1979). Effects of sterols, rat hematin and coproporphyrin on the free-living stages of *Nematospiroides dubius* and *Nippostrongylus brasiliensis. International Journal for Parasitology* **9**, 421–427.

Miegeville, M., Bouillard, C., Marjolet, M., Vermeil, C. and Avranche, P. (1981). Nouvelle contribution a l'étude de *Dipetalonema viteae* maintien en survie des adultes *in vitro. Bulletin de la Societé de Pathologie Exotique* **74**, 206–216.

Morakote, H., Thammasonthi, W., Maleewong, W. and Khamboonruang, C. (1985). Short term maintenance of *Gnathostoma* larvae *in vitro* without losing infectivity. *Southeast Asian Journal of Tropical Medicine and Public Health* **16**, 124–125.

Mossinger, J. and Wenk, P. (1986). Fecundity of *Litomosoides carinii* (Nematoda, Filarioidea) *in vivo* and *in vitro. Zeitschrift für Parasitenkunde* **72**, 121–131.

Murrell, K.D., Anderson, W.R., Schad, G.A., Hanbury, R.D., Kazacos, K.R., Gamble, H.R. and Brown, J. (1986). Field evaluation of the enzyme-linked immunosorbent assay for swine trichinosis: efficacy of the excretory-secretory antigen. *American Journal of Veterinary Research* **47**, 1046–1049.

Nelson, P.D. and Douvres, F.W. (1984). *In vitro* development of *Trichostrongylus colubriformis* from fourth-stage larvae to young adults in a chemically defined medium. *Proceedings of the Helminthological Society of Washington* **51**, 62–65.

Nelson, P.D., Weiner, D.J., Stromberg, B.E. and Abraham, D. (1982). *In vitro* cultivation of 3rd stage larvae of *Litomosoides carinii* to the 4th stage. *Journal of Parasitology* **68**, 971–973.

Ogura, N., Kobayashi, M. and Yamamoto, H. (1981). Studies on Filariasis IV. Critical period of molting from the 2nd to the 3rd stage in filarial larvae of *Brugia pahangi. Dokkyo Journal of Medical Science* **8**, 74–82.

Oshima, T., Oya, S. and Wakai, R. (1982). *In vitro* cultivation of *Anisakis* Type I and Type II larvae collected from fishes caught in Japanese coastal waters and their identification. *Japanese Journal of Parasitology* **31**, 131–134.

Platzer, E.G. (1977). Culture media for nematodes. *In* "Handbook Series in Nutrition and Food. Section G: Diets, Culture Media, and Food Supplements", Vol. II. (M. Rechcigl, ed.), pp. 29–59. CRC Press.

Pudney, M. and Varma, M.G.R. (1980). Present state of knowledge of *in vitro* cultivation of filariae. *In* "The *In Vitro* Cultivation of the Pathogens of Tropical Diseases, pp. 367–389. Schwabe, Basel.

Rapson, E.B., Jenkins, D.C. and Topley, P. (1985). *Trichostrongylus colubriformis*: *in vitro* culture of parasitic stages and their use for the evaluation of anthelmintics. *Research in Veterinary Science* **39**, 90–94.

Rew, R.S., Smith, C. and Colglazier, M.L. (1982). Glucose metabolism of *Haemonchus contortus* adults: effects of thiabendazole on susceptible versus resistant strains. *Journal of Parasitology* **68**, 845–850.

Rew, R.S., Douvres, F.W. and Saz, H.J. (1983). *Oesophagostomum radiatum* glucose metabolism of larvae grown *in vitro* and adults grown *in vivo. Experimental Parasitology* **55**, 179–187.

Rew, R.S., Urban, J.F.Jr. and Douvres, F.W. (1986). An *in vitro* screen for anthelmintics using larvae of *Ascaris suum. American Journal of Veterinary Research* **47**, 869–873.

Rose, J.H. (1973). The development *in vitro* of some gastro-intestinal nematodes of sheep, cattle and pigs. *Research in Veterinary Science* **14**, 326–333.

Rose, J.H. (1976). Preliminary results using metabolites and *in vitro* grown larvae of *Ostertagia circumcincta* to immunise lambs against oral challenge. *Research in Veterinary Science* **21**, 76–78.

Rose, J.H. (1978). Further attempts to immunise lambs using metabolites and *in vitro* grown larvae of *Ostertagia circumcincta. Research in Veterinary Science* **24**, 61–64.

Sakamota, T. (1979). Development and behaviour of adult and larval *Trichinella spiralis* cultured *in vitro. Memorial Faculty of Agriculture of Kagoshima University* **15**, 107–111.

Schiller, E.L., Bueding, E., Turner, V.M. and Fisher, J. (1975). Aerobic and anaerobic carbohydrate metabolism and egg production of *Schistosoma mansoni in vitro. Journal of Parasitology* **61**, 385–389.

Schiller, E.L., Turner, V.M., Marroquin, H.F. and D'Antonio, R. (1979). The cryopreservation and *in vitro* cultivation of larval *Onchocerca volvulus. American Journal of Tropical Medicine and Hygiene* **28**, 997–1009.

Sneller, V.P. and Weinstein, P.P. (1982). *In vitro* development of *Dirofilaria immitis* microfilariae: selection of culture media and serum levels. *International Journal for Parasitology* **12**, 233–238.

Sommerville, R.I. and Davey, K.G. (1976). Stimuli for cuticle formation and ecdysis *in vitro* of the infective larva of *Anisakis* sp. (Nematoda: Ascaridoidea). *International Journal for Parasitology* **6**, 433–439.

Stringfellow, F. (1984). Effects of bovine heme on development of *Haemonchus contortus in vitro. Journal of Parasitology* **70**, 989–990.

Stringfellow, F. (1986). Cultivation of *Haemonchus contortus* (Nematoda: Trichostrongylidae) from infective larvae to the adult male and egg-laying female. *Journal of Parasitology* **72**, 339–345.

Stromberg, B.E. (1979). The isolation and partial characterization of a protective antigen from developing larvae of *Ascaris suum. International Journal for Parasitology* **9**, 307–311.

Stromberg, B.E. (1980). Potentiation of the reaginic (IgE) antibody response to ovalbumin in the guinea pig with a soluble metabolic product from *Ascaris suum. Journal of Immunology* **125**, 833–836.

Stromberg, B.E., Khoury, P.B. and Soulsby, E.J.L. (1977). Development of larvae of *Ascaris suum* from the third to the fourth stage in a chemically defined medium. *International Journal of Parasitology* **7**, 149–151.

Swanson, J.A. and Bone, L.W. (1984). Response of *Nippostrongylus brasiliensis* (Nematoda) to bile. *Proceedings of the Helminthological Society of Washington* **51**, 332–333.

Sylk, S.R., Stromberg, B.E. and Soulsby, E.J.L. (1974). Development of *Ascaris suum* larvae from the third to fourth stage *in vitro. International Journal for Parasitology* **4**, 261–265.

Tamashiro, W.K. and Palumbo, N.E. (1985). Diethylcarbamazine *in vitro* inhibitory effect on microfilarial production by *Dirofilaria immitis. Journal of Parasitology* **71**, 381–383.

Tanner, M. (1981). *Dipetalonema viteae* (Filaroidea): development of the infective larvae *in vitro. Acta Tropica* **38**, 241–249.

Tanner, M. and Weiss, N. (1981). *Dipetalonema viteae* (Filarioidea): evidence for a serum-dependent cytotoxicity against developing third and fourth stage larvae *in vitro. Acta Tropica* **38**, 325–328.

Taylor, A.E.R., and Baker, J.R. (1968). "Methods of Cultivating Parasites *In Vitro*." Academic Press, London and New York.

Townson, S. and Connelly, C. (1986). Long-term survival of *Onchocerca gutturosa* adult worms *in vitro. Tropenmedizin ünd Parasitologie* **37** [Abstract No. 39]. [Proceedings from the Wellcome Filariasis Seminar '85 held on November 27, 1985, in London, England, p. 94.]

Townson, S. and Ham, P.J. (1986). Successful cryopreservation of adult male *Onchocerca gutturosa* in liquid nitrogen. *Tropenmedizin ünd Parasitologie* **37**, 117–120.

Townson, S., Connelly, C. and Ham, P.J. (1986). Cultivation of *Onchocerca lienalis* developing larvae *in vitro. Tropenmedizin ünd Parasitologie* **37**, 94 [Abstract No. 40].

Uga, S. and Matsumura, T. (1982). *In vitro* cultivation of *Angiostrongylus cantonensis* eggs. *Japanese Journal of Parasitology* **31**, 59–66.

Uga, S., Araki, K. and Matsumura, T. (1984). Deposition and development of eggs of *Angiostrongylus cantonensis in vitro. Japanese Journal of Parasitology* **33**, 257–261.

Urban, J.F.Jr. and Douvres, F.W. (1981). *In vitro* development of *Ascaris suum* from third- to fourth-stage larvae and detection of metabolic antigens in multiwell culture systems. *Journal of Parasitology* **67**, 800–806.

Urban, J.F. Jr. and Romanowski, R.D. (1985). *Ascaris suum*: Protective immunity in pigs immunized with products from eggs and larvae. *Experimental Parasitology* **60**, 245–254.

Urban, J.F. Jr. and Tromba, F.G. (1982). Development of immune responsiveness to *Ascaris suum* antigens in pigs vaccinated with ultraviolet-attenuated eggs. *Veterinary Immunology and Immunopathology* **3**, 399–409.

Urban, J.F. Jr., Douvres, F.W. and Xu, S. (1984). Culture requirements of *Ascaris suum* larvae using a stationary multi-well system: increased survival, development and growth with cholesterol. *Veterinary Parasitology* **14**, 33–42.

Van Banning, P. (1971). Some notes on a successful rearing of the herring-worm, *Anisakis marina* L. (Nematoda: Heterocheilidae). *Journal du Conseil International pour l'Exploration de la Mer* **34**, 84–88.

Varma, M.G.R. and Pudney, M. (1969). The growth and serial passage of cell lines from *Aedes aegypti* (L.) larvae in different media. *Journal of Medical Entomology* **6**, 432–439.

Ward, J.B. and Bone, L.W. (1983). *Nippostrongylus brasiliensis*: environmental influences on pheromone production by females. *International Journal of Parasitology* **13**, 499–502.

Ward, P.F.V. and Huskisson, N.S. (1978). The energy metabolism of adult *Haemonchus contortus in vitro. Parasitology* **77**, 255–271.

Ward, J.B., Nordstrom, R.M. and Bone, L.W. (1984). Chemotaxis of male *Nippostrongylus brasiliensis* (Nematoda) to some biological compounds. *Proceedings of the Helminthological Society of Washington* **51**, 73–77.

Weinstein, P.P. (1986). Filariasis: Problems and challenges. *The American Journal of Tropical Medicine and Hygiene* **35**, 221–233.

Weinstein, P.P. and Jones, M.F. (1956). The *in vitro* cultivation of *Nippostrongylus muris* to the adult stage. *Journal of Parasitology* **42**, 215–236.

Wenk, P., Illgen, B. and Seitz, H.M. (1978). Experiments *in vitro* with *Litomosoides carinii* (Nematoda, Filarioidea). Part 1. Maintenance of adult females and microfilariae as well as release of microfilariae in different culture media. *Zeitschrift für Parasitenkunde* **55**, 63–72.

Wong, M.M., Knighton, R., Fidel, J. and Wada, M. (1982). *In vitro* culture of infective-stage larvae of *Dirofilaria immitis* and *Brugia pahangi*. *Annals of Tropical Medicine and Parasitology* **76**, 239–242.

Xianxiang, H. and Jiajun, H. (1985). Studies on cultivation of adult *Necator americanus in vitro. Journal of Parasitology and Parasitic Diseases* **3**, 35–37.

TABLE II. Nematoda parasitic in vertebrates: summary of culture studies since 1977 (excluding techniques given in detail)

Species	Culture conditions[a,b,c]	Initial stage (source)[d]	Application or product of the techniques[d]	Reference
		Superfamily: Ancylostomoidea		
Ancylostoma caninum	SDS, U-Def: HI young dog serum, pH 6.0–6.8, air, 37 °C; 2–12 worms in 10 ml media changed daily	YA & MA (puppy dog)	Maintenance system: max. survival not given. YA grew but did not mature after 24 d; MA oviposited for 20 d	Fenglin *et al.* (1979)
A. duodenale	Same as for *A. caninum*	YA & MA (puppy dog)	Maintenance system: max. survival 29 & 22 d for YA & MA, respectively; YA grew but did not mature; MA ♀ oviposited for 235 d	Fenglin *et al.* (1979)
Necator americanus	SFS, S-Def: HI young dog serum + 5–50% M199 (Sm), pH 7.4, air, 37 °C; 2–10 worms in 1.5 ml or 30 worms in 3 ml, media changed daily	MA (dog)	Maintenance system: max. survival of 2 weeks; ♀ oviposited	Xianxiang and Jiajun (1985)
		Superfamily: Ascaridoidea		
Anisakis simplex	STS, same as described by VanBanning (1971; see Hansen & Hansen, 1978, p. 242).	L3 (hake from Pacific Ocean, Chile)	L3 "reached adult stage"	Carvajal *et al.* (1981)

TABLE II. *Continued*

Species	Culture conditions[a,b,c]	Initial stage (source)[d]	Application or product of the techniques[d]	Reference
An. simplex & *Anisakis* sp.	2-step STS, Def: (1) M199 (So), pH 7.0, 5% CO_2 in air, 35°C (Sommerville & Davey, 1976, in Hansen & Hansen, 1978, p. 242); U-Def (2): BLE (VanBanning, 1971; see Hansen & Hansen, 1978, p. 242) + 25% M199 + "two drops of fresh bovine blood" daily, pH 4.5, air, 35°C; 1 worm in 2 ml; media changed every 1–3 d	L3 (mackerel & pollack)	*An. simplex*: L3 to L4 in medium 1 after 3–7 d, then, in medium 2 L4 to YA after 7–14 d; "adults" after 22–29 d *Anisakis* sp: L3 to early L4 in medium 1; no development in medium 2	Oshima *et al*. (1982)
An. simplex	RBS, S-Def: medium API-1 + 24 μg bovine haemin ml^{-1}, pH 6.8, in 85% N_2, 5% O_2, 10% CO_2, 35°C (see Section II.B.1.1°); 25–50 worms in 40 ml or 80 ml; media changed every 5–7 d	L3 (rockfish)	L3 to MA & ovipositing ♀ after 28 & 31 d, respectively;	Bier and Douvres (unpublished)
	STS: Tyrode's solution, pH 7.2, air, 37°C; 1 larva in 10 ml; no media changes	L3 (herring, mackerel & whiting)	Maintenance system: ES with protease activity	Mathews (1982)

TABLE II. *Continued*

Species	Culture conditions[a,b,c]	Initial stage (source)[d]	Application or product of the techniques[d]	Reference
Ascaris suum	SMWS, Def: DMEM (Se), pH 7.0–7.2, 5% CO_2 in air, 37 °C. 10 000–15 000 larvae in 2.0 ml; no media changes	EE (pig)	L2 to L3 after 7–14 d; ES from medium immunogenic in pigs Bioassay for anthelmintics	Urban and Tromba (1982) Urban and Romanowski (1985) Rew *et al*. (1986)
	RBS, Def: M199 (Se), see Hansen & Hansen (1978, p. 244) + GHL (Sa) 20 ng ml^{-1}, pH 7.0–7.2, 90% N_2, 5% O_2, 5% CO_2, 37 °C; 2000 L3 in 20 ml; no media changes	Late-L3 (rabbit lung)	L3 to L4 after 4 d; ES recovered for immunological studies	Stromberg *et al*. (1977) Stromberg (1979, 1980)
	SMWS, DMEM (Se) or S-Def: DMEM + 10% HIFBS, pH 7.2, 5% CO_2 in air, 37.5 °C; 300 L3 in 2.0 ml; media changed every 7 d	Late-L3 (pig, rabbit lung)	L3 to L4 after 3 d; max. survival up to 28 d; sexual differentiation of L4 in S-Def medium; used for studies on moulting & metabolism; ES for immunological studies; larval nutritional requirements; anthelmintic screen.	Urban and Douvres (1981) Urban and Romanowski (1985) Fleming (1985a,b) Fetterer and Urban (1985) Urban *et al*. (1984) Rew *et al*. (1986) Komuniecki and Vanover (1986)
Phocanema decipiens	Same as for *An. simplex*, except BLE supplemented with	L3 (hake)	L3 to YA after 20–25 d	Carvajal *et al*. (1981)

TABLE II. *Continued*

Species	Culture conditions[a,b,c]	Initial stage (source)[d]	Application or product of the techniques[d]	Reference
	heparinized human blood not "citrated beef blood" (VanBanning, 1971; Hansen & Hansen, 1978, p. 242)			
	Superfamily: Filarioidea			
Brugia malayi	STFS: *Aedes malayensis* cell culture overlaid with U-Def medium: MM/MK (inorganic salts, glucose, LAH, yeast-olate, bovine albumin, HIFBS) (Sn) (after Varma & Pudney, 1969), air, 28 °C[b]	L1 (mf) (cat)	mf (artificially exsheathed) to sf after 6 d; max. survival 30 d	Devaney and Howells (1979)
B. pahangi	Same as for *B. malayi*	L1 (mf) (cat)	mf (artificially exsheathed) to sf after 6 d; max. survival for 30 d	Devaney and Howells (1979)
	SBS, U-Def: pieces of *Armigeres subalbatus* thorax floated on 2.0 ml of GITC (Se) + 20% HIFBS, air, 27 °C; no media changes[b]	L2 (mosquitoes)	L2 to M2 & L3 after 5 d; temperature important for ecdysis	Ogura *et al.* (1981)

TABLE II. *Continued*

Species	Culture conditions[a,b,c]	Initial stage (source)[d]	Application or product of the techniques[d]	Reference
	STFS: dog sarcoma cell line overlaid with S-Def M199 (So) + 10% NCS, 5% CO_2 in air, 37 °C; 50–100 larvae 5.0 ml^{-1}; no media changes[b]	L3 (mosquitoes) & Adv. L3 (jird peritoneum)	No development of L3, but growth & survival 20 d; 3- & 5-d-old Adv. L3 developed to L4 after 5 & 3 d, respectively	Chen and Howells (1979)
	STFS: primary or "low passage number" dog skeletal muscle cells overlaid with S-Def MEM (So) + 10% FBS, pH 6.8–7.4, 5% CO_2 in air, 37 °C; 200–300 larvae; fluid & cells changed every 4–6 d[b]	L3 (mosquitoes)	L3 to L4 after 9 d; survival for 33 d	Wong *et al.* (1982)
	SMWS: "early passage" dog kidney cells, overlaid with S-Def MEM (Sf) + 10% HIFBS, 5% CO_2 in air, 37 °C; 15 larvae in 2.0 ml; no media changes[b]	Adv. L3 (jird peritoneum)	4-d-old adv. L3 to L4 after 6 d; used as anthelmintic screen	Court and Lees (1983)
	SDS, S-Def: Click's medium (So) + 10% horse serum, 5% CO_2	MA ♀ (jird peritoneum)	MA survival up to 35 d; mf extruded & lectin-binding	Furman and Ash (1983)

TABLE II. *Continued*

Species	Culture conditions[a,b,c]	Initial stage (source)[d]	Application or product of the techniques[d]	Reference
	in air, 37 °C; worms "placed in 60 μm nylon mesh" in 12 ml; medium changed daily[b]		characteristics of mf analysed	
Dipetalonema viteae	STFS, U-Def: BHK-21 (Sg) + 10% TPB + 15% fresh jird serum, pH 7.3, 90% N_2, 5% O_2, 5% CO_2, 36 °C; 50–70 larvae in 5.0 ml; medium changed at 7 d	Adv. L3 (jirds)	"L3+" (recovered from micropore chambers implanted subcutaneously in jirds for 6 d) to L4 after 8–14 d.	Tanner (1981)
			Evaluated immune sera effects on L3 to L4	Tanner and Weiss (1981)
	STS, S-Def: RPMI 1640 (Sg) + 20% HIFBS, pH 7.1–7.3, 5% CO_2 in air, 36–37 °C; media changed daily[b]	L2 (ticks) MA female (hamster)	No development of L2, but survival for several d; maintenance system; max. survival of 100 d for MA, mf extruded for 75 d	Miegeville *et al.* (1981)
Dirofilaria immitis	Same as for *B. malayi*	L1 (mf) (dog blood)	"small number" of mf to "occasional" sf survival mf for 10–15 d	Devaney and Howells (1979)
	SBS, S-Def: NCTC 109 (So) + 10% FBS, air, 27 °C; 200 larvae per ml; no media changes[b]	L1 (mf) (dog blood)	mf to sf: survival up to 16 d	Ando *et al.* (1980)
			Osmotic pressure of medium adjusted to 330–390 mOsm	Ando and Kitamura (1982)

TABLE II. *Continued*

Species	Culture conditions[a,b,c]	Initial stage (source)[d]	Application or product of the techniques[d]	Reference
			kg^{-1}; increased yield of sf	
	STS, U-Def: mosquito organ culture in Schneider's *Drosophila* medium (Sg) + 20% HIFBS, air, 27°C; 67–115 larvae in 2.0 ml; no media changes[b]	L1 (mosquito Malpighian tubules)	L1 (24 h in Malpighian tubules) to L2 after 7 d	Devaney (1981)
	STS, U-Def: *Ae. albopictus* cell line in Ham's F-12 medium (Se), 5% CO_2 in N_2, 28°C; 50 larvae in 1.1 ml; no media changes[b]	L1 (mf) (dog blood)	mf to sf; various other Def & S-Def media, & cells tested	Sneller and Weinstein (1982)
	STS, S-Def: (1) Ham's F-12(K) (Sj) + 5% FBS, 5% CO_2 in air, 27°C; (2) NCTC 135 (Se) + 10% FBS, 5% CO_2 in air, 37°C[b]	L1 (mf) (dog blood) L3 (mosquito)	Fresh & cryopreserved mf to sf in medium 1; fresh & cryopreserved L3 to L4 in medium 2	Lok *et al.* (1983)
	STS, Def: (1) Ham's F-12(K) medium; S-Def; (2) Ham's F-12(K) + 5% FBS (Sj), 5% CO_2 in air, 37°C;	L1 (mf) (dog blood)	mf survival (50%) for 4 weeks in medium 1 & 6 weeks in medium 2	Abraham *et al.* (1986)

TABLE II. *Continued*

Species	Culture conditions[a,b,c]	Initial stage (source)[d]	Application or product of the techniques[d]	Reference
	30 000 larvae in 2.0 ml; media changes every 8 d[b]			
	Same as for *B. pahangi*, but gerbil skeletal muscle cells were used	L3 (mosquitoes)	L3 to L4 after 2–3 d; survival for 33 d	Wong *et al.* (1982)
	STS, S-Def: Ham's F-12(K) medium (Sj) + 20% HIFBS, pH 7.2, 5% CO_2 in air, 37 °C; 30–50 larvae per ml; media changed every 2 d	L3 (mosquitoes)	L3 to L4; various Def media, serum levels & serum substitutes examined	Lok *et al.* (1984a)
	STS, S-Def: RPMI 1640 (So) + 10% HIHuS, pH 7.2–7.4, in candle jar dessicator, 37 °C; 10–20 larvae in 3–4 ml; media changed every other d	L3 (mosquitoes)	L3 to L4 after 3 d; max. survival 26 d	Kumar *et al.* (1984)
	STFS, U-Def: dog sarcoma feeder cell line in ML-15 medium (=L-15 + 10% TPB) +20% HIFBS (Sc), pH 7.2–7.4, air, 37 °C; 100 larvae in 15 ml no media changes	L3 (mosquitoes)	L3 to L4 after 3–4 d; ultrastructural studies on cuticle	Devaney (1985)

TABLE II. *Continued*

Species	Culture conditions[a,b,c]	Initial stage (source)[d]	Application or product of the techniques[d]	Reference
	STFS, S-Def: (1) NCTC/IMDM (se) + 20% HIFBS; SMWS, Def: (2) NCTC/IMDM (se) pH 7.2, 5%Co_2 in air, 37°C; 10 larvae per ml	L3 (mosquitoes)	L3 in medium 1 for 2 d; L3 to L4 in medium 2. Albumin can substitute for HIFBS; no growth at 27°C	Abraham *et al.* (1987)
	STS, S-Def: RPMI 1640 (So) + 10% CS, pH 7.4, air, 37°C; 1 worm in 20 ml; media changed daily	MA ♀ (dogs)	Survival of MA for 25–36 d, mf extruded for 11–17 d; used as anthelmintic screen	Tamashiro and Palumbo (1985)
Litomosoides carinii	SMWS: medium L-15 + 2 μg ml^{-1} GHL or 1.1% glucose + 10% FBS or 10% agammaglobulinaemic horse serum (Se), pH 7.0–7.1, air 37°C; 10 larvae in 2.5 ml; media changed daily	Adv. L3 (multi-mammate rats)	Adv. L3 to L4; max. survival 4–6 d; lost infectivity for rats after 1 d	Nelson *et al.* (1982)
	STS, S-Def: M199 (So) + 33% CRS, air, 37°C; 1 worm in 2.0 ml; media changes daily[b]	L1 (mf), MA (cotton rat)	L1 (mf) no development, survival 6 d; MA survival 8 d & ♀ extrude max. number mf for 5–6 d	Wenk *et al.* (1978)
	STS, S-Def: M199 (So) + 33% CRS, air, 37°C; 1 ♀ or paired ♂ & ♀ in 2.0 ml; media changed daily[b]	YA (cotton rat)	Maintenance system: 4-week-old adult pairs survive for 45–46 d; no mating, ♀ grow	Illgen (1982)
	STS, S-Def: RPMI 1640 + 20% MEM + 10%	Ma (cotton rats)	Maintenance system: examined medium	Mossinger and Wenk (1986)

TABLE II. *Continued*

Species	Culture conditions[a,b,c]	Initial stage (source)[d]	Application or product of the techniques[d]	Reference
	FBS (Si, Sh) 5% CO_2 in air, 35°–37°C; 1 worm in 2.0 ml; media changes daily[a]		effects on embryogenesis; mf extruded; fecundity assay	
Onchocerca gutturosa	SMWS, S-Def: *Rhipicephalus appendiculatus* cell cultures in RPMI 1640 (So) + 20% HIFBS 28°C; 30–50 mf per well; no media changes[b]	L1 (mf)	mf to sf after 4–5 d	Pudney and Varma (1980)
	SMWS, S-Def: TS + 20% HuS, air, 11°C; cultures of 5–20 mf $well^{-1}$; no media changes	L1 (mf) (bovine skin)	Maintenance system: mf survival for 11 d	McDiarmid and Mathews (1982)
	SMWS, S-Def: RPMI 1640 (So) + 10% HIFBS, 5% CO_2 in air, 37°C; 5 worms in 1.8 ml; no media changes[b]	MA ♂ (bovine connective tissue)	Maintenance system: max. survival 7 d for both fresh & cryopreserved MA	Townson and Ham (1986)
	SMWS, S-Def: LLMCK2 monkey kidney cells in MEM + 10% HIFBS (Sf) pH 7.2–7.4, 5% CO_2 in air: one worm in 1.8 ml no media changes[b]	MA ♂ (bovine connective tissues)	MA max. survival 6–7 months	Townson and Connelly (1986)

TABLE II. *Continued*

Species	Culture conditions[a,b,c]	Initial stage (source)[d]	Application or product of the techniques[d]	Reference
O. lienalis	SMWS, S-Def: *Ae. malayenis* in RPMI 1640 + 20% HIFBS or ML-15 + 10% TPB + 20% HIFBS (Se, Sg) pH 7.2, air, 28°C; 40–60 mf in 0.25 ml; no media change	L1 (mf) (bovine umbilical skin)	mf to sf; survival 30 d	Devaney and Howells (1983)
	SMWS, S-Def: NI-20FBS (see Section II, C, 1, 2°) 95% N_2, 5% CO_2, 37°C; 6–10 larvae per ml; no media change[b]	L3 (mosquitoes)	L3 to L4 after 2–5 d; max. survival for 43 d; used as anthelmintic screen	Lok *et al.* (1984b, 1986)
	SMWS, S-Def: MEM + 10% HIFBS (So), 5% CO_2 in air, 37°C; 15 larvae in 2.0 ml; no media change[b]	L3 (black flies)	L3 to L4 after 4 d; max. survival for 10 d; used as anthelmintic screen	Court *et al.* (1985)
	SMWS, S-Def: LLMCK2 monkey kidney cells in NCTC 135 + 10% HIFBS (So), 5% CO_2 in air; 10 larvae in 1.8 ml; media changed every 3–4 d[b]	L3 (black flies)	L3 to "late" L4; survival of L4 for "longer than 60 d"	Townson *et al.* (1986)
O. volvulus	Same as for *O. gutturosa*	L1 (mf)	mf to sf after 4–5 d	Pudney and Varma (1980)

TABLE II. *Continued*

Species	Culture conditions[a,b,c]	Initial stage (source)[d]	Application or product of the techniques[d]	Reference
	Diphasic medium: blood-agar base overlaid with HBSS, 5 ml^{-1}; pH 7.5 (Schiller *et al.*, 1975, see Fried, 1978, p. 164) in	L1 (mf) (black flies & excised skin nodules from humans)	mf to "late" sf, ". . . 13 larvae developed to an early juvenile stage, averaging 11 × 1100 μm", after 30 d	Schiller *et al.* (1979)
	Erlenmeyer flask cultures 90% N_2, 5% O_2, 5% CO_2 — shaker bath, 37 °C[b]	MA (skin nodules)	Maintenance system: 12 immature & 1 MA ♀ survived for 7 d; 2 MA ♂ survived for 22 & 81 d	Schiller *et al.* (1979)
	Same as for *O. lienalis*	L3 (black flies)	L3 to L4, after 5–10 d; max. survival 25 d	Lok *et al.* (1984b)
	STS, S-Def: M199 + 10% HuS, air, 28–30 °C or 35 °C; 1 worm in 2.0 ml, media changed daily[b]	MA (skin nodules from humans, enzymatic isolation)	Maintenance system: max. survival of 28 d for ♂ & 42 d for ♀	Englebrecht and Schulz Key (1984)
Wuchereria bancrofti	STC, S-Def: M199 (So) + organic acids & sugars of GITC (see Taylor & Baker, 1968, p. 312) + 10% HIHuS, pH 7.4, air, 28 °C; media changed daily[b]	L1 (mf) (human blood)	mf to sf after 8 d; survival for 14 d	Kharat *et al.* (1980)

TABLE II. *Continued*

Species	Culture conditions[a,b,c]	Initial stage (source)[d]	Application or product of the techniques[d]	Reference
		Superfamily: Gnathostomatoidea		
Gnathostoma sp.	SDS, Def: RPMI 1640 (Se), pH 7.2, used "candle jar", 37 °C; 10 larvae in 4.0 ml; media changed daily[b]	L3 (eel's liver)	Maintenance system: L3 survived for 30 d; infective to mice; potential for ES collection	Morakote *et al*. (1985)
		Superfamily: Heligmosoidea		
Nematospiroides dubius (=*Heligmosomides polygyrum*, Durette-Desset, 1968)	SFS, U-Def: formalin-killed *Escherichia coli*: various sterols & porphyrins, air, 24 °C; 2000–5000 eggs per ml; no media changes[b]	Eggs (mouse faeces)	L1 to L3 does not require exogenous sterol; porphyrins enhance growth of L3	Mauro and Weinstein (1979)
	SFS, S-Def; CbMM (Se) + formalin-killed *E. coli*, air, 23 °C; 2000 larvae in 2.0 ml; no media changes[b]	Eggs (mouse faeces)	L1 to L3 after 6 d; development inhibited by azasteroids	Bottjer *et al*. (1984)
	Same as for *Nippostrongylus brasiliensis* (Jenkins *et al*., 1980)	Late L4 (mouse intestine)	L4 to YA after 7 d; used as anthelmintic screen	Jenkins and Ibarra (1984)

TABLE II. *Continued*

Species	Culture conditions[a,b,c]	Initial stage (source)[d]	Application or product of the techniques[d]	Reference
	Superfamily: Metastrongyloidea			
Angiostrongylus cantonensis	STS, S-Def: NCTC 109 (Sb) + 50% RS HoS, or CS, air, 37°C; 2000–35 000 eggs in 0.3 ml; media changes every 3 d[b]	Eggs (uteri of adults from rat lungs)	Eggs embryonated by 5 d; hatching of L1 after 8 d; slight gowth, viable for 32 d; infective for snails	Uga and Matsumura (1982)
	STS, S-Def: NCTC 109 (Sb) + 50% HoS, air, 37°C; 1 worm per ml; media changes every 3 d[b]	MA (rat lungs)	Maintenance system: max. survival 2 weeks ♀, 1 month ♂; ♀ laid eggs for 5 d, but only those deposited within 30 min embryonated	Uga *et al.* (1984)
Metastrongylus apri	STS, S-Def: NCTC 109 + 20% CS (Se), air, 37–38°C; 100–300 worms in 2.0 ml; 2 media changes per week[b]	L3 (M2) (earthworms)	L3 (M2) (artificially exsheathed) to YA after 14 d; YA maintained for 70 d; expressed sex specific antigen; effects of various antibiotics on L3 to YA	Hata *et al.* (1980, 1981, 1983)
	Superfamily: Strongyloidea			
Oesophagostomum radiatum	See Section II.G.1.1°	L3 (M2) (bovine faeces culture)	Examined glucose metabolism of developing M3, L4, & YA; labelled ES	Rew *et al.* (1983)

TABLE II. *Continued*

Species	Culture conditions[a,b,c]	Initial stage (source)[d]	Application or product of the techniques[d]	Reference
			demonstrated in Def & S-Def media	
			L4 grown from L3 expressed cuticular aberrations; examined by scanning electron microscopy	Douvres and Wergin (1984)
			ES derived from L3 to L4 acts as a potent inhibitor of bovine lymphoblastogenesis *in vitro*	Gasbarre *et al.* (1985)
			ES derived from L3 to L4 protected calves from parasite-induced weight loss	Gasbarre and Douvres (1986)
		Superfamily: Trichinelloidea		
Trichinella spiralis (derived from Polar Bear)	STFS, S-Def: (1) NCTC 135 + 0.5% LAH+0.1% yeast extract + 20% HICS (So), (2) medium 1 + 20% CEE (So), pH 7.4, with a "complete sheet" of embryonic mouse muscle cells; 37°C; 90% N_2, 5%	L1 (guinea pig muscle) M4, YA (rat intestine) NBL (derived from adults obtained from rat intestine)	L1 appear to undergo 1 moult in medium 1, after 8 h "many muscle larvae began to form a thin sheath . . . became loose . . ." & completed ecdysis to "free worms" that showed differentiation of	Sakamoto (1979)

TABLE II. *Continued*

Species	Culture conditions[a,b,c]	Initial stage (source)[d]	Application or product of the techniques[d]	Reference
	O_2, 5% CO_2; 1000 L1 in 5.0 ml; media changed every other d; modified Berntzen (1965, see Taylor & Baker, 1968, pp. 251–255) 2-step enzymic procedure for removal of L1 from muscle tissue & his "de-sheathing apparatus" were used. Rinaldini's solution was substituted for TS & sodium tauroglyocholate was deleted in step 2		genital organs after 48 h; larvae survived for 3–4 weeks, no further development. Ensheathed larvae, from 3-d-old cultures, were subjected to agitation (magnetic stirrer) in Rinaldini's solution containing 0.5% trypsin & 1% pancreatin for 30 min then "de-sheathing apparatus"; "exsheathed" ♂ & ♀ had adult genital systems without sperm & embryos; after a "few days", "spermatogenesis & embryogenesis were recognized . . .", "copulation was occasionally seen . . ." MA survived for 3 weeks. YA developed to MA in medium 1 after 36 h;	

TABLE II. *Continued*

Species	Culture conditions[a,b,c]	Initial stage (source)[d]	Application or product of the techniques[d]	Reference
			♀ larvaposited at 4 d; extruded NBL viable for 9 d. *In vivo* derived MA shed NBL that penetrated cells in medium 2; stichocytes & intracellular coiling of L1 visible.	
	SMWS, U-Def: same as for *N. brasiliensis* (Jenkins *et al.*, 1980); 500 larvae in 2.0 ml, no media changes	L1 (mouse muscle)	L1 (excysted in pepsin-HCl) changes from "coiled" phase to "serpentine" phase. Used as an anthelmintic screen	Jenkins and Carrington (1981)
	SMWS & SFS, S-Def: as described by Dennis *et al.* (1970, see Hansen & Hansen, 1978, p. 260)	MA (mouse intestine)	MA extruded NBL; *in vitro* fecundity assay used to correlate host protective response with production of NBL	Despommier *et al.* (1977)
	SFS, Def: L15 + 2.0 mM glutamine + 25 mM HEPES, pH 7.2, or DMEM + 2.0 mM glutamine + 10 mM HEPES (Se), 10% CO_2 in air, 37°C; 10 000 larvae per ml; no media changes[b]	L1 (rat muscle)	L1 (excysted in pepsin-HCl) maintained 72 h; ES for serodiagnostics, production of monoclonal antibodies, source of affinity purified antigens	Gamble *et al.* (1983, 1985) Murrell *et al.* (1986)

TABLE II. *Continued*

Species	Culture conditions[a,b,c]	Initial stage (source)[d]	Application or product of the techniques[d]	Reference
	Superfamily: Trichostrongyloidea			
Cooperia oncophora	Modified Douvres (1980) for *Trichostrongylus colubriformis* (see Section II.I.1,4°): RBS, S-Def medium API-16+0.2% pepsin +10 mM glut for 2 d & thereafter API-16 + 0.2% pepsin, pH 6.8, 85% N_2, 5% O_2, 10% CO_2, 39 °C; 80 000 larvae in 40 ml for 2 d, in 80 ml thereafter; media changed d 2, then every 3–7 d	L3 (M2) (bovine faeces cultures)	L3 (M2) (artificially exsheathed) to MA after 24 d; oviposited segmented eggs; survival 75% 24–35 d	Douvres, unpublished
Dictyocaulus viviparus	SMWS, Def: RPMI 1640, 4% CO_2 in air, 40 °C[b]	L3 (M2) (bovine faeces)	L3 (M2) (artificially exsheathed) to L4 within 48 h	Bos and Panhuijzen (1982)
Haemonchus contortus	RBS, Def: (1) EBSS, plus "antibiotics", saturated with 100% CO_2, 37 °C, for 16 h; 10 000 L3 (M2) ml^{-1} (modified from Hansen *et al.*, 1966); (2) PBS,	L3 (M2) (ovine faeces culture)	L3 (M2) exsheathed in medium 1 then developed to early L4 in media 2 & 3 in 2 d; larvicidal & inhibitory effects on development of 23	Boisvenue *et al.* (1977)

TABLE II. *Continued*

Species	Culture conditions[a,b,c]	Initial stage (source)[d]	Application or product of the techniques[d]	Reference
	pH 7.2, 5% CO_2 in air, 37 °C; 5000 larvae per ml; (3) same as medium 2, but + additional test compounds at 10, 50 & 100 ppm; no media changes		compounds with juvenile-hormone-like activity assayed	
	RBS, S-Def: API-27 or API-28 (see Section II.I.1.1°) + 24 μg bovine haemin ml^{-1}, pH 6.8, 85% N_2, 5% O_2, 10% CO_2, 39 °C; 100 or 250 worms in 20 ml; no media changes	MA (sheep abomasum)	Maintenance system: max. survival 3 d; viable 7 d; oviposited fertilized eggs that developed to L1 & hatched by 1 d; some L1 developed to M1 & L2	Douvres and Fetterer (1986)
	SFS, Def: Ham's F10 (Se), pH 7.4, 5% CO_2 in air, 4–5 °C for up to 24 h, thereafter 39–40 °C; 5 worms in 3.0 ml medium changed every 4 d	MA "gravid ♀" (sheep abomasum)	Maintenance system: for first 2 d MA viable & laid eggs; motility diminished gradually, egg-laying stopped from 2–5 d. About 10% of eggs deposited hatched to L1; no further development of L1 occurred	Ayalew and Murphy (1986)

TABLE II. *Continued*

Species	Culture conditions[a,b,c]	Initial stage (source)[d]	Application or product of the techniques[d]	Reference
			Described occurrence of *in utero* egg hatching & development of L1 to L4 in MA that were apparently alive ". . . for at least 12 h" & moribund & immobile by 2–5 d; "large number of eggs within the uterus of most females" hatched to L1 & then developed to L2 by 24 h, L3 by 2–3 d & to L4 by 6–8 d	
Nippostrongylus brasiliensis	Same as for *Nem. dubius*	Eggs (rat faeces)	Eggs to L3 (yield 2%); yield increased to 89% after addition of 25 μg cholesterol ml^{-1}; growth of L3 affected by cholesterol & up to 25 μg coproporphyrin ml^{-1}	Mauro and Weinstein (1979)
	SFS, Def; Krebs-Ringer solution, 0.1 M Tris-HCl, pH 7.4: S-Def;	Inf. L3 (rat faeces cultures)	L3 to "parasitic phase" of L3; effect of temperature & culture	Bonner (1979)

TABLE II. *Continued*

Species	Culture conditions[a,b,c]	Initial stage (source)[d]	Application or product of the techniques[d]	Reference
	NCTC 135 + 20% HICS (Se), 5% CO_2 in air, 25 °C or 37 °C; 25 worms in 2.0 ml; no media changes[b]		medium on development & RNA biosynthesis.	
	RBS, Def; (1) M199 (So) + 1.0 μg ml^{-1} vitamin B-12: S-Def; (2) medium 1 + 20% HIHoS: (3) medium 2 + 10% rat red blood lysate; & combinations of media, either 1 or 2 with an unknown concentration of soypeptone, 8% CO_2 in air, 37 °C; 25 larvae in 2.0 ml; ended after 2 d[b]	Inf. L3 (exsheathed, from rat faeces cultures)	L3 to "parasitic phase" did not occur in medium 1 but did occur in all other media; cathepsin D-like activity induced by media containing proteins or peptides	Bolla and Weinstein (1980)
	SMWS, U-Def: medium NB (Jenkins & Carrington, 1982) consists of nutrient broth described by Diamond & Douvres (1962, see Taylor & Baker, 1968, p. 299) without L-cysteine &	L4 (rat intestine)	L4 to YA after 2–7 d; used as anthelmintic screen	Jenkins *et al.* (1980)

TABLE II. *Continued*

Species	Culture conditions[a,b,c]	Initial stage (source)[d]	Application or product of the techniques[d]	Reference
	ascorbic acid + 15% HICS + 2% bovine erythrocyte lysate, pH 7.2, air, 37 °C; 50 larvae in 2.0 ml; no media changes			
	SMWS, U-Def; (1) medium NB (Jenkins *et al.*, 1980): Def; (2) CbMM (Sg), pH 6.0: S-Def; (3) CbMM + 15% NBCS, pH 6.2: air, 37 °C; 40–60 larvae per ml; no media changes	L4 (rat intestine)	L4 to YA after 3–7 d; 100% survival in medium 2 for 5 d & in media 1 & 3 for 7 d; development best in media 1, 3 & 2, respectively; used as anthelmintic screens	Jenkins and Carrington (1982)
	Same as for *N. dubius*	Eggs (mouse faeces)	Eggs to L3; development inhibited by 5–25 $\mu g\ ml^{-1}$ azasteroids, but reversed by up to 100 $\mu g\ ml^{-1}$ cholesterol	Bottjer *et al.* (1984)
	Stationary chamber: TS pH 5.9–8.7, 37 °C; 1 worm per μl or 100–1000 in 2.0 ml	YA & MA (rat intestine)	Maintenance system for bioassay of pheromone activity & feeding responses	Ward and Bone (1983) Ward *et al.* (1984) Bottjer and Bone (1985)

TABLE II. *Continued*

Species	Culture conditions[a,b,c]	Initial stage (source)[d]	Application or product of the techniques[d]	Reference
Ostertagia circumcincta	RTS, U-Def: (1) WAe (Rose, 1973, 1976 see Hansen & Hansen, 1978, p. 256); Def: (2) EBSS, pH 7.2–7.4, air, 38.5°C; 30 000 larvae in 40 ml; media changes 3 per week	L3 (M2) (ovine faeces cultures)	L3 (artificially exsheathed) to L4 & YA after 3 weeks in medium 1; ES from worms in medium 2 in immunological studies	Rose (1978)
O. ostertagi	RBS, S-Def: (1) same as described by Douvres & Malakatis (1977, see Section II.B.1.1°); Def: (2) maintenance medium using API-salt solution consisting only of the salts used in medium API-1 (see Section II.I.1.1°), pH 6.0, air, 39°C; 200 000 larvae in 100 ml; media changed at 3 d, & thereafter at 6–8d	L3 (M2) (bovine faeces cultures)	L3 (M2) exsheathed & developed to M4 after 15–19 d in 2-step culture system; 8-d-old cultured L3 & M3 in (2) completed development to L4 after 1–3 d; populations of L3 to M4 & ES used to immunize calves (unsuccessfully)	Herlich and Douvres (1979)
	RBS, S-Def: same as described by Douvres & Malakatis (1977, see Section II.I.1.1°)	L3 (M2) (bovine faeces cultures)	L3 (M2) exsheathed & developed to MA after 25–28 d; affects of insect growth-	Dourves *et al.* (1980)

TABLE II. *Continued*

Species	Culture conditions[a,b,c]	Initial stage (source)[d]	Application or product of the techniques[d]	Reference
	except step 2 cultures were gassed with 85% N_2, 5% O_2, 10% CO_2; 50 000 larvae in 50 ml; medium changed at d 3 & thereafter every 6–7 d		disrupting amines & amides on nematode development tested	
	RBS, S-Def: modification of Douvres & Malakatis (1977, see Section II.I.1.1°) medium API-1 + 0.2% pepsin + 5 mM L-cys for 7 d; thereafter API-1 + 0.2% pepsin, pH 6.0, 85% N_2, 5% O_2, 10% CO_2, 39°C; 80 000 larvae in 40 ml for 7 d; thereafter in 80 ml; media changed every 7 d	L3 (M2) (bovine faeces culture)	L3 (M2) (artificially exsheathed) developed to MA after 22–23 d, ♀ laid fertilized & unfertilized eggs after 27–34 d; max. survival 40 d	Douvres (unpublished)
Trichostrongylus colubriformis	Same as for *C. oncophora* (Douvres, unpublished)	L3 (M2) (ovine faeces cultures)	L3 (M2) (artificially exsheathed) to YA after 16 d; max. survival 21–28 d; yield 1–5% YA	Douvres (unpublished)

TABLE II. *Continued*

Species	Culture conditions[a,b,c]	Initial stage (source)[d]	Application or product of the techniques[d]	Reference
	RBS, Def: RPMI 1640 (Sn), pH 6.8, 85% N_2, 5% O_2, 10% CO_2, 39 °C; 8000–34 000 worms in 20 ml; media not changed or changed at d 3	L4 (guinea pigs intestines) L4-YA (sheep intestines)	L4 (5 d in guinea pigs) to YA after 6–7 d; L4 & M4 (7–8 d in sheep) to YA after 3 & 1 d, respectively; max. survival for 7 d; YA from sheep did not grow *in vitro*	Nelson and Douvres (1984)
	SMWS, U-Def: medium NBS, a modification of medium NB (Jenkins *et al.*, 1980) in that NBS formulated without bovine erythrocyte lysate, pH 7.2, air, 37 °C; 30–40 worms in 2.0 ml; no media changes	Late L4 (gerbil intestine)	Late L4 (M4?) to YA after 1–7 d; used as anthelmintic screen	Rapson *et al.* (1985)
	RTS, U-Def: medium Tc5 25% NCTC 135 (Sc), 25% unfiltered CEE (after Weinstein & Jones, 1956, see Taylor & Baker, 1968, p. 345), 25% YE (Sn), 25% FCS (Se), pH 7.0–7.2, 5% CO_2 in air, 27 °C; 400 larvae	Eggs (guinea pig faeces)	L1 (artificially hatched) to L2 after 2 d; L3 (M2) by 6 d; infective for guinea pig	Khan and Dorsman (1978)

TABLE II. *Continued*

Species	Culture conditions[a,b,c]	Initial stage (source)[d]	Application or product of the techniques[d]	Reference
	in 2.0 ml; media changed at d 5 & thereafter every other day			
	RTS, S-Def: medium CH4 54.5% balanced salt solution "C"; 20% YE Type II (Sn); & each of following ml^{-1}: $KHCO_3$, 1160 μg; casein hydrolysate AN:TN 0.39, 40 mg; cytochrome C, 340 μg; cholesterol, 75 μg; Tween 80, 3.0 mg; phosphatidylcholine, 2.0 mg (Sp); hydroxyproline, 300 μg; acetylglucosamine, 18 μg (Sq); lipoic acid, 5.0 μg; folic acid, 8.0 μg; & vitamin B-12, 3.0 μg (Sr), pH 7.3–7.5, 5% CO_2 in air, 31–32°C; 500 larvae in 3.0 ml; media	Eggs (guinea pig faeces)	L1 (artificially hatched) to L3 (M2); no bacteria, animal tissue extract or serum; infective for guinea pigs	Dorsman and Bijl (1985)

TABLE II. *Continued*

Species	Culture conditions[a,b,c]	Initial stage (source)[d]	Application or product of the techniques[d]	Reference
	changed d 3, thereafter every other day			
	Same as Ward & Bone (1983) for *N. brasiliensis*	MA (goat intestines)	Maintenance system; a bioassay for pheromone & feeding responses	Bone and Bottjer (1984, 1985) Bottjer *et al.* (1985)

[a] BLE, bovine liver extract; CEE, chicken embryo extract; CRS, cotton rat serum; CS, calf serum; cys, L-cysteine; Def, defined medium; DMEM, Dulbecco's modified Eagle's medium; EBSS, Earle's balanced salt solution; FBS, foetal bovine serum; GHL, glycl-histidyl-lysine tripeptide; GITC, Grace's insect tissue culture medium (see Taylor and Baker, 1968, p. 312); glut, glutathione (reduced); HBSS, Hanks's balanced salt solution; HI, heat-inactivated; HoS, horse serum; HuS, human serum; MEM, minimum essential medium; NCS, newborn calf serum; PBS, phosphate buffered saline; RBS, roller bottle culture system; RS, rat serum; RTS, roller tube culture system; SBS, stationary bottle culture system; S-Def, semi-defined medium; SDS, stationary dish culture system; SFS, stationary flask (e.g., Erlenmeyer type) culture system; SMWS, stationary multi-well plate culture system; STFS, stationary tissue flask culture system; STS, stationary tube culture system; TPB, tryptose phosphate broth; TS, Tyrode's solution; U-Def, undefined medium; YE, 20% fresh baker's yeast.

[b] Symbol denotes that author(s) excluded one or more of the following culture conditions: pH; gas phase; incubation temperature; number of worms in inoculum; medium volume; media changes.

[c] Unless otherwise noted, the following symbols identify the sources of media or ingredients: Sa, Calbiochem-Behring Corporation; San Diego, California; Sb, Difco; Baltimore, Maryland; Sc, Flow Laboratories; Irvine, Scotland; Sd; Flow Laboratories, Rockville, Maryland; Se, Grand Island Biological Chemical Company (Gibco), Grand Island, New York; Sf, Gibco Ltd, United Kingdom; Sg, Gibco Bio-Culture, Paisley Scotland; Sh, Gibco, Karlsruhe, Federal Republic of Germany; Si, Flow Laboratories, Mechenheim, Federal Republic of Germany; Sj, K.C. Biological, Lenexa, Kansas; Sk, M.A. Bioproducts, Walkersville, Maryland; Sl, Nissui Seiyaku Company, Ltd, Japan; Sm, Wellcome Ltd, United Kingdom; Sn, medium prepared in authors' laboratory; and So, source not given by authors; Sp, Sigma Chemical Company, St. Louis, Missouri; Sq, Serva (location unlisted); Sr, Merck (location unlisted).

[d] Adv. L3, *in vivo*-grown third-stage larvae; EE, embryonated eggs; ES, excretory-secretory products; L1, L2, L3 and L4, larvae in first, second, third and fourth stage, respectively; L1 (mf), microfilarial phase of first-stage larva; L3 (M2), infective, free-living larva, enveloped in sheath; M1, M2, M3, and M4, larvae enveloped by intact sheath (old cuticle) in first, second, third and fourth moult phases, respectively; MA, mature adult stage; NBL, newly born larvae; sf, advanced phase or "sausage form" of first-stage larva; YA, young adult (or fifth stage) stage.

Chapter 14

Nematodes Parasitic in Insects

IRENE POPIEL

Biosis, 1057 East Meadow Circle, Palo Alto, California, USA

I. INTRODUCTION

The *in vitro* culture of nematodes parasitic in insects, reviewed by Hansen and Hansen (1978) in an earlier version of this book was restricted to axenic methods. This chapter comprises a more extensive review and includes xenic culture. Emphasis has been placed on research towards mass *in vitro* culture of representatives of three families of entomogenous nematodes that have the greatest potential for the biological control of insects. However, the section on the Mermithidae has been treated as a historical review since, because they have more fastidious culture requirements, the development of *in vitro* systems is at a much earlier stage.

Steinernematid and heterorhabditid nematodes are now routinely cultured in kilogram quantities. The published state of the art has been almost entirely developed by Bedding (1976, 1981, 1982, 1984). Companies in the USA, Europe and Australia are commercially producing steinernematid and heterorhabditid nematodes for the control of a wide range of insect pests such as the black vine weevil, cutworms, white grubs and some borer insects. Present research is aimed towards methods of liquid culture that will increase yields and reduce costs. The principal limitation in an aerated liquid system is sheer. However, it has already been shown that in axenic liquid culture this limitation can be overcome (Buecher and Hansen, 1971).

The Neotylenctid species, *Deladenus siricidicola*, has been cultured for use against insects, but only in the mycetophagous rather than the insect parasitic phase of the life-cycle (Bedding, 1974).

II. MERMITHIDAE

The duration of the parasitic phase of the mermithid life-cycle varies between species, but is relatively short compared with the free-living phase. Mermithids feed only during the parasitic phase, achieving up to a 100-fold increase in size. They lack a functional gut and depend on transcuticular uptake of nutrients (Poinar and Hess, 1976, 1977; Gordon and Webster, 1972; Rutherford and Webster, 1974). Since uptake by active transport is likely to be influenced by the relative concentration of components, the composition of the culture medium should approach that of host fluids (Hansen and Hansen, 1978).

Culture *in vivo* is the only mass production method presently available (Petersen and Willis, 1972; Fasuliotis and Creighton, 1982). For mermithids to be cost-effective as pest control agents, the development of *in vitro* production methods is required. Early *in vitro* culture attempts were performed in the absence of information on the physiological and biochemical aspects of the host–parasite relationship. Media and culture conditions have been adapted as such information has become available.

1. Filipjevimermis leipsandra

Filipjevimermis leipsandra is a parasite of chrysomelid beetles available in large quantities from infected laboratory colonies of *Diabrotica balteata*, the banded cucumber beetle. The nematodes reside initially in the cerebral and neural ganglia, then complete growth and maturation in the haemocoel. This implies discrete nutritional and physiological requirements for the two phases of development *in vivo*.

Fassuliotis and Creighton (1982) reported successful cultivation through the complete life-cycle. Cultivation of preparasites to pre-adults was achieved in Schneider's *Drosophila* medium (SDM; Gibco) plus 20% foetal bovine serum (FBS) at 25°C within 26 d. Since transfer of pre-adults to a solid substrate was found to be necessary for the induction of the final moult to the adult stage, they were transferred to a sand-blasting aggregate (Starblast; DuPont) for 14 d by which time 70% had moulted. Upon transfer to water, the remaining worms moulted and females began to oviposit 1–3 d later. Nematodes grown *in vitro* were shorter than those recovered from *in vivo* infections of *Diabrotica balteata* (12.5–4.1 mm versus 31.3–14 mm).

It is puzzling that a nematode that requires two quite different sites within the host for growth and development can be successfully cultured in a single medium, when other mermithids, which develop completely within the haemocoel, cannot (see below).

2. Hydromermis conopophaga

In a single attempt to culture the pre-parasite of *Hydromermis conopophaga* species (Poinar, 1975), a morphological change to the parasitic form was achieved in a medium comprising pond water, chick embryo extract (see Taylor and Baker, 1968: pp. 345–346; 1978, p. 232) and bovine serum (3 : 1 : 1) at 20 °C.

3. Isomermis *sp.*

Limited growth of pre-parasites of *Isomermis* was obtained in a combination of SDM and Grace's tissue culture medium (GTM; see Taylor and Baker, 1968, p. 312) plus foetal calf serum (FCS) at 10 °C (Finney, 1976). No moult was observed.

4. Neomesomermis flumanalis

Neomesomermis flumanalis, a mermithid parasite of *Simulium* larvae, was first cultured by Myers (1973), who obtained limited growth in GTM at 15 °C. Finney (1976) reported limited growth and reproductive development in a combination of SDM and GTM plus FCS at 10 °C.

5. Romanomermis culicivorax

The mermithid *Romanomermis culicivorax* is available in large quantities from infections of laboratory colonies of *Culex pipiens* (Petersen, 1973).

Chlorox-treated eggs, or preparasites surface-sterilized with antibiotics are used to initiate *in vitro* cultures. Using surface-sterilized preparasites, Sanders *et al.* (1973) found that limited growth occurred in SDM plus 10% FCS at 25 °C and pH 6.5–7.0: many of the nematodes were abnormal in appearance and none had moulted. Myers (1975) obtained similar results using SDM or GTM plus heat-inactivated FCS (4 : 1).

Roberts and Van Leuken (1973) first reported the use of insect tissue culture for culturing *R. culicivorax*. Preparasites were incubated either in direct contact with the tissue culture cells or separated from them by a membrane. The cell lines used were those of Hsu (*Culex quinqui-*

fasciatus) and Singh (*Aedes aegypti*); the media were that of Mitsuhashi and Maramosche (1964) and SDM. Survival was 30% and only limited growth was achieved.

Culture studies were continued by Finney (1977), who reported the best *in vitro* development of *R. culicivorax* to date. Her most successful results were obtained using GTM plus 10% heat-inactivated FCS, pH 6.4–6.5, at 26°C (osmotic pressure adjusted to 240 mosmol in culture tubes). Media changes were made every 3–4 d. After 6 weeks, worms measuring up to 14 mm were obtained but survival was only 30%. Reproductive morphogenesis took placc in females only; storage material in the trophosome was lacking. Finney (1981) improved the nutritional state of the trophosome by incorporating fatty acids into the medium. Nematode development time was reduced by the addition of myristic acid, squalene and cholesterol.

Further testing by Castillo *et al.* (1982) of a number of vertebrate, invertebrate and microbiological media, with and without insect tissue culture cells, did not yield an improved system. The most advanced development was achieved in GTM plus 20% foetal bovine serum; 1.5% bovine albumin (BA); bacto-tryptose broth (Difco) 1.5 mg ml^{-1}; liver digest (Panmede; Pabyrn Laboratories), 2.0 mg; yeastolate (source unknown), 2.0 mg; vitamin B_{12} 0.3 mg ml^{-1}; and trehalose 2.0 mg ml^{-1}. After 3–4 weeks of culture, nematode development was comparable to that at 4–5 d *in vivo*; two moults were observed and maximum length was 7.0 mm.

Giblin and Platzer (1986) were unable to repeat Finney's results, and chose the GTM plus 20% FBS and 1.0% BA of Castillo *et al.* (1982) as a basal medium for the evaluation of biochemical and physical modifications. The most advanced development, equivalent to 2.4 d *in vivo*, was achieved in GTM in which the salt ratio was altered to that of the haemolymph of the fourth instar of *Culex pipiens*, the amino acid ratio was changed to that based on hydrolysed female *R. culicivorax* post-parasitic stage, and reduced glutathione (200 mg) was added. During this study, a method for mass surface sterilization of preparasites was devised. This involved light-induced migration of preparasites up through agar containing an antibiotic–antimycotic mixture (penicillin, 200 iu ml^{-1}; streptomycin, 200 mg ml^{-1}; chlortetracycline, 25 mg ml^{-1}; Kanamycin, 200 iu ml^{-1}; fungizone, 1.25 mg ml^{-1}) into an overlay of sterile distilled water. The preparasites were collected and washed three times before being dispensed into culture medium.

The singular lack of success in the culture of *Romanomermis* and other mermithids following extensive media testing invites a reappraisal of approach. Knowledge is lacking in three notable areas. Firstly, little is known about the parasites' specific physiological and nutritional requirements, or of the mechanism of nutrient uptake across the body wall. As suggested by Castillo *et al.* (1982), the nematodes may not be able to absorb sufficient nutrients *in vitro* and growth factors may not be available in the proper form and concentration. Those attempting culture would benefit from further basic research providing such physiological information for the establishment of optimal *in vitro* conditions. Thus the differences in the biochemical composition of mosquito and blackfly larval haemolymph (Gordon *et al.*, 1979, 1980) suggest that the media requirements of mermithids of mosquito and blackfly larvae will not be the same.

Secondly, it is known that sex in mermithids is influenced *in vivo* by environmental factors such as host diet and intensity of infection (Peterson, 1977; Gordon *et al.*, 1981), whereas the *in vitro* production of female worms only, has yet to be explained.

Thirdly, researchers have relied upon a combination of length and width measurements as the criterion with which to determine parasite growth and evaluate the culture conditions. Although this is important for the preliminary screening of media, monitoring the biochemical status and morphological integrity of the parasite, using *in vivo* data as a baseline, would provide a more complete evaluation of *in vitro* performance. For example, Poinar and Hess (1977) observed that the cuticular pores of *R. culicivorax* became clogged with deposits when the parasites were incubated in a medium containing heat-treated serum. This may result in inefficient transcuticular transport and may explain the poor growth obtained *in vitro* (Castillo *et al.*, 1982). Several authors have noted abnormalities in the body wall of *Romanomermis* grown *in vitro*. Comparative ultrastructural studies on the body wall of nematodes developing *in vivo* and *in vitro* would provide a better understanding of its role in nutrient uptake and of its integrity in *in vitro* conditions. A study on the effect of *in vitro* culture on reproductive morphogenesis might also provide valuable information on improvements to, or limitations of, culture conditions.

III. DIPLOGASTERIDAE

The *in vitro* culture of diplogasterid nematodes has received little attention. Continuous cultures of *Diplogaster aerivora* were maintained for almost 4 years on a medium consisting of agar and plain beef extract plus the bacterium *Serratia marcescens* (Swain, 1945).

Pristionchus uniformis was cultured under xenic conditions by Fedorko and Stanuszek (1971) on agar overlaid by raw liver slice.

IV. STEINERNEMATIDAE AND HETERORHABDITIDAE

Steinernematid and heterorhabditid nematodes are produced commercially for the biological control of insect pests. Each species of steinernematid and heterorhabditid nematode is symbiotically associated with a single bacterial species of the genus *Xenorhabdus* (Thomas and Poinar, 1979), except *Steinernema kraussei*, which is associated with *Flavobacterium* sp. (Mracek, 1977). Each species of *Neoaplectana* is associated with a different species of *Xenorhabdus* (Akhurst, 1983), whereas all species of *Heterorhabditis* have *X. luminescens* as the symbiont. The bacteria are carried monoxenically within the ventricular region of the intestine of the third stage infective juvenile. After entering the host, the nematodes pass to the haemocoel, exsheath and release the bacteria into the haemolymph. The bacteria multiply, the insect dies of septicaemia and the nematodes utilize nutrients from the bacteria and decomposing insect for growth and reproduction. Symbiont-derived antibiotics inhibit the growth of other microorganisms (Paul *et al.*, 1981; Akhurst, 1982).

Mass production *in vivo* using *Galleria mellonella* larvae (Dutky *et al.*, 1964; Poinar, 1975) can be used, but the method is labour-intensive and costly.

A. AXENIC CULTURE

Axenic cultures have been achieved on enriched agar overlaid with liver or kidney slices and in various liquid culture media (reviewed by Hansen and Hansen, 1978). Axenic culture has played an important role in the study of the relationship between nematodes deficient in bacteria and their insect hosts (e.g., Burman, 1982; Boemare *et al.*, 1982, 1983), and new nematode–bacteria associations (e.g., Dunphy *et al.*, 1985).

Investigations of specific nutritional requirements of the nematodes require defined liquid media. However, low nematode yields and the high cost of defined media components leave little potential for axenic mass production.

B. MONOXENIC CULTURE

A useful feature of steinernematid and heterorhabditid nematodes is that they can be grown in monoxenic culture on a wide variety of media. Initial methods involved systems in which growth of the nematodes was achieved on the surface of the medium in tubes, Petri plates or trays. Glaser *et al.* (1940) described the culture of *N. glaseri* on fermented potato mash or veal infusion agar. House *et al.* (1965) used a dog-food medium for the culture of the DD-136 strain of *N. carpocapsae*, which was then used by the Nutrilite corporation for its mass production. This corporation also achieved good nematode development using wheat, wheat bran, corn, corn gluten, oatmeal, whole fish and various fly media as the basic constituent of the medium (Laumond *et al.*, 1979).

In response to the need for a more economical culture method, Bedding (1976) devised a system for *Neoaplectana* species in which Aspen wood-wool was coated with medium comprising homogenized kidney or chicken heart. This system provided a large surface area and ample interstitial space. Only *Xenorhabdus* was associated with the nematodes. Although yields of infective juveniles reached 10^8 per 2.0 litre container, production was highly variable and the method was not suitable for *Heterorhabditis* species.

Following these early attempts at monoxenic mass production, Akhurst (1980) found that the symbionts *Xenorhabdus* and *Flavobacterium* occur in two forms. These forms can be differentiated by colony characteristics and just two biochemical tests. Generally only one form, designated primary, is carried by infective stages. The more stable secondary form, which occurs with primaries in cultures initiated only with primaries, is of great significance as its presence is detrimental to the final yield. Bedding (personal communication cited in Akhurst, 1980) found that cultures of *N. feltiae* Agriotos and *H. heliothidis* NZ with the primary form bacteria yielded 1.6 and 2.9 times, respectively, as many infective juveniles as cultures with the secondary form.

More reliable techniques for the culture of both *Neoaplectana* and *Heterorhabditis* species were developed with this new knowledge of symbiont dimorphism and involved monoxenic culture on crumbed polyether polyurethane foam coated with medium (see below; Bedding, 1981, 1984; Wouts, 1981).

1. *Detailed methods*

1° *Neoaplectana* and *Heterorhabditis* species (Bedding, 1981)

(i) ISOLATION AND MAINTENANCE OF SYMBIOTIC BACTERIA

The symbiont can be selected from the bacterial colonies obtained from streaking (a) the haemolymph of recently infected *G. mellonella* larvae (Bedding, 1981), (b) homogenized surface-sterilized infective juveniles [infective juveniles are immersed in 0.1% w/v merthiolate (Ely Lilly and Co.) or 0.5% hyamine for 3 h and 15 min respectively, then washed three times in sterile water (Akhurst, 1980)] or (c) haemolymph from uninfected *G. mellonella* larvae in which surface-sterilized infective juveniles are allowed to exsheath (Poinar, 1966). In each case the selected colonies are added to yeast salts broth (YS: $NH_4H_2PO_4$, 0.5 g; K_2HPO_4, 0.5 g; $MgSO_4 \cdot 7H_2O$, 0.2 g; NaCl, 5.0 g; yeast extract, 5.0 g; water to 1.0 litre (Dye, 1968)). Stock cultures (about 2.5 ml) can be maintained on YS agar (15 g litre^{-1}) at 12°C and subcultured monthly. The bacteria can be stored at −18°C in 17% glycerol in nutrient broth (Bacto-beef extract, 3.0 g; Bacto-peptone, 5.0 g; H_2O to 1.0 litre) or freeze dried in 5% peptone/3% sucrose in YS broth (Akhurst, 1980).

(ii) ESTABLISHMENT AND MAINTENANCE OF MONOXENIC CULTURES

Surface-sterilized infective juveniles (see (i) above) are inoculated on to a 48 h growth of the symbiont on dog-food (a mixture of crude protein > 21%, crude fat > 6%, crude fibre > 6%; moisture < 12%), or nutrient agar slants (Bacto-beef extract, 3.0 g; Bacto-peptone, 5.0 g; agar, 15 g; water, 1.0 litre), and maintained at 20–28°C. Subculture, with a loop-full of nematodes, may be made after 2–4 weeks. When subculturing, it is not necessary to pre-inoculate the slants with the symbiont. For efficient development of *H. heliothidis* on nutrient agar, the nutrient broth content of the agar is doubled and 1% corn oil added (Wouts, 1981).

Preliminary investigations have yielded >60% survival of infective juveniles of *N. carpocapsae* following liquid nitrogen freezing (James *et al.*, unpublished). Procedures for the cryopreservation of steinernematid and heterorhabditid nematode stock cultures are being developed.

2° Mass culture of steinernematids and heterorhabditis
(Bedding, 1981, 1984; Wouts, 1981)

Parts of the technology for mass rearing these nematodes (described below) are protected by US patent 4 334 498 (Bedding, 1982).

Media with a high protein content are used:

		Neoaplectana spp.	*Heterorhabditis* spp.
(i)	Bedding (1981)	%	%
	Homogenized pig kidney	70.0	60.0
	Melted beef fat	10.0	20.0
	Tap water	20.0	20.0

(ii) Wouts (1981)		
Soy flour		25.45
Nutrient broth		1.1
Yeast extract		0.36
Corn oil		11.8
Tap water		61.0
(iii) Bedding (1984)		
Homogenized chicken offal	80.0	70.0
Melted beef fat	—	10.0
Tap water	20.0	20.0

The media components of Bedding (i and iii) are mixed in a blender and incorporated (by hand squeezing) into crumbed polyether polyurethane foam in a medium-to-foam ratio of 12 : 1 (w/w). The media components of Wouts (ii) are mixed and cooked to a smooth roux and then incorporated into foam in a medium-to-foam ratio of 4.4 : 1 (w/w).

Several possible containers have been described as detailed below.

500 ml Erlenmeyer flasks (Bedding, 1981). 700 g of foam plus medium [2°, (i)] are added to a flask and autoclaved for 3 h. Each flask is inoculated with 10 ml of a YS broth shake culture (20–30°C) of the symbiont and incubated at 20–30°C for 2–3 d. Each flask is then inoculated with approximately 4 g of foam from a mature culture flask or with the entire contents of a culture tube and incubated at 20–28°C. Maximum yields

TABLE I. Yields of infective nematodes from 500 ml flasks of kidney/fat homogenate on polyether polyurethane crumbed foam, pre-inoculated with primary form bacterial symbiont (from Bedding, 1981)

Nematode species	No. of flasks	Average yield	Range of yields
N. carpocapsae (Leningrad)	30	38×10^6	$29–55 \times 10^6$
N. bibionis (Tasmania)	73	29×10^6	$20–40 \times 10^6$
N. glaseri (N. Carolina)	11	8×10^6	$6.5–10 \times 10^6$
H. bacteriophora (S. Australia)	10	36×10^6	$23–55 \times 10^6$
H. heliothidis (New Zealand)	15	32×10^6	$26–35 \times 10^6$
H. heliothidis (Tasmania)	8	27×10^6	$23–35 \times 10^6$
Heterorhabditis sp.	9	49×10^6	$41–56 \times 10^6$

are obtained for *Neoaplectana* species in 2–3 weeks; the infective stages should be harvested within this time as mortality is high afterwards. Maximum yields for *Heterorhabditis* species are obtained in 4–5 weeks

Fig.1. (A) Homogenized chicken offal is added to crumbed polyether polyurethane foam. (B) Homogenate is thoroughly mixed with foam. (C) Foam/homogenate is added to bag via funnel to protect bag edges for sealing. (D) Autoclavable plastic bag is sealed with an impulse sealer. (E) Culture bags protected in sacks are autoclaved horizontally after evacuation of air. (From Bedding, 1984.)

(see Table I). *Heterorhabditis* infective stages survive well for several months in flask cultures kept at 12°C.

Wouts (1981) described a similar method for the culture of *H. heliothidis* in 250 ml Erlenmeyer flasks containing medium (ii) and obtained average yields of 10^6 infective juveniles per flask.

Aerated polypropylene culture bags (Bedding, 1984). This method was described for the culture of *N. bibionis*, but is applicable to other neoaplectanid and heterorhabditid species. The stages in the process from incorporation of medium into foam to harvest and storage are illustrated in Figures 1 and 2. Bags (1.0 × 0.5 m) are constructed from

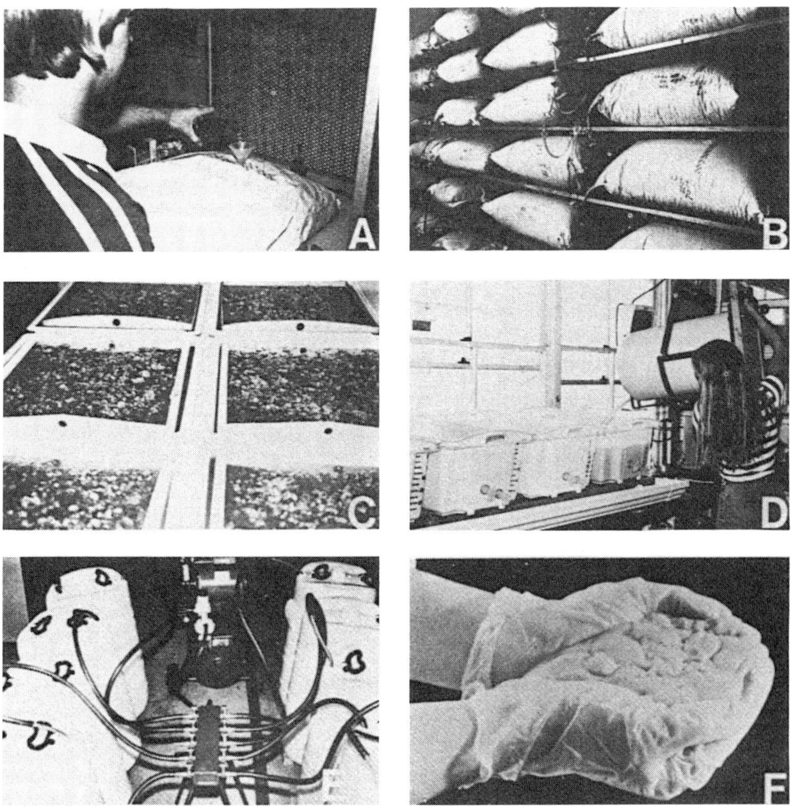

Fig.2. (A) Bacterial inoculum is added to sterilized culture bags in laminar flow cabinet. (B) Culture bags are inflated and aerated with sterile air during incubation. (C) Foam standing in sieves, immersed in water, for harvesting. (D) Nematodes are cleaned by sedimenting in tanks. Decanting is aided by taps and guttering. (E) Nematode suspensions are vigorously aerated in inclined drums to allow break-down of remaining medium. (F) Cleaned nematodes from a single culture bag.

autoclavable polypropylene tubing with air inlet and outlet tubes and an inoculation panel of autoclave tape. The bags are loaded with 3.0 or 5.0 kg of chicken offal medium [2°(iii)] on 240 g (for *Neoaplectana*), or 400 g (for *Heterorhabditis*), of crumbed foam. Sealed bags are autoclaved for 1 h. 100 ml of symbiont inoculum (2 d YS-broth shake culture) kg^{-1} is added to the bags and mixed with the foam. The bags are connected to a filtered air supply (1.0 litre min^{-1}) and incubated for 5 d at 23 °C. They are then inoculated with one flask culture for every 3.0 g of medium, reconnected to the air supply and incubated at 23 °C for 2 weeks (*Neoaplectana*) or 3–4 weeks (*Heterorhabditis*). Maximum yields achieved for *N. bibionis*, *N. glaseri*, *N. feltiae* (= *N. carpocapsae*) and *Heterorhabditis* sp. are 2000, 700, 1550 and 1700, respectively.

3° Harvesting and storage of steinernematids and heterorhabditids
(Bedding, 1984)

Foam cubes 5.0 cm deep are placed on stainless steel sieves (mesh size 0.6 mm) standing in trays and just covered with tap water. Within 2 h 95% of the nematodes will migrate out into the water and are washed twice by sedimentation and re-suspension in tap water. The final nematode suspension is aerated in a tilted drum for 1 week to allow for the breakdown of any remaining medium particles. Infective nematodes are concentrated to a paste and incorporated into crumbed polyether polyurethane foam and stored in aerated plastic bags at 4 °C.

ACKNOWLEDGEMENTS

Thanks are due to W. Lanier for valuable discussion during the preparation of this manuscript.

REFERENCES

Akhurst, R.J. (1980). Morphological and functional dimorphism in *Xenorhabdus* spp., bacteria symbiotically associated with the insect pathogenic nematodes *Neoaplectana* and *Heterorhabditis*. *Journal of General Microbiology* **121**, 303–309.

Akhurst, R.J. (1982). Antibiotic activity of *Xenorhabdus* spp., bacteria symbiotically associated with insect pathogenic nematodes of the families Heterorhabditidae and Steinernematidae. *Journal of General Microbiology* **128**, 3061–3065.

Akhurst, R.J. (1983). *Neoaplectana* species: specificity of association with bacteria of the genus *Xenorhabdus*. *Experimental Parasitology* **55**, 258–263.

Bedding, R.A. (1974). Five new species of *Deladenus* (Neotylenchidae) entomophagous mycetophagous nematodes parasitic in siricid wood-wasps. *Nematologica* **20**, 204–225.

Bedding, R.A. (1976). New methods increase the feasibility of using *Neoaplectana* spp. (Nematoda) for the control of insect pests. *In* "Proceedings of the 1st International colloquium of Invertebrate Pathology, Kingston, Canada", pp. 250–254.

Bedding, R.A. (1981). Low cost *in vitro* mass production of *Neoaplectana* and *Heterorhabditis* species (Nematoda) for field control of insect pests. *Nematologica* **27**, 109–114.

Bedding, R.A. (1982). Assemblage for rearing nematodes. US patent 4 334 498.

Bedding, R.A. (1984). Large scale production, storage and transport of the insect parasitic nematodes *Neoaplectana* and *Heterorhabditis* spp. *Annals of Applied Biology* **104**, 117–120.

Boemare, N., Laumond, C. and Luciani, J. (1982). Mise en évidence d'une toxicogenèse provoquée par le nématode axénique entomophage *Neoaplectana carpocapsae* Weiser chez l'insecte axenique *Galleria mellonella* L. *Comptes rendus hébdomonadaire des Séances de l'Academie des Sciences. Series III* **294**, 543–546.

Boemare, N., Bonifassi, E., Laumond, C. and Luciani, J. (1983). Etude expérimentale de l'action pathogène du nématode *Neoaplectana carpocapsae* Weiser; recherches gnotobiologiques chez l'insecte *Galleria mellonella* L. *Agronomie* **3**, 407–415.

Buecher, E.J. and Hansen, E.L. (1971). Mass culture of axenic nematodes using continuous aeration. *Journal of Nematology* **3**, 199–200.

Burman, M. (1982). *Neoaplectana carpocapsae*: toxin production by axenic insect parasitic nematodes. *Nematologica* **28**, 62–70.

Castillo, J.M., Chin, P. and Roberts, D.W. (1982). Growth and development of *Romanomermis culicivorax in vitro*. *Journal of Nematology* **14**, 476–485.

Dunphy, G.B., Rutherford, T.A. and Webster, J.M. (1985). Growth and virulence of *Steinernema glaseri* influenced by different subspecies of *Xenorhabdus nematophilus*. *Journal of Nematology* **1**, 476–482.

Dutky, S.J., Thompson, J.V. and Cantwell, G.E. (1964). A technique for the mass propagation of the DD-136 nematode. *Journal of Insect Pathology* **6**, 417–422.

Dye, D.W. (1968). A taxonomic study of the genus *Erwinia*. I. The "amylovora" group. *New Zealand Journal of Science* **11**, 590–607.

Fassuliotis, G., and Creighton, C.S. (1982). *In vitro* cultivation of the entomogenous nematode *Filipjevimermis leipsandra*. *Journal of Nematology* **14**, 126–131.

Fedorko, A. and Stanuszek, S. (1971). *Pristionchus uniformis* sp. n. (Nematoda, Rhabditida, Diplogasteridae) a facultative parasite of *Lepinitarsa decemlineata* Say and *Melolontha melolontha* L. in Poland. Morphology and biology. *Acta Parasitologica Polonica* **19**, 95–112.

Finney, J.R. (1976). The *in vitro* culture of mermithid parasites of blackflies and mosquitoes. *Journal of Nematology* **8**, 284.

Finney, J.R. (1977). The development of *Romanomermis culicivorax* in *in vitro* culture. *Nematologica* **23**, 479–480.

Finney, J.R. (1981). Mermithid nematodes: *In vitro* culture attempts. *Journal of Nematology* **13**, 275–280.

Giblin, R.M. and Platzer, E.G. (1986). Mass surface sterilization and *in vitro* culture of *Romanomermis culicivorax*. *Revue de Nématologie* **10**, 67–74.

Glaser, R.W., McCoy, E.E. and Girth, H.B. (1940). The biology and economic importance of a nematode parasitic in insects. *Journal of Parasitology* **26**, 479–495.

Gordon, R. and Webster, J.M. (1972). Nutritional requirements for protein synthesis during parasitic development of the entomophilic nematode *Mermis nigrescens. Parasitology* **64**, 161–172.

Gordon, R., Finney, J.R., Condon, W.J. and Rusted, T.N. (1979). Lipids in the storage organs of three mermithids and in the haemolymph of their hosts. *Comparative Biochemistry and Physiology* **64B**, 585–590.

Gordon, R., Condon, W.J. and Squires, J.M. (1980). Sterols in the trophosomes of the mermithid nematodes *Neomesomermis flumenalis* and *Romanomermis culicivorax* relative to sterols in the host haemolymph. *Journal of Parasitology* **66**, 585–590.

Gordon, R., Squires, J.M., Babie, S.J. and Burford, I.R. (1981). Effects of host diet on *Romanomermis culicivorax*, a mermithid parasite of mosquitoes. *Journal of Nematology* **13**, 285–290.

Hansen E.L. and Hansen, J.W. (1978). Nematoda parasitic in animals and plants. *In* "The Cultivation of Parasites *in vitro*." (A.E.R. Taylor and J.R. Baker, eds). Academic Press, London and New York.

House, H.L., Welch, H.E. and Cleugh, T.R. (1965). Food medium of prepared dog biscuit for the mass production of nematode DD-136 (Nematoda: Steinernematidae). *Nature* **4986**, 847.

Laumond, C., Mauleon, H. and Kernarrec, A. (1979). Donnes nouvelles sur le spectre d'hôtes et le parasitisme du nématode entomophage *Neoaplectana carpocapsae. Entomophaga* **24**, 13–27.

Mitsuhashi, J. and Maramosche, K. (1964). Aseptic cultivation of four virus-transmitting species of leafhoppers (Cicadellidae) *Contributions, Boyce Thompson Institute* **22**, 165–173.

Mracek, Z. (1977). *Steinernema kraussi*, a parasite of the body cavity of the saw fly *Cephalia abietis* in Czechoslovakia. *Journal of Invertebrate Pathology* **30**, 87–94.

Myers, R.F. (1973). Third biannual report, Research Unit on Vector Pathology, Memorial University, St Johns, Newfoundland, pp. 9–10.

Myers, R.F. (1975). Annual Report, Research Unit on Vector Pathology, Memorial University, St Johns, Newfoundland, pp. 9–10.

Paul, V.J., Frautschy, S., Fenical, W. and Nealson, K.H. (1981). Isolation and structure assignment of several new antibacterial compounds from the insect symbiotic bacteria *Xenorhabdus* spp. *Journal of Chemical Ecology* **7**, 589–598.

Petersen, J.J. (1973). Role of mermithid nematodes in biological control of mosquitoes. *Experimental Parasitology* **33**, 239–247.

Petersen, J.J. (1977). Effects of host size and parasite burden on sex ratio in the mosquito parasite *Octomyomermis muspratti. Journal of Nematology* **9**, 343–346.

Petersen, J.J. and Willis, O.R. (1972). Procedures for the mass rearing of a mermithid parasite of mosquitoes. *Mosquito News* **32**, 226–230.

Poinar, G.O. Jr (1966). The presence of *Achromobacter nematophilus* in the infective stage of *Neoaplectana* sp. (Steinernematidae: Nematoda). *Nematologica* **12**, 105–108.

Poinar, G.O. Jr (1975). "Entomogenous Nematodes. A manual and host list of insect nematode associations", pp. 22–28. E.J. Brill, Leiden.

Poinar, G.O. Jr. and Hess, R. (1976). Uptake of ferritin particles through the body wall of a mermithid nematode. *International Research Communications System/Medical Sciences* **4**, 296.

Poinar, G.O. Jr. and Hess, R. (1977). *Romanomermis culicivorax*: morphological evidence of transcuticular uptake. *Experimental Parasitology* **42**, 27–33.

Roberts, D.W. and Van Leuken, W. (1973). Limited growth of a mermithid nematode cultured in association with tissue-culture cells. *International Colloquium on Insect*

Pathology and Microbiological Control. Oxford, England, p. 91. [Abstract].

Rutherford, T.A. and Webster, J.M. (1974). Transcuticular uptake of glucose by the entomophilic nematode *Mermis nigrescens*. *Journal of Parasitology* **60**, 804–808.

Sanders, R.D., Stokstad, E.L.R. and Malatesta, C. (1973). Axenic growth of *Reesimermis nielseni* (Nematoda: Mermithidae) in insect tissue culture media. *Nematologica* **19**, 567–568.

Swain R.B. (1945). The association of nematodes of the genus *Diplogaster* with white fringed beetles. *Journal of Economic Entomology* **38**, 488–490.

Taylor, A.E.R. and Baker, J.R. (1968). "The Cultivation of Parasites *In Vitro*". Blackwell Scientific Publications, Oxford.

Taylor, A.E.R. and Baker, J.R. (1978). "Methods of Cultivating Parasites *In Vitro*". Academic Press, London and New York.

Thomas, G.M. and Poinar, G.O. Jr. (1979). *Xenorhabdus* gen. nov., a genus of entomopathogenic, nematophilic bacteria of the family Enterobacteriaceae. *International Journal of Systematic Bacteriology* **29**, 352–360.

Wouts, W.M. (1981). Mass production of the entomogenous nematode *Heterorhabditis heliothidis* (Nematoda: Heterorhabditidae) on artificial media. *Journal of Nematology* **13**, 467–469.

Chapter 15

Acanthocephala

D.W.T. CROMPTON and OLIVIA L. LASSIÈRE

Zoology Department, The University, Glasgow, Scotland

I. INTRODUCTION

All known species of the phylum Acanthocephala are dioecious endoparasitic worms with an indirect life-history involving an arthropod intermediate host and a vertebrate definitive host; sometimes paratenic hosts are found (see Crompton and Nickol, 1985). The adult worms attain sexual maturity in the small intestine (Crompton, 1985) and it may be assumed that their restriction to this habitat is related to their lack of a gut of their own and a dependence on the host's digestive processes for the supply of materials for tegumentary absorption (Starling, 1985). Three classes are recognized within the phylum. The Eoacanthocephala generally parasitize poikilothermic definitive hosts, the Palaeacanthocephala are found in both poikilothermic and homoiothermic definitive hosts, while the Archiacanthocephala tend to be restricted to homoiothermic hosts. Generalizations like these must be accepted with caution because, although only about 1000 species of Acanthocephala have been described, very few are known in detail.

Acanthocephalan development begins when mature eggs (containing acanthors) pass out of the definitive host. The acanthor is already infective to the intermediate host before the egg is discharged by the parent female worm, so no further development occurs until ingestion by the intermediate host. The acanthor burrows into the haemocoel and metamorphoses through a series of acanthella stages to the cystacanth or infective juvenile (Schmidt, 1985). Cystacanths of *Moniliformis moniliformis*, which develops in cockroaches, appear to retain their infectivity to rats for about as long as the cockroaches survive (Crompton *et al.*, 1984).

Acanthocephalan worms take a relatively long time to develop and mature, particularly in poikilothermic hosts living at low ambient temperatures, and this together with the complexity of the alimentary environment experienced in definitive hosts (Mettrick and Podesta, 1974; Crompton and Nesheim, 1976) and the remarkable sensitivity of the worms to host dietary changes (Crompton, 1987) makes these parasites particularly difficult subjects for *in vitro* culture. Nevertheless some progress has been made since Lackie's summary almost a decade ago (1978). Most of the impetus has come from detailed studies of acanthocephalan physiology leading to better techniques for their short-term maintenance *in vitro*. These recipes and methods are listed (Section II) together with information about attempts to achieve their long-term *in vitro* cultivation (Section III). The importance of establishing reliable criteria for the assessment of growth is well recognized. Thus measurements of changes in length or wet weight are perhaps less satisfactory than measurements of changes in dry weight: tissue analysis (Crompton, 1972), coupled with monitoring of morphological and cytological changes, is even better.

II. SHORT-TERM MAINTENANCE *IN VITRO*

Some information about a range of procedures for the *in vitro* maintenance of Acanthocephala for up to 12 h is summarized in Table I.

III. LONG-TERM MAINTENANCE *IN VITRO*

There have been numerous attempts to culture acanthocephalans, from all three classes, for extended periods *in vitro*. Successful culture would provide valuable information about the parasites' biology (e.g., attainment of maturity, fecundity, mating regime, effect of intraspecific competition for nutrients) and hence lead to a better understanding of field observations. Further to this, axenic culture in a medium of known composition would allow controlled metabolic experiments to be carried out. Detailed methods for *M. dubius* (= *M. moniliformis*) (adults: Nicholas and Grigg, 1965; larvae: Lackie and Rotheram, 1972; Lackie, 1975) and for *Pomphorhynchus bulbocolli* culture (Jensen, 1952) have already been documented by Lackie (1978, pp. 281–284).

TABLE I. Details of the short-term maintenance of *Acanthocephala in vitro*

Parasite[a]	Stage	Medium[b]	Conditions			Sterility[c]	Duration (h)	Objective	Reference
			Temp(°C)	pH	Gas				
Archiacanthocephala									
M *Moniliformis moniliformis*	Adult	Tyrode's saline (± glucose)	37.4	7.5	N_2	cbp 1 mg ml^{-1} ss 1 mg ml^{-1}	<2.5	Metabolic study	Ward and Crompton (1969)
	Adult	KRB	37.0	7.5	$N_2 : CO_2$ (95:5)	ppG 500 iu ml^{-1} ss 75 μg ml^{-1}	4	Metabolic study	Körting and Fairbairn (1972)
	Adult	Tyrode's saline + glucose + amino acids	37.0	6.88	Air	bp 1 mg ml^{-1} ss 1.7 mg ml^{-1}	3	Metabolic study	Crompton and Ward (1984)
	Larval	KRT	37.0	7.4	Air 5% CO_2 in air N_2 5% CO_2 in N_2	—	4	Glycogen metabolism	Hovarth (1971)
M *Macracanthorhynchus hirudinaceus*	Adult	30% artificial sea water + 0.1 M glucose and sucrose	30.0	7.0	Air	—	—	Electro-physiology	Wong *et al.* (1979)
Palaeacanthocephala									
B *Polymorphus minutus*	Adult	Hanks's saline + glucose	41.7±0.5	7.8	N_2	cbp 1 mg ml^{-1} ss 1 mg ml^{-1}	4	Metabolic study	Crompton and Ward (1967)
Eoacanthocephala									
F *Paulisentis fractus*	Adult	STG	22–25	—	Air	Washed in 0.75% NaCl with ppG 200 iu ml^{-1} ss 1 mg ml^{-1}	12	Metabolic study	Hibbard and Cable (1968)

[a] M = mammal definitive host
B = bird definitive host
F = fish definitive host

[b] KRB = Krebs-Ringer bicarbonate
KRT = Krebs-Ringer saline with 0.025M tris (hydroxymethylamino) methanemalate
STG = saline tritiated glucose solution

[c] bp = benzylpenicillin
cbp = crystapen benzylpenicillin
ppG = potassium penicillin G
ss = streptomycin sulphate
iu = international units

1. Detailed techniques

1° Acanthellae: *Moniliformis moniliformis* (Archiacanthocephala)

(A.M. Lackie, 1981; personal communication)

STERILIZATION AND MEDIUM

(i) *Sterilization*. Swab infected cockroach (*Periplaneta americana*) with 70% alcohol. Gently remove the crop from the abdomen and discard; this reduces contamination by gut contents. Open abdomen in a 3 cm Petri dish under Hills' saline (mg dl^{-1}: NaCl, 760.5; KCl, 74.5; $MgCl_2.6H_2O$, 40.7; $CaCl_2.2H_2O$, 43.8; $NaH_2PO_4.2H_2O$, 9.63; $NaHCO_3$, 33.6; pH 7.2). Collect larvae from bottom of dish and transfer to sterile Hills' saline in a sterile conical plastic tube (15 ml). Wash three times in Hills' saline allowing the larvae to settle under gravity. Handle the larvae carefully to minimize damage to the envelopes which surround them. To prevent "stickiness" add medium + 5% heat-inactivated foetal calf serum (FCS) in the last wash and transfer to a clean sterile conical tube. **Take sterile precautions from now on**. Wash three times in medium + 5% FCS, transfer to a clean tube and wash three times more.

(ii) *Medium*. Modified D73 (I.M. Huxham and A.M. Lackie, unpublished), which is based on a recipe devised by Quiot *et al*., 1985, for the maintenance of cells from the cockroach *Gromphadorhina laevigata*.

Amino acids	mg dl^{-1}
L-Arginine-HCl	30
L-Aspartic acid	25
L-Cysteine-HCl(BDH)	26
L-Glutamic acid	95
Glycine	15
L-Histidine-HCl	26
L-Isoleucine	24
L-Leucine	25
L-Methionine	50
L-Phenylalanine	20
L-Proline	67
L-Serine	8
L-Threonine	20
L-Tryptophan	20
L-Tyrosine	10
L-Valine	15

Salts	mg dl^{-1}
NaCl	600
KCl	80
$MgSO_4.H_2O$	18
Na_2HPO_4	5.0
KH_2PO_4	6.0
$CaCl_2.2H_2O$	14
Other components	
D-Glucose	100 mg dl^{-1}
HEPES	476 mg dl^{-1}
Stretomycin sulphate	10 mg dl^{-1}
Penicillin (sodium salt)	0.15 mg dl^{-1}
Cyanocobalamine	0.5 μg
Flow BME (basal medium Eagle) vitamins (× 100)	0.5 ml
Gibco medium 199* (× 10) without bicarbonate and glutamine	4.0 ml
Gentamicin sulphate	100 μg ml^{-1}

*See Taylor and Baker (1968, p. 355) for ingredients of medium 199.

Adjust pH to 7.2–7.3 with NaOH and make up to 100 ml with distilled water. Filter sterilize medium with a 0.22 μm Millipore filter. *Note*: all amino acids Sigma unless marked otherwise.

TECHNIQUE

Transfer sterilized larvae to wells in Nunc plastic culture dish (24-well multidish; supplied sterile by Gibco Ltd) as follows: stage I acanthellae — up to 50 per well; stage III acanthellae or cystacanths — 20–30 per well. Developmental stages are defined by King and Robinson (1967) and J.M. Lackie (1972). Each well contains 750 μl medium plus 10% FCS. Place culture tray in a plastic sandwich box and incubate in the dark at 25–28 °C. Replace half of the medium every 2–3 d.

RESULTS

Growth was assessed visually in terms of changes in size and morphology. Stage II acanthellae developed to stage V.

2° Cystacanths: *Moniliformis moniliformis* (Archiacanthocephala)

(Tobias and Schmidt, 1977)

STERILIZATION AND MEDIUM

(i) *Sterilization*. Dissect cystacanths from infected *P. americana*, momentarily wash in 0.0625% Zephiran chloride (Roberts, 1973) and hold in 0.85% NaCl. Remove the transparent envelope surrounding the cystacanth with *"minuten nadeln"*. Quickly rinse "uncovered" cystacanths individually in Zephiran chloride and then immerse in sterile 0.85% NaCl. Activate cystacanths by exposure to 0.05% solution of sodium taurocholate in Krebs–Ringer saline at pH 8.5 (adjust with 0.1 M glycine-NaOH) at 37 °C for 15 min as described by Graff and Kitzman (1965).

Sterilize all media with 300 mg ml^{-1} streptomycin sulphate, 300 iu ml^{-1} penicillin G (Voge, 1967) and 50 mg ml^{-1} tetracycline-HCl; adjust pH to 7.5 with 10% $NaHCO_3$.

(ii) *Media*. (a) "NCTC 199 with 10% inactivated calf serum (ICS) and 1% glucose". (b) Medium (a) supplemented with 20% hen's egg yolk (see Berntzen and Macy, 1969 in Taylor and Baker, 1978, p. 178 for method of preparation). (c) Medium (a) supplemented with 10% rat intestinal extract (RIE). Prepare RIE from the anterior half of the small intestine of necropsied rats. Open the intestinal pieces lengthwise, remove the contents and wash five times in sterile 0.85% NaCl. Homogenize 1 part of washed intestinal pieces with 4 parts of sterile triple-distilled water (1 : 4, w/v) for 60 s. Centrifuge (speed not given) the homogenate for 1 h at room temperature. Remove the supernatant RIE and place in sterile culture tubes and store at 4 °C. Mix medium a with RIE in a 9 : 1 ratio (v/v).

TECHNIQUE

Place pairs of activated cystacanths into 16 × 125 mm Kimax culture tubes, with Teflon-lined screw caps, containing 12 ml of medium pre-incubated to 37 °C. Gas the tubes for 5 min with 90% N_2, 10% CO_2, seal with parafilm and incubate at a slight incline at 37 °C. Change medium three times a week by replacing 5.0 ml of old medium.

RESULTS

In NCTC 199 with 10% ICS, 1% glucose and 20% egg yolk, the worms maintained normal body form and motility throughout the 55 d experimental period and showed a 135% average change in standard length[1]. Growth occurred in all three media but at a slower rate than *in vivo*. No growth of ovaries or testes was seen.

COMMENT

Tanaka and MacInnis (1980) analysed the body fluid of adult *Moniliformis moniliformis* with a view to the *in vitro* culture of the ovaries. Their findings may be helpful for future attempts to grow or maintain this species *in vitro*.

3° Cystacanth: *Corynosoma constrictum* (Palaeacanthocephala)
(Young and Lewis, 1977)

MEDIUM

Chick chorioallantoic membrane (CAM).

TECHNIQUE

Cystacanth is implanted on to CAM.

[1]Standard length = length of trunk not including the presoma or the protruding male bursa.

RESULTS

Beginning of somatic and germinal growth was seen after 3 d on CAM. The standard proboscis length[2] increased 116% and 32–78% increases in other somatic measurements were observed. After 3 d the lengths and widths of the testes increased by 80% and 91% respectively.

COMMENT

Surprisingly, this study seemed to involve a solitary cystacanth of *C. constrictum*; much more work is needed since the technique seems promising. Cystacanths of *Polymorphus paradoxus* and *P. marilis* survived, but failed to grow, on the CAM for 9 d and 7 d respectively.

4° Cystacanths: *Pomphorhynchus laevis* • (Palaeacanthocephala)
(Broughton, 1974)

STERILIZATION AND MEDIUM

(i) *Sterilization*. Dissect cystacanths from infected *Gammarus pulex* (Amphipoda) and rinse thoroughly before transferring to sterile Medium 199 (see Taylor and Baker, 1968, pp. 355–356) containing Fungizone (Squibb). Make up a stock solution of Fungizone by adding 1 phial, containing 50 mg amphotericin B solubilized with 41 mg sodium deoxycholate with phosphate buffer, to 100 ml with sterile distilled water.

(ii) *Medium*. Medium 199 (Wellcome Reagents Ltd: available in 100 ml sterile bottles).

TECHNIQUE

(i) In a sterile cabinet, add 12 ml aliquots of sterile Medium 199 to 30 ml sterile Falcon plastics culture flasks with a solid phase of Lab-Lemco agar (Difco) to provide a substrate for the worms to embed into. Aseptically transfer 10 sterilized cystacanths, using a sterile Pasteur pipette, to each culture flask and close caps tightly. Incubate at 10 °C. Change the medium, as follows, every 3–4 d **in a sterile cabinet**:

Place the parasites and old medium in a sterile plastic Petri dish.
Draw off the old medium with a sterile Pasteur pipette.
Wash the parasites twice in fresh sterile medium.

[2]Standard proboscis length = an average of measurements of the proboscides of cystacanths.

Add fresh medium to a new culture flask and add the washed parasites using flame-sterilized fine forceps.

Return the tightly capped culture flask to a cooled incubator at 10 °C. After 28 d collect all the worms that are unaffected by fungal growth and wash several times in sterile Medium 199. Place in a fresh culture flask containing fresh sterile Medium 199 only. Incubate flasks at 6 °C and change the medium weekly as described.

(ii) Alternatively, dissect cystacanths from *Gammarus* and incubate in non-sterile Medium 199 at 6 °C and change the medium at weekly intervals.

RESULTS

With culture technique (i) the parasites remained active and normal in appearance throughout the 63 d experimental period. Dry weight readings indicated that no increase in body mass had taken place. In the females, the ovaries showed no sign of breaking down into ovarian balls and no male worms were seen to evert the bursa.

With culture technique (ii) the parasites remained active and normal throughout the 33 d experimental period but no growth was recorded.

COMMENT

Broughton's procedure was based initially on that devised by Jensen (1952) which allowed *P. bulbocolli* to grow *in vitro* from acanthellae to prepatent adults. However, Broughton found the type of glassware used by Jensen to be unsatisfactory, so she used Falcon Plastics' culture flasks instead.

5° Adults: *Neoechinorhynchus emydis* and *N. pseudemydis* (Eoacanthocephala)

(Dunagan, 1962)

STERILIZATION AND MEDIUM

(i) *Sterilization*. Remove worms from the turtle definitive hosts (*Pseudemydis scripta*, *Graptemys geographica* or *G. pseudogeographica*) and wash three times in 0.7% NaCl by shaking in a test tube and removing fluid. Transfer to a Petri dish and separate the worms according to sex and species. Discard any worms which fail to evert the proboscis (to avoid contamination by organisms that could be retained within the retracted proboscis). Separate the mature worms from the immature and "old" worms. Pass the mature worms through three changes of sterile Tyrode's balanced salt solution (T-BSS; see Taylor and Baker, 1968, p. 350) containing 0.9% NaCl. The first two changes also contain 200 iu ml^{-1} of penicillin G potassium and 1 mg ml^{-1} of streptomycin sulphate.

(ii) *Media.*

(a) *N. emydis*: Eagle's HeLa medium (see Taylor and Baker, 1968, p. 354) at pH 8.0 plus 20% calf serum.

(b) *N. pseudemydis*: use T-BSS at pH 8.0 with 0.1% glucose and 11% turtle serum. Turtle serum is obtained aseptically from the ventricle of the heart of *Pseudemydis scripta* (available from commercial suppliers; address not given) after a hole has been drilled through the plastron. 10–15 ml of blood can be collected from a 10-inch turtle at weekly intervals and the hole in the plastron can be closed with a collodion-coated cork stopper. If not used immediately, serum can be stored at −20°C.

TECHNIQUE

After the third wash in sterile T-BSS place worms, in groups of four, into culture tubes. Two types of culture tubes can be used: one is a modification of that used by Smyth (1946, 1955) containing dialysing tubing (24° Å holes) within which the worms are placed with the T-BSS; the other is a screw-capped 8.0 ml test tube in which the worms are in direct contact with 2.0 ml of medium. Incubate the culture tubes at 22–23°C without agitation except for brief shaking before daily examination and change the medium every 7 d. When the worms fail to move within 10 min of being shaken, on two consecutive days, cultures are considered to be at an end.

RESULTS

In medium (a), *N. emydis* remained motile for between 26 and 96 d (average 61 d). No growth was observed, but copulation and egg maturation occurred. Female worms, examined on 51st and 73rd days of survival, contained motile sperm attached to the developing ovarian balls and also eggs with no obvious abnormalities were seen.

In medium (b), *N. pseudemydis* remained motile for between 65 and 88 d (average 75 d).

Over the period of culture, all worms in either medium gradually swelled up and became immobile. The cuticle became opaque and blisters appeared on the body surface.

6° Adults: *Octospinifer macilentis* (Eoacanthocephala) (Harms, 1965)

STERILIZATION AND MEDIUM

(i) *Sterilization*. Dissect adult worms (those 3–5 mm long appear most suitable) from the common sucker (*Catostomus commersoni*) and place them in sterile 0.85% NaCl (pH adjusted to 7.8 with $NaHCO_3$) for 30 min to allow them to free themselves of debris. Disinfect the adults by washing them in sterile buffered saline containing the following: penicillin G, 1000 iu ml^{-1}; streptomycin sulphate, 1000 μg ml^{-1}; myostatin, saturated solution, 0.015 ml^{-1}; and Zephiran chloride-Benzalkonium chloride, 1 in 1000 solution, 0.015 ml^{-1}. Using a sterile wire loop to manipulate the worms, pass them through three changes of solution allowing 20 min in each.

(ii) *Medium*. Sterilize the media with 500 iu ml^{-1} penicillin G and 500 mg ml^{-1} streptomycin sulphate.

(a) Basal medium (medium 8 of Jensen, 1952): Difco Neopeptone, 1.8 g; Difco yeast extract, 0.2 g; glucose, 2.0 g; horse serum, 40 ml; NaCl, 1.04 g; KCl, 0.04 g; $CaCl_2$, 0.048 g; and H_2O, 166 ml.

(b) To 8 parts of basal medium add 2 parts of minimal essential medium (MEM)-vitamin mixture (Microbiological Associates).

(c) Mix 7 parts of basal medium with 2 parts of MEM-vitamin mixture and 1 part of MEM-amino acid mixture (Microbiological Associates).

Add sodium bicarbonate solution to media (b) and (c) to adjust the pH to the range where phenol red changes from red to yellow (about pH 7.8). Media can be stored at 4°C in 20 ml screw-cap tubes for up to 3 weeks. Check sterility of the media in thioglycolate broth prior to use.

Note. The two most successful media of the 28 tested by Harms (1965) are given here.

TECHNIQUE

Place a single worm into a 1 dram homeopathic vial with a rubber serum stopper containing 2.0 ml of either medium. Change the medium every 3 d using a syringe with a 23-gauge needle and incubate at 20°C. Use the apparatus described by Harms (1965) to observe and measure the worms *in vitro*.

RESULTS

The worms maintained normal motility and appearance in both media. In medium (b) the worms survived for up to 58 d and an 18% increase in standard length[3] was recorded. In medium (c) the worms survived for up to 43 d with a 45% increase in standard length.

ACKNOWLEDGEMENTS

We thank Dr A.M. Lackie for kindly allowing us to include details of her unpublished method for the *in vitro* culture of *M. moniliformis* and Dr C.R. Kennedy for providing information about Broughton's work on the *in vitro* culture of *P. laevis*.

REFERENCES

Berntzen, A.K. and Macy, R.W. (1969). *In vitro* cultivation of the digenetic trematode *Sphaeridiotrema globulus* (Rudolphi) from the metacercarial stage to egg production. *Journal of Parasitology* **55**, 136–139.

Broughton, P.F. (1974). Specificity of *Pomphorhynchus laevis* (Acanthocephala) with special reference to growth in trout and *in vitro*. MSc dissertation, University of Exeter.

Crompton, D.W.T. (1972). The growth of *Moniliformis dubius* (Acanthocephala) in the intestine of male rats. *Journal of Experimental Biology* **56**, 19–29.

Crompton, D.W.T. (1985). Reproduction. *In* "Biology of the Acanthocephala" (D.W.T. Crompton and B.B. Nickol, eds), pp. 213–271. Cambridge University Press.

Crompton, D.W.T. (1987). Host diet as determinant of parasite growth, reproduction and survival. *Mammal Review* (in press).

Crompton, D.W.T. and Nesheim, M.C. (1976). Host–parasite relationships in the alimentary tract of domestic birds. *Advances in Parasitology* **14**, 96–194.

Crompton, D.W.T. and Nickol, B.B. (1985), "Biology of the Acanthocephala". Cambridge University Press.

Crompton, D.W.T. and Ward, P.F.V. (1967). Lactic and succinic acids as excretory products of *Polymorphus minutus* (Acanthocephala) *in vitro*. *Journal of Experimental Biology* **46**, 423–430.

Crompton, D.W.T. and Ward, P.F.V. (1984). Selective metabolism of L-serine by *Moniliformis* (Acanthocephala) *in vitro*. *Parasitology* **89**, 133–144.

Crompton, D.W.T., Keymer, A.E. and Arnold, S.E. (1984). Investigating over-dispersion: *Moniliformis* (Acanthocephala) and rats. *Parasitology* **88**, 317–331.

Dunagan, T.T. (1962). Studies on *in vitro* survival of Acanthocephala. *Proceedings of the Helminthological Society of Washington* **2**, 131–135.

Graff, D.J. and Kitzman, W.B. (1965). Factors influencing the activation of acanthocephala cystacanths. *Journal of Parasitology* **51**, 424–429.

[3]Standard length = length of trunk not including the presoma or the protruding male bursa.

Harms, C.E. (1965). *In vitro* cultivation of an Acanthocephalan, *Octospinifer macilentis*. *Experimental Parasitology* **17**, 41–45.

Hibbard, K.M. and Cable, R.M. (1968). The uptake and metabolism of tritiated glucose, tyrosine, and thymidine by adult *Paulisentis fractus* Van Cleave and Bangham, 1949 (Acanthocephala: Neoechinorhynchidae). *Journal of Parasitology* **54**, 517–523.

Hovarth, K. (1971). Glycogen metabolism in larval *Moniliformis dubius* (Acanthocephala). *Journal of Parasitology* **57**, 132–136.

Jensen, T. (1952). The life cycle of the fish acanthocephalan, *Pomphorhynchus bulbocolli* (Linkins) Van Cleave, 1919, with some observations on larval development *in vitro*. PhD dissertation, University of Minnesota.

King, D. and Robinson, E.S. (1967). Aspects of the development of *Moniliformis dubius*. *Journal of Parasitology* **53**, 142–149.

Körting, W. and Fairbairn, D. (1972). Aerobic energy metabolism in *Moniliformis dubius* (Acanthocephala). *Journal of Parasitology* **58**, 45–50.

Lackie, A.M. (1978). Acanthocephala. *In* "Methods of Culturing Parasites *In Vitro*" (A.E.R. Taylor and J.R. Baker, eds), pp. 279–85. Academic Press, London and New York.

Lackie, A.M. (1981). The specificity of the serum agglutinins of *Periplaneta americana* and *Schistocerca gregaria* and its relationship to the insects' immune response. *Journal of Insect Physiology* **27**, 138–143.

Lackie, J.M. (1972). The course of infection and growth of *Moniliformis dubius* (Acanthocephala) in the intermediate host *Periplaneta americana*. *Parasitology* **64**, 95–106.

Lackie, J.M. (1975). The host specificity of *Moniliformis dubius* (Acanthocephala), a parasite of cockroaches. *International Journal for Parasitology* **5**, 301–307.

Lackie, J.M. and Rotheram, S. (1972). Observations on the envelope surrounding *Moniliformis dubius* (Acanthocephala) in the intermediate host, *Periplaneta americana*. *Parasitology* **65**, 303–308.

Mettrick, D.F. and Podesta, R.B. (1974). Ecological and physiological aspects of helminth-host interactions in the mammalian gastrointestinal canal. *Advances in Parasitology* **12**, 183–278.

Nicholas, W.L. and Grigg, H. (1965). The *in vitro* culture of *Moniliformis dubius* (Acanthocephala). *Experimental Parasitology* **16**, 332–340.

Quiot, J.M., Paradis, S. and Louis, C. (1985). Establishment of a cell line (SP[Gl]13) from dorsal vessels of the cockroach *Gromphadorhina laevigata*, Saurssum and Zentner (Dictyoptera: Blattidae). *In Vitro* **21**, 603–605.

Roberts, L.S. (1973). Modifications in media and surface sterilization methods for *in vitro* cultivation of *Hymenolepis diminuta*. *Journal of Parasitology* **59**, 474–479.

Schmidt, G.D. (1985). Development and life cycles. *In* "Biology of the Acanthocephala" (D.W.T. Crompton and B.B. Nickol, eds), pp. 273–286. Cambridge University Press.

Smyth, J.D. (1946). Studies on tapeworm physiology. I. Cultivation of *Schistocephalus solidus in vitro*. *Journal of Experimental Parasitology* **23**, 47–70.

Smyth, J.D. (1955). Problems relating to the *in vitro* cultivation of Pseudophyllidean cestodes from egg to adult. *Revista Iberica de Parasitologica. Granada (Espana)*. Tomo Extraordinario. Marzo. pp. 65–86.

Starling, J.A. (1985). Feeding, nutrition and metabolism. *In* "Biology of the Acanthocephala" (D.W.T. Crompton and B.B. Nickol, eds), pp. 125–212. Cambridge University Press.

Tanaka, R.D. and MacInnis, A.J. (1980). Analyses of the pseudocoelomic fluid from *Moniliformis dubius*. *Journal of Parasitology* **66**, 354–355.

Taylor, A.E.R. and Baker, J.R. (1968). "The Cultivation of Parasites *In Vitro*". Blackwell Scientific Publications, Oxford.

Taylor, A.E.R. and Baker, J.R. (1978). "Methods of Cultivating Parasites *In Vitro*". Academic Press, London and New York.

Tobias, R.C. and Schmidt, G.D. (1977). *In vitro* cultivation of *Moniliformis moniliformis* juveniles. *Journal of Parasitology* **63**, 588–589.

Voge, M. (1967). Development *in vitro* of *Mesocestoides* from oncosphere to young tetrathyridium. *Journal of Parasitology* **53**, 78–82.

Ward, P.V.F. and Crompton, D.W.T. (1969). The alcoholic fermentation of glucose by *Moniliformis dubius* (Acanthocephala), *in vitro*. *Proceedings of the Royal Society B* **172**, 65–88.

Wong, B.S., Miller, D.M. and Dunagan, T.T. (1979). Electrophysiology of acanthocephalan body wall muscles. *Journal of Experimental Parasitology* **82**, 273–280.

Young, B.W. and Lewis Jr., P.D. (1977). Growth of an acanthocephalan on the chick chorioallantois. *Proceedings of the Montana Academy of Sciences* **37**, 88.

Chapter 16

Helminth Vaccines

M.D. RICKARD* and M.J. HOWELL†

**University of Melbourne, Veterinary Clinical Centre, Princes Highway, Werribee, Victoria, Australia*

and

†*Department of Zoology, Australian National University, Canberra, A.C.T, Australia*

I. INTRODUCTION

The enormous impact of helminth parasites on the health and productivity of man and his domesticated animals has resulted in substantial investment in research on their eradication or control. Advances in chemotherapy and increased understanding of the factors governing transmission of these parasites have played a major practical role in progress toward this goal. However, there are few instances where total control or eradication has been achieved, often because the parasite is able to alter its physiology or behaviour to cope with the injuries sustained at the hand of man. The clear evidence for host-protective immunity during infection with many parasites has, for over half a century, prompted scientists to search for means of stimulating this immunity artificially so that parasites can be controlled by vaccination of the host.

Interest in the use of *in vitro* cultivation methods as a means for producing parasite antigens for immunoprophylaxis was stimulated by several factors. Poor results were generally achieved by the parenteral administration of dead parasites, or parasite extracts, as vaccines. Early experiments demonstrated precipitates around the various natural orifices of *Nippostrongylus muris* larvae incubated in immune serum *in vitro* (Sarles, 1937, 1938). This observation, together with the much more promising results achieved when living attenuated helminths were used as vaccines, or when infective stages were administered into abnormal sites in the body, led to speculation that metabolic or excreted-secreted (ES) products of living parasites were likely to be the most effective antigens. Early vaccination experiments with ES antigens from nematodes (Thorson, 1953; Campbell, 1955; Chute, 1956) reinforced this notion. Another important factor which encouraged research on *in vitro* culture methods was the often limited quantities of living helminth parasites, or their infective stages, available from infections in man or animals.

This chapter describes methods that have been employed experimentally with at least some success. Because of limitations of space only the more recent experiments are described, but reference is made to earlier works. In many instances the processes described are not strictly examples of *in vitro* cultivation, but are rather the maintenance of parasites alive *in vitro* for short periods. The last part of the chapter (Section V) describes some new approaches which are being examined as alternatives to the traditional culture of whole living parasites, but *in vitro* methods such as recombinant DNA technology are not considered to fall within the scope of this book.

II. TREMATODA

Host-protective immunity against the important trematode infections of man and his domesticated animals is generally not as clear-cut as that which occurs in infection with metacestode stages. Despite this, a substantial amount of work has been carried out to examine the feasibility of developing vaccines against these parasites (Clegg and Smith, 1978; Taylor, 1980; Lloyd, 1981; Rickard and Howell, 1982). The majority of this work has focused on *Fasciola hepatica* and *Schistosoma* spp., because they are important parasites of many domesticated animals and man.

A. FASCIOLIDAE

Adult or juvenile worms have in all cases been recovered from the bile duct or liver parenchyma of infected mice or rats, usually by dissection or a combination of dissection and incubation. Original papers should be consulted for details.

ES products of either metacercariae or juvenile or adult worms have been collected from *in vitro* cultures and injected into hosts with a view to stimulating resistance to challenge infection.

1. *Detailed techniques*

1° *Fasciola hepatica* : metacercariae
(Howell, 1979; Howell and Sandeman, 1979; Sandeman *et al.*, 1980)

(i) EXCYSTMENT PROCEDURE (HOWELL *ET AL.*, 1977)

Metacercariae (mc) were recovered from experimentally infected *Lymnaea tomentosa* and excysted essentially as described by Wikerhauser (1961): 200–500 mc were incubated in a Leighton tube containing 5.0 ml 0.25% pepsin (Calbiochem) in 0.85% NaCl, adjusted to pH 2.0 with 1M HCl, for 2 h at 37 °C. They were washed several times by sedimentation in 0.85% NaCl until the pH of the washings reached 7.0. The final rinse was decanted and 0.85% NaCl containing 0.4%, trypsin (ICN Life Sciences Group, catalogue no. 101179), 1% $NaHCO_3$, and 20% bile (centrifuged, filtered, pH 7.0) from fluke-free sheep, was added. Incubation was carried out for 3–4 h at 37 °C. Periodic and vigorous shaking of the Leighton tube by hand promoted excystment. 50–80% of the mc excysted within 3–4 h in the bile/trypsin solution. There appeared to be variation between batches of cysts and the chemical constituents varied in their ability to induce excystment — particularly the trypsin; the type described above gave consistently good results.

(ii) PREPARATION OF CULTURES

Excysted mc were washed four times by sedimentation in Medium 199 (see Taylor and Baker, 1968, p. 356) containing 100 iu ml^{-1} penicillin, 100 μg^{-1} streptomycin sulphate and 2.5 μg ml^{-1} amphotericin B. The mc were transferred to a sterile Leighton tube with a minimum of medium and again washed four times by sedimentation as above. Following transfer to another sterile Leighton tube, the mc were washed once and the rinsing solution removed immediately prior to the addition of culture medium. **Throughout the washing process, aseptic procedures were followed.** All tubes and Pasteur pipettes were siliconized with either Ajax silicone glass treatment solution or Sigmacote® to prevent mc from adhering to them. Groups of about 50–100 excysted mc were placed in sterile Leighton tubes with 5.0 ml of culture medium. This consisted of 20% "immune" serum (see (iii) below) from sheep or rats in Medium 199 plus antibiotics as described above. The tubes were gassed with 5% CO_2 in air to maintain a pH of around 7.2–7.4. Cultures were incubated at 37°C in a shaker water bath operating at 120 agitations min^{-1}. Irrespective of the type of serum used, almost half the mc remained alive and active and survived for at least 7 d.

(iii) PREPARATION OF CULTURE ANTIGENS

The immune rat (Wistar) serum used in the cultures was obtained from rats given a primary infection of 10 mc of *F. hepatica* at 5 weeks of age, then challenged 7 weeks later with 20 mc, and the serum prepared one week after challenge (Howell *et al.*, 1977).

The immune sheep (Merino/Border Leicester cross) serum was obtained from lambs 10 weeks after infection with 100 mc, when antibody levels have attained peak values (Sandeman and Howell, 1980). In the presence of immune rat or sheep serum, a precipitate of antigen–antibody complexes formed on the teguments of the mc and in the culture medium. From a single culture containing about 100 mc in 5.0 ml culture medium, between 50 and 100 μg of precipitate could be collected after a week. The culture medium containing the antigen–antibody complexes in suspension was decanted from the mc and the precipitate was recovered by centrifugation (4000*g*, 30 min). The pellet was washed three times with phosphate buffered saline (PBS) by centrifugation as above. If not used immediately, the complex was stored at −20°C.

(iv) VACCINATION EXPERIMENTS

(a) Five-week-old Wistar rats were each given either an injection of antigen–rat antibody complexes equivalent to 45–50 μg protein suspended in 0.5 ml PBS emulsified with an equal volume of Freund's complete adjuvant (FCA) [0.2 ml intramuscularly (i.m.) and 0.8 ml intraperitoneally (i.p.)] or 100–250 μg protein in 0.4 ml PBS emulsified with an equal volume of FCA (0.4 ml divided between hind foot pads). Booster injections were given 3 weeks or 5 weeks later. The rats were then challenged with mc after 9 d. Necropsies were performed 7–8 weeks later and the success of the vaccination was determined by comparing the numbers of adult flukes recovered from test animals and controls. Significant levels of protection (ranging from 50% to 87%) were obtained in 3 of 4 separate experiments. The greatest levels of protection were associated with the larger amount of complex administered, but there is insufficient evidence to exclude the route of injection as an important factor in stimulating protection. Further studies on the nature of the immune complex have been published (Irving and Howell, 1982).

(b) Five-month-old worm-free Merino lambs were used. Animals were vaccinated with an initial i.m. injection of 1.4 mg of precipitated antigen/sheep antibody complexes suspended in 1.0 ml PBS and emulsified with an equal volume of FCA. The injection was repeated after 4 weeks and the animals were challenged 1 week later with 200 mc. Although the vaccinated animals produced high levels of antibody in response to the injection of complexes, no protection against challenge was evident at necropsy.

(v) COMMENTS

Davies *et al.* (1979) failed to confer any protection on rats using antigens collected during *in vitro* culture of metacercariae in contrast with the results of the above experiments. However, there were some substantial differences in the vaccination protocols: Freund's incomplete adjuvant (FIA) rather than FCA was used; antigens rather than antigen–antibody complexes were used; only a single sensitizing injection was administered and this was given subcutaneously (s.c.) rather than i.p., i.m. or into the footpads.

2° *Fasciola hepatica*: juvenile worms

(Lang, 1974, 1976; Lang and Hall, 1977; Hall and Lang, 1978)

(i) CULTURE METHOD

Sixteen-day-old worms, collected from mouse livers, were washed twice (10 min, 37.5 °C) in sterile PBS. Groups of 40–50 were transferred to 25 ml glass flasks containing 0.5 ml per worm of Medium 199 (without antibiotics: see Taylor and Baker, 1968, pp. 355–356) and incubated at 37.5 °C in a water bath shaker set at the slowest oscillation. The flasks were loosely capped and maintained in equilibrium with air; relatively short-term incubations presumably obviated the need for antibiotics.

(ii) VACCINATION EXPERIMENTS

A single i.p. injection of 1.0 ml culture medium from a 4 h culture failed, in terms of worm recovery, to protect white inbred mice (unspecified strain) against challenge with 2 *F. hepatica* mc, but host mortality was significantly reduced. However, twenty 12-week-old male mice (inbred, but again strain unspecified) given a total of 1.0 ml of culture medium i.p. in three 0.2 ml aliquots taken from 2, 7 and 12 h cultures plus a 0.4 ml aliquot from a 24 h culture, and challenged 30 d later with 2 mc, showed an 83.3% reduction in worm recovery (Lang, 1976). Similar degrees of worm reduction were achieved by giving a single dose of either 2.0 ml of culture medium from a 24 h culture or 4.0 or 6.0 ml from a 12 h culture. Alternatively, the 2.0 ml could be spread over several equal doses from 2, 7, 12 and 24 h cultures, or 2, 12 and 24 h cultures or 12 and 24 h cultures. Estimates of total protein administered to each of the successfully vaccinated mice ranged from 670 to 800 μg.

In an experiment involving one cow, three injections (at 1 d, 24 d, 42 d) of culture medium (volume not specified) were given i.m. in the neck region. The animal was challenged at 42 d with 200 mc. Eighty-two days later, the vaccinated animal had 79.1% fewer flukes than a control cow.

3° *Fasciola hepatica*: juvenile worms
(Rajasekariah *et al.*, 1979)

(i) CULTURE METHOD

Four-week-old *F. hepatica* were removed from the livers of infected rats, washed repeatedly in 0.85% NaCl at 37°C to remove debris then rinsed twice in Hedon-Fleig (HF) saline. Worms were transferred to Costar tissue culture flasks and incubated (37°C, 30 min) to allow regurgitation of caecal contents and identification of the most active worms. These were transferred to flasks containing 1.0 ml HF saline per worm and incubated for 48 h (37°C) with the medium changed at 6–8 h intervals. The pooled, used, incubation medium was centrifuged (4000 rpm, 30 min) and the supernatant was stored at − 20°C until required: it was then concentrated by ultrafiltration using Amicon PM-10 filters with high purity nitrogen as the propellant.

(ii) VACCINATION EXPERIMENTS

Outbred Wistar rats (5-week-old males) were vaccinated by giving each 1.0 ml of medium (1.6 mg fluke protein) emulsified in 1.0 ml of FCA s.c. on two occasions, 14 d apart, then challenged with 30 mc 14 d later. At necropsy, 9 weeks after challenge, there was a 66% reduction in worm burden compared with the controls.

Male CBA/H mice (8–9 weeks old) were given an initial injection of medium containing 50 μg fluke protein emulsified with FCA into the hind footpads and i.p. Precise details of volumes injected and respective amounts given i.p. and into the footpads were not given. Three and 6 weeks later, medium containing 20 μg protein was injected into the same sites; 14 d after the last injection, the mice were challenged with 3 mc; in terms of mortality or worm recoveries, the vaccination protocol afforded no protection.

(iii) COMMENTS

Results of vaccination studies using antigens derived from juvenile flukes cultured *in vitro* have been quite variable and conflicting. Lang (1976) and Lang and Hall (1977) have stimulated strong resistance against challenge infection in mice and Rajasekariah *et al.* (1979) have successfully vaccinated rats but not mice. Burden *et al.* (1982) were unable to vaccinate rats or mice using culture antigens from juvenile flukes. If standard vaccination protocols are to be developed it would seem, from the limited studies done so far, that such factors as host strain and the age of flukes donating antigens could be important.

4° *Fasciola hepatica*: adult worms
(Dragneva, 1972)

Adult flukes were cultured for 20–34 h in Tyrode's solution according to the method of Rohrbacher (see Taylor and Baker, 1978, p. 169). Rabbits, immunized at weekly intervals for 5 weeks with 30–40 mg of ES antigens injected s.c. into the hip region showed about 50% protection when challenged with 15 mc 1 week after the last injection.

COMMENT

Except in this one instance (Dragneva, 1972), it has not been possible to stimulate any degree of resistance to *F. hepatica* using, as vaccines, antigens recovered from culture media in which adult worms have been maintained. Unsuccessful attempts have been described by Lehner and Sewell (1979) and Rajasekariah *et al.* (1979). Such negative results using culture antigens from adult flukes tie in with the observation that metacercariae and juvenile flukes implanted s.c. into rats stimulate resistance to challenge but implanted adult flukes do not (Rajasekariah and Howell, 1978).

B. SCHISTOSOMATIDAE

1. *Detailed techniques*

1° Schistosoma japonicum
(Sadun and Lin, 1959)

(i) CULTURE METHOD

Adult *S. japonicum* were recovered from a variety of infected hosts (mice, rabbits and monkeys). Between 50 and 100 worms were placed in a Carrel flask containing 5.0 ml serum ultrafiltrate, 10 ml Simms X-6 solution (ingredients and source unspecified) plus 10 iu ml^{-1} penicillin and 100 μg ml^{-1} streptomycin. The flask was shaken gently and incubated at 37 °C for 72 h. The worms were then removed and the culture media was lyophilized.

(ii) VACCINATION EXPERIMENT

Forty mice were given three i.p. injections each of 0.5 ml adult *S. japonicum* culture antigen (protein content unspecified) 3, 2 and 1 week(s) before infection with 50 cercariae. Thirty days later the number of worms recovered from these animals was about 20% less than from control mice.

2° *Schistosoma mansoni*
(Levine and Kagan, 1960)

(i) CULTURE METHOD

Cercariae of a Puerto Rican strain of *S. mansoni*, collected by exposing infected *Australorbis glabratus* in small beakers, were concentrated by centrifugation (320*g*, time unspecified) and groups of 400 000 were held overnight in 5–10 ml water in a Petri dish at room temperature. The contents of the Petri dish, presumed to contain cercarial antigens, were centrifuged as above and the supernatant was frozen and stored at −20 °C. The final antigen preparation represented material derived from 1.6×10^6 cercariae.

Adult metabolic antigen was made in a similar manner using 300 adult male and female worms (recovered by perfusing infected hamsters) for each incubation (in saline ?).

A third antigen consisted of equal amounts of the cercarial and adult antigen preparations.

(ii) **VACCINATION EXPERIMENT**

Antigen preparations were emulsified with an equal volume of FCA. Groups of 15 mice (18 g white female, strain not indicated) were each given 0.5 ml of FCA containing one of the antigen preparations weekly for 5 weeks. At 6 weeks the mice were challenged with 300 cercariae. There was a slight, but significant increase in the longevity of the mice.

3° *Schistosoma mansoni*

(Murrell and Clay, 1972)

(i) **CULTURE METHOD**

Adult *S. mansoni* were recovered from NMRI mice and groups of 40 flukes incubated in 5.0 ml medium NCTC-135 (containing 400 iu ml^{-1} of a penicillin–streptomycin mixture) for 36–48 h at 37 °C with a gas phase of 5% CO_2 in air. The culture medium was then removed, pooled and concentrated 250 times by ultrafiltration using an Amicon PM-10 membrane. Further details of maintenance of cultures and storage of antigen are not given.

(ii) **VACCINATION EXPERIMENT**

Male NMRI, 10–11 weeks old mice were given five injections of concentrated culture medium as follows: s.c. injection of 0.25 ml of medium emulsified with FIA; 10 d later 0.4 ml of 6% medium in alum i.p.; 5 d later, a repeat of the s.c. injection; 5 and 10 d later repeats of the i.p. injection. The mice were then exposed to 100 cercariae of *S. mansoni* and necropsied 8 weeks later. Vaccinated mice showed a 40% reduction ($0.02 < P < 0.05$) in the number of worms recovered compared with the controls.

2. *Comments*

A study using secretions from cercariae to immunize mice has been undertaken (Minard *et al.*, 1977) but no protection was obtained. Clearly, adult *S. mansoni* and *S. japonicum* secrete substances during *in vitro* culture that confer partial protection on mice. It is surprising that no further work has been carried out in this area, although many attempts have been made to characterize the antigenic material produced by schistosomes *in vitro* (e.g. Kusel and MacKenzie, 1975; Kusel *et al.*, 1975a, b). The observation that adult schistosomes appear to produce host-protective antigens is in marked contrast to the situation with *Fasciola hepatica*.

III. CESTODA

Host-protective immunity is a striking feature of repeated infection with metacestode stages in mammalian intermediate hosts and in many cases plays a major role in regulating natural transmission of these parasites (Williams, 1979; Rickard and Williams, 1982). It is hardly surprising, therefore, that substantial research effort has been directed toward developing vaccines against these parasites using *in vitro* culture, especially the more important parasites in the family Taeniidae. The evidence for host-protective immunity against infection with the important adult cestodes is less encouraging for development of vaccines (Hopkins, 1980; Rickard, 1983), although some production of vaccines via *in vitro* culture has been attempted.

A. *TAENIA* SPP.

1. *Detailed techniques*

1° ***Taenia ovis*** **metacestodes**
(Rickard and Bell, 1971a; Rickard *et al.*, 1976, 1977a)

(i) PREPARATION OF ACTIVATED ONCOSPHERES AND CULTIVATION METHOD

Eggs were collected from the terminal proglottides of worms purged from experimentally infected dogs, washed several times in physiological saline containing 1.0×10^6 iu litre^{-1} penicillin G, 1.0 g litre^{-1} streptomycin sulphate and 1×10^4 iu litre^{-1} of mycostatin and stored at 4°C until required. The methods for hatching eggs, activating oncospheres, and *in vitro* culture were those of Heath and Smyth (1970; see Taylor and Baker, 1978, p. 213).

Medium 858 (Table I; obtained from Commonwealth Serum Laboratories) was supplemented with 20% heat-inactivated (56°C, 30 min) serum from 3-week-old colostrum-deprived helminth-free lambs. Approximately 40 000 activated oncospheres were added to each Leighton tube (Bellco Biological Glassware) containing 10 ml of pre-warmed (37.5°C) culture medium. Culture tubes were rotated on a roller drum apparatus (4 rev min^{-1}). Medium was replaced every 2 d after the developing oncospheres had settled on standing: the total culture period was 8 d. The culture medium removed at each change was centrifuged (16 000*g*, 4°C, 1 h), and the supernatant was stored at −20°C.

TABLE I. Tissue culture medium 858 (Healey *et al.*, 1955)

	mg litre^{-1}		mg litre^{-1}
L-Arginine	70	Diphosphopyridine nucleotide (95% pure)	7.0
L-Histidine	20	Triphosphopyridine nucleotide (80% pure)	1.0
L-Lysine	70	Coenzyme A (75% pure)	2.5
L-Tyrosine	40	Cocarboxylase (88% pure)	1.0
L-Tryptophane	10	Flavin adenine dinucleotide (60% pure)	1.0
L-Phenylalanine	25	Uridine triphosphate (90% pure)	1.0
L-Cystine	20	Glutathione	10.0
L-Methionine	15		
L-Serine	25	Adenine desoxyriboside	10
L-Threonine	30	Guanine desoxyriboside	10
L-Leucine	60	Cytosine desoxyriboside	10
L-Isoleucine	20	5-Methyldesoxycytidine	0.1
L-Valine	25	Thymidine	10
L-Glutamic acid	75		
L-Aspartic acid	30	Sodium acetate	50
L-Alanine	25	Sodium glucuronate	4.2
L-Proline	40	L-Glutamine	100
L-Hydroxyproline	10	*d*-Glucose	1000
L-Cysteine	260	Phenol red (pH indicator)	20
Glycine	50	Ethanol (as an initial solvent for fat-soluble constituents)	16
Pyridoxine	0.025	NaCl	6800
Pyridoxal	0.025	KCl	400
Biotin	0.01	$CaCl_2$	200
Folic acid	0.01	$MgSO_4.7H_2O$	200
Choline	0.5	$NaH_2PO_4.H_2O$	140
Inositol	0.05	$NaHCO_3$	2200
p-Aminobenzoic acid	0.05	Fe, as $Fe(NO_3)_3$	0.1
Vitamin A	0.1		
Calciferol	0.1	Sodium penicillin G	1.0
α-Tocopherol phosphate	0.01	Dihydrostreptomycin sulphate	100
Menadione	0.01	n-Butyl parahydroxybenzoate	0.2
Ascorbic acid	50		
Tween 80 (oleic acid)	5.0		
Cholesterol	0.2		

(ii) **VACCINATION EXPERIMENTS**

The total culture supernatant collected during the culture period was thawed, pooled and concentrated 8-fold by dialysis in cellulose tubing (Visking) against polyethylene glycol. Merthiolate was added to a final concentration of 1 in 10 000 and the antigen stored at 4 °C. For vaccination, the concentrated culture medium was emulsified in an equal volume of FCA, and each lamb received 2 ml by i.m. injection in the thigh (Rickard and Bell, 1971a). In other experiments the culture medium, emulsified in FIA, was equally effective (Rickard *et al.*, 1976, 1977a). Several laboratory and field experiments showed that these antigens could be used (a) to protect young lambs against experimental infection with 2000 *T. ovis* eggs given 5 weeks after immunization; (b) to protect lambs against natural infection acquired by grazing on contaminated pasture; (c) to cause the passive transfer of immunity to the lamb via colostrum of ewes vaccinated before parturition; (d) to allow lambs to be vaccinated while under cover of colostral immunity; (e) to give 1 year of protection from naturally acquired infection with a single vaccination. Vaccination had no effect on the survival of established metacestodes, which contrasted with the destruction of metacestodes in sheep that had been previously naturally infected (Rickard *et al.*, 1976).

2° *Taenia ovis* metacestodes
(Rickard and Adolph, 1977)

These authors modified the previous method (Section III.A.1°) as follows.

(i) Oncospheres were incubated in medium 858 with or without the addition of 20% lamb serum.
(ii) The duration of incubation was reduced to a total of either 24 or 48 h.
(iii) Culture media were centrifuged at only 500*g*.
(iv) Culture supernatants were concentrated 10-fold by dialysis against polyethylene glycol.
(v) Culture medium equivalent to supernatant from 20 000 oncospheres was used as the dose for each lamb, and homogenized in an equal volume of FIA.

The antigens from oncospheres incubated in medium 858 supplemented with serum gave excellent protection at both 24 and 48 h incubations. The antigens from medium 858 without added serum gave excellent protection at 48 h of incubation and lesser, but significant, protection at 24 h.

3° Immunization against metacestode stages of *T. ovis* using *Taenia hydatigena* antigens
(Rickard and Bell, 1971a)

Taenia hydatigena oncospheres were cultured in a manner identical to that already described for *T. ovis* oncospheres (Section III.A.1°), and culture antigens emulsified in FCA were injected (i.m.) into helminth-free lambs, followed 5 weeks later by challenge with *T. ovis* eggs. Necropsy examination 7 weeks later showed that the *T. hydatigena* antigens had stimulated a high degree of protection against infection with *T. ovis*.

4° Immunization against *Taenia hydatigena* metacestodes using *T. hydatigena* antigens
(Onamunwi and Coles, 1980)

Taenia hydatigena eggs were collected from gravid proglottides of worms obtained from experimentally infected beagles, rinsed in saline and stored at 4°C. The eggs were hatched and cultured according to the method described earlier for *T. ovis* (Section III.A.1°) except that medium CMRL 1066 (see Taylor and Baker, 1968, p. 356) plus antibiotics (0.4×10^5 iu ml^{-1} penicillin G, 40.0 mg ml^{-1} streptomycin sulphate and 10^4 iu ml^{-1} mycostatin respectively) without added serum was used as the culture medium. Methods of treating the antigen, preparation of the vaccine and inoculation of the lambs were as described for *T. ovis* (Section III.A.1°). Lambs, challenged with *T. hydatigena* eggs orally 3 weeks after vaccination and examined at necropsy 6 weeks later for metacestode stages, showed a high level of immunity.

5° *Taenia hydatigena* adult worms
(Heath *et al.*, 1980)

T. hydatigena worms were maintained in Eagle's minimum essential medium (see Taylor and Baker, 1968, p. 355) for an unspecified period of time, and the medium was changed every 2 d. The medium was concentrated by ultrafiltration. Antigen (7 mg protein) was incorporated into 4.0 ml FCA and 1.0 ml of mixture injected into each of two s.c. and two deep i.m. sites in dogs. Dogs developed circulating antibodies to the antigen, but showed no resistance to infection with *T. hydatigena* worms.

6° Immunization against metacestode stages of *Taenia saginata* using *T. hydatigena* antigens
(Rickard and Adolph, 1976; Rickard *et al.*, 1981, 1982)

In vitro culture antigens from *T. hydatigena* have been used to immunize calves against experimental and naturally acquired infection with *T. saginata* metacestodes. Rickard and Adolph (1976) and Rickard *et al.* (1981) used preparation and culture techniques as described previously for *T. ovis* (Section III.A.1°). Rickard *et al.* (1982) used the same procedure to prepare one batch of antigen tested, but modified the procedure for a second batch. In the latter case, 20% heat-inactivated (56°C, 30 min.) foetal calf serum (FCS) was used to supplement medium 858 instead of colostrum-deprived lambs' serum, and 1% glucose was also added.

The *in vitro* culture antigens from *T. hydatigena* oncospheres gave significant protection against experimental infection of calves with *T. saginata*, although the level of protection achieved was not as great as with the homologous *T. saginata* antigen.

7° *Taenia saginata* metacestodes

(Rickard and Adolph, 1976; Rickard *et al.*, 1977b, 1982; Rickard and Brumley, 1981)

(i) PREPARATION OF ACTIVATED ONCOSPHERES

T. saginata eggs were collected from the terminal segments of fresh tapeworms recovered from patients treated with the anthelmintic niclosamide and given a saline purge. The oncospheres, teased from the gravid proglottides, were washed and stored as for *T. ovis* eggs (Section III.A.1°). *T. saginata* eggs were hatched and activated using Rickard and Bell's (1971b) modification of Silverman's (1954) method. Eggs were first incubated for 1 h at 37.5°C in 10 ml screw-capped test tubes containing artificial gastric fluid (AGF: 1% pepsin + 1% HCl in physiological saline). This was followed by incubation in 10 ml of an artificial intestinal fluid (AIF) each 100 ml of which contained 1.0 g $NaHCO_3$, 1.0 g pancreatin, 13 mg cholesterol, 40 mg trypsin, and 200 mg each of Na taurocholate, Na glycocholate and Na tauroglycocholate in distilled water. The AIF required stirring at 37.5°C to allow most components to dissolve, followed by filtration to clarify it. Either a small amount of AGF could be left in the tube when AIF was introduced, so that CO_2 released from the $NaHCO_3$ could be trapped in the tube, or alternatively, the tube was gassed with a 5% CO_2, 10% O_2, 85% N_2 mixture. Incubation in AIF was for 1 h on a roller-drum apparatus (4 rev min^{-1}). Activated oncospheres were washed three times with warm culture medium using gentle centrifugation.

(ii) CULTURE METHOD

This followed that described for *T. ovis* (Section III.A.1°) except that heat inactivated (56°C, 30 min) FCS replaced lamb serum in the culture medium. There was minimal development of *T. saginata* oncospheres during the culture period, and some oncospheres degenerated.

(iii) VACCINATION EXPERIMENTS

Animals were vaccinated once i.m. with culture antigens derived from 20 000 oncospheres emulsified in an equal volume of FCA (total volume 5.0 ml). Rickard and co-workers showed that (a) calves could be protected against experimental infection with *T. saginata* eggs, (b) calves could be protected by passive transfer of immunity via the colostrum of mothers vaccinated with culture antigens in late pregnancy; (c) vaccination protected calves against infection acquired by grazing infected pasture; (d) vaccination could be implemented as part of a farm management system to allow at least partial control of *T. saginata* infection.

8° *Taenia saginata* metacestodes
(Lloyd, 1979)

(i) PREPARATION OF ACTIVATED ONCOSPHERES AND CULTURE METHOD

Eggs were hatched and oncospheres activated using the techniques described previously for *T. ovis* (Section III.A.1°) and by Gallie and Sewell (1970). The latter technique added trypsin at 40 mg 100 ml^{-1} and substituted 20% ox bile for sheep bile in AIF (Section III.A.7°(i)). Culture antigens were collected by modification of Rickard and Adolph's (1977) techniques for *T. ovis* (Section III.A.2°). The medium used was RPMI 1640 supplemented with HEPES buffer, glutamine, antibiotics (Gibco) and 10% foetal bovine serum. 20 000 oncospheres were cultured in 3.0 ml of medium for 48 h. The culture medium was centrifuged, concentrated 10 or 30 times, and frozen at − 20°C (no further details were given).

(ii) VACCINATION EXPERIMENTS

Pre-parturient heifers were immunized either i.m. (10 ml of 30 times concentrated antigen emulsified in FCA twice at a 23-day interval) or by the intramammary route (15 ml of 10 times concentrated antigen infused via the teat canal on three occasions at 2 week intervals). Calves that were fed immunoglobulin prepared from colostrum from the immunized heifers had high levels of resistance to challenge infection with *T. saginata* eggs. Three-month-old calves, which were immunized i.m. with 5.0 ml of 10 times concentrated antigen in FCA and challenged with 34 000 eggs of *T. saginata* 3 weeks after immunization, also showed significant immunity to the challenge infection.

(iii) COMMENT

Publications by Wikerhauser *et al.* (1978) and Mitchell and Armour (1980) described unsuccessful attempts to vaccinate cattle against *T. saginata* infection using cultured oncosphere antigens prepared in a similar manner to that described by Rickard and Adolph (1976, 1977). Rickard *et al.* (1982) discussed some of the possible reasons for failure in these cases, which included the removal of important particulate antigens by centrifugation at high speeds and variations in sex and breed of the animals used for experiments. A substantial number of factors can effect the ability of animals to develop resistance and these are reviewed by Rickard and Williams (1982).

9° *Taenia pisiformis* metacestodes
(Rickard and Outteridge, 1974)

Methods for the preparation and culture of *T. pisiformis* oncospheres were as described earlier for *T. ovis* (Section III.A.1°) except that (i) the AIF described by Rickard and Bell (Section III.A.7°(i)) was used to hatch and activate oncospheres; (ii) the lamb serum was replaced with 20% heat-inactivated (56°C, 30 min) rabbit serum collected from 6-week-old cestode-free rabbits. Rabbits were vaccinated twice by i.m. injection of 0.05 ml antigen (equivalent to 10 000 oncospheres) emulsified in an equal volume of FCA. Rabbits were challenged with 1500 *T. pisiformis* eggs each, 3 weeks after the second injection of antigen, and killed 6 weeks later. Vaccinated animals demonstrated high levels of resistance against the challenge infection.

10° *Taenia pisiformis*
(Heath, 1973, 1976)

(i) PREPARATION OF ACTIVATED ONCOSPHERES

Eggs teased from washed gravid proglottides were sieved and allowed to settle overnight at 4°C. The supernatant was removed and the eggs were treated with 1 : 500 Hibitane® (chlorhexidine gluconate) for 20 min, rinsed three times in sterile distilled H_2O and stored at 4°C in sterile saline (0.85% NaCl). The procedure for hatching and activation was a modification of that of Heath and Smyth (1970; see Section III.A.1°) except that a tissue-culture roller was not used, and both artificial gastric and intestinal solutions were passed through 0.2μm Millipore filters before use.

(ii) CULTURE TECHNIQUE

Major additives to the culture medium included heat-inactivated FCS obtained commercially, and rabbit blood cells. The rabbit blood cells were collected from the central ear artery into sterile heparinized syringes, washed three times in PBS and stored at 4°C in PBS containing 300 mg dl^{-1} of dextrose. The culture medium was contained in flat plastic culture flasks (30 ml or 250 ml) and consisted of Medium NCTC-135 (Gibco) with dextrose added to a final level of 400 mg dl^{-1}, penicillin G 100 iu ml^{-1}, streptomycin sulphate 100 $\mu g\ ml^{-1}$, and mycostatin 100 iu ml^{-1}. If Gram negative organisms became apparent during culture, gentamicin sulphate was also added; 1% of washed, packed rabbit blood cells plus FCS (50% for the first 7 d, 20% for the next 14 d) were added to the medium. The gas phase was N_2 and culture vessels were held stationary at 37–39°C, with medium changed twice weekly.

Experiments were carried out to collect culture antigens free from macromolecules and with incorporated [^{14}C]amino acids. One million *T. pisiformis* oncospheres were dispensed equally in 4 × 250 ml culture flasks and cultivated *in vitro* for 10 d. At 6 d of culture, two flasks received 2.0 ml of a protein hydrolysate labelled with 0.125 mCi of ^{14}C which had been passed through a PM-10 ultrafiltration membrane to remove macromolecules. At 10 d, all developing larvae were washed free from all except adsorbed macromolecules by repeated changes of PBS, pH 7.2 at 37°C. Larvae were pipetted into two other 1 litre culture vessels containing culture medium filtered through a PM-10 membrane to remove macromolecules. Approximately 200 000 larvae were contained in each flask. The medium was changed every 3 d for 9 d; at each change the used medium was immediately concentrated to 25 ml and washed three times with 200 ml of autoclaved PBS in an Amicon 402 stirred cell using a PM-10 ultrafiltration membrane. All manipulations involving the culture media were carried out under either N_2 (unlabelled larvae) or air (labelled larvae) to assess the sensitivity of culture antigens to oxidation.

(iii) VACCINATION EXPERIMENTS

Rabbits were each immunized once by s.c. injection of 1.0 ml of the concentrated larval antigens combined with 2.0 ml of colloidal aluminium phosphate as adjuvant. Rabbits were challenged by oral administration of 2500 *T. pisiformis* eggs 14 d after vaccination, and autopsied 30 d later. The culture antigens collected under nitrogen stimulated a high level of protective immunity in the rabbits whereas the "oxidized" media were not protective. This paper also presented data on amino acid analysis of the culture antigen.

11° *Taenia taeniaeformis*: vaccination with oncosphere antigens

(Lloyd, 1979)

The preparation of oncospheres and culture method are given in Section

III.A.8° but 10% normal mouse serum was substituted for FBS. CFI mice, 6 weeks old, were immunized twice at 2-week intervals by i.m. injection of 0.5 ml of 10 times concentrated antigen emulsified in FCA (volume unspecified). Challenge infection was with 400 *T. taeniaeformis* eggs 2 weeks after the last injection. Immunized mice, killed 4 weeks after challenge, showed high levels of protection (94%) against *T. taeniaeformis* infection.

12° *Taenia taeniaeformis*: vaccination with oncosphere antigens

(Rajasekariah *et al.*, 1980a)

(i) PREPARATION OF ACTIVATED ONCOSPHERES

Oncospheres were hatched and activated using a modification of the method described by Rickard and Bell (Section III.A.7°(i)). Eggs were incubated initially for 3–4 h in pepsin-HCl followed by 6–8 h in AIF (Section III.A.7°(i)) supplemented with 0.1% trypsin. Oncospheres were washed in 0.85% NaCl, centrifuged (200*g*) and the embryophoric blocks and unhatched eggs separated from hatched oncospheres by centrifugation on Percoll (Rajasekariah *et al.*, 1980b): a 1.0 ml suspension of 100 000 freshly hatched oncospheres in 0.85% NaCl was layered over 5.0 ml of Percoll in 10 ml polycarbonate tubes. The tubes were spun (4500*g*, 30 min, 4°C) using a swing-out head. The sample at the interface of the saline and Percoll was carefully removed and washed twice (centrifuged at 500*g*) in a large excess of ice-cold sterile saline. Care was taken not to carry over excess Percoll into the saline for the first wash, because increased density of the saline resulted in loss of oncospheres when the supernatant was removed.

(ii) PREPARATION OF CULTURE ANTIGENS

Hatched and activated oncospheres were washed in warm sterile culture medium 858 (Section III.A.1°) supplemented with 1% glucose. Culture tubes containing 200 000 oncospheres each in 12 ml of medium were incubated for 72 h at 37°C and rotated at 4 rev min^{-1}. Culture medium was collected either by standing the tubes upright for 2 h and removing the supernatant or by centrifugation (3500*g*, 1 h).

(iii) VACCINATION EXPERIMENTS

Groups of CBA/H mice were injected simultaneously both s.c. and i.p. with 0.1 ml antigen either with no adjuvant or in an equal volume of FCA. Second antigen injections were given by the same route 2 weeks later without adjuvant; the total oncosphere equivalent dose of culture supernatant was 8000 per mouse. Mice were challenged with 450 *T. taeniaeformis* eggs 2 weeks after the last injection and killed 30 d later. The gravity sedimented supernatant gave levels of protection ranging from 73–87%, whilst the centrifuged supernatant gave only 34–45% protection.

(iv) COMMENT

The results suggest that some of the antigens collected in culture are insoluble particles, which has important implications for purification studies.

13° *Taenia taeniaeformis*: vaccination using metacestode antigens
(Kwa and Liew, 1977)

(i) CULTURE METHOD

Strobilocerci, collected from livers of infected rats, were immediately washed several times in saline and then twice in PBS (pH 7.2). They were gently rotated in culture tubes (37 °C, 24 h) containing Eagle's minimum essential medium (see Taylor and Baker, 1968, p. 355) supplemented with 1 g litre^{-1} glucose, 5 iu ml^{-1} penicillin and 5 μg ml^{-1} streptomycin. The supernatant was decanted, the protein concentration was adjusted to 1.35 mg ml^{-1}, and it was stored at -10 °C.

(ii) VACCINATION EXPERIMENT

Groups of outbred albino rats were immunized by s.c. injection of 1.0 mg culture antigen protein (volume not specified) in FCA. Rats were challenged with 1000 eggs orally and killed 3 weeks later. Immunized rats had levels of protection around 95% compared with control animals.

14° *Taenia taeniaeformis*: vaccination with metacestode antigens
(Ayuya and Williams, 1979)

(i) PREPARATION OF ANTIGENS

Strobilocerci were collected from rats and washed in distilled water followed by sterile physiological saline. Fifty strobilocerci were placed in culture vials containing 150 ml of Hanks's BME (Gibco) with 1.0 mg ml^{-1} streptomycin, 1000 iu ml^{-1} penicillin, 500 iu ml^{-1}

polymixin B, and incubated (37 °C, 24 h). The culture medium was collected and dialysed (48 h against distilled water and 12 h against PBS). The medium was concentrated by vacuum dialysis or by dialysis against polyethylene glycol. If the latter method was used, the culture medium was further dialysed against PBS for 48 h. It was frozen at – 20 °C until used.

(ii) VACCINATION EXPERIMENTS

Groups of six rats were immunized with culture antigens using a variety of routes (i.m., *per os*, i.p.) and with a variety of adjuvants (FCA, *Bordetella pertussis*, aluminium hydroxide) or without adjuvant. Oral and i.p. administration without adjuvant were effective in stimulating resistance against challenge with homologous eggs. Adjuvants were necessary for i.m. immunization; both *B. pertussis* and aluminium hydroxide were effective adjuvants, but not FCA.

15° *Taenia taeniaeformis*: vaccination with metacestode antigens

(Rajasekariah *et al.*, 1980c)

(i) PREPARATION OF CULTURE ANTIGEN

Culture antigen was prepared from mature (6 month) strobilocerci as well as from 3-week-old metacestodes. The latter were collected from mouse livers by digestion in 1.0% trypsin in 0.85% NaCl at 37 °C for 1 h. Larvae were dissected from the liver tissue and washed in several changes of warm (37 °C) medium 858 (Section III.A.1°) supplemented with 1% glucose. 20 three-week-old larvae were incubated (37 °C) in tubes containing 12 ml of the above medium for 4–6 d with continuous rotation (4 rev min^{-1}). Culture medium was changed daily. The total culture medium collected was pooled and centrifuged (500*g*, 5 min) and the supernatant was stored at – 20 °C.

Fully developed strobilocerci were washed in several changes of warm (37 °C) sterile 0.85% NaCl and incubated in 25 ml of the above medium in 30 ml culture flasks for 6 d. Medium was changed daily and the pooled medium was centrifuged (either 4500*g*, 30 min or 500*g*, 5 min) and the supernatants were stored at – 20 °C.

(ii) VACCINATION EXPERIMENTS

Several different vaccination protocols were used and the original reference should be consulted for the details. Mice vaccinated with concentrated culture antigens centrifuged (4500*g*) from 6-month-old strobilocerci did not develop protective immunity against homologous infection with eggs, irrespective of the adjuvant used. When mice were vaccinated twice with concentrated antigen (centrifuged at 500*g*) from either 3-week-old or 6-month-old metacestodes (*B. pertussus* adjuvant), a significant level of protection was achieved.

(iii) COMMENT

These results again suggest that particulate antigens are important components of those collected in culture medium during *in vitro* incubation of these parasites.

B. *MESOCESTOIDES*

1° *M. corti*
(Kowalski and Thorson, 1972)

(i) ANTIGEN PREPARATION

Tetrathyridia were removed aseptically from the peritoneal cavity of an infected mouse, washed six times in sterile Krebs–Ringer solution (K-R) containing 250 μg ml^{-1} streptomycin and 250 iu ml^{-1} penicillin, and then washed three times in sterile K-R without antibiotics. Larvae were incubated in 50 ml of sterile K-R for 24 h at 37°C. The supernatant fluid was stored at −20°C. Before immunization, the protein content (Lowry method) of the antigen was adjusted with sterile K-R to 90 μg of protein ml^{-1}.

(ii) VACCINATION EXPERIMENT

Lobund strain mice, 3–4 weeks old, were immunized by i.p. inoculation of antigen according to the following schedule. Days 0, 2, 4, 6, 0.1 ml; days 8, 10, 12, 14, 0.2 ml; days 16, 18, 20, 22, 0.3 ml. On day 24 the mice were challenged with 100 tetrathyridia of *M. corti* either orally or i.p., and necropsied 10 d later. Mice receiving antigen via either route had approximately 60% fewer larvae compared to the controls.

C. *ECHINOCOCCUS*

1° *E. granulosus* adult worms
(Herd *et al.*, 1975)

These experiments describe attempts to use antigens from *in vitro* cultured *E. granulosus* adults to immunize dogs against infection with protoscoleces.

(i) PREPARATION OF ANTIGENS

Adult *E. granulosus* were allowed to develop in dogs for 33–39 d before harvesting for culture. The culture medium consisted of 1 litre of 10 times concentrated, Medium 199 (see Taylor and Baker, 1968, pp. 355–356), a dialysate of 2 litres of the inactivated serum of new born calves, 150 ml 30% glucose, 100 ml 2% potassium chloride, 10^7 iu sodium benzylpenicillin, 10 g streptomycin sulphate, 2.5×10^6 iu polymyxin B sulphate, 2×10^6 iu nystatin, 200 ml 4.4% sodium bicarbonate and deionized distilled water to 10 litres. The final pH was 7.4. The culture fluid, except for the nystatin which was added last, was passed through an HICP-10 membrane in an Amicon CH3 concentrator to remove macromolecules, then sterilized by filtration. Worms were cultured (at 38–39 °C) in batches of about 500 in 50 ml of culture fluid in sterile Falcon T flasks, gassed daily with 10% O_2 and 5% CO_2 in N_2 at 38–39 °C. The secretions and culture fluid were collected daily and pooled and fresh culture medium was added to the flasks. The secretory antigens were concentrated and separated from other culture components using an Amicon HICP-10 membrane and freeze-dried.

(ii) VACCINATION EXPERIMENT

Dogs were given 4 s.c. injections of culture antigens each containing 5.0 mg of protein emulsified in equal volumes of FCA (first injection) or FIA (subsequent injections). Dogs were challenged by oral administration of 80 000 protoscoleces each. Antibodies to culture antigens were demonstrated in the sera of immunized dogs by immunoprecipitation in gels, and there was a significant depression of worm egg production by *E. granulosus* worms in the immunized dogs.

(iii) COMMENT

Herd (1977) described a further experiment using the above techniques to immunize dogs against *E. granulosus* infection. The experiment was largely unsuccessful owing to a high level of natural resistance to infection in some of the control dogs. This emphasizes the need for more information on the genetic factors controlling both natural resistance and acquired immunity.

IV. NEMATODA

This section outlines only the most recent successful experiments using *in vitro* cultured antigens to immunize against nematode infection. Readers interested in early experiments with *Nippostrongylus muris* (Thorson, 1953), *N. brasiliensis* (Denham, 1968), *Trichinella spiralis* (Campbell, 1955; Chute, 1956), *Ascaris lumbricoides* (Soulsby, 1957), *Dictyocaulus viviparus*, *Trichostrongylus colubriformis*, *Strongyloides papillosus* (Silverman *et al.*, 1962), or *Haemonchus contortus* (Silverman, 1965) should consult the original papers for details of their methods.

A. IMMUNIZATION USING LIVING LARVAE CULTURED *IN VITRO*

1° *Oesophagostomum radiatum*
(Herlich *et al.*, 1973)

(i) CULTURE METHOD

Larvae were cultured according to the method of Douvres (1970) using medium KW-2, a modification of medium KW-1 (Douvres *et al.*, 1966, detailed in Taylor and Baker 1968, p. 284, and 1978, p. 255). Medium NCTC-135 was substituted for NCTC-109 and the pH was adjusted to 6.8. The other ingredients (bovine serum, peptone, yeast extract, dextrose and antibiotics) remained the same. Roller system cell production bottles, containing 100 or 200 ml of medium, were rotated on a roller apparatus (1 rev every 1.5 min). Inocula of up to 2000 larvae ml^{-1} of medium were used, and all cultures were incubated at 39°C. It is not clear from the description whether the larvae were grown with the addition of EBK or EBTr cells, or without the addition of any cells.

(ii) VACCINATION EXPERIMENTS

Calves were given a single i.p. inoculation of either advanced parasitic third-stage larvae (L_3) only (58, 670 larvae) or a mixture of 13% L_3, 17% third-moult larvae and 70% fourth-stage larvae (mixture called L_4, approximately 50 000–60 000 larvae). Calves were challenged at various intervals with oral inoculation of 15 000 normal infective larvae and killed 50–54 d after challenge. Several calves underwent a mild to severe febrile reaction within the first 10 d after inoculation. Vaccination with L_3 gave 44–90% protection against the challenge infection and with L_4, 36–83% protection. Many nodular lesions were produced in the peritoneal cavity of immunized calves.

2° *Ostertagia circumcincta*
(Rose, 1976)

(i) CULTURE METHOD

O. circumcincta were grown *in vitro* using the method of Leland (1963) and Rose (1973) detailed in Taylor and Baker (1968, pp. 293–295, 1978, pp. 255–256) except that the culture vessel was a 100 ml glass tube fitted with a screw cap. Approximately 30 000 larvae were grown in 40 ml of medium which was changed three times a week. Cultures were maintained for 5.5 weeks during which time development took place from the infective third stage larva (L_3) to the adult worm.

(ii) VACCINATION EXPERIMENT

Two lambs were injected s.c. with a mixture of living L_3, fourth-stage larvae (L_4) and adult stages, on three occasions at 7 d intervals. The total number of worms injected was 50 000. Only a limited degree of protection was achieved against challenge infection with 2000 infective larvae of *O. circumcincta* given by mouth five times weekly for 3 weeks.

B. IMMUNIZATION USING WORM METABOLITES COLLECTED *IN VITRO*

1° *Ostertagia circumcincta*
(Rose, 1976, 1978)

(i) CULTURE METHOD

For the culture method see Section IV.A.2°. Medium was collected during both a 5.5 week culture and a 3 week culture. In addition, metabolites were collected in Earle's balanced salt solution (EBSS, see Taylor and Baker, 1968, p. 350) by transferring worms at the fourth and early fifth stages from the culture medium into EBSS with antibiotics. Worms were maintained for 10 d at 38.5 °C on the roller drum and the EBSS changed three times a week. Pooled culture media were concentrated either by lyophilization or by ultrafiltration with Amicon PM-10 membranes.

(ii) VACCINATION EXPERIMENTS

In the first experiment (Rose, 1976), three cross-bred lambs were injected i.m. with 5.0 ml of concentrated medium emulsified in an equal volume of FCA on 5 consecutive days each week for 11 weeks, and two lambs were given lyophilized medium in distilled water by mouth 5 days each week for 2 weeks. Each lamb received 300 000 larval equivalents of metabolites. Four weeks after the last treatment lambs were given 2000 infective larvae of *O. circumcincta* by mouth 5 days each week for 3 weeks. These immunization procedures gave marginal protection against the challenge infections, as shown by lower worm burdens, reduced length of worms and lower faecal egg counts compared with the control lambs.

In the later experiments (Rose, 1978), lyophilized antigens and antigens concentrated by ultrafiltration were incorporated into FCA and used to immunize 9- , 6-, and 3-month-old lambs by i.m. injection. The total antigen dose per lamb was derived from culture of 300 000 infective larvae of which about 10% developed to late fourth and early fifth stages. Both the culture antigens and the antigens collected in EBSS stimulated a degree of protective immunity against challenge infection in the 6- and 3-month-old lambs, but not the 9-month-old lambs. Worm burdens and worm egg counts were lower in the immunized animals than in the controls. It was suggested that an age resistance resulting in low worm burdens in the control lambs obscured any positive effect of the vaccination in the older lambs.

(iii) COMMENT

Herlich and Douvres (1979) failed to demonstrate any protective immunity in calves vaccinated with culture media in which *O. ostertagi* had been grown.

2° *Haemonchus contortus*
(Ozerol and Silverman, 1970)

(i) CULTURE METHOD

The method is detailed by Silverman *et al.* (1966) and Ozerol and Silverman (1969) (see also Taylor and Baker, 1968, pp. 290–292). A suspension of *H. contortus* L_3 larvae (50 000 ml^{-1}) in 10 ml Earle's balanced salt solution (EBSS, see Taylor and Baker, 1968, p. 350) without glucose was gassed with CO_2 and incubated on a slowly rotating roller drum for 24 h at 38°C. The exsheathed L_3 were removed from EBSS and cultured in *Caenorhabditis briggsae* medium (see Table II) reinforced with five times the normal quantity of vitamins. Metabolites were harvested at 5–8 d when over 70% of the larvae were in L_4 or early fifth stage (L_5) and mortality was less than 10%. The supernatant fluids were separated from the living worms by sedimentation and filtration, dialysed against deionized water at room temperature for several hours, and concentrated by lyophilization.

TABLE II. Chemically defined basal medium[a] for axenic cultivation of *Caenorhabditis briggsae* (Sayre *et al.*, 1963)

	M%	mM	mg litre^{-1}
1. Amino acids			
"Essential" (EAA)			
L-Arginine	10.4	7.45	1300
L-Histidine	3.4	2.43	377
L-Isoleucine	12.1	8.74	1148
L-Leucine	20.3	14.60	1918
L-Lysine	13.0	9.37	1372
L-Methionine	4.8	3.47	518
L-Phenylalanine	6.9	5.03	830
L-Threonine	11.2	8.03	956
L-Tryptophane	1.7	1.20	245
L-Valine	16.2	11.62	1360
Total	100.0	71.94	
"Non-essential" (NEAA)			
L-Alanine	23.4	20.8	1860
L-Aspartic	18.2	16.2	2160
L-Cysteine	0.2	0.21	26
L-Glutamic (Na)	4.8	4.32	733
L-Glutamine	14.5	13.0	1950
L-Glycine	14.4	12.81	962
L-Proline	8.5	7.55	870
L-Serine	11.3	10.0	1051
L-Tyrosine	3.9	2.0	363
L-Phenylalanine[b]		1.45	240
α,ϵ-Diaminopimelic acid	0.8	0.72	137
Total	100.0	89.06	
Total amino acids		161	
Molar ratio (NEAA/EAA) = 1.24			
2. Nucleic acid substituents			
Adenylic acid		1.0	347
Cytidylic acid		1.0	323
Guanylic acid		1.0	363
Uridylic acid		1.0	324
Thymine		1.0	126
3. Vitamins and growth factors			
N-Acetylglucosamine		6.7×10^{-3}	15.0
DL-Alanine		168.0×10^{-3}	15.0
p-Aminobenzoic acid[b]		54.0×10^{-3}	7.5
Biotin[b]		15.3×10^{-3}	3.75
Cyanocobalamine		2.8×10^{-3}	3.75
Folinate (Ca)		7.3×10^{-3}	3.75
Niacin[b]		61.0×10^{-3}	7.5
Niacinamide		61.1×10^{-3}	7.5
Pantetheine		13.3×10^{-3}	3.75

TABLE II. *Continued*

Pantothenate (Ca)	29.0×10^{-3}	7.5
Pteroylglutamic acid[b]	16.9×10^{-3}	7.5
Pyridoxal.HCl	18.3×10^{-3}	3.75
Pyridoxamine.2HCl	15.5×10^{-3}	3.75
Pyridoxine.HCl	36.4×10^{-3}	7.5
Riboflavin 5.PO_4(Na)	14.6×10^{-3}	7.5
Thiamine.Cl.HCl	22.2×10^{-3}	7.5
DL-Thioctic acid[a]	18.8×10^{-3}	3.75
4. Glutathione, reduced	1.33	408
5. D-Glucose	7.3	1315
6. Choline chloride	6.34	885
i-Inositol.$2H_2O$	3.59	648
7. KH_2PO_4	9.0	1225.5
KOH	16.5	925.5
Mg.H.citrate.$5H_2O$	3.0	915.0
$CaCl_2.2H_2O$	1.5	220.0
K_3 citrate	1.5	486.0
$Fe(NH_4)_2(SO_4)_2.6H_2O$	0.15	58.8
$MnCl_2.4H_2O$	0.113	22.2
$ZnCl_2$	0.075	10.2
$CuCl_2.2H_2O$	0.038	6.5

[a]pH of medium = 5.8.
[b]Solvent: triethanolamine, 6.5 mg $litre^{-1}$.

(ii) VACCINATION EXPERIMENT

Culture media were separated into components of different molecular weight by column chromatography using Sephadex G-200. Two serologically active fractions, A and B, containing metabolites of mol.wt 650 000, 240 000 and 35 000 protected lambs against *H. contortus*. Three immunization injections were given to each lamb, but no other details are specified. Medium from L_3, L_4 and adults stimulated immunity that suppressed the egg production of the challenge infection by 72–82%.

(iii) COMMENT

Neilson (1975) describes experiments in which he failed to immunize lambs against infection with *H. contortus* using metabolic products from larvae cultured *in vitro*.

3° *Trichostrongylus colubriformis*
(Rothwell and Love, 1974)

(i) CULTURE METHOD

T. colubriformis L_4 larvae were collected from the anterior small intestine of sheep 8 d after infection with 100 000 L_3. The intestine was removed and opened longitudinally immediately after slaughter; food particles were rinsed off in 0.85% NaCl and the intestine was placed on a Baermann apparatus in saline at 38 °C. After 1.5–2 h the worms were removed from any underlying debris by pipette and rinsed in warm saline. Larvae were allowed to settle in a centrifuge tube and approximately 5.0 ml incubated, without agitation, under air in a conical flask in 20 ml of saline at 38 °C for 6 h. The supernatant fluid was removed, lightly centrifuged and then filtered (0.45 μm pore size Millipore filter). Some incubation fluid was concentrated to 1/5 volume by ultrafiltration using Amicon UM-2 membranes.

(ii) VACCINATION EXPERIMENTS

Guinea pigs were injected with either 10 μg of incubation fluid protein in 0.15 ml FCA (10 animals) or 100 μg of incubation fluid protein per 100 g body weight (7 animals) and then challenged with 2000 *T. colubriformis* infective larvae. In both groups the injection of the incubation fluid stimulated strong protective immunity against the challenge infection when compared with sham-vaccinated controls.

(iii) COMMENT

Herlich and Douvres (1979) described experiments in which they failed to immunize calves against infection with *T. colubriformis* and *T. axei* larvae grown *in vitro*.

4° *Ascaris suum*
(Guerrero and Silverman, 1971)

(i) CULTURE METHODS

The preparation of second (L_2), third (L_3) and fourth (L_4) stage larvae and the culture methods are described in detail by Levine and Silverman (1969) and Guerrero and Silverman (1969) (see Taylor and Baker, 1978, p. 243).

L_2 and early L_3 larvae were cultured in sterile bottles in a CO_2 environment using two media. One batch was cultured in 20 ml of a 1 : 1

dilution of *Caenorhabditis briggsae* medium (Table II) 2 × with 5 × vitamins in sterile water (Gibco) supplemented with 4.0 ml sterile pig serum and 1.0 ml of antibiotic mixture. The other batch was cultured in 20 ml of Eagle's minimum essential medium (Gibco) and received the same supplement as described above. On day 5, the bottles were removed from the incubator and received an additional 2.0 ml of the *C. briggsae* culture medium, 1.0 ml of a 25% solution of bovine serum albumin, and 0.15 ml of Fungizone (Squibb). On day 12, the bottles were removed from the incubator, opened, sampled, and the larvae counted; their viability and degree of development were also determined. The metabolic antigens from L_2 and L_3 were separated into 4.0 ml aliquots and then freeze-dried and stored at 4°C.

To collect metabolic antigens from late L_3 and early L_4, 500 mice were each inoculated with a single dose of 10 000 embryonated *A. suum* eggs, and 7 d later larvae were collected from the lungs of these animals by the Baermann method. The recovered larvae were axenized for 30 min in a 1 : 20 solution of antibiotic mixture (50 g streptomycin, 55 × 10^6 iu penicillin, 2.5 g neomycin in 500 ml of sterile saline) at 38°C.

The antibiotic-treated larvae were transferred to sterile silicon-stoppered test tubes containing 3.0 ml of Eagle's minimum essential medium. The medium was supplemented with 4% of a 25% solution of bovine serum albumin, 2% pig serum, 0.5% Fungizone and 4% of the previously described antibiotic solution. The silicone stoppers were flamed and then cooled and the plugged tubes were wrapped with parafilm before being put on a roller drum at 38°C. These cultures were left undisturbed for 3 d. On day 3, the tubes were opened and gassed with CO_2 for 3 s in the liquid medium and 5 s in the tube environment. CO_2 was supplied from a 500 ml Erlenmeyer flask containing three or four small chips of dry ice and equipped with a rubber tube and a sterile Pasteur pipette. After gassing, the tubes were allowed to remain two more days on the roller drum at 38°C. On day 5, the tubes were removed from the incubator, the metabolic antigens separated and treated as for L_2 and L_3 antigens.

(ii) VACCINATION EXPERIMENTS

Groups of 11 mice were vaccinated i.p. with the various culture antigens with 4% sodium alginate (Algivant; Colab Laboratories) as adjuvant in doses of 1.0 ml per animal of 1 : 1 mixture of antigen and adjuvant, and challenged with a single dose of 10 000 *A. suum* eggs orally. Efficacy of vaccination was assessed according to a lung scoring technique on a scale of 0 to 5 (Guerrero and Silverman, 1969). Metabolic antigens from L_2, L_3 and L_4 were effective in immunizing recipient mice.

5° *Ascaris suum*
(Stromberg and Soulsby, 1977)

(i) CULTURE METHODS

Culture antigens associated with moulting of L_3 to L_4 were collected. 2000 L_3 were obtained by the Baermann technique from the lungs of rabbits 7 d after oral infection with 100 000 embryonated *A. suum* eggs. 2000 L_3 were cultured in Medium 199 (see Taylor and Baker, 1968, pp. 355–356) supplemented with 0.0225 mg ml^{-1} glucose, 300 iu ml^{-1} penicillin, 0.3 mg ml^{-1} streptomycin, 100 iu ml^{-1} mycostatin and 0.04 ml swine serum ml^{-1} under an atmosphere of 5% O_2, 5% CO_2, 90% N_2 in 15 ml conical centrifuge tubes incubated at 37°C. The culture fluid was collected 5 d later, when 95% of the L_3 had moulted to the L_4, and concentrated 20 times by ultrafiltration using Amicon PM-10 membranes (details of storage not given).

(ii) VACCINATION EXPERIMENT

Guinea pigs were immunized with single i.m. injection of antigens, in FCA, equivalent to 0.75 mg of protein. The antigens stimulated a significant degree of protection against the challenge with 5000 artificially hatched L_2 injected into the mesenteric vein. In further similar experiments Stromberg (1979) demonstrated that the *A. suum* culture fluid contained a single vaccinating antigen of mol. wt 67 000 with about 79% protein and 22% carbohydrate.

2. *Comments on 4° and 5°*

Interestingly the results of Guerrero and Silverman (1971) and Stromberg and Soulsby (1977) differed in that the latter authors were unable to stimulate protective immunity in guinea pigs with culture antigens from non-moulting L_2, L_3 or L_4 stages of *A. suum*; this could be partly accounted for by differences in culture times and in the interpretation of the various stages of development, since Maung (1978) has shown that the infective larva of *A. suum* is a double-ensheathed L_3, rather than an L_2, unless the second moult is delayed.

6° *Toxocara canis*
(Nicholas *et al.*, 1984)

(i) CULTURE METHOD

60 000 L_3, hatched by Fairbairn's (1961) technique, were suspended in

100 ml of modified Eagle's medium (Dulbecco) and 5.0 ml aliquots dispensed in screw-capped tubes gassed with 10% CO_2 in air. The medium was changed weekly and the removed medium concentrated 10 times and washed free of low-molecular-weight substances with PBS (Dulbecco) by filtration through an Amicon YM-10 filter.

(ii) VACCINATION EXPERIMENT

CBA mice were immunized by three s.c. injections of 0.05 ml culture antigen (approximately 1.0 μg protein) in an equal volume of FCA. Mice were challenged orally 3 weeks later with 500 infective *T. canis* eggs and killed after 21 d. The immunized mice showed a high degree of protection against homologous challenge infection.

7° *Angiostrongylus cantonensis*

(Uahkowithchai *et al.*, 1977; Techasoponmani and Sirisinha, 1980)

(i) CULTURE METHOD

Actively motile adult worms were collected from the pulmonary arteries of infected adult albino rats and washed several times in Hanks's balanced salt solution (see Taylor and Baker, 1968, p. 350) containing 100 or 300 iu ml^{-1} penicillin, and 200 or 500 μg ml^{-1} streptomycin. Two different culture media were used: either NCTC-109 or Eagle's basal medium (see Taylor and Baker, 1968, pp. 351–352 and 354) in the proportion of 0.1 ml per worm. The loosely-stoppered 20 ml cylindrical glass culture vials were incubated at 37 °C in an atmosphere of 5% CO_2 and the medium was changed daily. In the first experiment (NCTC-109), the male and female worms were cultured separately (24 h) and the supernatant of the centrifuged (speed not given) culture medium was used as antigen. In the second experiment, 2 days' changes of culture medium were pooled, centrifuged (300*g*, 30 min), dialysed at 4 °C in excess 0.01 M NH_4HCO_3 and lyophilized. For immunization, the medium was reconstituted in PBS, pH 7.2, and centrifuged (speed not given) to remove insoluble materials.

(ii) VACCINATION EXPERIMENTS

Outbred albino rats and Swiss mice were immunized by s.c. injection of antigen from either male or female worms. Adult rats (250–300 g) were given either two injections (0.65 mg antigen in FCA followed by 0.50 mg antigen in FIA a week later) or three injections (the third dose of 0.30 mg antigen in PBS a week later). The volumes of adjuvant used were not specified. Mice (30 g) were given three injections of antigen (0.10, 0.05 and 0.05 mg) in a similar protocol to that used for the rats. The animals immunized with antigens from female worms were largely protected from what would otherwise have been a lethal challenge infection with L_3. Immunity was expressed as reduced mortality (especially in rats), reduced parasite development and reduced output of L_1 in the faeces.

8° *Trichuris muris*
(Jenkins and Wakelin, 1977; 1983)

(i) CULTURE METHOD

Male and female worms, recovered from the large intestine of NIH inbred mice, were washed in cold Krebs–Ringer solution and 400 worms were incubated in 8.0 ml Krebs-Ringer (3 h at 37 °C). Worms were filtered from the medium which was then centrifuged (60*g*, time unspecified) to remove worm eggs. The protein concentration of the medium was 1.0 mg ml^{-1} (determined spectrophotometrically, as in Waddell, 1956).

(ii) VACCINATION EXPERIMENTS

In the first experiments (Jenkins and Wakelin, 1977), mice were injected once (s.c.) with 200 μg of the culture antigen protein incorporated into an equal volume of FIA, and challenged 10 d later with 400 *T. muris* eggs. Approximately 70% fewer worms were recovered from immunized mice compared with the controls. In their later experiments (Jenkins and Wakelin, 1983) it was shown that: (a) 10 μg of culture antigen stimulated highly effective immunity; (b) 100 μg of antigen was effective without adjuvant; (c) FIA or alum were the best adjuvants; (d) antigen–antibody complexes were immunogenic. Further studies on chromatographic separation of the antigen were also described (Jenkins and Wakelin, 1983).

9° *Trichinella spiralis*
(Gamble, 1985)

(i) CULTURE METHOD

Gamble (1985) stated that *T. spiralis* muscle larvae, from muscle of female Sprague–Dawley rats were maintained in Dulbecco's minimum essential medium for periods of 18–20 h as described by Gamble *et al.* (1983). However, the description in the latter paper varies from Gamble (1985) in the culture medium used, Leibovitz medium (L-15; Gibco), and in the duration of incubation (72 h). The method described by Gamble *et al.* (1983) is as follows. Following recovery from rats, worms were washed three times in Earle's balanced salt solution (pH 7.4) containing penicillin and streptomycin (1000 iu ml^{-1} and 1000 $\mu g\ ml^{-1}$, respectively). Larvae were inoculated (10 000 worms ml^{-1}) into medium L-15 (pH 7.2) containing 2 mM L-glutamine, 25 mM HEPES and penicillin–streptomycin as before. The medium was collected 72 h later by filtration through a 0.20 μm membrane (Nalgene) to remove larvae. The filtrate was dialysed against saline, concentrated 100-fold on a PM-10 membrane (Amicon) at 4 °C and stored at − 20 °C (no details of temperature of culture, the vessels used or the system used were given).

(ii) VACCINATION EXPERIMENTS

Outbred female Swiss-Webster mice, 2–3 months old, were immunized three times at 2-week intervals by i.p. injection of antigen emulsified in FCA (volume not specified). Animals were challenged with 150 infective larvae 1 week after the final immunizing dose. Experiments showed that 10 μg, but not 1.0 μg, of antigen protein stimulated significant immunity, as measured by the number of larvae recovered from the muscles. Immunized mice had fewer (but not significantly so) adult worms in the intestines 6 d after challenge, and the fecundity of adult worms was significantly reduced.

V. ALTERNATIVE METHODS OF OBTAINING HELMINTH ANTIGENS *IN VITRO*

A. CELL CULTURES

The aim of this approach is the establishment of helminth cell lines in continuous culture, but few have attempted it so far.

1° *Echinococcus multilocularis*
(Sakamoto, 1978)

Daughter cysts were collected aseptically from the livers of cotton rats infected with *E. multilocularis* of Alaskan origin. They were initially treated with 0.2% trypsin (presumably in isotonic saline) for 5 min then washed three times with Hanks's saline (see Taylor and Baker, 1968, p. 350). The tissue was cut finely and trypsinized again on a magnetic stirrer. Larger pieces of debris were removed by passage through a sieve (230 mesh) and the sieved material allowed to stand for 5 min. The supernatant was transferred to a fresh tube and centrifuged (5 min, 800 rpm, *g* force not indicated but probably about 100–150). The pellet was dispersed in Hanks's saline and the centrifugation and washing procedure were repeated three times before inoculating cultures (apparently no quantitation of the cells was carried out and the number of cells added to cultures was not specified).

The culture medium comprised NCTC-135 containing 0.5% lactalbumin hydrolysate, 0.1% yeast extract and 20–30% inactivated FCS (sera from other species including calf, cow, cotton rat, Wistar rat, horse and ICR/JCL mouse gave less satisfactory results). In some experiments, cotton rat embryo or liver extract, or chick embryo extract (mode of preparation not given) was added to the culture medium in unspecified amounts. Cells (unspecified numbers in unspecified volumes of culture medium) were cultured in 30 ml Falcon flasks with a gas phase of 5% O_2, 5% CO_2, 90% N_2.

So-called germinal cells were released from the germinal layer of trypsin-treated cysts and attached to the bottom of the flask. They tended to appear spindle-shaped initially but then became asteroid, interconnecting with one another by means of long processes. Cells clearly multiplied and were subcultured (presumably by standard trypsinization procedures) at 10 d intervals, but they failed to proliferate after the third subculture. Cotton rat liver extract stimulated cell division. Embryo-extract promoted differentiation of what appeared to be multilocular vesicles, but these gradually degenerated and became detached from the bottom of the flask.

2° *Schistosoma mansoni*
(Capron and Dupas, 1972)

Explants of adult *S. mansoni* were cultured in plasma clots with a liquid overlay. Adult worms (25–30 d) were recovered from infected hamsters and washed in Hanks's saline (see Taylor and Baker, 1968, p. 350) containing penicillin (200 iu ml^{-1}) and streptomycin (200 $\mu g\ ml^{-1}$). Worms were cut into fragments and treated with a solution of 0.5% trypsin for 15 min. The fragments were fixed in position with the aid of a plasma clot formed from 2 parts plasma and 1 part embryo extract. Medium (4.5–5.0 ml) was then added and changed twice weekly. The best results were obtained with Eagle's basal medium (see Taylor and Baker, 1968, p. 354) containing 10% calf serum together with penicillin (100 iu ml^{-1}) and streptomycin (50 $\mu g\ ml^{-1}$). It is inferred that all incubations were carried out at 37 °C.

After 8 d *in vitro*, migration of fibroblast-like cells from the tissue fragments into the plasma clots had taken place from 41 of 188 explants. Explants were considered to be alive only if contractile.

3° *Schistosoma mansoni*
(Weller and Wheeldon, 1982)

The above reference reviews in detail earlier attempts by the authors to culture schistosome cells (Weller, 1976). The culture medium used is referred to as Medium 1779 (see Table III), which consistently yielded plaques of schistosome cells *in vitro*. Schistosomes (Puerto Rican strain) were recovered from CD-1 mice infected 4 weeks previously with 300 cercariae using a perfusion technique (consult the original paper for details). The principal solution used for perfusion and many subsequent

TABLE III. Composition of medium 1779

Inorganic salts	mg litre^{-1}		mg litre^{-1}
$CaCl_2.2H_2O$	400	Cytosine	0.5
KCl	1000	Guanine-HCl	1.0
$MgSO_4.7H_2O$	400	Thymine	0.5
NaCl	3000	Uracil	1.0
$NaHCO_3$	2500	Xanthine Na Salt	1.0
NaH_2PO_4	10	Orotic acid Na salt	50.0
NaOH	200	**Organic acid**	
		α-Ketoglutaric acid Na	
Buffer		salt	200
HEPES	2380		
Trace metals		**Vitamins**	
Trager metals[a]	1.0 ml litre^{-1}	DL-Carnitine-HCl	5.0
Amino acids		Choline chloride	20.0
L-Alanine	1000	Menadione Na	
L-Glutamine	150	bisulphfite/albumin	1.0
Glycine	250	Vitamin B^{-12}	10.0
L-Histidine	1000	BME vitamin block	5.0 ml litre^{-1}
Hydroxy-L-proline	500	**Hormones**	
L-Lysine-HCl	100	Dexamethasone	392.5 ng
L-Ornithine-HCl	10	Insulin	5.0
L-Serine	250	Triiodo-L-thyronine	10 μg
BME essential amino	1.0 ml litre^{-1}	**Protein fractions**	
acids		Cohn's fraction IV	100
Peptides		Foetal calf serum	20.0 ml litre^{-1}
Glutathione	10	Gelatin	1000
L-Histidyl-L-alanine	25	Yeast extract/serum	1.0 ml litre^{-1}
Carbohydrates		**Others**	
Glucose	12 000	Phenol red	20
Trehalose	500	Serotonin creatinine	
Purines and pyrimidines		sulphate	0.4
Adenine	2.5	Cholesterol/albumin	3 drops ml^{-1} culture

[a]Trager (1957).

manipulations is referred to as schistosome salt solution (SSS), the composition of which is as follows:

	mg $litre^{-1}$
$CaCl_2 \cdot H_2O$	400
KCl	1000
$MgSO_4 \cdot 7H_2O$	400
NaCl	3000
$NaHCO_3$	2500
NaH_2PO_4	10

Worms recovered by perfusion with SSS + glucose (12 g $litre^{-1}$), HEPES (2.38 g $litre^{-1}$) and penicillin G (100 μg ml^{-1}), were cooled for 10–15 min at 5 °C and transferred to a tube (8.5 × 1.5 cm) for mincing. The worms were rinsed once with SSS containing 0.25% trypsin (pancreatic; Sigma T-8128) and minced in 0.25 ml of the same solution with fine scissors. The volume of SSS + 0.25% trypsin in the tube was increased to 3.0 ml and the fragments were trypsinized for 10 min at room temperature. One or two changes of supernatant were made to reduce the amount of finely granular caecal material present in the suspension. The trypsin solution was removed and the cell harvest centrifuged and resuspended three times in 3.0 ml aliquots of Medium 1779. Leighton tubes and coverslips were washed thoroughly before use and preconditioned by incubating them with 1.25 ml medium for 24–48 h at 35 °C before use. This apparently enhances cell attachment to coverslips at the base of the tubes.

Sedimented cells and fragments were resuspended in 2.0 ml of medium and 0.1 ml amounts were added to 1.0 ml of medium in preconditioned Leighton tubes. Loosely capped cultures were placed in racks and incubated in a horizontal position at 35 °C in 5% CO_2 in air. The culture medium was changed twice weekly, but if the pH fell to 6.8 (assessed by phenol red colouration) three changes were made each week. The medium was changed by standing the tubes vertically, removing the fluid over the sedimented material and replacing it with 1.0 ml of fresh medium.

The original reference should be consulted for details of the many experiments that led to the evolution of Medium 1779 and additional information on its preparation.

The worm fragments appeared semi-transparent and refractile and sustained their muscular activity for several weeks. Phenol red accumulated in parts of the intact excretory system, indicating that functional activity was retained. Cyst-like structures also developed and these often contained brick-red inclusions. Exposed cut surfaces of fragments underwent repair, and fragments frequently reassociated.

COMMENTS

Cell proliferation apparently takes place from trypsin-treated explanted tissue of schistosomes and multilocular hydatids. However, long-term cell cultures have not been established. To achieve this may require transformation of the cells to a malignant state (by virus or carcinogen?) as with many mammalian cell lines. Given that few attempts have been made to culture parasitic flatworm cells, the promising results noted above suggest that more research in this area is very much warranted.

B. CELL FUSION

The objective was to use an established cell line as a carrier of parasite genetic material with a view to obtaining continuous synthesis of parasite antigens *in vitro*.

1° *Fasciola hepatica*
(Howell, 1981)

Cells of adult *F. hepatica* were fused with a rat cell line that cannot synthesize the enzyme hypoxanthine-guanine phosphoribosyl transferase (HGPRT). This genetic defect provides a means of selecting fused or hybrid cells in a medium containing hypoxanthine, aminopterin and thymidine (HAT medium). The assumption is that the *F. hepatica* HGPRT gene in any hybrid cells that form will correct the deficiency in the rat cells and the hybrid cells will continue to grow and possibly express other parasite genes that code for antigens of interest.

(i) FUSION AND CULTURE METHODS

Adult *F. hepatica* were recovered from the bile ducts of naturally infected sheep. Twenty flukes were finely minced in 1.0 ml Dulbecco's modification of Eagle's medium (DMM) plus 1.0 ml trypsin/versene (Commonwealth Serum Laboratories) and shaken in a McCartney bottle (27 ml universal container) for 20 min at 37 °C. The cells were decanted, centrifuged at 600*g* and resuspended in 1.0 ml DMM. About 2×10^7 cells were obtained with over 90% viability as determined by trypan blue exclusion. *F. hepatica* cells (2×10^7) in 1.0 ml DMM were mixed with 10^7 HGPRT negative rat fibroblasts in 4.0 ml DMM in a 15 ml glass centrifuge tube. The mixture was spun (600*g*, 5 min) and the supernatant was discarded. Fusion was effected as follows (method of Galfre *et al.*, 1979). The cell pellet was disrupted by tapping the tube. All further steps were carried out at about 37°C. 0.8 ml of 50% (w/w) polyethylene glycol (PEG, mol.wt 1500) in DMM (pH 7.6) was added over 1 min to resuspend the cells. The cell suspension was held at 37°C for 1 min and 1 ml of DMM was added over a further 1 min. 10 ml of DMM was then added over the next 5 min. The cells were immediately centrifuged as above and gently resuspended in 10 ml DMM containing 10% FCS. Fused cells (2×10^6 in about 0.6 ml) were inoculated into Falcon flasks (25 cm^2 growth area) containing 10 ml of 10% FCS in DMM. One day later, half the medium was removed and replaced by 10% FCS in HAT medium. This was again repeated 24 h later, then every second day. Cultures were regularly examined for the appearance of colonies of cells.

(ii) RESULTS

About 2 weeks after fusion with PEG, growing colonies of cells could be distinguished in the flasks. The hybridization frequency was about 1 in 5×10^5 cells treated. When the cells had grown to higher densities, they were cloned and karyotypic, biochemical and immunological analyses were performed. The HGPRT in the hybrids had the characteristics of fluke rather than rat enzyme. However, no fluke chromosomes could be detected with any confidence in the hybrids, and no fluke antigens could be identified in the cells using a rabbit antiserum to soluble somatic antigens of *F. hepatica*.

(iii) COMMENTS

Clearly, cell fusion between helminth and mammalian cells is possible and stable hybrid cell lines can be derived. However, since there are no known selection pressures that can be applied to the hybrids for the continued synthesis of parasite antigens, it may well be that the technique is as yet of limited application. In the present case, preferential loss of parasite chromosomes seemed to take place. The use of other host cell lines and separation of hybrids soon after fusion may avoid this problem.

VI. CONCLUSIONS

Unfortunately, studies on the *in vitro* culture of helminths for the production of vaccines have not lived up to the early expectations. Although there have been several promising advances, which are described in this chapter, the goal of a commercially viable process of *in vitro* production of a vaccine has not yet been achieved. It is probably fair to concede at this stage that *in vitro* cultivation of helminth parasites, as such, is not likely to be a practical avenue for vaccine production. Conceivably, approaches such as the *in vitro* culture of parasite cell lines and cell fusion techniques, such as described here, may reach practical reality, but it is more likely that recombinant DNA methods utilizing bacteria, yeasts or mammalian cells will provide commercial quantities of parasite antigens. None the less, *in vitro* cultivation methods have contributed, and will continue to contribute, extensively to the analysis of parasite antigens. None of the methods referred to above is likely to succeed until we can clearly identify the antigens we seek to produce *in vitro*.

REFERENCES

Ayuya, J.M. and Williams, J.F. (1979). The immunological response of the rat to infection with *Taenia taeniaeformis*. VII. Immunization by oral and parenteral administration of antigens. *Immunology* **36**, 825–834.

Burden, D.J., Harness, E. and Hammet, N.C. (1982). *Fasciola hepatica*: attempts to immunise rats and mice with metabolic and somatic antigens derived from juvenile flukes. *Veterinary Parasitology* **9**, 261–266.

Campbell, D.H. (1955). The antigenic role of the excretions and secretions of *Trichinella spiralis* in the production of immunity in mice. *Journal of Parasitology* **41**, 483–491.

Capron, A. and Dupas, H. (1972). Essais d'obtention de cultures cellulaires a partir d'explants de *Schistosome mansoni* (Sambon, 1907). *Comptes Rendus Héboromonadaires des Séances de l'Académie de Sciences, Seri. D* **275**, 1243–1246.

Chute, R. M. (1956). The dual antibody response to experimental trichinosis. *Proceedings of the Helminthological Society of Washington* **23**, 49–58.

Clegg, J.A. and Smith, M.A. (1978). Prospects for the development of dead vaccines against helminths. *Advances in Parasitology* **16**, 165–218.

Davies, C., Rickard, M.D., Smyth, J.D. and Hughes, D.L. (1979). Attempts to immunise rats against infection with *Fasciola hepatica* using *in vitro* culture antigens from newly excysted metacercariae. *Research in Veterinary Science* **26**, 259–260.

Denham, D.A. (1968). *Bordetella pertussis* vaccine as an adjuvant for helminth antigens. *Journal of Parasitology* **54**, 68.

Douvres, F.W. (1970). Cultivation of *Oesophagostomum radiatum* developmental stages in cell culture systems. *Journal of Parasitology* **56**, Sect. II, Pt 1, 83–84.

Douvres, F.W., Tromba, F.G. and Doran, J.D. (1966). The influence of NCTC 109, serum and swine kidney cell cultures on the morphogenesis of *Stephanurus dentatus* to fourth stage *in vitro*. *Journal of Parasitology* **52**, 875–889.

Dragneva, S. (1972). Vaccination against *Fasciola hepatica* with somatic and metabolic antigens. *Izvestiya Na Tsentralanya Khelminlogicha Laboratoriya* **15**, 81–87. [In Russian.]

Fairbairn, D. (1961). The *in vitro* hatching of *Ascaris lumbricoides* eggs. *Canadian Journal of Zoology* **39**, 153–162.

Galfre, G., Howe, S.C., Milstein, C., Butcher, G.W. and Howard, J.C. (1977). Antibodies to major histocompatibility antigens produced by hybrid cell lines. *Nature, London* **266**, 550–552.

Gallie, G.J. and Sewell, M.M.H. (1970). A technique for hatching *Taenia saginata* eggs. *Veterinary Record* **86**, 749.

Gamble, H.R. (1985). Comparison of immune effects in mice immunized with *Trichinella spiralis* adult and larval antigens. *Journal of Parasitology* **71**, 680–682.

Gamble, H.R., Anderson, W.R., Graham, C.E. and Murrell, K.D. (1983). Diagnosis of swine trichinosis by enzyme-linked immunosorbent assay (ELISA) using an excretory-secretory antigen. *Veterinary Parasitology* **13**, 349–361.

Guerrero, J. and Silverman, P.H. (1969). *Ascaris suum*: immune reactions in mice. I. Larval metabolic and somatic antigens. *Experimental Parasitology* **26**, 272–281.

Guerrero, J. and Silverman, P.H. (1971). *Ascaris suum*: immune reactions in mice. II. Metabolic and somatic antigens from *in vitro* cultured larvae. *Experimental Parasitology* **29**, 110–115.

Hall, R.F. and Lang, B.Z. (1978). The development of an experimental vaccine against *Fasciola hepatica* in cattle. *Proceedings of the US Animal Health Association, 82nd annual general meeting 1978*, pp. 55–60.

Healy, G.M., Fisher, D.C. and Parker, R.C. (1955). Nutrition of animal cells in tissue culture. X. Synthetic medium No. 858. *Proceedings of the Society for Experimental Biology and Medicine* **89**, 71–77.

Heath, D.D. (1973). An improved technique for the *in vitro* culture of taeniid larvae. *International Journal for Parasitology* **3**, 481–484.

Heath, D.D. (1976). Resistance to *Taenia pisiformis* larvae in rabbits: Immunization against infection using non-living antigens from *in vitro* culture. *International Journal for Parasitology* **6**, 19–24.

Heath, D.D. and Smyth, J.D. (1970). *In vitro* cultivation of *Echinococcus granulosus*, *Taenia hydatigena*, *T. ovis*, *T. pisiformis* and *T. serialis* from oncosphere to cystic larva. *Parasitology* **61**, 329–343.

Heath, D.D., Parmeter, S.N. and Osborn, P.J. (1980). An attempt to immunise dogs against *Taenia hydatigena*. *Research in Veterinary Science* **29**, 388–389.

Herd, R.P. (1977). Resistance of dogs to *Echinococcus granulosus*. *International Journal for Parasitology* **7**, 135–138.

Herd, R.P., Chappel, R.J. and Biddell, D. (1975). Immunization of dogs against *Echinococcus granulosus* using worm secretory antigens. *International Journal for Parasitology* **5**, 395–399.

Herlich, H. and Douvres, F.W. (1979). Gastrointestinal nematode immunization trials in cattle. *American Journal of Veterinary Research* **40**, 1781–1782.

Herlich, H., Douvres, F.W. and Romanowski, R.D. (1973). Vaccination against *Oesophagostomum radiatum* by injecting killed worm extracts and *in vitro*-grown larvae into cattle. *Journal of Parasitology* **59**, 987–993.

Hopkins, C.A. (1980). Immunity and *Hymenolepis diminuta*. *In* "Biology of the Tapeworm *Hymenolepis diminuta*" (H. Arai ed.), pp. 551–614. Academic Press, London and New York.

Howell, M.J. (1979). Vaccination of rats against *Fasciola hepatica*. *Journal of Parasitology* **65**, 817–819.

Howell, M.J. (1981). An approach to the production of helminth antigens *in vitro*: the formation of hybrid cells between *Fasciola hepatica* and a rat fibroblast cell line. *International Journal for Parasitology* **11**, 235–242.

Howell, M.J. and Sandeman, R.M. (1979). *Fasciola hepatica*: some properties of a precipitate which forms when metacercariae are cultured in immune rat serum. *International Journal for Parasitology* **9**, 41–45.

Howell, M.J., Sandeman, R.M. and Rajasekariah, G.R. (1977). *In vivo* and *in vitro* studies of the effect of immune rat serum on *Fasciola hepatica*. *International Journal for Parasitology* **7**, 367–371.

Irving, D.O. and Howell, M.J. (1982). Characterization of excretory-secretory antigens of *Fasciola hepatica. Parasitology* **85**, 179–188.

Jenkins, S.N. and Wakelin, D. (1977). The source and nature of some functional antigens of *Trichuris muris. Parasitology* **74**, 153–161.

Jenkins, S.N. and Wakelin, D. (1983). Functional antigens of *Trichuris muris* released during *in vitro* maintenance: their immunogenicity and partial purification. *Parasitology* **86**, 73–82.

Kowalski, J.C. and Thorson, R.E. (1972). Immunization of laboratory mice against tetrathyridia of *Mesocestoides corti* (cestoda) using a secretory and excretory antigen and a soluble somatic antigen. *Journal of Parasitology* **58**, 732–734.

Kusel, J.R. and Mackenzie, P.E. (1975). The measurement of the relative turnover rates of proteins of the surface membranes and other fractions of *Schistosoma mansoni* in culture. *Parasitology* **71**, 261–273.

Kusel, J.R., Mackenzie, P.E. and McLaren, D.J. (1975a). The release of membrane antigens into culture by adult *Schistosoma mansoni. Parasitology* **71**, 247–259.

Kusel, J.R., Sher, F.A., Perez, H., Clegg, J.A. and Smithers, S.R. (1975b). The use of radioactive isotopes in the study of specific schistosome membrane antigens. *In* "Isotopes and Radiation in Parasitology IV," pp. 127–143. International Atomic Energy Agency, Vienna.

Kwa, B.H. and Liew, F.Y. (1977). Immunity in taeniasis-cysticercosis. I. Vaccination against *Taenia taeniaeformis* in rats using purified antigen. *Journal of Experimental Medicine* **146**, 118–131.

Lang, B.Z. (1974). Host–parasite relationship of *Fasciola hepatica* in the white mouse. VI. Studies on the effects of immune and normal sera on the viability of young worms transferred to normal recipients. *Journal of Parasitology* **60**, 925–929.

Lang, B.Z. (1976). Host–parasite relationships of *Fasciola hepatica* in the white mouse. VII. Effects of anti-worm incubate sera on transferred worms and successful vaccination with a crude incubate antigen. *Journal of Parasitology* **62**, 232–236.

Lang, B.Z. and Hall, R.F. (1977). Host–parasite relationships of *Fasciola hepatica* in the white mouse. VIII. Successful vaccination with culture incubate antigens and antigens from sonic disruption of immature worms. *Journal of Parasitology* **63**, 1046–1049.

Lehner, R.P. and Sewell, M.M.H. (1979). Attempted immunisation of laboratory animals with metabolic antigens of *Fasciola hepatica. Veterinary Science Communications* **2**, 337–340.

Leland, S.E. Jr. (1963). Studies on the *in vitro* growth of parasitic nematodes. I. Complete or partial parasitic development of some gastrointestinal nematodes of sheep and cattle. *Journal of Parasitology* **49**, 600–611.

Levine, D.M. and Kagan, I.G. (1960). Studies on the immunology of schistosomiasis by vaccination and passive transfer. *Journal of Parasitology* **46**, 787–792.

Levine, H.S. and Silverman, P.H. (1969). Cultivation of *Ascaris suum* larvae in supplemented and unsupplemented chemically defined media. *Journal of Parasitology* **55**, 17–21.

Lloyd, S. (1979). Homologous and heterologous immunization against the metacestodes of *Taenia saginata* and *Taenia taeniaeformis* in cattle and mice. *Zeitschrift für Parasitenkunde* **60**, 87–96.

Lloyd, S. (1981). Progress in immunization against parasitic helminths. *Parasitology* **83**, 225–242.

Maung, M. (1978). The occurrence of the second moult of *Ascaris lumbricoides* and *Ascaris suum. International Journal for Parasitology* **8**, 371–378.

Minard, P., Murrell, K.D. and Stirewalt, M.A. (1977). Proteolytic, antigenic and immunogenic properties of *Schistosoma mansoni* cercarial secretion material. *American Journal of Tropical Medicine and Hygiene* **26**, 491–499.
Mitchell, G.B.B. and Armour, J. (1980). Failure to protect calves against *Taenia saginata* using antigens prepared from *in vitro* cultivation of the larval stage. *Research in Veterinary Science* **29**, 373–377.
Murrell, K.D. and Clay, B. (1972). *In vitro* determination of cytotoxic antibodies to *Schistosoma mansoni* schistosomules. *American Journal of Tropical Medicine and Hygiene* **21**, 569–577.
Neilson, J.T.M. (1975). Failure to vaccinate lambs against *Haemonchus contortus* with functional metabolic antigens identified by immunoelectrophoresis. *International Journal for Parasitology* **5**, 427–430.
Nicholas, W.L., Stewart, A.C. and Mitchell, G.F. (1984). Antibody responses to *Toxocara canis* using sera from parasite-infected mice and protection from toxocariasis by immunisation with ES antigens. *Australian Journal of Experimental Biology and Medical Science* **62**, 619–626.
Onawunmi, O.A. and Coles, G.C. (1980). Attempts to immunise sheep with culture antigens of *Taenia hydatigena*. *Research in Veterinary Science* **29**, 122–123.
Ozerol, N.H. and Silverman, P.H. (1969). Partial characterization of *Haemonchus contortus* exsheathing fluid. *Journal of Parasitology* **55**, 79–87.
Ozerol, N.H. and Silverman, P.H. (1970). Further characterization of active metabolites from histotropic larvae of *Haemonchus contortus* cultured *in vitro*. *Journal of Parasitology* **56**, 1199–1205.
Rajasekariah, G.R. and Howell, M.J. (1978). *Fasciola hepatica*: role of developmental stages in the rat's resistance to challenge. *Experimental Parasitology* **44**, 233–238.
Rajasekariah, G.R., Mitchell, G.F., Chapman, C.B. and Montague, P.E. (1979). *Fasciola hepatica*: attempts to induce protection against infection in rats and mice by injection of excretory/secretory products of immature worms. *Parasitology* **79**, 393–400.
Rajasekariah, G.R., Mitchell, G.F. and Rickard, M.D. (1980a). *Taenia taeniaeformis* in mice: protective immunization with oncospheres and their products. *International Journal for Parasitology* **10**, 155–160.
Rajasekariah, G.R., Rickard, M.D. and Mitchell, G.F. (1980b). Density-gradient separation of *Taenia pisiformis* oncospheres. *Journal of Parasitology* **66**, 355–356.
Rajasekariah, G.R., Rickard, M.D. and Mitchell, G.F. (1980c). Immunization of mice against infection with *Taenia taeniaeformis* using various antigens prepared from eggs, oncospheres, developing larvae and strobilocerci. *International Journal for Parasitology* **10**, 315–324.
Rickard, M.D. (1983). Immunity. *In* "Biology of the Eucestoda" (C. Arme and P.W. Pappas, eds). Vol. 2, pp. 539–579. Academic Press, London and New York.
Rickard, M.D. and Adolph, A.J. (1976). Vaccination of calves against *Taenia saginata* infection using a "parasite-free" vaccine. *Veterinary Parasitology* **1**, 389–392.
Rickard, M.D. and Adolph, A.J. (1977). Vaccination of lambs against infection with *Taenia ovis* using antigens collected during short-term *in vitro* incubation of activated *T. ovis* oncospheres. *Parasitology* **75**, 183–188.
Rickard, M.D. and Bell, K.J. (1971a). Successful vaccination of lambs against infection with *Taenia ovis* using antigens produced during *in vitro* cultivation of the larval stages. *Research in Veterinary Science* **12**, 401–402.
Rickard, M.D. and Bell, K.J. (1971b). Immunity produced against *Taenia ovis* and *T. taeniaeformis* infection in lambs and rats following *in vivo* growth of their larvae

in filtration membrane diffusion chambers. *Journal of Parasitology* **57**, 571–575.

Rickard, M.D. and Brumley, J.L. (1981). Immunisation of calves against *Taenia saginata* infection using antigens collected by *in vitro* incubation of *T. saginata* oncospheres or ultrasonic disintegration of *T. saginata* and *T. hydatigena* oncospheres. *Research in Veterinary Science* **30**, 99–103.

Rickard, M.D. and Howell, M.J. (1982). Comparative aspects of immunity in fascioliasis and cysticercosis in domesticated animals. *In* "Biology and Control of Endoparasites" (L.E.A. Symons, A.D. Donald and J.K. Dineen, eds), pp. 343–373. Academic Press, Sydney.

Rickard, M.D. and Outteridge, P.M. (1974). Antibody and cell-mediated immunity in rabbits infected with the larval stages of *Taenia pisiformis. Zeitschrift für Parasitenkunde* **44**, 187–201.

Rickard, M.D. and Williams, J.F. (1982). Hydatidosis/cysticercosis: immune mechanisms and immunization against infection. *Advances in Parasitology* **21**, 229–296.

Rickard, M.D., White, J.B. and Boddington, E.B. (1976). Vaccination of lambs against infection with *Taenia ovis. Australian Veterinary Journal* **52**, 209–214.

Rickard, M.D., Boddington, E.B. and McQuade, N. (1977a). Vaccination of lambs against *Taenia ovis* infection using antigens collected during *in vitro* cultivation of larvae: passive protection via colostrum from vaccinated ewes and the duration of immunity from a single vaccination. *Research in Veterinary Science* **23**, 368–371.

Rickard, M.D., Adolph, A.J. and Arundel, J.H. (1977b). Vaccination of calves against *Taenia saginata* infection using antigens collected during *in vitro* cultivation of larvae: passive protection via colostrum from vaccinated cows and vaccination of calves protected by maternal antibody. *Research in Veterinary Science* **23**, 365–367.

Rickard, M.D., Arundel, J.H. and Adolph, A.J. (1981). A preliminary field trial to evaluate the use of immunisation for the control of naturally acquired *Taenia saginata* infection in cattle. *Research in Veterinary Science* **30**, 104–108.

Rickard, M.D., Brumley, J.L. and Anderson, G.A. (1982). Field trial to evaluate the use of antigens from *Taenia hydatigena* oncospheres to prevent infection with *Taenia saginata* in cattle grazing on sewage-irrigated pasture. *Research in Veterinary Science* **32**, 189–193.

Rose, J.H. (1973). The development *in vitro* of some gastro-intestinal nematodes of sheep, cattle and pigs. *Research in Veterinary Science* **14**, 326–333.

Rose, J.H. (1976). Preliminary results using metabolites and *in vitro* grown larvae of *Ostertagia circumcincta* to immunise lambs against oral challenge. *Research in Veterinary Science* **21**, 76–78.

Rose, J.H. (1978). Further attempts to immunise lambs using metabolites and *in vitro* grown larvae of *Ostertagia circumcincta. Research in Veterinary Science* **24**, 61–64.

Rothwell, T.L.W. and Love, R.J. (1974). Vaccination against the nematode *Trichostrongylus colubriformis*. I. Vaccination of guinea-pigs with worm homogenates and soluble products released during *in vitro* maintenance. *International Journal for Parasitology* **4**, 293–299.

Sadun, E.H. and Lin, S.S. (1959). Studies on the host–parasite relationships to *Schistosoma japonicum*. IV. Resistance acquired by infection, by vaccination and by the injection of immune serum in monkeys, rabbits and mice. *Journal of Parasitology* **45**, 543–548.

Sakamoto, T. (1978). Development of echinococcal tissue culture *in vitro* and *in vivo*. *Memoirs of the Faculty of Agriculture, Kagoshima University* **14**, 109–115.

Sandeman, R.M. and Howell, M.J. (1980). *In vitro* studies of the response of sheep to infection with *Fasciola hepatica. Veterinary Parasitology* **6**, 347–357.

Sandeman, R.M., Howell, M.J. and Campbell, N.J. (1980). At attempt to vaccinate sheep against *Fasciola hepatica* using a juvenile fluke antigen–sheep–antibody complex. *Research in Veterinary Science* **29**, 255–259.

Sarles, M.P. (1937). The *in vitro* action of immune rat serum on *Nippostrongylus muris* (Nematoda). *Journal of Parasitology* **23**, 560–561.

Sarles, M.P. (1938). The *in vitro* action of immune rat serum on the nematode *Nippostrongylus muris. Journal of Infectious Diseases* **62**, 337–348.

Sayre, F.W., Hansen, E.L. and Yarwood, E.A. (1963). Biochemical aspects of the nutrition of *Caenorhabditis briggsae. Experimental Parasitology* **13**, 98–107.

Silverman, P.H. (1954). Studies on the biology of some tapeworms of the genus *Taenia*. I. Factors affecting hatching and activation of taeniid ova, and some criteria of their viability. *Annals of Tropical Medicine and Parasitology* **48**, 207–215.

Silverman, P.H. (1965). Some immunologic aspects of parasitic helminth infections. *American Zoologist* **5**, 153–163.

Silverman, P.H., Poynter, D. and Podger, K.R. (1962). Studies on larval antigens derived by cultivation of some parasitic nematodes in simple media: protection tests in laboratory animals. *Journal of Parasitology* **48**, 562–571.

Silverman, P.H., Alger, N.E. and Hansen, E.L. (1966). Axenic helminth cultures and their use for the production of antiparasitic vaccines. *Annals of the New York Academy of Sciences* **139**, 124–142.

Soulsby, E.J.L. (1957). Immunization against *Ascaris lumbricoides* in the Guinea Pig. *Nature, London*, **179**, 783–784.

Stromberg, B.E. (1979). The isolation and partial characterization of a protective antigen from developing larvae of *Ascaris suum. International Journal for Parasitology* **9**, 307–311.

Stromberg, B.E. and Soulsby, E.J.L. (1977). *Ascaris suum*: immunization with soluble antigens in the guinea pig. *International Journal for Parasitology* **7**, 287–291.

Taylor, A.E.R. and Baker, J.R. (1968). "The Cultivation of Parasites *In Vitro*". Blackwell Scientific Publications, Oxford.

Taylor, A.E.R. and Baker, J.R. (eds) (1978). "Methods of Cultivating Parasites *In Vitro*". Academic Press, London and New York.

Taylor, M.G. (1980). Vaccination against trematodes. *Symposia of the British Society for Parasitology* **18**, 115–140.

Techasoponmani, R. and Sirisinha, S. (1980). Use of excretory and secretory products from adult female worms to immunize rats and mice against *Angiostrongylus cantonensis* infection. *Parasitology* **80**, 457–469.

Thorson, R.E. (1953). Studies on the mechanism of immunity in the rat to the nematode, *Nippostrongylus muris. American Journal of Hygiene* **58**, 1–15.

Trager, W. (1957). Nutrition of a hemoflagellate (*Leishmania tarentolae*) having an interchangeable requirement for choline or pyridoxal. *Journal of Protozoology* **4**, 269–276.

Uahkowithchai, V., Techasoponmani, R. and Sirisinha, S. (1977). Effect of immunization with metabolic antigens on experimental infection of rats with *Angiostrongylus cantonensis. Southeast Asian Journal of Tropical Medicine and Public Health* **8**, 486–493.

Waddell, W.J. (1956). A simple ultraviolet Spectrometric method for the determination of protein. *Journal of Laboratory and Clinical Medicine* **48**, 311–314.

Weller, T.H. (1976). Manson's schistosomiasis; frontiers *in vivo, in vitro* and in the body politic. *American Journal of Hygiene and Tropical Medicine* **25**, 208–216.

Weller, T.H. and Wheeldon, S.K. (1982). The cultivation *in vitro* of cells derived from adult *Schistosoma mansoni*. I. Methodology; criteria for evaluation of cultures; and development of media. *American Journal of Tropical Medicine and Hygiene* **31**, 335–348.

Wikerhauser, T. (1961). Immunobiological diagnosis of fascioliasis. II. The *in vitro* action of immune serum on the young parasitic stage of *F. hepatica* — a new precipitin test for fascioliasis. *Veterinarski Archiv, Zagreb* **31**, 71–80.

Wikerhauser, T., Brglez, J., Dzakula, N., Asaj, R. and Matić-Piantanida, D. (1978). Eksperimentalno istraživanje vrijednosti metabolickih proizvoda aktiviranih onkosfera *Taenia saginata* za vakcinaciju teladi protiv misićne cisticerkoze. *Acta Parasitologica Jugoslavica* **9**, 51–56.

Williams, J.F. (1979). Recent advances in the immunology of cestode infections. *Journal of Parasitology* **65**, 337–349.

Appendix

NAMES AND ADDRESSES OF SUPPLIERS MENTIONED IN THE TEXT

This list is as complete as possible, but no responsibility can be accepted for errors or omissions. Alternative products may be equally satisfactory [Eds].

Abbott Laboratories Ltd, Queenborough, Kent ME11 5EL, UK.
Abbott Laboratories, Abbott Park, North Chicago, IL 60064, USA.
Amicon Corporation, Lexington, MA 02173, USA.
Anachem Ltd, 20 Charles Street, Luton, Bedfordshire LU2 OEB, UK.
Armour Pharmaceutical Co. Ltd, Hampden Park, Eastbourne, East Sussex, UK.
Armour Pharmaceutical Co., 303 S. Broadway, Trarrytown, New York NY 10591, USA.
Baird & Tatlock Ltd, PO Box 1, Romford, Essex RM1 1HA, UK.
Baltimore Biological Laboratories (BBL), Division of Bioquest, PO Box 175, Cockeysville MD 21030, USA. [UK agency: Becton Dickinson UK Ltd]
Bayer UK Ltd, Bayer House, Newbury, Berkshire RG13 1JA, UK.
BBL Microbiology Systems, PO Box 243, Cockeysville, MD 21030, USA.
BDH Chemicals Ltd, Freshwater Road, Dagenham, Essex RM8 1RZ, UK.
Beaudoin, 1 & 3 rue Rataird, Paris V^{e}, France. [UK agent: Camlab Ltd, Nuffield Road, Cambridge, CB4 1TH, UK.]
Becton Dickinson UK Ltd, Between Towns Road, Cowley, Oxford OX 4 3LY, UK.
Bellco Glass Inc., PO Box B340, Vineland, NJ 08360, USA.
British Drug Houses, Broom Road, Poole, Dorset BH12 4NN, UK.
British Oxygen Co. Ltd (BOC), Special Gases, Deer Park Road, London SW19 3UF, UK.
Calbiochem-Behring Corporation, PO Box 12087, San Diego, CA 92112, USA.
Camlab Ltd, Nuffield Road, Cambridge CB4 1TH, UK.
Colab Laboratories Inc., Chicago Heights, Ilinois, USA.
Collaborative Research, Waltham, MA 02154, USA.
Collagen Corporation, Palo Alto, CA 94303, USA.
Commonwealth Serum Laboratories (Australia), 212 Willoughby Road, Narembum, NSW 2065, Australia.
Costar, 205 Broadway, Cambridge, MA 02139, USA.
Coulter Electronics Ltd, Northwell Drive, Luton, Bedfordshire LU3 3RH, UK.

Difco Laboratories Ltd, PO Box 14B, Central Avenue, West Moseley, Surrey KT8 OSE, UK.

Difco Laboratories Inc., PO Box 1058, Detroit, MI 48232, USA.

Dow Corning Medical Products, South Saginaw Road, Midland, MI 48640, USA.

Falcon Plastics, UK agent: Becton Dickinson UK Ltd.

Fisher Chemical Co., 7722 Fenton Street, Silver Spring, MD 20910, USA.

Flow Laboratories Ltd, PO Box 17, Second Avenue Industrial Estate, Irvine AA12 8NB, Scotland.

Flow Laboratories Inc., 7655 Old Springhouse Road, McClean, VA 22102, USA.

Fluka AG, Buchs, Switzerland.

Gailard-Schlesinger Inc., 584 Mineola Avenue, Carle Place, Long Island, NY 11514, USA.

Gallenkamp Ltd, PO Box 290, Technico House, Christopher Street, London EC2P 2ER, UK.

Gibco Ltd, Unit 4, Cowley Mill Trading Estate, Longbridge Way, Uxbridge, Middlesex UB8 2XG, UK.

Gibco Laboratories, 3175 Staley Road, Grand Island, NY 14072, USA.

Hopkins & Williams Ltd, Millbruck Industrial Estate, Moston Road, Sandbach, Cheshire CW11 9YA, UK.

ICN Nutritional Biochemicals (ICN Pharmaceutical Inc.), PO Box 28050, Cleveland, OH 44128, USA.

K.C. Biological, Lenexa, KA 66215, USA.

Kimax (USA), Owens-Illinois Inc., 1 Sea Gate, Toledo, OH 43666, USA.

Laboratoire Roger Bellon, Neuilly, France.

Laboratory Glass Inc., PO Box B610, Vineland, New Jersey, 08360, USA.

Lee Engineering Co., Milwaukee, Wisconsin, USA.

Lilly, Eli, International Corporation, 307 East McCarthy Street, Indianapolis, IN 46285, USA.

Linbro Chemicals, New Haven, Connecticut, USA.

M.A. Bioproducts, Biggs Ford Road, Walkersville, MD 21793, USA.

Mallinckrodt Inc., PO Box 5840, St Louis, MO 63134, USA.

Merck & Co., Inc., Rahway, New Jersey, USA.

Merck, Sharpe & Dohme Ltd, Hertford Road, Hoddesdon, Hertfordshire, EN11 9BU, UK.

Miles Scientific Co., 30W475 North Aurora Road, Naperville, IL 60566, USA.

Millipore Corporation, Ashby Road, Bedford, MA 01730, USA.

Millipore (UK) Ltd, Millipore House, Abbey Road, London NW10 7SP, UK.

Nalgene, PO Box 365, Rochester, NY 14602, USA.

New Brunswick Scientific Ltd, 6 Colonial Way, Watford, Hertfordshire WD2 2PT, UK.

Nissui Seiyaku Co. Ltd, Japan (further details not known).

Nunc — see Gibco.

Nycomed (UK) Ltd, 2111 Coventry Road, Sheldon, Birmingham BZ6 3EA, UK.

Nyegaard & Co., Nycoveien 2, Postbox 4220, Torshov, NO401, Oslo 4, Norway. [US & Canadian distributor, Accurate Chemical & Scientific Corp., 28 Tec Street, Hicksville, NY 11801, USA].

Olympus Optical Co., Tokyo, Japan. [UK agent: Gallenkamp.]

Oxoid Ltd, Wade Road, Basingstoke, Hampshire RG24 OPW, UK.

Paines & Byrne Ltd, Pabyrn Laboratories, 177–179 Bilton Road, Perivale, Greenford, Middlesex UB6 7HG, UK. [US agent: Samrak Trading Co., Division of Harrison's Trading Co. Inc., PO Box 310, One Stone Place, Bronxville, NY 10708, USA].

Pharmacia Ltd, Pharmacia House, Midsummer Boulevard, Central Milton Keynes, Buckinghamshire MK9 3HP, UK.

Pharmacia Fine Chemicals, 800 Centennial Avenue, Piscataway, NJ 08854, USA.

Planer Products Ltd, Windmill Road, Sunbury-on-Thames, TW16 7HD UK.

Prewett, W., Ltd, Stone Flour Mills, Worthing Road, Horsham, West Sussex RH12 1ST, UK.

Schering (Pharmaceutical Division), Burgess Hill, West Sussex RH15 9NE, UK.

Scientific Industries Inc., 70 Orville Drive, Bohemia, NY 11716, USA.

Serva-Feinbiochemica GmbH & Co., D-6900 Heidelberg 1, PO Box 105260, Federal Republic of Germany. [UK agent: Uniscience Ltd, 12–14 St Ann's Crescent, London SW18 2LS, UK.]

Serva-Fine Biochemicals, Inc., 200 Shames Drive, Westbury, NY 11590, USA.

Shandon Southern Products Ltd, Chadwick Road, Astmoor, Runcorn, Cheshire WA7 1PR, UK.

Sigma Chemical Co. Ltd, Fancy Road, Poole, Dorset BH17 7NH, UK.

Sigma Chemical Co., PO Box 14508, St Louis, MO 63178, USA.

Squibb & Sons, Squibb House, Staines Road, Hounslow, Middlesex TW3 3JB, UK.

Squibb, E.R. & Sons, Inc., Princeton, NJ 08540, USA.

Sterilin Ltd, 43–45 Broad Street, Teddington, Middlesex TW11 8QZ, UK.

Technicon Instruments Co. Ltd, Hamilton Close, Houndsmill, Basingstoke, Hampshire RG21 2YE, UK.

Travenol Laboratories Ltd, Caxton Way, Thetford, Norfolk IP24 3SE, UK.

Union Carbide agent: Jencons Scientific Ltd, Cherry Court Way Industrial Estate, Stanbridge Road, Leighton Buzzard, Bedfordshire LU7 8UA, UK.

United States Biochemical Corp., PO Box 22400, Cleveland, OH 44122, USA.

Watson-Marlow Ltd, Falmouth, Cornwall TR11 4RU, UK.

Wellcome Reagents Ltd, Wellcome Research Laboratories, Langley Park, Beckenham, Kent BR3 3BS, UK.

Whatman Lab Sales Ltd, Unit 1, Coldred Road, Parkwood, Maidstone, Kent ME15 9XN, UK.

Winthrop Laboratories Ltd, Winthrop House, Surbiton-upon-Thames, Surrey KT6 4PH, UK.

Winthrop Laboratories, 1450 Broadway, New York 18, New York, USA. [or Winthrop-Breon Laboratories, 90 Park Avenue, New York, NY 10016, USA.]

Index